FRONTIERS IN
RELIABILITY

SERIES IN QUALITY, RELIABILITY & ENGINEERING STATISTICS

Series Editors: M. Xie (*National University of Singapore*)
                  T. Bendell (*Nottingham Polytechnic*)
                  A. P. Basu (*University of Missouri*)

VOL 4   Series on Quality, Reliability & Engineering Statistics

FRONTIERS IN RELIABILITY

A volume commemorating the first 25 years of the Indian Association for Productivity, Quality and Reliability

Editors

Asit P. Basu
University of Missouri-Columbia, USA

Sujit K. Basu
Indian Institute of Management, Calcutta, India

Shyamaprasad Mukhopadhyay
Calcutta University, India

World Scientific
Singapore • New Jersey • London • Hong Kong

Published by

World Scientific Publishing Co. Pte. Ltd.

P O Box 128, Farrer Road, Singapore 912805

USA office: Suite 1B, 1060 Main Street, River Edge, NJ 07661

UK office: 57 Shelton Street, Covent Garden, London WC2H 9HE

Library of Congress Cataloging-in-Publication Data
Frontiers in reliability / edited by Asit P. Basu, Shyamaprasad
    Mukhopadhyay, Sujit K. Basu.
        p.   cm. -- (Series on quality, reliability, and engineering
    statistics ; vol. 4)
        Includes bibliographical references and index.
        ISBN 9810233604
        1. Reliability (Engineering)   2. Reliability (Engineering) -
- Statistical methods.   I. Basu, Asit P.   II. Mukhopadhyay,
Shyamaprasad.   III. Basu, Sujit K.   IV. Series.
TA169.F75   1998
620'.00452'015195--dc21   97-44020
                                                           CIP

British Library Cataloguing-in-Publication Data
A catalogue record for this book is available from the British Library.

Printed in Singapore by Uto-Print

PREFACE

The Indian Association for Productivity, Quality, and Reliability (IAPQR), established on February 5, 1972 to promote productivity through quality and reliability, is celebrating its 25th year. The membership of the Association includes researchers from India and abroad as well as memberships from academic institutions and industries.

The Association conducts a number of courses on Statistical Process Control (SPC) and offers the Post-Graduate Diploma in SPC. It also publishes the biannual journal *IAPQR Transactions*. A number of research monographs on topics related to quality and reliability are planned to celebrate the Silver Jubilee year.

This present volume, *Frontiers in Reliability*, offers a survey of important topics, reviews recent developments in reliability theory, and seeks to promote further research in the field. The volume consists of refereed invited papers from leading researchers from several countries on a broad spectrum of topics in reliability. The subjects covered include Bayesian reliability, Bayesian reliability modeling, confounding in a series system, DF tests, Edgeworth approximation to reliability, estimation under random censoring, fault tree reduction for reliability, inference about changes in hazard rates, information theory and reliability, mixture experiment, mixture of Weibull distributions, queueing network approach in reliability theory, reliability estimation, reliability modeling, repairable systems, residual life function, software, spare allocation systems, stochastic comparisons, stress-strength models, system-based component test plans, and TTT-transform.

Names of authors and the titles of their papers are given in the "Contents".

This volume will be of interest to both users and researchers in reliability, statisticians, engineers, probabilists, and graduate students interested in reliability.

We would like to thank a number of persons for their help in completing this volume. We are grateful to the contributors of all the articles for accepting our invitations, for their valuable contributions, and for their encouragement, support, and patience.

Two Graduate Assistants at the University of Missouri, Patricia Jones and Mary Richardson, helped at all stages of editing. Tiffany McBee, secretary at the Missouri Department of Statistics also helped in the editing. We are grateful to the U.S. Air Force Office of Scientific Research for providing research support.

September 1997
Columbia, Missouri

PURNENDU KUMAR BOSE

A Brief Biography

Professor P. K. Bose, Professor of Statistics at the University of Calcutta, India, was an eminent scholar, educator, humanitarian, and patriot. His scholarship advanced the field of Statistics, his leadership shaped the Department of Statistics of Calcutta University, and his vision influenced higher education and economic planning for the entire country of India.

Purnendu Kumar Bose was born on July 16, 1916 at Sonarpur, a village near Calcutta. He lost his father at an early age, and it was his mother and grandmother who raised him, along with his brothers and sisters. He had his elementary education in the village school followed by secondary education in Calcutta. He took intermediate and bachelors degrees from City College and Presidency College, respectively, both in Calcutta. In 1939 he received the M.Sc. degree in Applied Mathematics from the University of Calcutta.

In India during those early stages, statistics as a discipline was being developed by scholars from other disciplines such as physics and mathematics. Notable among these scholars was Professor P. C. Mahalanobis, the founder of the Indian Statistical Institute, who was head of the Physics Department of Presidency College in Calcutta.

A strong influence in Professor Bose's life was his elder brother, Subhendu Sekhar Bose. Subhendu was a brilliant student of Physics who had joined the first band of workers under Professor P. C. Mahalanobis. Subhendu's untimely death in 1938 was a turning point in Professor Bose's life. He sought to fill his brother's place in the pursuit of a new subject and joined the Indian Statistical Institute in 1939. There he worked closely with Professor Mahalanobis and other early workers in Statistics in India.

In 1942 Professor Bose joined the newly created Department of Statistics in Calcutta University. This department was the first in India to offer a full-fledged postgraduate teaching program in Statistics. Teaching Statistics as an honors subject at the undergraduate level was begun in 1944 in the new Department of Statistics at the Presidency College of the University of Calcutta under the leadership of Professor Mahalanobis. Professor Bose was among the first batch of teachers in this department.

The Calcutta University Department was making significant contributions to statistical inference, multivariate analysis, survey sampling, and design and analysis of experiments. Professor Bose played a leading role by constructing tables facilitating the computation of confidence intervals and significance levels for the classical D-square statistic, as well as extending tables of hyper-geometric functions and numerically evaluating the power functions of F, t, and r statistics.

Professor Bose earned the D.Phil. degree in Statistics in 1950 - the first to receive this degree in Statistics from Calcutta University - on the basis of his dissertation entitled, "Parametric Relations and Statistical Tables for the Distribution Functions of Classical and Studentised D-Square Statistic and the Use of these Distribution Functions as Graduating Populations for Frequency Data".

A second phase in the life of the Calcutta University Statistics Department began in early 1951 when Professor Bose became Head of the Department. Though his research continued, Professor Bose plunged into the task of expanding and strengthening the department as well as promoting the cause of statistics in India. He was active in various statistical organizations of India including those of universities, research organizations, and government agencies. Dr. Bose was appointed the first Centenary Professor of Statistics in 1962 and continued as Head of the Department until 1969. He was appointed the first Pro-Vice-Chancellor for Academic Affairs of Calcutta University in 1968 and remained in that office until 1976, acting as Vice-Chancellor for a few months during the time as well. In this capacity Professor Bose initiated reforms in teaching and research in the post-graduate departments that benefited the entire university. In the midst of the student violence that rocked the university in the early seventies his wisdom and leadership helped ensure the continued functioning of the university.

Outside of Calcutta University, Professor Bose was deeply involved in the Indian education system and served on various commissions on education and planning for the Government of India. He authored a book on his experiences and perceptions entitled "Higher Education at Cross-Roads".

Professor Bose contributed significantly to the field of Statistical Quality Control and its application for the development of Indian industry and wrote profusely on the subject in numerous Indian and foreign journals. He was Chairman of the Quality Control and Industrial Statistics Sectional Committee of the Indian Standards Institution (subsequently the Bureau of Indian Standards) from its inception until his death. He was one of the earliest members of the Indian Society for Quality Control formed in 1948 at the initiative of Professor Mahalanobis. When this Society became defunct in 1972, Professor Bose led a group to establish the Indian Association for Productivity, Quality, and Reliability. Initially the Vice President, Professor Bose became the Chairman of the Association in 1976 and inspired and guided its activities until he died.

Professor Bose had an abiding interest in the development of Indian Agriculture. He wrote on the subject and served on government commissions in regard to agriculture. He actively promoted the cause of Science in India, was a life member of the Indian Science Congress Association, and a member of its council. He chaired the National Committee on Mathematics Education and Research set up in 1991.

Apart from his professional activities, Professor Bose was a great lover of games and sports. Having played football (soccer) in his youth, he retained his interest in the sport by holding office in local sports associations and in 1992 was honored by the West Bengal Sports Council as one of the great sports administrators in the state. A great humanist, he would go out of his way to help students, friends, and relatives and to offer hospitality. He was an active office-bearer of the National Committee for Linguistic and Communal Harmony, organizing national debates on the social issues addressed by the committee. He was involved in relief work for the displaced and distressed and was closely connected with the cultural and welfare activities of the Ramakrishna Mission. He had a love for literature and a flair for writing both in English and Bengali and wrote many expository and review articles on a wide range of topics. Plain living and high thinking was the motto of his life.

He was a fellow of the Royal Statistical Society (London, U. K.), National Academy of Science of India, Institute of Quality Assurance (U. K.), and Indian Standards Institution. The Indian Society of Agricultural Statistics conferred on him the title of Sankhyiki Bhusan, and the Indian Science Congress Association honored him with the Asutosh Mookerjee Memorial Award.

His was a life of scholarship and service to his profession, his country, and his fellow man.[*]

---

[*]The foregoing article has been abstracted from a longer paper, "Homage to a Departed Leader", by S. P. Mukherjee, published in *IAPQR Transactions*, volume 20, 1995.

CONTENTS

A BAYESIAN APPROACH USING NONHOMOGENEOUS POISSON PROCESSES FOR SOFTWARE RELIABILITY MODELS

J. A. ACHCAR

Department of Computer Sciences and Statistics, ICMSC,
University of Sao Paulo, C.P. 668, 13560-970,
Sao Carlos, S.P., Brazil
E-mail: jorge@taba.icmsc.sc.usp.br

D.K. DEY, M. NIVERTHI

Department of Statistics, Box U-120,
University of Connecticut,
Storrs, CT 06269, USA
E-mail: dey@stat.uconn.edu, niverthi@stat.uconn.edu

Modeling the interfailure times between successive failures, or the number of failures up to a given time are the two most widely used strategies in software reliability. This paper looks at the latter of the two modeling strategies. Let $M(t)$ be the cumulative number of failures of the software that are observed during time (0,t]. M(t) is modeled by a Nonhomogeneous Poisson Process (NHPP) with mean value function m(t) or equivalently an intensity function $\lambda(t)$, the first derivative of m(t). We present a Bayesian inference procedure for two special cases of the NHPP-I (where the number of bugs is finite) models; a supermodel given by the generalized gamma order statistics model which incorporates some of the standard models discussed in the literature, and the log normal order statistics model. These two classes give better flexibility for the shape of the intensity function $\lambda(t)$. Bayesian model adequacy and model comparison techniques are used to validate our model choice. Our methods are illustrated on a well known software reliability data set introduced by Jelinski and Moranda (1972) using Markov Chain Monte Carlo methods.

1 Introduction

Software reliability models help us describe the software debugging process and measure the quality of software. Failures caused by bugs or faults in the program are observed. Different stochastic models have been proposed to describe the reliability of software. Traditionally, one either models the successive times, generally CPU times, between failures or the number of failures of the software upto a given time. Interest generally lies in the inferential aspects of the models: predicting the mean time to failures (MTTF) and estimating the number of residual faults and the survival function.

In the Jelinski Moranda model, the failure rate is a linear function of the number of bugs yet to be detected in the software. Schick and Wolverton (1973, 1978) also look at models where the failure rate is a function of the number of bugs in the software yet to be detected. Similarly Xie and Bergman (1988) generalize the Jelinski-Moranda model to cases where the failure rate is a function of the number of bugs yet to be detected, the power-type and the exponential failure rate models.

For inferential purposes both maximum likelihood and Bayesian methods have been extensively used. Jelinski and Moranda (1972) and Littlewood and Verrall (1973) are examples of well known models that look at interfailure times. Jelinski and Moranda (1972) and Joe and Reid (1985) resort to maximum likelihood methods to estimate the number of bugs in the software. The Jelinski Moranda estimates were found to be unstable with many local maxima due to a flat likelihood. This motivated the use of Bayesian methods which have the additional advantage of allowing software engineers to utilize information from similar testing.

Raftery (1987) did an empirical Bayes study for a General Order Statistics (GOS) model with unknown population size. Singpurwalla and Soyer (1992) developed the Kalman filter algorithm for the AR(1) time series model. Chen and Singpurwalla (1994) employ a Markov Chain Monte Carlo method for inference on the non-Gaussian Kalman filter model. Campodonico and Singpurwalla (1995) study Bayesian inference on the Musa-Okumoto process using expert opinions.

There have been some attempts at model unification due to the numerous model proposals in software reliability. Al-Mutairi, Chen, and Singpurwalla (1996) unified concatenated hazard function models using the shot-noise process. Chen and Singpurwalla (1996) unified software reliability models by self excitation point processes. Singpurwalla and Wilson (1994) and Kuo (1996) give a comprehensive review of the research done in software reliability so far.

A more direct approach, as compared to modeling the interfailure times, is to model the number of failures M(t) discovered in the interval (0,t]. The counting process most commonly used to model software errors is the Nonhomogenuous Poisson Process (NHPP), though there is no reason other counting processes may not be used. M(t) is modeled by a NHPP with the mean function given by m(t) = E[M(t)], a non decreasing function of t. The intensity function is given by $\lambda(t)$ = $\frac{dm(t)}{dt}$. The probability function is given by P[M(t)=n] = $\frac{[m(t)]^n}{n!} e^{-m(t)}$, where n=0,1, ... For a constant $\lambda(t)$, M(t) reduces to a Homogenuous Poisson Process.

Suppose there are an unknown number of faults N, at the beginning of the debugging stage. When the software failures are modeled as the first n order statistics taken from N i.i.d. observations from a common density f and cdf F, with support in R^+, the statistical model is referred to as a General Order Statistics (GOS) Model. In addition if N is distributed as a Poisson variable with mean θ, then M(t) is NHPP-I and m(t) = θ F(t). Goel and Okumoto (1979) with $F(t) = 1 - e^{-\beta t}$ and Goel (1983) with $F(t) = 1 - e^{-\beta t^\alpha}$ are common examples of these processes. Observe that $m(t) \to \theta$ as $t \to \infty$.

When new faults are introduced during debugging, the Record Value Statistics (RVS) model replaces the GOS setup. The software failures are modeled as record breaking statistics of unobserved i.i.d. outcomes from a density f. Here $m(t) \to \infty$ as $t \to \infty$. A special case of the mean function is m(t) = -ln[1-F(t)] which yields the Musa-Okumoto process (1984) with $\lambda(t) = \frac{\alpha}{(t+\beta)}$, the Duane process(1964) with $\lambda(t) = \alpha\beta t^{\alpha-1}$, and the Cox and Lewis process(1966) with $\lambda(t) = \exp(\alpha + \beta t)$.

Kuo and Yang (1996) developed a unified approach for the software reliability growth models incorporating both the GOS and the RVS cases. They proposed the use of Markov chain Monte Carlo methods to model the software failures. Kuo, Lee, Choi, and Yang (1996) also developed Bayesian inference for the Ohba et al (1982) process by considering a further extension given by the NHPP - Gamma-k model with F(t)=$1-e^{-\beta t} \sum_{j=0}^{k-1} \frac{(\beta t)^j}{j!}$.

We present a Bayesian inference procedure for two special classes of GOS models: a supermodel given by the generalized gamma order statistics model which unifies several standard order statistics software reliability models and the log-normal order statistics model. The use of these two classes of models give greater flexibility for the shape of the intensity function $\lambda(t)$, which implies a better fit for software reliability data. We also explore some Bayesian model checking and model selection criteria.

The format of the paper is as follows. Section 2 introduces the likelihood and prior structure for the generalized gamma model and some of its special cases. The lognormal model is discussed in section 3. Some inferential tools and model selection criteria are explained in section 4. Section 5 discusses the data set, choice of hyperparameter, convergence diagnostics, and the results from our data analysis. Finally we make some concluding remarks in section 6.

2   Bayesian Inference for NHPP-I Software Reliability Models

Let us assume that the mean value function m(t) is indexed by the unknown parameters θ and ψ, where ψ is possibly vector-valued. There can be two possible scenarios here. Time truncated testing refers to the case when the testing is continued till a given time. On the other hand when we monitor the software till a specified number of failures occur we are looking at the failure truncated situation. Given the time truncated model testing until time t, the ordered epochs of the observed n failure times are denoted by $x_1, x_2, ..., x_n$. The likelihood function is given by

$$L_{NHPP}(\theta, \psi \mid D_t) = \prod_{i=1}^{n} [\lambda(x_i)] e^{-m(t)} \tag{1}$$

where $D_t = \{n; x_1, x_2, ..., x_n; t\}$ is the data set (Cox and Lewis, 1966 or Lawless, 1982). For the failure truncated model a similar expression can be applied with t replaced by x_n.

For Bayesian inference of such software reliability models, we consider the use of the Gibbs Sampler (Gelfand and Smith, 1990 and Casella and George, 1992), and the Metropolis-Hastings algorithm (Chib and Greenberg, 1995). The presence of the expression $m(t) = \theta F(t)$ in the likelihood function for NHPP-I in equation (1) usually prevents us from specifying a convenient form for the conditional density of ψ and θ given D_t, needed in the Gibbs sampling. Therefore we introduce a latent variable $N' = N - n$ which has a Poisson distribution with parameter $\theta[1 - F(t \mid \psi)]$, where ψ indexes all the parameters involved in the cdf $F(t \mid \psi)$ (Kuo and Yang, 1996). Now the posterior distribution $p(\theta, \psi \mid D_t)$ can be obtained from the joint density $p(\theta, \psi, N' \mid D_t)$ by marginalization. The joint posterior density $p(\theta, \psi, N' \mid D_t)$ is approximated from the Gibbs samplers drawn from the following conditional densities: $p(N' \mid \theta, \psi, D_t)$, $p(\theta \mid N', \psi, D_t)$, and $p(\psi \mid N', \theta, D_t)$.

2.1 Generalized Gamma Order Statistics Model

Let $F(t)=I_k(\beta t^\alpha)$ be the cdf of a generalized gamma distribution. Therefore, the corresponding mean function is given by

$$m_{GG}(t) = \theta I_k(\beta t^\alpha), \tag{2}$$

where $I_k(s)$ is the incomplete gamma integral given by

$$I_k(s) = \frac{1}{\Gamma(k)} \int_0^s x^{k-1} e^{-x} dx \tag{3}$$

and m_{GG} refers to the mean function for the generalized gamma model. The intensity function becomes

$$\lambda_{GG}(t) = m'_{GG}(t) = \frac{1}{\Gamma(k)} \theta \beta^k \alpha t^{\alpha k - 1} e^{-\beta t^\alpha}. \tag{4}$$

Observe that several usual order statistics models (NHPP-I) which are already considered in the literature are special cases of the generalized gamma order statistics model:

(i) Plugging k=1 in the expression for $I_k(\beta t^\alpha)$ yields the Goel process (1983) or the Weibull order statistics model.

(ii) If k=1, α=1, equation (4) gives the Goel and Okumoto (1979) process or the exponential order statistics model.

(iii) The NHPP-Gamma-k process or the gamma-k order statistics model which was considered in Ohba etal (1982) and Kuo etal (1996), is obtained by plugging in α =1 in equation (2).

(iv) If $k \to \infty$, we have a normal order statistics model (Lawless (1982)), not considered in this paper.

For the generalized gamma order statistics model, the likelihood function for θ, α, β, and k is given by,

$$L_{NHPP}(\theta, \alpha, \beta, k \mid D_t) \propto \frac{\theta^n \alpha^n \beta^{kn}}{\Gamma(k)^n} \prod_{i=1}^n x_i^{\alpha k - 1} e^{-[\beta \sum_{i=1}^n x_i^\alpha - \theta I_k(\beta t^\alpha)]}.$$

For Bayesian inference, introduce a latent variable $N' = N - n$ with a Poisson density and the following independent prior densities for θ, α, β and k

$$N' \sim P[\theta(1 - I_k(\beta t^\alpha))], \quad \theta \sim \Gamma(a, b), \quad \beta \sim \Gamma(c, d),$$

$$\alpha \sim \pi_1(\alpha), \text{ and } \quad k \sim \pi_2(k).$$

The prior structure is chosen similarly for the special cases of the generalized gamma model. Here, $P(\lambda)$ denotes a Poisson distribution with parameter λ and $\Gamma(a, b)$ denotes a gamma distribution with mean $\frac{a}{b}$ and variance $\frac{a}{b^2}$. The joint posterior density is $P(N', \alpha, \beta, k, \theta \mid D_t)$ which is proportional to

$$\frac{\theta^{N'+n+a-1}\alpha^n\beta^{kn+c-1}}{N'!\{\Gamma(k)\}^n} \times [\prod_{i=1}^{n} x_i^{\alpha k-1}][1-I_k(\beta t^\alpha)]^{N'} e^{-[(b+1)\theta+(d+\sum_{i=1}^{n} x_i^\alpha)\beta]}\pi_1(\alpha)\pi_2(k).$$

Again for the failure truncated model, similar expressions can be obtained with t replaced by x_n. The marginal posterior densities for the Gibbs algorithm are given by,

$$(N' \mid \alpha, \beta, k, \theta, D_t) \sim P[\theta(1-I_k(\beta t^\alpha))], \qquad (\theta \mid N', \alpha, \beta, k, D_t) \sim \Gamma[a+n+N', b+1],$$

$$(\alpha \mid \theta, N', \beta, k, D_t) \propto \alpha^n[\prod_{i=1}^{n} x_i^{\alpha k-1}]e^{-\beta\sum_{i=1}^{n} x_i^\alpha}[1-I_k(\beta t^\alpha)]^{N'}\pi_1(\alpha),$$

$$(\beta \mid N', \alpha, k, \theta, D_t) \propto \beta^{kn+c-1}e^{-\beta[\sum_{i=1}^{n} x_i^\alpha+d]}[1-I_k(\beta t^\alpha)]^{N'},$$

and

$$(k \mid N', \alpha, \beta, \theta, D_t) \propto \frac{\beta^{kn}}{\Gamma(k)^n}[\prod_{i=1}^{n} x_i^{\alpha k-1}][1-I_k(\beta t^\alpha)]^{N'}\pi_2(k).$$

The variables α, β, and k should be generated using Metropolis-Hastings algorithm (Chib and Greenberg, 1995).

2.2 Special Cases of the Generalized Gamma Order Statistics Model

For k=1, $I_k(\beta t^\alpha) = 1 - e^{-\beta t^\alpha}$ which is the cdf for the Weibull order statistics model (Goel (1983) process). Here the joint posterior density for $(N', \alpha, \beta, \theta \mid D_t)$ is proportional to,

$$\frac{1}{N'!}\theta^{N'+n+a-1}\alpha^n\beta^{n+c-1} \times [\prod_{i=1}^{n} x_i^{\alpha-1}]e^{[-(b+1)\theta+\sum_{i=1}^{n} x_i^\alpha+N't^\alpha+d]\beta}\pi_1(\alpha).$$

The marginal posterior densities are given by

$$(N' \mid \theta, \alpha, \beta, D_t) \sim P(\theta e^{-\beta t^\alpha}), \qquad (\theta \mid N', \alpha, \beta, D_t) \sim \Gamma(a+n+N', b+1),$$

$$(\beta \mid N', \alpha, \theta, D_t) \sim \Gamma(n+c, N't^c + d + \sum_{i=1}^{n} x_i^\alpha),$$

and

$$(\alpha \mid N', \theta, \beta, D_t) \propto \alpha^n[\sum_{i=1}^{n} x_i^\alpha]e^{-[N't^\alpha+\sum_{i=1}^{n} x_i^\alpha]\beta}\pi_1(\alpha).$$

Observe that when k=1, we only need to use the Metropolis-Hastings algorithm to generate the variable α.

Assuming $\alpha = 1$, we have a gamma-k order statistics model. The joint posterior density for N', β, θ and k is given by $P[N', \beta, \theta, k \mid D_t]$ which is proportional to

$$\frac{1}{N'!\{\Gamma(k)\}^n}\theta^{N'+n+a-1}\beta^{kn+c-1} \times [\prod_{i=1}^{n} x_i^{k-1}]e^{-[(b+1)\theta]}e^{-(d+\sum_{i=1}^{n} x_i)\beta}[1 - I_k(\beta t)]^{N'}\pi_2(k).$$

The marginal posterior densities for the Gibbs algorithm are given by

$$(N' \mid \beta, \theta, k, D_t) \sim P[\theta(1 - I_k(\beta t))], \qquad (\theta \mid N', \beta, k, D_t) \sim \Gamma[a + n + N', b + 1],$$

$$(\beta \mid N', \theta, k, D_t) \propto \beta^{kn+c-1}e^{-(\sum_{i=1}^{n} x_i+d)\beta}[1 - I_k(\beta t)]^{N'},$$

and

$$(k \mid N', \beta, \theta, D_t) \propto \frac{\beta^{kn}}{\{\Gamma(k)\}^n}[\prod_{i=1}^{n} x_i^{k-1}][1 - I_k(\beta t)]^{N'}\pi_2(k).$$

When $\alpha = 1$, we need to use the Metropolis-Hastings algorithm to generate the variables β and k.

When $\alpha = 1$ and k=1 (Goel and Okumoto (1979) process), we have the exponential order statistics model. In this case, the joint posterior density for $(N', \theta, \beta \mid D_t)$ is proportional to

$$\frac{1}{N'!}\theta^{N'+n+a-1}\beta^{n+c-1}e^{-[(b+1)\theta+(\sum_{i=1}^{n} x_i+N't+d)\beta]}.$$

With $\alpha = 1$ and k=1, the marginal posterior densities for the Gibbs algorithm are given by

$$(N' \mid \theta, \beta, D_t) \sim P(\theta e^{-\beta t}), \qquad (\theta \mid N', \beta, D_t) \sim \Gamma[a + n + N', b + 1],$$

and

$$(\beta \mid N', \theta, D_t) \sim \Gamma[n + c, d + N't + \sum_{i=1}^{n} x_i].$$

3   The Log-Normal Order Statistics Model

So far software reliability counting process models includes the exponential, Weibull, Pareto, and extreme value distributions as special cases. The lognormal model is also quite popular in reliability modeling. Here we want to investigate whether a lognormal model gives a better fit for the intensity function $\lambda(t)$. Let us assume that F(t) is the c.d.f. of a log-normal distribution, that is, the mean function is

$$m_{LN}(t) = \theta\Phi_z(\frac{\ln t - \mu}{\sigma}), \qquad (5)$$

where $\Phi_z(\cdot)$ is the cdf of a standard normal distribution. $m_{LN}(t)$ refers to the mean function. The intensity function is given by

$$\lambda_{LN}(t) = \frac{\theta}{t\sqrt{2\pi}\sigma} e^{-\frac{(\ln t - \mu)^2}{2\sigma^2}}. \tag{6}$$

The likelihood function for θ, μ, and σ is given by

$$L_{NHPP}(\theta, \mu, \sigma) \propto \frac{\theta^n}{(2\pi)^{n/2}\sigma^n}[\prod_{i=1}^{n}\frac{1}{x_i}] \times e_2^{-[\frac{\sum_{i=1}^{n}\frac{(\ln x_i - \mu)^2}{2\sigma^2}}{}+\theta\Phi_z(\frac{\ln t - \mu}{\sigma})]}.$$

As before we introduce a latent variable $N' = N - n$ and assume the following density for N', and independent priors on θ, μ, and σ :

$$N' \sim P[\theta(1 - \Phi_z(\frac{\ln t - \mu}{\sigma}))], \qquad \theta \sim \Gamma[a, b],$$

$$\mu \sim \pi_3(\mu) \qquad -\infty < \mu < \infty,$$

and

$$\sigma \sim \pi_4(\sigma) \qquad \sigma > 0.$$

$\pi_3(\mu)$ is chosen to be proportional to 1 and $\pi_4(\sigma)$ proportional to $\frac{1}{\sigma}$. The joint posterior density for $N', \theta, \mu, \sigma \mid D_t$ is proportional to

$$\frac{1}{\sigma^n N'!}\theta^{n+N'+a-1}e^{-\theta(b+1)} \times [1 - \Phi_z(\frac{\ln t - \mu}{\sigma})]^{N'}e^{-\frac{\sum_{i=1}^{n}\frac{(\ln x_i - \mu)^2}{2\sigma^2}}{}} \pi_3(\mu)\pi_4(\sigma).$$

In this case, the marginal posterior densities are given by

$$(N' \mid \theta, \mu, \sigma, D_t) \sim P[\theta(1 - \Phi_z(\frac{\ln t - \mu}{\sigma}))], \qquad (\theta \mid N', \mu, \sigma, D_t) \sim \Gamma[a+n+N', b+1],$$

$$(\mu \mid N', \theta, \sigma, D_t) \propto e^{-\frac{\sum_{i=1}^{n}\frac{(\ln x_i - \mu)^2}{2\sigma^2}}{}} \times [1 - \Phi_z(\frac{\ln t - \mu}{\sigma})]^{N'} \pi_3(\mu),$$

and

$$(\sigma \mid N', \mu, \theta, D_t) \propto e^{-\frac{\sum_{i=1}^{n}\frac{(\ln x_i - \mu)^2}{2\sigma^2}}{}} \times \sigma^{-n}\{1 - \Phi_z(\frac{\ln t - \mu}{\sigma})\}^{N'} \pi_4(\sigma).$$

Observe that, we again need to use the Metropolis-Hastings algorithm to generate the variables μ and σ.

4   Bayesian Inference and Model Determination

We use the Monte Carlo samples to infer about the parameters or functions of these parameters. For example, the expected number of remaining errors in the software is given by $\epsilon(t) = m(\infty) - m(t)$. For the generalized gamma order statistics model, we have $\epsilon(t) = \theta[1 - I_k(\beta t^\alpha)]$ and a Bayes estimator of $\epsilon(t)$ with respect to the squared error loss function is given by

$$E[\epsilon(t) \mid D_t] = E[\theta(1 - I_k(\beta t^\alpha)) \mid D_t]. \tag{7}$$

Let $\theta^{(s)}$, $k^{(s)}$, $\alpha^{(s)}$, and $\beta^{(s)}$ denote the sth iterates of $\theta, k, \alpha,$ and β drawn from their respective sampling distributions. We monitor convergence of the chains using the Gelman and Rubin (1992) convergence diagnostics. We consider every 10th iterate starting from the 110th iterate for each chain of 1000 iterations. The chain is thinned to obtain independence among the sample points chosen from the complete conditionals of the parameters. This gave us in all S=900 points from the sampling distribution for each parameter. Then a Monte Carlo estimate of $\epsilon(t)$ can be obtained by

$$\hat{\epsilon}(t) = \frac{1}{S} \sum_{s=1}^{S} \theta^{(s)} (1 - I_{k^{(s)}} (\beta^{(s)} t^{\alpha^{(s)}})). \tag{8}$$

Similarly one can infer about $\epsilon(t)$ for other order statistics models. For model checking, observe that, if the model is correct, apriori $\frac{m(t)}{\theta}$=F(t) has a standard uniform, U(0,1), distribution. Therefore, at each failure time we consider empirical Q-Q plots of the Monte Carlo estimates of $\frac{m(t)}{\theta}$ versus quantiles of the Uniform(0,1) distribution for each failure time and for each model. Departure from a straight line indicates model inadequacy.

For model selection, we could consider the prequential conditional predictive ordinate (PCPO) as suggested by Dawid (1984), for the future epoch x_{i+1} defined by $c_i = p(x_{i+1} \mid D_{x_i})$, the conditional density of X_{i+1} evaluated at the future observed time x_{i+1} given $(x_1, x_2, ..., x_i)$. The sequence $\{X_i\}_{i>1}$ is a Markov chain (Yang, 1994, pp 52). Therefore, we can write $c_i = p(x_{i+1} \mid x_i)$. The PCPO can be computed by

$$p(x_{i+1} \mid x_i) = \int p(x_{i+1} \mid \psi, D_{x_i}) p(\psi \mid D_{x_i}) d\psi$$

$$= \int \lambda(x_{i+1}) \exp\{-m(x_{i+1}) + m(x_i)\} p(\psi \mid D_{x_i}) d\psi, \tag{9}$$

where $m(t) = \theta F(t)$ is the mean value function of a NHPP-I process and $\lambda(t)$ is the intensity function.

For the generalized gamma order statistics model with $F(t) = I_k(\beta t^\alpha)$ we have $m(t) = \theta I_k(\beta t^\alpha)$ and $\lambda(t) = \frac{\alpha \beta^k t^{\alpha k - 1} e^{-\beta t^\alpha}}{\Gamma(k)}$. Thus, the PCPO for the future epoch x_{i+1}, is given by

$$p(x_{i+1} \mid x_i) =$$

$$\int \int \int \frac{\theta \alpha \beta^k x_{i+1}^{\alpha k-1} e^{-\beta x_{i+1}^{\alpha}}}{\Gamma(k)} e^{-\theta[I_k(\beta x_{i+1}^{\alpha}) - I_k(\beta x_i^{\alpha})]} p(\alpha, \beta, k \mid D_{x_i}) d\alpha d\beta dk. \quad (10)$$

PCPO's for the special cases of the generalized gamma order statistics model are easily obtained. For example, if k=1 (a Weibull order statistics model), we have $1 - I_k(\beta x_i^{\alpha}) = e^{-\beta x_i^{\alpha}}$ and,

$$p(x_{i+1} \mid x_i) = \int \int \theta \alpha \beta x_{i+1}^{\alpha-1} e^{-\beta x_{i+1}^{\alpha}} e^{-\theta[e^{-\beta x_i^{\alpha}} - e^{-\beta x_{i+1}^{\alpha}}]} p(\alpha, \beta \mid D_{x_i}) d\alpha d\beta. \quad (11)$$

We can approximate the above quantity by its Monte Carlo estimate

$$\hat{p}(x_{i+1} \mid x_i) = \frac{1}{S} \sum_{s=1}^{S} (\theta^s \alpha^{(s)} \beta^{(s)}{}^{k^{(s)}} x_{i+1}^{\alpha^{(s)} k^{(s)} - 1} e^{-\beta^{(s)} x_{i+1}^{\alpha^{(s)}}} e^{-\theta^{(s)}}). \quad (12)$$

For the log-normal order statistics model with mean value function $m(t) = \theta \Phi_z(\frac{\ln t - \mu}{\sigma})$ and $\lambda(t) = m'(t) = \theta \frac{1}{t\sigma} \phi_z(\frac{\ln t - \mu}{\sigma})$. Here ϕ refers to the normal pdf. In this case, the PCPO for the future epoch x_{i+1} considering the log-normal order statistics model is given by

$$p(x_{i+1} \mid x_i) = \int \int \theta \frac{1}{t\sigma} \phi_z(\frac{\ln t - \mu}{\sigma}) e^{-\theta[\Phi_z(\frac{\ln x_{i+1} - \mu}{\sigma}) - \Phi_z(\frac{\ln x_i - \mu}{\sigma})]} p(\mu, \sigma \mid D_{x_i}) d\mu d\sigma. \quad (13)$$

Now we can use the obtained PCPO, $c_i = p(x_{i+1} \mid x_i)$, in model selection. In this way, we could consider plots of $c_i(\ell)$ versus i ($i = 1, 2, ..n$) for $\ell = 1, \ldots$,5, different models considered in this exercise. n refers to the number of errors observed. Larger values of $c_i(\ell)$ (on the average) indicates the better model. For a single summary measure we also choose the model for which $c(\ell) = \prod_{i=2}^{n} c_i(\ell)$ is maximum (ℓ indexes models). We are leaving out the case i = 1 (as the first failure may not be predicted too well).

5   A Numerical Illustration

We study a software reliability data set introduced by Jelinski and Moranda (1972). The data consists of the number of days between the 26 failures that occurred during the production phase of a software (NTDS data - Naval Tactical Data System).

From the data, we observe that n=26, $\sum_{i=1}^{n} ln(x_i) = 112.4776$ and $x_n = x_{26} = 250$.

We discuss the parameter choice in detail for the generalized gamma model. For the generalized gamma order statistics model consider the failure truncated model with t replaced by $x_{26} = 250$, we consider the latent variable $N' \sim P[\theta(1-$

Table 1: NTDS data for the error times in the production of a software

	interfailure	actual	number	interfailure	actual
1	9	9	14	9	87
2	12	21	15	4	91
3	11	32	16	1	92
4	4	36	17	3	95
5	7	43	18	3	98
6	2	45	19	6	104
7	5	50	20	1	105
8	8	58	21	11	116
9	5	63	22	33	149
10	7	70	23	7	156
11	1	71	24	91	247
12	6	77	25	2	249
13	1	78	26	1	250

$I_k(\beta(250)^\alpha))]$ and priors $\theta \sim \Gamma(60,2)$, $\beta \sim \Gamma(10,2)$, and noninformative prior densities for α and k given by $\pi_1(\alpha) \propto 1/\alpha$, $\alpha > 0$ and $\pi_2(k) \propto 1/k$, $k > 0$. The choice of the hyperparameters is done to make the priors diffuse so that the inference is driven by the data. From the marginal posterior densities for $N', \theta, \alpha, \beta$, and k given, we generated 10 separate Gibbs chains each of which ran for 1000 iterations and monitored the convergence of the Gibbs samplers using the Gelman and Rubin (1992) method. We have obtained summaries for the posterior distributions of the parameters θ, α, β, and k from S = 900 sample points. It is interesting to observe that the maximum likelihood estimators for θ, α, β, and k obtained by using the software package SAS are given by $\hat{\theta} = 29.4301$, $\hat{k} = 18.2010$, $\hat{\alpha} = 0.2748$, and $\hat{\beta} = 5.1279$.

Usually, software engineers consider special cases of the supermodel to analyze software reliability data. For the k=1, Goel (1983) process (failure truncated model) assume $N' \sim P(\theta e^{-(250)^\alpha \beta})$ and priors $\theta \sim \Gamma(90,3)$, $\beta \sim \Gamma(26,160)$, and a noninformative prior density for α given by $\pi_2(\alpha) = 1/\alpha$, $\alpha > 0$. The posterior summaries for the parameters N', θ, α, and β using S=900 samples are obtained from 10 Gibbs chains each of 1000 iterations. The same procedure is followed for the Goel (1983), Gamma-k order statistics model, and the Goel-Okumoto (1979) models. The results are displayed in Tables 2-6. Next we approximate the Bayes estimates for the mean value function $m(t)$, shown in Table 8 with respect to the squared error loss for the different models. From the PCPO plot shown in Figure 1 for c_i against i, observe that the generalized gamma order statistics model gives the largest PCPO values. The values of $c(l) = \prod_{i=2}^n c_i(l)$, $l = 1,2,...,5$, are also displayed in Table 7.

$RE(\ell) = \sum_{i=1}^{26} \frac{[n_i - \hat{m}(x_i)]^2}{\hat{m}(x_i)}$, the sum of relative errors is another model selection criteria. Here n_i is the observed number of bugs in the interval $(0, x_i]$, $\hat{m}(x_i)$ is the Bayesian estimator for the mean value function, and ℓ indexes models. We observe that the generalized gamma order statistics model fits the NTDS data the best among the proposed models (bigger value for $c(\ell)$ and smaller value for $RE(\ell)$). Q-Q plots of the Monte Carlo estimates of $\frac{m(t)}{\theta}$ against the corresponding quantiles of the Uniform(0,1) model, shown in Figure 2, come closest to a straight line for the generalized gamma order statistics model.

Table 2: Posterior Summaries for Generalized Gamma Order Statistics Model

Parameter	Mean	Median	S.D.	95% Credible Interval
θ	28.78	28.21	5.190	(19.91, 39.58)
α	0.303	0.308	0.056	(0.195, 0.399)
β	5.188	4.732	2.107	(2.096, 10.170)
k	18.93	18.88	5.131	(10.42, 31.25)

Table 3: Posterior Summaries for Weibull Order Statistics Model

Parameter	Mean	Median	S.D.	95% Credible Interval
θ	31.02	30.89	2.93	(26.00, 37.37)
α	0.479	0.478	0.062	(0.356, 0.605)
β	0.111	0.111	0.023	(0.071, 0.159)

Table 4: Posterior Summaries for Gamma-k Order Statistics Model

Parameter	Mean	Median	S.D.	95% Credible Interval
θ	29.61	29.41	2.780	(24.65, 35.56)
β	0.017	0.0169	0.0042	(0.0099, 0.0262)
k	1.890	1.854	0.404	(1.188, 2.840)

Table 5: Posterior Summaries for Exponential Order Statistics Model

Parameter	Mean	Median	S.D.	95% Credible Interval
β	0.0059	0.0057	0.0016	(0.0032, 0.0094)
θ	31.242	31.074	3.484	(25.09, 38.79)

Table 6: Posterior Summaries for Lognormal Order Statistics Model

Parameter	Mean	Median	S.D.	95% Credible Interval
θ	30.12	29.73	4.405	(22.68, 40.21)
μ	4.57	4.48	0.413	(4.13, 5.96)
σ	0.852	0.779	0.299	(0.576, 1.694)

Table 7: Values for $c(\ell)$ and RE(ℓ) for different models

Models	$c(\ell)$	R.E.(ℓ)
Generalized Gamma Order Statistics Model	9.576×10^{-50}	6.775
Weibull Order Statistics Model (k = 1)	9.874×10^{-58}	68.589
Gamma-k Order Statistics Model ($\alpha = 1$)	1.325×10^{-51}	23.376
Exponential Order Statistics Model ($\alpha = 1$, k = 1)	3.587×10^{-57}	221.450
Lognormal Order Statistics Model	1.130×10^{-51}	7.972

Table 8: Bayes Estimators for m(x_i), i=1, ... ,26

i	x_i	t_i	Generalized Gamma	Gamma-k ($\alpha = 1$)	Weibull ($k = 1$)	Exponential ($k = 1, \alpha = 1$)	Log-Normal
1	9	9	0.3846	0.5818	8.4327	1.5897	0.1411
2	21	12	2.0812	2.1213	11.7309	3.5758	1.1853
3	32	11	4.3854	3.9483	13.6585	5.2713	3.0620
4	36	4	5.31023	4.6694	14.2275	5.8599	3.8828
5	43	7	6.9691	5.9725	15.1072	6.8558	5.4090
6	45	2	7.4451	6.3508	15.3360	7.1326	5.8561
7	50	5	8.6263	7.3020	15.8716	7.8099	6.9785
8	58	8	10.4612	8.8234	16.6368	8.8515	8.7491
9	63	5	11.5574	9.7633	17.0676	9.4772	9.8186
10	70	7	13.0122	11.0534	17.6201	10.3218	11.2487
11	71	1	13.2119	11.2346	17.6947	10.4396	11.4459
12	77	6	14.3654	12.3034	18.1224	11.1314	12.5889
13	78	1	14.5502	12.4782	18.1905	11.2443	12.7726
14	87	9	16.1163	14.0042	18.7678	12.2302	14.3377
15	91	4	16.7573	14.6534	19.0055	12.6515	14.9828
16	92	1	16.9124	14.8128	19.0633	12.7552	15.1393
17	95	3	17.3656	15.2839	19.2329	13.0628	15.5975
18	98	3	17.8009	15.7441	19.3972	13.3649	16.0392
19	104	6	18.6197	16.6318	19.7107	13.9530	16.8745
20	105	1	18.7497	16.7755	19.7612	14.0490	17.0077
21	116	11	20.0653	18.2748	22.2848	15.0679	18.3662
22	149	33	22.9720	21.9213	21.5821	17.7547	21.4633
23	156	7	23.4314	22.5449	21.8159	18.2606	21.9703
24	247	91	26.7075	27.3343	24.0499	23.2699	25.8712
25	249	2	26.7454	27.3909	24.0870	23.3534	25.9213
26	250	1	26.7640	27.4187	24.1054	23.3948	25.9460

6   Concluding Remarks

This paper proposes two families of models to model the number of software failures in a given time interval. The generalized gamma model, so far not used in software literature, provides the best fit among the models proposed. The Q-Q plot of $\frac{m(t)}{\theta}$ versus quantiles of the Uniform(0,1) distribution may be criticized as it may give a straight line apriori, not aposteriori. Moreover, the model comparison technique, using the PCPO criterion, does not penalize for complexity of the model. But the paper does provide a useful way of looking at software failures. In a separate paper Niverthi, Nayak, and Dey (1997) look at the same family of models in a different light. They propose recapture debugging, where each error is allowed to recur, as a useful tool to help in inference, in particular to estimate the number of errors in a software. They also examine the role of N, the number of errors, in some cases as a parameter and in other cases like NHPP, where N is a latent variable introduced to facilitate computation.

Acknowledgments

J. A. Achcar thanks the Brazilian institution, FAPESP, for providing financial support to visit the Department of Statistics at the University of Connecticut, Storrs, CT, USA.

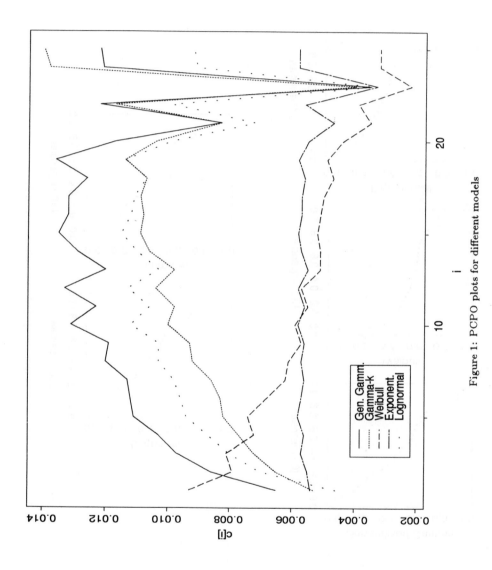

Figure 1: PCPO plots for different models

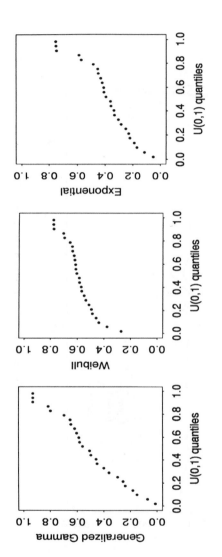

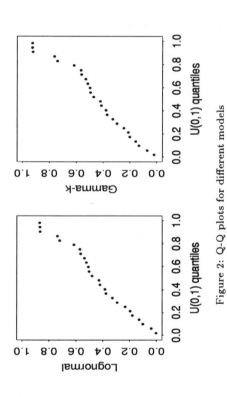

Figure 2: Q-Q plots for different models

References

1. Al-Mutairi, Y. Chen and N.D. Singpurwalla, "Software Reliability models whose concatenated failure rates are shot-noise processes." *Preprint*, (1996).
2. G. Casella and E.I. George, "Explaining the Gibbs Sampler." *American Statistician*, 46, 167-174, (1992).
3. Y. Chen and N.D. Singpurwalla, "Unification of software reliability models by self-exciting point processes." *Advances in Applied Probability*, in press, (1996).
4. S. Chib and E. Greenberg, "Understanding the Metropolis-Hastings Algorithm." *American Statistician*, 49, 327-335, (1995).
5. D.R. Cox and P.A. Lewis, "Statistical Analysis of Series of Events." Methuen, (1966).
6. A.P. Dawid, "Statistical Theory: The Prequential Approach." *Journal of the Royal Statistical Society*, Series A, 147, 278-292, (1984).
7. J.T. Duane, "Learning curve approach to reliability monitoring." *IEEE Transaction on Aerospace*, AS-2(2), 563-566, (1964).
8. A.E. Gelfand and A.F.M. Smith, "Sampling-based approaches to calculating marginal densities." *Journal of the American Statistical Association*, 85, 398-409, (1990).
9. A. Gelman and D.B. Rubin, "Inference from iterative simulation using multiple sequences." (with discussion), *Statistical Science*, 7, 457-511, (1992).
10. A.L. Goel, "A guidebook for software reliability assessment." *Technical Report* RADC-TR-83-176, (1983).
11. A.L.Goel and K. Okumoto, "Time dependent error detection rate model for software reliability and other performance measures." *IEEE Transactions on Reliability*, 28, 206-211, (1979).
12. Z. Jelinski and P.B. Moranda, "Software Reliability Research. In Statistical Computer Performance Evaluation." ed. W.Freiburger, New York: Academic Press, 465-497 (1972).
13. H. Joe and N. Reid, . "Estimating the number of faults in a system." *Journal of the American Statistical Association*, 80, 222-226, (1985).
14. L. Kuo, "Software Reliability." *Technical Report*, No. 96-27, Department of Statistics, University of Connecticut. (1996).
15. L. Kuo and T.Y. Yang, "Bayesian Computation for Nonhomogenous Poisson Processes in Software Reliability." *Journal of the American Statistical Association*, 91, 763-773, (1996).
16. L. Kuo, J. Lee, K. Choi and T.Y. Yang, "Bayesian inference for S-shaped software reliability growth models." *Technical Report* No. 96-05, Department of Statistics, University of Connecticut, (1996).
17. J.F. Lawless, "Statistical Models and Methods for Lifetime Data." New York: John Wiley & Sons, (1982).
18. B. Littlewood and J.L. Verrall, "A Bayesian reliability growth model for computer software." *Applied Statistics*, 22, 332-346, (1973).
19. J.D. Musa and K. Okumoto, "A logarithmic Poisson execution time model for software reliability measurement." In *Proceedings of Seventh International*

Conference on Software Engineering, Orlando, pp. 230-238, (1984).

20. M. Niverthi, T.K. Nayak and D.K. Dey, "A Bayesian Approach for estimating the number of errors in software reliability modeling." *Technical Report*, Department of Statistics, University of Connecticut, (1997).

21. M. Ohba, S. Yamada, K. Takeda, and S. Osaki, "S-shaped software reliability growth curve: how good is it?" *COMPSAC* 82, 38-44, (1982).

22. A.E. Raftery, "Inference and Prediction for a General Order Statistic Model with Unknown Population Size." *Journal of the American Statistical Association* 82, 1163-1168, (1987).

23. G.J. Schick and R.W. Wolverton, "Assessment of Software Reliability." *Proceedings of Operations Research* Wiezberg-Wien: Physica-Verlag, 395-422, (1973).

24. G.J. Schick and R.W. Wolverton, "An analysis of computing software reliability models." *IEEE Transactions on Software Engineering*, 4, 104-120, (1978).

25. N.D. Singpurwalla and R. Soyer, "Non Homogenuous autoregressive processes for tracking (software) reliability growth, and their Bayesian analysis." *Journal of the Royal Statistical Society*, B, 54, 145-156, (1992).

26. N.D. Singpurwalla and S.P. Wilson, "Software Reliability Modeling." *International Statistical Review*, 62, 289-317, (1994).

27. M. Xie and B. Bergman, "On modeling reliability growth for software." *Proceedings of the 8th IFAC Symposium on Identification and System Parameter Estimation* (1988).

28. T.Y. Yang, "Computational approaches to Bayesian inference for software reliability." *Ph.D. Thesis*, Department of Statistics, University of Connecticut, Storrs (1994).

BAYESIAN ESTIMATION OF THE NUMBER OF UNDETECTED ERRORS WHEN BOTH REVIEWERS AND ERRORS ARE HETEROGENEOUS

SANJIB BASU

Division of Statistics, Northern Illinois University, De Kalb, IL 60115

E-mail: basu@niu.edu

The error detection problem involves estimation of the unknown size N of a population based on multiple capture-recapture samples. In the context of software review, the capture-recapture samples represent errors detected by k reviewers who inspect the software in parallel. The interest here is to estimate the number of undetected errors remaining in the software. Another application area is in estimation of animal abundance where the capture-recapture samples are obtained by k successive trappings. In typical capture-recapture studies, the reviewers are non-identical and errors are heterogeneous. Earlier works in this area separately addressed either the heterogeneity among the reviewers or among the errors. In this article, a flexible Bayesian model is proposed which can simultaneously incorporate both heterogeneities under a unified setup. A Markov Chain Monte Carlo method is described for obtaining Bayesian point and interval estimates of N. The method is illustrated in the software review data of AT&T 5 ESS switches.

1 Introduction

Complex systems of computer software sometimes fail due to software faults. Root cause analysis shows that many faults are actually due to errors in the requirements or design stage of the development process. For the purpose of detecting faults in the design stage, software engineers prepare documents describing feature design and requirements. Several reviewers individually read these documents. A tabulation of detected defects provides information regarding the number of faults undetected. If this latter number can be judged to be small the process can proceed to the next step, otherwise further document refinement is necessary in order to avoid expensive fixes later in the process. Our main goal in this paper is to estimate the number of faults remaining in the software which are not detected by the members of the review team.

As an example, Table 1 list data from Eick et al (1993) on faults identified for a particular feature of the AT&T 5 ESS Switch. Altogether, 43 distinct faults were detected by 6 reviewers. The problem is to estimate the number of faults still remaining undetected.

The general statistical problem described above has different ramifications in different application areas. The systems reliability problem mentioned above is one such application. Another important application is in estimating the size of a closed biological population based on the capture-recapture methodology which takes multiple trapping samples from the population, identifying the individuals which appear more than once. First used by Laplace to estimate the population size of France, this problem has an extensive literature, see Seber (1982), Burnham and Overton (1978), and more recent Bayesian works by George and Robert (1992) and Madigan and York (1997).

19

For simplicity of exposition, we use software reliability terminology in this article rather than attempt to simultaneously use terminologies of both software inspection and population size estimation. Moreover, we use errors/defects/faults/issues to describe software errors.

In the most basic formulation of the software review problem, one assumes that both the faults and the reviewers are homogeneous and independent. A Binomial model results with unknown detection probability p and the total number of faults $= N$ including those discovered and those remaining undetected. Several methods are proposed in the literature for estimation of N under this simple model including likelihood and Bayesian methods.

However, there are several features of the software review process which are not addressed by this basic model. They include :

(i) **Inhomogeneous reviewers** It may be that some reviewers detect more faults than others. Reviewers may have areas of specializations making them more likely to detect certain issues. The model thus should allow for different detection probabilities for the reviewers.

(ii) **Non-independent reviewers** The simple model assumes that the engineers detect the faults independently. In reality, engineers may discuss issues among themselves and thus discover more errors in common.

(iii) **Inhomogeneous errors** The basic model treats all errors as probabilistically identical. In reality, some issues are often more difficult to detect than others.

Vander Wiel and Votta (1993) perform a simulation study where they simulate data from probability distributions mimicking the above three scenarios and then they apply statistical estimators (maximum likelihood and Jackknife) which are obtained without considering the above issues. They find that ignoring the above issues can lead to moderately poor to very poor performances for the estimators. Bickel and Yahav (1988), however, comment that asymptotically (number of errors and number of reviewers $\to \infty$), "while the inhomogeneity of the readers is not important the inhomogeneity of the errors is crucial."

Eick et al. (1993) show how maximum likelihood estimation can be carried out when the errors are homogeneous but the reviewers are heterogeneous. George and Robert (1992) propose a Bayesian model for the same scenario and describe how Gibbs sampling based methodology can be applied. They concentrate only on the biological population size estimation problem. This heterogeneous reviewers model is extended by Ebrahimi (1996) and Basu and Ebrahimi (1997) who further allow the reviewers to be dependent on one another.

The diagonally opposite scenario is when reviewers are homogeneous but errors are heterogeneous. Burnham and Overton (1978) describe a jackknife based method here. They again concentrate exclusively on animal trapping studies. Basu and Ebrahimi (1997) recently propose a flexible Bayesian model for the heterogeneous errors case and describe how Markov Chain Monte Carlo (MCMC) methods can be applied to obtain point and interval estimates for the number of undetected errors. Madigan and York (1997) use Bayesian graphical models to include covariate informations.

In this article, we develop a Bayesian model where we assume that reviewers and errors act independently between each other and also within themselves. The unique feature of this model is that it can simultaneously incorporate the heterogeneity of the reviewers and the heterogeneity of the errors in a flexible way. The existing models in the literature can only incorporate heterogeneity of one kind. To our knowledge, this is the first attempt to incorporate both heterogeneity under a unified setup.

The article is organized as follows. In section 2, we introduce the notations and develop the preliminaries. In section 3, we briefly outline the existing heterogeneous reviewers model and the heterogeneous errors model. In sections 4.1 and 4.2, we describe our Bayesian model which can incorporate both heterogeneities under a unified umbrella. Our proposed method is flexible enough to incorporate a wide variety of priors for N including some improper priors. For these improper priors, we provide sufficient conditions (see Result 1) guaranteeing the properness of the posteriors. The posterior analysis and the Gibbs sampling algorithm is described in 4.3. In the context of Gibbs sampling, two full conditional distributions turn out to be not of common parametric forms. We show that these conditional distributions have log-concave densities and use the adaptive rejection sampling method of Gilks and Wild (1992) to simulate from them. Application of our proposed model and analysis in the AT&T 5 ESS Switch review data is described in section 5.

Our proposed Bayesian model provides flexibility to the users and provides avenues to build-in inherent characteristics of the detection process directly into the model. Another major advantage is that our model readily provides interval estimates (credible intervals) for N in addition to point estimates. This is because the complete posterior distribution of N is available to the user. On the other hand, the maximum likelihood based methods are mostly geared towards obtaining only point estimates for N.

We briefly note here the connection between estimation of the number of undetected errors and the capture-recapture sampling problem which we are focusing on and the broader software reliability models. In software reliability, the goal is not only to estimate the number of faults in the software but also to provide an estimate of the reliability of the software after removal of detected faults. We refer the reader to Achcar et al. (1997), Nayak (1997), Ries and Basu (1997), and Sen (1997) for different statistical approaches to software reliability modeling.

2 Preliminaries

Let N be the unknown number of faults. Suppose k reviewers inspect in parallel to detect these faults. Let $X_{ij} = 1$ if fault i is detected by j-th reviewer, and $= 0$ otherwise, $i = 1, \ldots, N$, $j = 1, \ldots, k$. The basic data would be the $N \times k$ matrix $x = ((x_{ij}))$, however, we do not observe this matrix. Note that we use the standard terminology of using upper case letters to denote random variables and lower case letters to denote observed realizations of the random variables.

Suppose, in the review process, n distinct errors are discovered by k reviewers. These n discovered errors represent a sample from the population of N errors, what we do not know is which ones among the $\{1, \ldots, N\}$ are discovered. The observed data informs us that $N - n$ rows of the $\boldsymbol{x}$ matrix are zero-rows. For the other $\{i_1, \ldots, i_n\}$ non-zero x-rows, we observe $\underset{\sim}{x}_{i_l} = (x_{i_l,1}, \ldots, x_{i_l,k})$, $l = 1, \ldots, n$. However, we do not know which ones among the N possible rows are the non-zero rows. There are $\binom{N}{n}$ such possibilities.

Let $p_{ij} = P(X_{ij} = 1)$, i.e., p_{ij} is the detection probability of the i-th error by the j-th reviewer, $i = 1, \ldots, N$, $j = 1, \ldots, k$. Let $\boldsymbol{p}$ be the $n \times k$ matrix $((p_{ij}))$. Then, under the assumption of independence among the k reviewers and independence among the N errors, the likelihood of N and $\boldsymbol{p}$ is

$$L(N, \boldsymbol{p} \,|\, \text{data}) \propto \binom{N}{n} \prod_{i=1}^{N} \prod_{j=1}^{k} p_{ij}^{x_{ij}} (1 - p_{ij})^{1 - x_{ij}}. \tag{1}$$

This likelihood will play a central role in our development, we will refer back to it time and again.

Let n_j be the number of errors detected by the j-th reviewer. Thus, $n_j = \sum_{i=1}^{N} x_{ij}$, $j = 1, \ldots, k$. Note that n_j is the j-th column sum of the $\boldsymbol{x}$ matrix. The quantity $\sum_{j=1}^{k} n_j$ gives the total number detections.

Similarly, let $y_i = \sum_{j=1}^{k} x_{ij}$, $i = 1, \ldots, N$. y_i is the i-th row sum of the $\boldsymbol{x}$ matrix and denotes how many reviewers have detected the i-th error. Note that each y_i can only take values $0, 1, \ldots, k$. Let $f_j = \#\{y_l : y_l = j\}$, $j = 0, \ldots, k$. Thus f_j is the number of errors which were detected by exactly j reviewers. Note $f_0 = N - n$. Also, $f_1, \ldots, f_k$ are observed in data, whereas f_0 is not observed.

3   Existing models

3.1   IID model

The simplest model is when all k reviewers are independent and identical and so also are the N errors. Under these assumptions $p_{ij} = p$, $j = 1, \ldots, k$, $i = 1, \ldots, N$. From (1), the likelihood for this iid model is,

$$L(N, p) \propto \binom{N}{n} p^{\sum_{j=1}^{k} n_j} (1 - p)^{Nk - \sum_{j=1}^{k} n_j}. \tag{2}$$

Note that N is the parameter of interest here; p is a nuisance parameter. Maximum likelihood estimation of N for this model can pursued in a straightforward way by first finding $\widehat{p}_{mle}$ for a fixed N and then maximizing $L(N, \widehat{p}_{mle})$ over N. Bayesian analysis can also be obtained with equal ease by assigning a prior distribution $\pi(N, p)$ on the parameters. An independent prior structure $\pi(N, p) = \pi(N) \cdot \pi(p)$ is a common choice.

3.2 Heterogeneous reviewers model

The homogeneity assumption among the reviewers is often not practical. Reviewers sometimes have areas of specializations making them more likely to detect certain faults. This possible inhomogeneity among the reviewers is modeled here by assuming that the detection probability p_{ij} varies from reviewer to reviewer but remains identical over the different errors. Thus, $p_{ij} = p_j$, $i = 1, \ldots, N$, $j = 1, \ldots, k$. Let $\underline{p} = (p_1, \ldots, p_k)$. The likelihood for this heterogeneous reviewers model is then given by

$$L(N, \underline{p}) \propto \binom{N}{n} \prod_{j=1}^{k} p_j^{n_j} (1 - p_j)^{N - n_j}. \tag{3}$$

Eick et al (1993) consider this model and show that the maximum likelihood estimate of p_j is $n_j / \widehat{N}$, where the maximum likelihood estimate $\widehat{N}$ is obtained as a solution of a profile likelihood equation. They also propose a separate conditional maximum likelihood estimator.

For Bayesian analysis, N and $\underline{p}$ are typically assumed to be a priori independent, i.e., $\pi(N, \underline{p}) = \pi(N)\pi(\underline{p})$. The marginal likelihood of N then is $L(N) = \int L(N, \underline{p}) d\pi(\underline{p})$. If $p_1, \ldots, p_k$ are further assumed to be independent, i.e., $\pi(\underline{p}) = \prod_{j=1}^{k} \pi(p_j)$, we then have $L(N) \propto \binom{N}{n} \prod_{j=1}^{k} \int p_j^{n_j} (1 - p_j)^{N - n_j} d\pi(p_j)$. These k univariate integrals can often be obtained easily. For example, if each individual $\pi(p_j)$ is Beta(α_j, β_j), the marginal $L(N)$ involves product of k complete-Beta functions.

The posterior distribution of N can be obtained as $\pi(N \mid \text{data}) \propto L(N) \, \pi(N)$. In some cases, (for example, if $\pi(N)$ has finite support) the posterior distribution of N can be obtained with relative ease.

Obviously not all priors allow as clean a solution as above. Moreover, without the prior independence of $p_1, \ldots, p_k$, the k-dimensional integration $\int L(N, \underline{p}) d\pi(\underline{p})$ for obtaining $L(N)$ can be quite formidable. However, as shown in George and Robert (1992), if one can simulate easily from the conditional posteriors $\pi(N \mid \underline{p}, \text{data})$ and $\pi(\underline{p} \mid N, \text{data})$, one can implement Markov Chain Monte Carlo (MCMC) sampling methodology to estimate the marginal posterior $\pi(N \mid \text{data})$. If the conditional posteriors are not of the form of common distributions, one can choose from a whole list of available simulation strategies including adaptive rejection, Metropolis-Hastings, and many others. For earlier Bayesian works in this area (including empirical Bayes work and reference prior based results) see Smith (1991, 1988) and Castledine (1981).

A generalization of this model is proposed by Ebrahimi (1996) and Basu and Ebrahimi (1997). They consider the situation when the reviewers can be dependent on one another in addition to being heterogeneous. Without independence among the reviewers, the basic likelihood of (1) is no longer valid. Instead, suppose we define the joint probabilities : $p(x_1, \ldots, x_k) = P(X_{i1} = x_1, \ldots, X_{ik} = x_k)$, $i = 1, \ldots, N$, where $x_j = 0$ or 1, $j = 1, \ldots, k$. There are 2^k such $p(x_1, \ldots, x_k)$ which add up to 1. Similarly, let $n(x_1, \ldots, x_k) = \#\{X_{i1} = x_1, \ldots, X_{ik} = x_k, \ i = 1, \ldots, N\}$, where $x_j = 0$ or 1, $j = 1, \ldots, k$. The multinomial likelihood for this dependent reviewer model then has the form

$$L(N, \underset{\sim}{p}) \propto \frac{N!}{(N-n)!} \prod_{\{x_j=0 \text{ or } 1, \ j=1,\ldots,k\}} p(x_1, \ldots, x_k)^{n(x_1,\ldots,x_k)}. \tag{4}$$

The likelihood has $2^k - 1$ nuisance parameters $p(x_1, \ldots, x_k)$ in addition to the parameter of interest N. Ebrahimi (1996) uses a nonparametric kernel method to obtain estimates $\hat{p}(x_1, \ldots, x_k)$ for fixed N. He then obtains the maximum likelihood estimate of N based on the estimated likelihood. Basu and Ebrahimi (1997) describe a Bayesian model which achieves a dramatic dimension reduction. In this model, prior specification and posterior analysis are needed only for N and $p(0, \ldots, 0)$, the other $2^k - 2$ parameters can be completely arbitrarily specified. We refer the reader to these two papers for further details.

3.3 Heterogeneous errors model

In this model, the reviewers are assumed to be independent and homogeneous, whereas the errors are now treated as heterogeneous. Inhomogeneity of the errors is, in fact, typical in software inspection as reported by Vander Wiel and Votta (1993) based on their personal experience of reviews in Bell Laboratories. In population size estimation, Burnham and Overton (1982) also report that capture probabilities are generally heterogeneous among individuals.

The development of the model begins from the general likelihood structure given in (1). The heterogeneity of the errors is modeled by assuming that the detection probability p_{ij} varies from error to error but remains the same over different reviewers. Thus, $p_{ij} = p_i$, $i = 1, \ldots, N$, $j = 1, \ldots, k$, and the resulting likelihood is

$$L(N, \underset{\sim}{p}) \propto \binom{N}{n} \prod_{i=1}^{N} \binom{k}{i} p_i^{y_i} (1 - p_{ij})^{k-y_i}. \tag{5}$$

Here $\underset{\sim}{p} = (p_1, \ldots, p_N)$ and $y_i = \sum_{j=1}^{k} x_{ij}$ as defined earlier. The complication with this model stems from the fact that the dimension of the parameter vector $\underset{\sim}{p} = (p_1, \ldots, p_N)$ is not known, but is determined by another unknown parameter N. Further complexity is added since the data matrix $((x_{ij}))$ has only n non-zero rows corresponding to n of the p_i's; the other $N - n$ are zero rows.

Typically, $p_1, \ldots, p_N$ are assumed to be i.i.d from a probability distribution F_p with support on $[0, 1]$. Burnham and Overton (1982) use this model and propose a jackknife based estimator for the parameter of interest N. Recently, Basu and Ebrahimi (1997) propose a Bayesian model for this heterogeneous errors case. They assume that F_p is a discrete distribution with finite support. The resulting Bayesian model is analytically intractable. A direct implementation of a Markov Chain Monte Carlo method is also problematic since the dimension of the nuisance parameter $\underset{\sim}{p} = (p_1, \ldots, p_N)$ depends on the parameter of interest N. Basu and Ebrahimi (1997) develop an intelligent Gibbs sampling algorithm which circumvents this problem.

4 Heterogeneous reviewers & heterogeneous errors

4.1 Introduction

In the previous section, we briefly described many existing models in software review and capture-recapture animal trapping literature. These models either incorporate the heterogeneity of the reviewers or incorporate the heterogeneity of the errors. However, in real-life applications, both heterogeneities are generally simultaneously present in data. For example, Table 1 shows data from Eick et al (1993) on AT&T 5 ESS Switch review by six reviewers. The column sums, n_j's, clearly suggest presence of inhomogeneity among the six reviewers. On the other hand, the row sums, y_i's, indicate presence of some inhomogeneity among the errors, though not very strongly. The analysis of Basu and Ebrahimi (1997) show that there are significant differences among the results obtained under the IID model, the heterogeneous reviewers model, and the heterogeneous errors model. A summary of the their results is shown in Table 1. The presence of both kind of heterogeneities is evident in this data set, and hence there is need for a model which simultaneously incorporate both heterogeneities in a unified setup.

As another example, Burnham and Overton (1982) estimate the size of a cottontail rabbit population based on multiple capture-recapture trappings. They consider a heterogeneous errors/animals model and use a jackknife estimate of N, whereas Basu and Ebrahimi (1997) use a Bayesian estimate. Both find that "there is clear evidence of time variation in the capture probabilities" (which we call heterogeneity among the reviewers). Again, these earlier models only addressed the heterogeneity among the errors/animals when the data provide evidence for both kind of heterogeneities.

In section 1, we mentioned the comments of Vander Wiel and Votta (1993) based on their personal experience of over 50 reviews in Bell Laboratories and their interview of over 100 reviewers. They observe that reviewers are often heterogeneous, and additionally, "faults occur in at least two categories, easy to observe (not requiring much thought) and hard to observe (requiring much thought)". Vander Wiel and Votta (1993) carry out a simulation study where they include a few specialists among the reviewers (heterogeneity) and they include two kinds of faults, type A and type B, with different detection probabilities. They find that both the maximum likelihood and the jackknife estimates can severely misestimate the number of undetected errors. Another simulation study with similar results was performed by Otis et al. (1978).

In our model, we reflect the simultaneous heterogeneities of the reviewers and the errors through the detection probabilities p_{ij}. Recall that $p_{ij} = P(i$-th error is detected by j-th reviewer). We let p_{ij} depend on both error i and reviewer j, but assume a product structure : $p_{ij} = e_i \cdot r_j$, $i = 1, \ldots, N$, $j = 1, \ldots, k$. In a sense, we are assuming that both error i and reviewer j have effects on the detection probability p_{ij}, but these two effects act independently. With this structure, the likelihood of (1) changes to

$$L(N, \underset{\sim}{e}_N, \underset{\sim}{r}_k) \propto \binom{N}{n} \prod_{i=1}^{N} \prod_{j=1}^{k} (e_i r_j)^{x_{ij}} (1 - e_i r_j)^{1-x_{ij}}. \tag{6}$$

Here $\underset{\sim}{e}_N = (e_1, \ldots, e_N)$ are the error effects, and $\underset{\sim}{r}_k = (r_1, \ldots, r_k)$ are the reviewer effects. One major complication of the analysis stems from the fact that the dimension of the parameter vector $\underset{\sim}{e}_N$ is not known, but is determined by another unknown parameter N.

The above model is overparameterized, and further model structure is required for any fruitful statistical inference. We assume that the error effects $e_1, \ldots, e_N$ are a random sample from a probability distribution F_e with support on $[0, 1]$. We make a similar assumption about the reviewer effects : $r_1, \ldots, r_k$ are i.i.d. from a distribution F_r which also has support on $[0, 1]$. We thus assume that the individual error effects are exchangeable and so also are the individual reviewer effects.

4.2   The prior model

Here we develop prior structures on F_e and F_r. First we describe the prior model for F_e. In this development, we directly address the issue raised by Vander Wiel and Votta (1993) based on their experience at Bell Laboratories. They mention that the errors are typically of two basic types, hard to detect and easy to detect. We model this heterogeneity of the errors by assuming that $F_e = \alpha \delta_{\{\varepsilon_1\}} + (1 - \alpha)\delta_{\{\varepsilon_2\}}$ where $\delta_{\{\varepsilon\}}$ denotes the degenerate distribution at ε, $\varepsilon_1 \leq \varepsilon_2$ are respectively the effects for the hard and easy errors, and α is the proportion of hard errors in the software.

This model can easily be generalized to more than two type of errors. For example, Montgomery (1991, p185) describes a demerit system which has four error types : very serious, serious, moderately serious, and minor. In general, suppose there are V errors types with effects $\varepsilon_1, \ldots, \varepsilon_V$. Then, $F_e = \sum_{v=1}^{V} \alpha_v \delta_{\{\varepsilon_v\}}$, $0 \leq \alpha_v \leq 1$, $\sum_{v=1}^{V} \alpha_v = 1$, and there is a possible ordering among the ε_v's, for example, $\varepsilon_1 \leq \ldots \leq \varepsilon_V$.

First consider the case when there are only two types of errors: hard and easy. In some reviews, the supervisor can assign specific values to ε_1 and ε_2 based on past experience. However, the value of α is typically unknown. We assume that as a result of prior elicitation, the supervisor will specify ranges $[L_e^v, U_e^v]$, $v = 1, 2$, for possible values of ε_1 and ε_2. The two ranges may or may not overlap, but $L_e^1 \leq L_e^2$ and $U_e^1 \leq U_e^2$. For example, the supervisor may specify $5\% \leq \varepsilon_1 \leq 20\%$ for the hard errors and $15\% \leq \varepsilon_2 \leq 40\%$ for the easy errors. We propose an uniform prior model: $(\varepsilon_1, \varepsilon_2) \sim \text{Uniform}([L_e^1, U_e^1] \bigotimes [L_e^2, U_e^2] \cap \{\varepsilon_1 \leq \varepsilon_2\})$. We further assume that α has a $\text{Beta}(\kappa_e \gamma_e, \kappa_e(1 - \gamma_e))$ prior. The hyperparameter $0 \leq \gamma_e \leq 1$ represents one's prior belief about the proportion of hard errors in the software. Large γ_e implies large α values are more likely and vice versa. The concentration parameter $\kappa_e > 0$ represents how strongly one believes in the value of γ_e. If κ_e is large, then an α generated from the above prior will be close to γ_e with high probability.

In the general case of V types of errors, we assume that $\underset{\sim}{\varepsilon} = (\varepsilon_1, \ldots, \varepsilon_V) \sim \text{Uniform}(\bigotimes_{v=1}^{V} [L_e^v, U_e^v] \cap \{\varepsilon_1 \leq \ldots \leq \varepsilon_V\})$. We also assume that $\underset{\sim}{\alpha} = (\alpha_1, \ldots, \alpha_V)$ has a $\text{Dirichlet}(\kappa_e \underset{\sim}{\gamma}_e)$ prior distribution, where $\kappa_e > 0$ is the concentration parameter and $\gamma_e = (\gamma_e^1, \ldots, \gamma_e^V)$ with $0 \leq \gamma_e^v \leq 1$ and $\sum_{v=1}^{V} \gamma_e^v = 1$.

We next proceed to describe the prior model for the reviewer effect distribution F_r. This model is similar to the model for F_e. Suppose there are W different types of reviewers with reviewer effects $\rho_1, \ldots, \rho_W$. We then assume that $F_r = \sum_{w=1}^{W} \beta_w \delta_{\{\rho_w\}}$, where $0 \leq \beta_w \leq 1$ and $\sum_{w=1}^{W} \beta_w = 1$. There could be an ordering among the ρ_w's reflecting the varying degree of expertise of different reviewer types, for example, $\rho_1 \leq \ldots \leq \rho_W$. The β_w's represent the proportion of each type of reviewers in the study. The prior structures on ρ_w and β_w are also similar to the F_e case. We assume that $\rho = (\rho_1, \ldots, \rho_W) \sim \text{Uniform}(\bigotimes_{w=1}^{W} [L_r^w, U_r^w] \cap \{\rho_1 \leq \ldots \leq \rho_W\}$. We also assume that $\beta = (\beta_1, \ldots, \beta_W)$ has a Dirichlet $(\kappa_r \gamma_r)$ prior distribution, where $\kappa_r > 0$ is the concentration parameter and $\gamma_r = (\gamma_r^1, \ldots, \gamma_r^W)$ with $0 \leq \gamma_r^w \leq 1$ and $\sum_{w=1}^{W} \gamma_r^w = 1$.

In the following, we describe a Markov Chain Monte Carlo (MCMC) algorithm for posterior analysis of this model. We remark here that our analysis is, in fact, more general than the specific prior model we describe here. For example, our sampling algorithm can be easily modified if instead of assuming Uniform priors on ε and ρ, we assume independent truncated Beta priors.

We complete the prior description by specifying a prior distribution for N. Here, our model is flexible enough to incorporate a wide variety of prior choices. We show in the following subsection that the conditional posterior of N is of a common parametric form for any of the following prior choices, and hence Gibbs sampling can be routinely implemented. The choices are: (i) Poisson prior on N, (ii) Gamma-mixed-Poisson prior on N, (iii) Jeffreys prior $\pi_1(N) = 1/N$, $N = 1, 2, \ldots$, and (iv) discrete uniform prior $\pi_0(N) \equiv 1$, $N = 1, 2, \ldots$. Other priors for N are also possible, however, then the conditional posterior of N may not be of a common form, and one may need to use specialized sampling methods (such as Metropolis algorithm).

The priors in (iii) and (iv) mentioned above are improper, and it is crucial to verify that the resulting posteriors for N are proper before MCMC sampling is attempted. We provide sufficient conditions for properness of the marginal posteriors of N.

Result 1 *Suppose $L_e^v > 0 \; \forall \; 1 \leq v \leq V$ and $L_r^w > 0 \; \forall \; 1 \leq w \leq W$. Then the marginal posteriors $\pi_0(N \mid data)$ and $\pi_1(N \mid data)$ correponding to the discrete uniform prior $\pi_0(N)$ and Jeffreys' prior $\pi_1(N)$ are both proper.*

Proof : The proof is similar to the proof of Result 2 in Basu and Ebrahimi (1997) ∎

4.3 Conditional distributions

A closed form posterior analysis of N for the proposed model is intractable. We take recourse to Gibbs sampling to estimate the posterior distribution of N and other quantities of interest. We refer the reader to Gelfand (1994) and the book by

Gilks et al (1996) for broad description of Gibbs sampling, its convergence issues and the convergence diagnostics.

The Gibbs sampler simulates alternately and iteratively from the conditional distribution of one parameter given the values of the remaining parameters and the data. In our setup, the data is the x matrix whose last $N - n$ rows are zero-rows. The parameter of interest is N. Other parameters are α, $\underset{\sim}{\varepsilon}$, β, ρ, $\underset{\sim}{e}_N$, and $\underset{\sim}{r}_k$.

There is, however, a catch in a direct implementation of the Gibbs sampler. The dimension of the latent vector $\underset{\sim}{e}_N = (e_1, \ldots, e_N)$ is determined by N. Hence, given $\underset{\sim}{e}_N$, the value of N is completely determined, thus resulting in a reducible Markov chain.

A solution to the above problem can be suggested if we assume, a priori, that N is bounded above by M, i.e., the possible values of N are $1, 2, \ldots, M$ where M is a large positive integer. We can then assume that there are actually M many $e_1, \ldots, e_M$. These e_i's are, as before, i.i.d. from the distribution F_e. However, this distribution specification is no longer conditional on N.

With this formulation, a Gibbs sampling based method can be outlined. At each iteration of the Gibbs sampler, $e_1, \ldots, e_N$ given $\{N$, other parameters, and data$\}$ will be generated from their full conditional posterior distribution. On the other hand, given $\{N$, other parameters, and data$\}$, the full conditional distributions of $e_{N+1}, \ldots, e_M$ are i.i.d. $\sim F_e$, i.e., same as their prior distribution.

The N generation step of the Gibbs sampler will also be straightforward since the dimension of $\underset{\sim}{e}$ is no longer determined by N. Thus, the Gibbs sampler can be implemented in a routine manner.

There is a possible shortfall of the above mentioned strategy. At each iteration of the above Gibbs algorithm, $M - N$ many $e_{N+1}, \ldots, e_M$ will be generated from F_e, their so-called prior distribution. That is, the data will not influence this step of the generation in any fashion. If M is much larger than where most of the posterior of N is concentrated (for example, if $M = 300$ whereas the posterior of N is mostly concentrated at $N \leq 100$), this may cause the Gibbs sampler to be slow mixing and converge very slowly. The problem being that the sampled $e_{N+1}, \ldots, e_M$ from the so-called prior F_e will enter the full conditional distributions of the other parameters. For example, the full conditional distribution of $\underset{\sim}{\alpha}$ depends crucially on the generated values of these latter $e_{N+1}, \ldots, e_M$.

We here suggest an alternative strategy for implementation of the Gibbs sampler. Recall that the involved parameters are N, $\underset{\sim}{\alpha}$, $\underset{\sim}{\varepsilon}$, β, ρ, $\underset{\sim}{r}_k$, and $\underset{\sim}{e}_N$. In this alternative strategy, each parameter is drawn from its full conditional distribution except for N and $\underset{\sim}{e}_N$. For these latter two, we generate them jointly from the joint conditional distribution of $N, \underset{\sim}{e}_N \mid \text{data}, \underset{\sim}{\alpha}, \underset{\sim}{\varepsilon}, \beta, \rho, \underset{\sim}{r}_k$. This last step is achieved by first generating N from the conditional distribution of $N \mid \text{data}, \underset{\sim}{\alpha}, \underset{\sim}{\varepsilon}, \beta, \rho, \underset{\sim}{r}_k$ and then generating $\underset{\sim}{e}_N$ from $\underset{\sim}{e}_N \mid \text{data}, N, \underset{\sim}{\alpha}, \underset{\sim}{\varepsilon}, \beta, \rho, \underset{\sim}{r}_k$.

The full conditional distributions required for the Gibbs iterations are described below.

- The following definitions are used in the conditional distributions of $\underset{\sim}{\alpha}, \underset{\sim}{\varepsilon}, \beta$, and ρ. These definitions are conditional on $N, \underset{\sim}{e}_N$, and $\underset{\sim}{r}_k$, i.e., we assume that their values are available from a previous iteration.

Let $\mathcal{E}_v = \{i : e_i = \varepsilon_v, \ i = 1, \ldots, N\}$ and let $\eta_v = \#\mathcal{E}_v = $ cardinality of $\mathcal{E}_v$, $v = 1, \ldots, V$. Note that $\sum_{v=1}^{V} \eta_v = N$. Also, define z_{vj} to be the sum of only those elements in the j-th column of the x matrix for which $i \in \mathcal{E}_v$. Thus, $z_{vj} = \sum_{i \in \mathcal{E}_v} x_{ij}, \ v = 1, \ldots, V, \ j = 1, \ldots, k$.

The corresponding definitions for $\underset{\sim}{r}_k$ and $\underset{\sim}{\rho}$ follow. Let $\mathcal{R}_w = \{j : r_j = \rho_w, \ j = 1, \ldots, k\}$, and let $\mu_w = \#\mathcal{R}_w$. We have $\sum_{w=1}^{W} \mu_w = k$. Finally, define $y_{wi} = \sum_{j \in \mathcal{R}_w} x_{ij}, \ w = 1, \ldots, W, \ i = 1, \ldots, N$.

Now, we describe the full conditional distributions.

$\underset{\sim}{\alpha}$: The full conditional distribution of $\underset{\sim}{\alpha} = (\alpha_1, \ldots, \alpha_V)$ is given by $\pi(\underset{\sim}{\alpha} \mid \text{data}, N, \underset{\sim}{\varepsilon}, \beta, \underset{\sim}{\rho}, \underset{\sim}{e}_N, \underset{\sim}{r}_k) = \text{Dirichlet}(\kappa_e \underset{\sim}{\gamma}_e \oplus \underset{\sim}{\eta})$ where $\underset{\sim}{\eta} = (\eta_1, \ldots, \eta_V)$ and $\oplus$ implies coordinate-wise addition. Note that this conditional distribution is independent of $\beta, \underset{\sim}{\rho}$, and $\underset{\sim}{r}_k$.

$\underset{\sim}{\varepsilon}$: Let $\pi(\varepsilon_1, \ldots, \varepsilon_V)$ denote a generic prior on $\underset{\sim}{\varepsilon}$. Then, a series of simple calculations shows that the full conditional distribution of $\underset{\sim}{\varepsilon}$ is given by $\pi(\underset{\sim}{\varepsilon} \mid \text{data}, N, \underset{\sim}{\alpha}, \beta, \underset{\sim}{\rho}, \underset{\sim}{e}_N, \underset{\sim}{r}_k) \propto \pi(\varepsilon_1, \ldots, \varepsilon_V) \prod_{v=1}^{V} \{\prod_{j=1}^{k} r_j^{z_{vj}} \varepsilon_v^{z_{vj}} (1 - \varepsilon_v r_j)^{\eta_v - z_{vj}}\}$. Note that: (i) this conditional distribution is independent of $\underset{\sim}{\alpha}, \beta$, and $\underset{\sim}{\rho}$; and (ii) if $\varepsilon_1, \ldots, \varepsilon_V$ are independent a priori, i.e., $\pi(\varepsilon_1, \ldots, \varepsilon_V) = \prod_{v=1}^{V} \pi(\varepsilon_v)$, then they are also independent conditionally due to the product structure of the conditional density.

When $\underset{\sim}{\varepsilon}$ has the Uniform$(\bigotimes_{v=1}^{V} [L_e^v, U_e^v] \cap \{\varepsilon_1 \leq \ldots \leq \varepsilon_V\}$ of section 4.2, we generate them sequentially, first ε_1, then ε_2, and so on. We have, $\pi(\varepsilon_v \mid \text{data}, \text{other parameters, and } \varepsilon_1, \ldots, \varepsilon_{v-1}) \propto \prod_{j=1}^{k} \{\varepsilon_v^{z_{vj}} (1 - \varepsilon_v r_j)^{\eta_v - z_{vj}}\}$ on the support interval $[\max(\varepsilon_{v-1}, L_e^v), U_e^v]$.

The above conditional density is not of a common parametric form. However, in Result 2 below, we show that it is, in fact, a log-concave density. We use the adaptive rejection sampling method of Gilks and Wild (1992) to simulate from this log-concave density.

$\underset{\sim}{\beta}$: This distribution is similar to $\underset{\sim}{\alpha}$. We have $\pi(\beta \mid \text{data}, N, \underset{\sim}{\alpha}, \underset{\sim}{\varepsilon}, \underset{\sim}{\rho}, \underset{\sim}{e}_N, \underset{\sim}{r}_k) = \text{Dirichlet}(\kappa_r \underset{\sim}{\gamma}_r \oplus \underset{\sim}{\mu})$, where $\underset{\sim}{\mu} = (\mu_1, \ldots, \mu_W)$. Note that this conditional distribution is independent of $\underset{\sim}{\alpha}, \underset{\sim}{\varepsilon}$, and $\underset{\sim}{e}_N$.

$\underset{\sim}{\rho}$: This distribution is similar to $\underset{\sim}{\varepsilon}$. For a generic $\pi(\rho_1, \ldots, \rho_W)$ prior on $\underset{\sim}{\rho}$, we have $\pi(\underset{\sim}{\rho} \mid \text{data}, N, \underset{\sim}{\alpha}, \underset{\sim}{\varepsilon}, \beta, \underset{\sim}{e}_N, \underset{\sim}{r}_k) \propto \pi(\rho_1, \ldots, \rho_W) \prod_{w=1}^{W} \{\prod_{i=1}^{N} e_i^{y_{wi}} \rho_w^{y_{wi}} (1 - e_i \rho_w)^{\mu_w - y_{wi}}\}$.

For $\pi(\rho_1,\ldots,\rho_W) = \text{Uniform}(\bigotimes_{w=1}^{W}[L_r^w,U_r^w] \cap \{\rho_1 \leq \ldots \leq \rho_W\}$, we generate $\rho_1,\ldots,\rho_W$ sequentially. We have, $\pi(\rho_w \,|\, \text{data, other parameters, and } \rho_1,\ldots,$ $\rho_{w-1}) \propto \prod_{i=1}^{N}\{\rho_w^{y_{wi}}(1-e_i\rho_w)^{\mu_w - y_{wi}}\}$ on the support interval $[\max(\rho_{w-1},L_r^w),U_r^w]$. Result 2 below shows that this is a log-concave density, and hence adaptive rejection sampling can again be applied.

$\underset{\sim}{r}_k$: It is easy to see that $r_1,\ldots,r_k$ are conditionally independent due to the product structure of the likelihood in (6). Also, note that each r_j is a discrete variable taking values $\rho_1,\ldots,\rho_W$. For each $1 \leq j \leq k$, we have

$$\pi(r_j \,|\, \text{data}, N, \underset{\sim}{\alpha}, \underset{\sim}{\varepsilon}, \beta, \rho, \underset{\sim}{e}_N) = \sum_{w=1}^{W} \beta_{wj}^* \,\delta_{\{\rho_w\}} \text{ where } \beta_{wj}^* = c\cdot\beta_w \prod_{i=1}^{N}\rho_w^{x_{ij}}(1-$$

$e_i\rho_w)^{1-x_{ij}}$, $w = 1,\ldots,W$, and the constant of proportionality c is such that $\sum_{w=1}^{W}\beta_{wj}^* = 1$. Note that the conditional distribution is independent of $r_1,\ldots,r_{j-1},r_{j+1},\ldots,r_k$, $\underset{\sim}{\alpha}$, and $\underset{\sim}{\varepsilon}$.

N and $\underset{\sim}{e}_N$: We mentioned earlier that we simulate N and $\underset{\sim}{e}_N$ jointly from their joint conditional distribution. Notice, $\pi(\underset{\sim}{e}_N, N \,|\, \text{data}, \underset{\sim}{\alpha}, \underset{\sim}{\varepsilon}, \beta, \rho, \underset{\sim}{r}_k) = \pi(N \,|\, \text{data}, \underset{\sim}{\alpha}, \underset{\sim}{\varepsilon}, \beta, \rho, \underset{\sim}{r}_k) \cdot \pi(\underset{\sim}{e}_N \,|\, \text{data}, N, \underset{\sim}{\alpha}, \underset{\sim}{\varepsilon}, \beta, \rho, \underset{\sim}{r}_k)$.

(i) We first describe $\pi(\underset{\sim}{e}_N \,|\, \text{data}, N, \underset{\sim}{\alpha}, \underset{\sim}{\varepsilon}, \beta, \rho, \underset{\sim}{r}_k)$. This is similar to the conditional distribution of $\underset{\sim}{r}_k$. We have $e_1,\ldots,e_N$ conditionally independent,

and for every $1 \leq i \leq N$, $\pi(e_i \,|\, \text{data}, N, \underset{\sim}{\alpha}, \underset{\sim}{\varepsilon}, \beta, \rho, \underset{\sim}{r}_k) = \sum_{v=1}^{V} \alpha_{vi}^* \,\delta_{\{\varepsilon_v\}}$,

where $\alpha_{vi}^* = c\cdot\alpha_v \prod_{j=1}^{k}\varepsilon_v^{x_{ij}}(1-\varepsilon_v r_j)^{1-x_{ij}}$, $v = 1,\ldots,V$, and the constant

of proportionality c is such that $\sum_{v=1}^{V} \alpha_{vi}^* = 1$.

(ii) Next, we state the distribution of N, conditional on $\{\text{data}, \underset{\sim}{\alpha}, \underset{\sim}{\varepsilon}, \beta, \rho, \underset{\sim}{r}_k\}$ but not conditioned on $\underset{\sim}{e}_N$'s. Note that N is a discrete variable taking positive integer values. Let $\pi(N)$ denote a generic prior on N.

It is easy to see that only the last $N - n$ zero rows of the x matrix contribute to the conditional distribution of N. It follows that $\pi(N \,|\, \text{data}, \underset{\sim}{\alpha}, \underset{\sim}{\varepsilon}, \beta, \rho, \underset{\sim}{r}_k) \propto \pi(N)\,\{N!/(N-n)!\}\,[p(\underset{\sim}{\alpha}, \underset{\sim}{\varepsilon}, \underset{\sim}{r}_k)]^{N-n}$ where $p(\underset{\sim}{\alpha}, \underset{\sim}{\varepsilon}, \underset{\sim}{r}_k) = \sum_{v=1}^{V} \alpha_v\{\prod_{j=1}^{k}(1 - \varepsilon_v r_j)\}$. Note that this distribution is independent of β and ρ.

This completes the description of the conditional distribution. The Gibbs sampler simulates iteratively and sequentially from these distribution. The conditional distributions of ε_v and ρ_w stated above are not of a common parametric form. The following result shows that these densities are log-concave, and hence we can simulate from them by using the adaptive rejection sampling of Gilks and Wild (1992).

Result 2 (i) $\pi(\varepsilon_v \,|\, \text{data}, N, \underset{\sim}{\alpha}, \beta, \rho, \underset{\sim}{e}_N, \underset{\sim}{r}_k, \varepsilon_1,\ldots,\varepsilon_{v-1})$ is log-concave in ε_v.

(ii) $\pi(\rho_w \mid data, N, \underset{\sim}{\alpha}, \underset{\sim}{\varepsilon}, \beta, , \underset{\sim}{e}_N, \underset{\sim}{r}_k, \rho_1, \ldots, \rho_{w-1})$ is log-concave in ρ_w.

Proof : We prove the result for ε_v. The proof for ρ_w is similar. For $\varepsilon_v \in$ its support interval, we have $\log \pi(\varepsilon_v \mid$ data and other parameters$)$ = constant term +

$$\sum_{j=1}^{k} \{z_{vj} \log \varepsilon_v + (\eta_v - z_{vj}) \log(1 - \varepsilon_v r_j)\} \text{ and } \tfrac{\partial^2}{\partial \varepsilon_v^2} \log \pi(\varepsilon_v \mid \ldots) = \sum_{j=1}^{k} \{-z_{vj}/\varepsilon_v^2 - (\eta_v - z_{vj}) r_j^2/(1 - \varepsilon_v r_j)^2 \} < 0.$$

This completes the proof ∎

We next comment on simulation from the conditional distribution of $N \mid$ data, $\underset{\sim}{\alpha}$, $\underset{\sim}{\varepsilon}, \beta, \rho, \underset{\sim}{r}_k$. If the support of the prior $\pi(N)$ is on a bounded set $\{0, \ldots, M\}$, then conditionally, N is a discrete random variable supported on $\{n, \ldots, M\}$ and hence can be easily simulated.

Even when the support for $\pi(N)$ is not bounded, the conditional posteriors of N are of common forms for the four priors mentioned in section 4.2. For the Poisson prior $\pi(N) = \exp(-\lambda) \lambda^N/N!$, $N = 0, 1, \ldots$, considered by George and Robert (1992), the conditional distribution of $N - n$ is again Poisson($\lambda p(\underset{\sim}{\alpha}, \underset{\sim}{\varepsilon}, \underset{\sim}{r}_k)$). The Gamma-mixed-Poisson prior is considered by Jewel (1985). Here, $\pi(N \mid \lambda) =$ Poisson(λ) where $\lambda \sim$ Gamma(a, b). For this prior, the conditional posterior of λ is Gamma($a+n, (1-p(\underset{\sim}{\alpha}, \underset{\sim}{\varepsilon}, \underset{\sim}{r}_k)+1/b)^{-1}$) whereas $N-n \mid \{\lambda, $ data, and other parameters$\}$ $\sim$ Poisson($\lambda p(\underset{\sim}{\alpha}, \underset{\sim}{\varepsilon}, \underset{\sim}{r}_k)$). For the Jeffreys' prior $\pi_1(N) = 1/N$, the conditional distribution of $N - n$ is Negative Binomial with parameters n and $p(\underset{\sim}{\alpha}, \underset{\sim}{\varepsilon}, \underset{\sim}{r}_k)$. The Jeffreys' prior was considered by Castledine (1981), Smith (1991), George and Robert (1992). Finally, for the improper discrete uniform prior $\pi_0(N) \equiv 1$, the conditional distribution of $N - n$ is again Negative Binomial with parameters $n + 1$ and $p(\underset{\sim}{\alpha}, \underset{\sim}{\varepsilon}, \underset{\sim}{r}_k)$.

5 Applications

Our proposed heterogeneous reviewers and errors model uses each individual entry of the x data matrix. Most reports of either software review or animal trapping surveys, however, do not report the complete data. Instead, only a summary of the data is reported, for example, n_j = the number of errors detected by each reviewer and/or y_i = the number of times each error was detected. These are only the column and row sums of the x matrix, whereas we need the individual x_{ij} entries for our model.

A complete description of the x matrix is available for the software review data listed in Eick et al (1993) on errors identified for a particular feature of the AT&T 5 ESS switch. Six reviewers, $k = 6$, found a total of 47 distinct errors. Out of these 47 errors, 4 errors were detected during the review meeting and are not included in our analysis. The x_{ij} entries for the remaining 43 errors are shown in Table 1. Recall that $x_{ij} = 1$ if i-th errors is detected by j-th reviewer and $x_{ij} = 0$ otherwise. Eick et al (1993) estimate the total number of errors to be $\widehat{N}_{\text{mle}} = 65$ based on a maximum likelihood analysis assuming independent but non-identical reviewers. Ebrahimi (1996) assumes that reviewers can be dependent. He obtains $\widehat{N} = 71$, based on a maximum likelihood analysis after kernel based estimation. Basu and Ebrahimi (1997) perform Bayesian analyses for the models described in section 3. A summary of their results from the IID model, the heterogeneous reviewers model and the heterogeneous errors models are shown in Table 2.

Table 1: AT&T 5 ESS Switch review

Error	Reviewer j							Error	Reviewer j						
i	1	2	3	4	5	6	y_i	i	1	2	3	4	5	6	y_i
1	1	0	0	0	0	0	1	23	0	0	0	0	0	1	1
2	0	0	0	1	0	1	2	24	0	0	0	0	1	0	1
3	0	0	0	0	1	0	1	25	0	0	0	1	1	0	2
4	1	0	0	0	0	0	1	26	1	0	0	0	0	0	1
5	0	0	0	1	0	0	1	27	0	0	0	0	1	0	1
6	0	0	0	1	0	0	1	28	0	0	1	0	0	0	1
7	0	0	0	1	0	0	1	29	1	0	0	0	0	0	1
8	1	0	0	0	0	0	1	30	1	0	0	1	1	0	3
9	1	0	0	0	0	0	1	31	1	0	0	1	0	0	2
10	0	0	0	1	0	0	1	32	1	0	0	0	0	0	1
11	1	0	0	1	0	0	2	33	1	0	0	0	0	0	1
12	1	0	0	0	0	0	1	34	0	0	1	0	0	0	1
13	1	0	0	1	0	0	2	35	0	0	1	0	0	0	1
14	1	0	0	0	0	0	1	36	0	0	0	0	0	1	1
15	1	0	0	1	0	0	2	37	0	0	1	0	0	0	1
16	1	0	0	1	0	0	2	38	1	0	0	0	0	1	2
17	1	1	0	1	1	1	5	39	1	0	0	0	0	0	1
18	1	0	0	0	0	1	2	40	1	0	0	0	0	0	1
19	0	1	0	0	0	0	1	41	0	0	0	0	1	0	1
20	0	1	0	0	0	0	1	42	1	0	0	0	0	0	1
21	1	0	0	1	0	0	2	43	1	0	0	0	0	0	1
22	1	0	0	1	0	0	2	n_j	25	3	4	13	9	6	

The variation of the n_j values clearly suggests heterogeneity among the six reviewers. On the other hand, the row sums y_i indicate presence of some heterogeneity among the errors. We analyze the data using our proposed heterogeneous reviewers and errors model. For simplicity, we chose $V = 2$ and $W = 2$, i.e., there are two types of errors and two types of reviewers. The following priors are used : (i) $\kappa_e = \kappa_r = 2$, $\gamma_e = \gamma_r = 1/2$, (These result in Uniform$(0, 1)$ priors for α_1 and β_1.), (ii) $(\varepsilon_1, \varepsilon_2)$ and $(\rho_1, \rho_2) \sim \mathrm{U}([0.01, 0.5] \times [0.3, 1])$, and (iii) $N \sim$ improper discrete uniform distribution on $\{1, 2, \ldots\}$. The hyperparameters are chosen empirically based on data and results of earlier analysis.

The posterior point and interval estimates of N based on our model are shown in the last row of Table 2. We find that our heterogeneous reviewers and errors model yield significantly higher values of posterior mean and posterior median of N than the other models. We have also tried $\pi(N) =$ the Jeffreys prior $1/N$. The posterior mean of N dropped to 101, and the posterior median was 84. This is expected as the prior $\pi(N) = 1/N$ clearly prefers smaller values of N.

The kernel estimate of the posterior of N is shown in Figure 1. This estimate is right skewed. In fact, very large values of N, even $N \geq 1000$, are occasionally drawn in the Gibbs iteration. This right skewness resulted in a high value for the posterior mean of N. On the other hand, even for the outlier resistant posterior median, the value from our heterogeneous reviewers and errors model is again significantly higher than the other models.

Figure 2 shows the posterior density estimates of $\alpha_1, \varepsilon_1, \varepsilon_2, \beta_1, \rho_1$, and ρ_2. The posterior summaries of these parameters are shown in Table 3. The posterior density of α_1 is left skewed. In fact, the posterior mean and median of α are 0.729 and 0.815, respectively. The analysis thus suggests that there is a higher proportion of "hard" errors, but there are some "easy" errors as well (we cannot really say that α is 1). On the other hand, the posterior density estimate for β_1 is only slightly left skewed with a posterior mean of 0.644 and posterior median of 0.656. Thus, the analysis suggests that both types of reviewers are present in significant proportions.

We also obtain credible intervals for N in Table 2. Note that these are equal posterior tail area intervals and may not be HPD intervals. The credible intervals obtained under our model are wider than the earlier models. They suggest that modeling simultaneous heterogeneities incorporates large variation of N values into our model which was disregarded in the earlier models. This variation is also evident in the long right tail of the posterior density of N. The kernel density plots and the posterior summaries are obtained using the CODA software (Best et al. (1995)).

In summary, the data provide evidence for heterogeneity among the reviewers and also among the errors. If both heterogeneity are modeled, then the estimate of N can be significantly higher than estimates obtained under earlier models.

References

1. Achcar, J.A., Dey, D.K. and Niverthi, M., A Bayesian approach using non-homogeneous Poisson process for software reliability models, in *this volume* (1997).

2. Basu, S. and Ebrahimi, N., Bayesian methods for error detection and size of population, Tech. Rept., Northern Illinois University (1997).

3. Best, N., Cowles, M.K. and Vines, K., *CODA: Convergence Diagnosis and Output Analysis Software for Gibbs sampling output*, version 0.30 (1996).

4. Bickel, P.J. and Yahav, J.A., On estimating the number of unseen species and system reliability, in *Stat. decision Theory and Related Topics IV*, (S.S. Gupta, J.O. Berger Eds.), Springer-Verlag, Vol. 2, 265-272 (1988).

5. Burnham, K.P. and Overton, W.S., Estimation of the size of a closed population when capture probabilities vary among animals, *Biometrika*, 65, 625-633 (1978).

6. Castledine, B., A Bayesian analysis of multiple-recapture sampling from a closed population, *Biometrika*, 67, 197-210 (1981).

7. Ebrahimi, N., On the statistical analysis of the number of errors remain in a software design after inspection, Tech. Rept., Northern Illinois Univ. (1996).

8. Eick, S.G., Loader, C. R., Vander Wiel, S. A., Votta, L. G., How many errors remain in a software design document after inspection?, in *Proc. of the 25th symposium of the interface*, San Diego (1993).

9. Gelfand, A.E., Gibbs Sampling (a contribution to the Encyclopedia of Statistical Sciences), To appear in *the Encyclopedia of Statistical Sciences* (1994).

10. George E.I. and Robert, C.P., Capture-recapture estimation via Gibbs sampling, *Biometrika*, 79, 677-683 (1992).
11. Gilks, W.R., Full conditional distributions, In *Markov Chain Monte Carlo in Practice* (W.R. Gilks, S. Richardson, and D.J. Spiegelhalter Eds.), Chapman and Hall, London (1996).
12. Gilks, W.R. and Wild, P., Adaptive rejection sampling for Gibbs sampling, *Appl. Statist.*, 41, 337-348 (1992).
13. Jewel, W.S., Bayesian estimation of undetected errors, in *Bayesian Statistics 2* (Bernardo et al eds.), Elsevier Science Publishers, 663-672 (1982).
14. Madigan, D. and York, J.C., Bayesian methods for estimation of the size of a closed population, *Biometrika*, 84, 19-31 (1997).
15. Montgomery, D.C., *Introduction to Statistical Quality Control*, 2nd Ed., Wiley, New York (1991).
16. Nayak, T.K., Statistical approaches to modeling and estimating software reliability, in *this volume* (1997).
17. Otis, D.L., Burnham, K.P., White, G.C. and Anderson, D.R., Statistical inference for capture data on closed animal populations, *Wildlife Monographs*, 62, 1-135 (1978).
18. Ries, L.D. and Basu, A.P., Parameter estimation in software reliability growth models, in *this volume*, (1997).
19. Seber, G.A.F., *The Estimation of Animal Abundance and Related Parameters*, 2nd Ed. MacMillan, New York (1982).
20. Sen, A., Analysis of repairable systems – past, present and future, in *this volume* (1997).
21. Smith, P.J., Bayesian analysis for a multiple capture-recapture model, *Biometrika*, 78, 399-408 (1991).
22. Smith, P.J., Bayesian methods for multiple capture-recapture surveys, *Biometrika*, 44, 1177-89 (1988).
23. Vander Wiel, S.A. and Votta, L.G., Assessing software design using capture-recapture methods, *IEEE Trans. on Software Engg.*, 19, 1045-1054 (1993).

Table 2: AT&T Switch: Posterior Means, Medians and approximate credible intervals for N

Model	Mean	Median	95% interval	80% interval
IID model	77.014	75	(54,105)	(59,89)
Heterogeneous Reviewers	66.424	65	(52,82)	(56,74)
Heterogeneous Errors	89.800	84	(58,158)	(65,121)
Heterogeneous Reviewers & Errors	140.000	98	(66.254)	(59,518)

Table 3: Estimated posterior mean (1st entry) and posterior median (2nd entry)

α_1	ε_1	ε_2	β_1	ρ_1	ρ_2
0.729, 0.815	0.184, 0.163	0.694, 0.699	0.644, 0.656	0.194, 0.189	0.726, 0.736

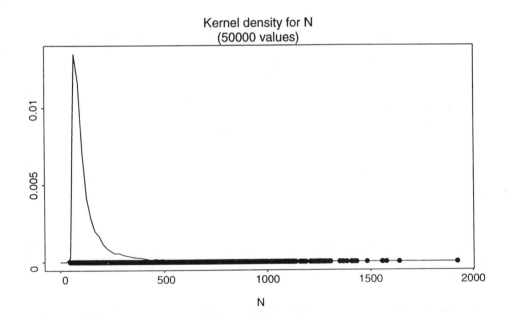

Figure 1: Posterior distribution of N for heterogeneous reviewers and errors model

S. Basu

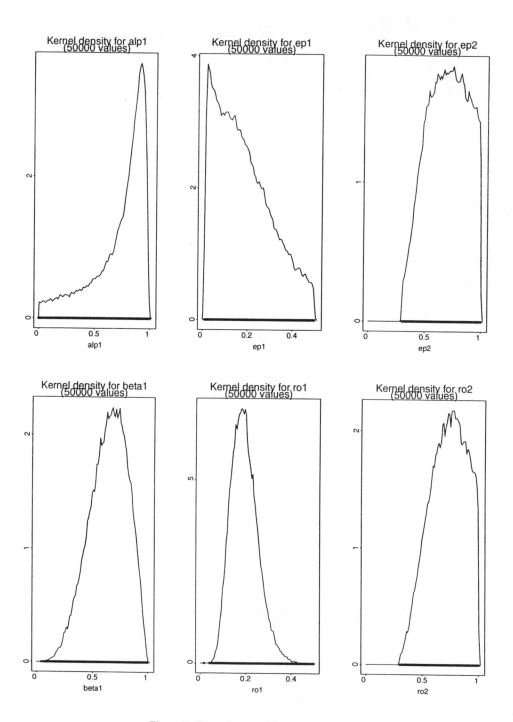

Figure 2: Posterior densities of parameters

THE CLASS OF LIFE DISTRIBUTIONS THAT ARE LAPLACE ORDER DOMINATED BY THE EXPONENTIAL LAW AND ITS RAMIFICATIONS

SUJIT K. BASU

Indian Institute of Management Calcutta

MURARI MITRA

Bengal Engineering College (Deemed University)

The class of life distributions that are Laplace order dominated by the exponentials, commonly called the $\mathcal{L}$-class, is known to be the largest among the important aging families and was introduced by Klefsjö. This article reviews various aspects of this class as available in the literature so far.

1 Introduction

The concept of `aging' is central to typical problems of reliability and survival analysis. Various attempts to capture the phenomenon of aging of equipment or biological species have led to various life distribution classes, of which the important and most researched ones are the classes of increasing failure rate (IFR), increasing failure rate on average (IFRA), decreasing mean residual life (DMRL), new better than used (NBU), new better than used in expectation (NBUE), and harmonic new better than used in expectation (HNBUE) life distributions and their duals (DFR, DFRA, IMRL, NWU, NWUE, and HNWUE classes of life distributions). The well-known sources of details regarding such classes of life distributions are Barlow and Proschan (1975), Rolski (1975), and Klefsjö (1982).

One of the important characteristics common to life distributions belonging to these families is that of Laplace order domination by exponentials with identical means; specifically, if a life distribution F with finite mean μ belongs to any of the above classes, then

$$L_F(s) \le L_G(s) \tag{1}$$

for all $s \ge 0$, where

$$L_F(s) := \mathcal{E}\, e^{-sX} = \int_0^\infty e^{-sT}\, dF(t)$$

and $G(t) = 1 - e^{-t/\mu}$.

Definition 1.1 (Klefsjö (1983)). *A distribution F with support $[0, \infty)$ and mean μ belongs to the $\mathcal{L}$-class of life distributions if its Laplace transform $L_F(\cdot)$ satisfies the above inequality (1) for all $s \ge 0$; its dual i.e. the class of life distributions for which the reverse inequality holds for all $s \ge 0$ is called the $\overline{\mathcal{L}}$ class.*

It is well-known that the inequality (1) is satisfied by the Laplace transforms of all life distributions belonging to the important families mentioned above; thus, the $\mathcal{L}$-class is hierarchically the largest in relation to these classes (see Section 3 of this article).

A corresponding class of discrete life distributions is called the G-class.

Definition 1.2 (Klefsjö (1983)). *A strictly positive integer-valued random variable N or equivalently, its distribution $\{p_k, \ k = 1, 2, \ldots \}$, is said to belong to the $G(\overline{G})$-class of distributions if*

$$\mathcal{E} Z^N \leq (\geq) \mathcal{E} Z^{N^*}, \quad \forall \, 0 \leq Z \leq 1,$$

where N^* is a geometric random variable such that $\mathcal{E} N^* = \mathcal{E} N$, i.e. the probability mass function of N^* is given by

$$P(N^*) = \frac{1}{m} \left(1 - \frac{1}{m} \right)^{k-1}, \quad k = 1, 2, \ldots$$

where $m = \mathcal{E} N \, (>1)$.

One of the basic reasons for the importance of the $\mathcal{L}$-class lies in the fact that in some reliability studies, it may be easier to determine the Laplace transform than the corresponding survival function explicitly. Various interpretations of the inequality (1) in the context of aging and/or other aspects are given in Klefsjö (1983) and Alzaid, Kim, and Proschan (1991).

In this paper, we propose to present a review of various aspects of the $\mathcal{L}$-class of life distributions. In Section 2, we summarize the properties of this class of distributions so far available in the literature; Section 3 deals with some interesting ideas that emerged from the concept of the $\mathcal{L}$-class.

2 Basic Properties of the $\mathcal{L}$-class of Life Distributions

In this section, we highlight some basic properties of life distributions of the $\mathcal{L}$-class.

2.1 *Closure Properties*

Some of the earliest works in the $\mathcal{L}$-class due to Klefsjö (1983) focused on issues concerning preservation of the $\mathcal{L} \left(\overline{\mathcal{L}} \right)$-property under typical reliability operations like formation of coherent structures, mixtures, and convolution. Such issues regarding the IFR, IFRA, NBU, NBUE, DMRL, and HNBUE families (and their duals) had already been resolved (see Barlow and Proschan (1975), Klefsjö (1982), and Bondesson (1981)). The results obtained by Klefsjö in this context for the $\mathcal{L}$ and $\overline{\mathcal{L}}$ classes are summarized in the following table.

Table 1

Reliability Operations	$\mathcal{L}$-Class	$\overline{\mathcal{L}}$-Class
Formation of coherent structures	Not preserved	Not preserved
Mixture	Not preserved	Preserved
Convolution	Preserved	Not preserved

Chaudhuri (1995) proved the following theorem to conclude that the $\mathcal{L}$-class is closed under weak convergence:

Theorem 2.1 (Chaudhuri (1995)). *Let $\{F_n, n \geq 1\}$ be a sequence of life distribution functions such that F_n converges weakly to F as $n \to \infty$. If $F_n \in \mathcal{L}$ for all $n \geq 1$, then the limiting distribution F also has the $\mathcal{L}$ property. In such a situation, $\lim_{n \to \infty} \mathcal{E}_{F_n} X^r = \mathcal{E}_F X^r$ for all $r \geq 1$.*

2.2 Finiteness of Moments etc.

Unlike in cases of other well-known life distribution families mentioned above, not much is known about the existence of moments of various orders of distributions in the $\mathcal{L}$-class. For example, it is known that distributions in the HNBUE and hence in all hierarchically smaller classes of life distributions have finite moments of *all* (positive) orders (Klefsjö (1982)); but, as far as the $\mathcal{L}$-class is concerned, the most that is known so far in this regard is that for $F \in \mathcal{L}$,

$$\mathcal{E}_F X_r < \infty, \tag{2}$$

for all $-1 < r \leq 2$ (Bhattacharjee and Sengupta (1996), Basu and Mitra (1997)). Bhattacharjee and Sengupta (1996) further showed that as in the HNBUE class,

$$\mathcal{E}_F X^2 \leq 2 (\mathcal{E}_F X)^2 \tag{3}$$

so that the coefficient of variation of $F \in \mathcal{L}$ cannot exceed unity; we propose to come back to this theme once again in Section 3 while discussing some results concerning characterization of the exponential distribution in the $\mathcal{L}$-class.

In this context, it may also be relevant to present a result involving expectation of completely monotone functions of random variables whose distributions possess the $\mathcal{L}$-property.

Definition 2.1 *A non-negative function ϕ on $[0, \infty)$ is completely monotone if it possesses derivatives $\phi^{(n)}$ of all orders and $(-1)^n \phi^{(n)} (t) \geq 0$ for $t \geq 0$ and $n = 1, 2, \dots$.*

The following result now follows from a result of Alzaid, Kim and Proschan (1989):

Theorem 2.2 *$F \in \mathcal{L}$ if and only if $E_F (\phi(X)) \geq E_G (\phi(x))$ for each non-negative function ϕ having a completely monotone derivative and for which the expectations exists, G being the exponential distribution having mean $\mu = \int_0^\infty \{1 - F(t)\} \, dt$.*

2.3 Lower Bound for Survival Function

Recently, Chaudhuri, Deshpande, and Dharmadhikari (1996) proposed a lower bound for survival functions belonging to the $\mathcal{L}$-class as follows.

Theorem 2.3 *Let F be a life distribution with mean μ. If $F \in \mathcal{L}$, then $1 - F(t) \leq \dfrac{t}{\mu} e^{\left(1 - \frac{t}{\mu}\right)}$*

for $0 \leq t \leq \mu$.

Such a bound is of importance whenever a survival function is not completely known, but known to have the $\mathcal{L}$-property.

2.4 Shock Model Formulations

Another important aspect of the study concerning the class of $\mathcal{L}$-distributions has been to investigate how such distributions arise in the context of shock models. Interesting results in this regard are discussed in Klefsjö (1983), Alzaid, Kim, and Proschan (1989), and Bhattacharjee and Basu (1994).

Consider the standard framework of general shock model distribution S with survival probability

$$1 - S(t) = \mathcal{E}\overline{P}_{N(t)} = \sum_{k=0}^{\infty} \overline{P}_k \, P\{N(t) = k\}, \quad t \geq 0, \tag{4}$$

where $\overline{P}_k$ is the probability of surviving k shocks and $N(t)$ is the number of shocks that arise till time $t \geq 0$ according to some specified process. Klefsjö (1983) gave sufficient conditions for the survival function of a device subject to shocks governed by a birth process to have an $\mathcal{L}$-distribution. It is to be noted that if all the birth intensities λ_k equal λ, then one has the homogeneous Poisson shock model studied by Esary, Marshall, and Proschan (1973). An application of Klefsjö's sufficient conditions to this special case would lead to the following important theorem.

Theorem 2.4 (Klefsjö (1983)). *Suppose the shocks arise according to a Poisson process with intensity parameter* λ *and let* $\{p_k \ k = 1, 2,...\}$ *be the probability distribution of the number* N *of shocks to failure so that* $\overline{P}_k := P(N > k) = \sum_{k+1}^{\infty} p_j$. *If* $\{p_k \ k \geq 1\} \in G$ $(\overline{G})$, *then the survival function S given in (2) belongs to the* $\mathcal{L}(\overline{\mathcal{L}})$-*class.*

Bhattacharjee and Basu (1994) later extended the above theorem in the case where the shocks arise in accordance with a renewal process driven by a life distribution function F which belongs to the $\mathcal{L}$ family. Such results indicate that the G property, which is analogous to the $\mathcal{L}$ property in the discrete case, gets translated into the $\mathcal{L}$ property for the survival function of a device subjected to shocks arising according to a renewal process. Counterparts of various shock model results (including those for a cumulative damage model) for NBUE and HNBUE families by Block and Savits (1978) and Klefsjö (1981) have also been obtained for the $\mathcal{L}$-class.

3 Ramification of the $\mathcal{L}$-class of Life Distributions

We now summarize the ramifications of the $\mathcal{L}$-property reported in the literature so far. Typically, these ramifications have resulted from efforts to verify, in the context of the $\mathcal{L}$-class of life distributions, the properties usually common to well-known aging classes of life distributions. One such ramification has its root in the observation of Bhattacharjee and Sengupta (1995) that, though, as in the cases of IFR (Basu and Simons (1984)), NBUE, or HNBUE (Basu and Bhattacharjee (1984)) classes, the coefficient of variation (c.v.) η_F of a life distribution with the $\mathcal{L}$-property cannot exceed 1, unlike in such classes, there can be distributions in the $\mathcal{L}$-class other than the exponential, for which the c.v. is equal to 1. Motivated by this observation, Mitra, Basu, and Bhattacharjee (1995) tried to locate a suitable sub-class of the $\mathcal{L}$-family, presumably larger than the HNBUE family, within which the exponential distribution can be characterized based on whether the corresponding c.v. is unity or less. Another kind of ramification has been initiated by Lin (1997) through his efforts to explore if, as in the HNBUE class, the property of equivalence of weak convergence and moment convergence would hold in the $\mathcal{L}$-class and even larger classes.

Definition 3.1 (Mitra, Basu, and Bhattacharjee (1995)). *A life distribution F with finite mean* μ *is said to belong to the* $\mathcal{L}_D$-*class if both F and* F_1 *belongs to* $\mathcal{L}$, *where* F_1 *is the defined as*

$$F_1(x) = \frac{1}{\mu} \int_0^x \{1 - F(t)\} \, dt.$$

Notice that, F_1 is the first derived distribution corresponding to F; more importantly, it is the asymptotic distribution of the *remaining life of the component in use* for a renewal process driven by F. In the context of aging relative to repairs (Bhattacharjee (1994)), consider a device with survival time distribution F, which is replaced every time it fails by a statistically identical new copy (perfect repairs strategy). The $\mathcal{L}_D$-property of F, therefore, describes, in an intuitively appealing way, that the aging profile of a new component is shared by its remaining life at great age under instantaneous and perfect repairs as both F and F_1 belong to $\mathcal{L}$.

Example 3.1. Consider a life d.f. F(x) which equals 0, 0.5, and 1 for $0 \le x < 1$, $1 \le x < 3$, and $x \ge 3$. Using the facts that (i) F is DMRL and hence has $\mathcal{L}$-property, and (ii) F DMRL implies that F_1 is IFR and hence $F_1 \in \mathcal{L}$; it can be shown that $F \in \mathcal{L}_D$ (Mitra, Basu, and Bhattacharjee (1995)).

The above example also establishes that the subclass $\mathcal{L}_D$ is strictly larger than the DMRL class. On the other hand, consider a life d.f. F(x) which equals 0, 0.3, and 1 for $0 \le x < 0.3$, $0.3 \le x < 3$, and $x \ge 3$, respectively; it can be shown that $F \in \mathcal{L}_D$ (Bhattacharjee and Basu (1994)), but F is not HNBUE (Klefsjö (1983)) and hence is *not* NBUE. Beyond this observation, however, whether any hierarchical relationship holds between $\mathcal{L}_D$ and the subclass HNBUE of $\mathcal{L}$ is still an open question. Based on results so far available, the hierarchical relationship of the $\mathcal{L}_D$-class to the standard aging classes is shown by the chain of implications:

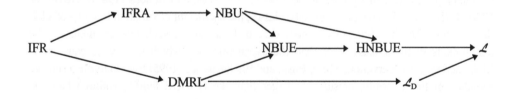

Figure 1

The closure properties of the $\mathcal{L}_D$-class are summarized as follows.

Theorem 3.1 (Bhattacharjee and Basu (1994)).
  (i) Neither $\mathcal{L}_D$ nor $\overline{\mathcal{L}}_D$ is closed under formation of coherent structures.
  (ii) $\mathcal{L}_D$ is closed under convolutions, but $\overline{\mathcal{L}}_D$ is not.
  (iii) $\mathcal{L}_D$ is not closed under mixtures, but $\overline{\mathcal{L}}_D$ is.

From definition 3.1, and Theorem 2.1, it is easy to observe that the $\mathcal{L}_D$-class is closed under weak convergence, and if $\mathcal{L}_D$ distributions $F_n \to F$ weakly as $n \to \infty$, then for all $r \leq 2$, $\lim_{n\to\infty} \mathcal{E}_{F_n} X^r = \mathcal{E}_F X^r$. It also follows from the definition of the $\overline{\mathcal{L}}_D$-class and (3) - (4) that for $F \in \overline{\mathcal{L}}_D$, $\mathcal{E}_F X^r < \infty$ for all $-1 < r \leq 3$ and

$$1 \geq \frac{\mathcal{E}_F X^2}{2(\mathcal{E}_F X)^2} \geq \frac{\mathcal{E}_F X^3}{3!(\mathcal{E}_F X)^3}. \tag{5}$$

Further, the following characterization of the exponential distribution in the $\mathcal{L}_D$-class holds.

Theorem 3.2 (Mitra, Basu, and Bhattacharjee (1995)). *Suppose $F \in \overline{\mathcal{L}}_D$. Then F is exponential if and only if its coefficient of variation is 1.*

Mitra, Basu, and Bhattacharjee (1995) also obtained characterizations of exponentials through crossing properties as well as the convergence of suitably scaled $\overline{\mathcal{L}}_D$-distributions to the unit exponential distribution. The latter result is useful for characterizing the limiting lifetime of a system having independent components with cold standby redundancy. However, the result does not seem to hold in the broader class of $\mathcal{L}$-distributions since their characterization through the unit value of the coefficient of variation fails there.

We now discuss $\mathcal{L}_D$ aging with shocks. Consider once more the standard framework of shock models already referred to in the previous section, and suppose the shocks arise according to a renewal process driven by the d.f. F. Notice that corresponding to the random variable N, there exists an *induced* random variable N_1 with distribution

$$P(N_1 = k) = \frac{P(N \geq k)}{\mathcal{E}N}, \quad k = 1, 2, \dots.$$

Definition 3.2 (Bhattacharjee and Basu (1994)). *The random variable N or equivalently $\{p_k, k = 1, 2, \dots\}$ is said to belong to the G_D $(\overline{G}_D)$-class of distributions if both N and N_1 belong to the G $(\overline{G})$-class.*

Theorem 3.3 (Bhattacharjee and Basu (1994)). *If $F \in \overline{\mathcal{L}}_D$ $(\overline{\mathcal{L}}_D)$ and $\{p_k, k \geq 1\} \in G_D$ $(\overline{G}_D)$, then $S \in \overline{\mathcal{L}}_D$ $(\overline{L}_D)$.*

While the $\mathcal{L}_D$ is a contraction of the $\mathcal{L}$-class, a query in Lin (1997) leads to an expansion of the latter. Basu and Bhattacharjee (1984) proved the equivalence of weak convergence and moment convergence for life distributions in the HNBUE class. Lin (1997) investigated if this property would hold for any larger class. He succeeded in showing that the property does hold for the classes Π_α $(\alpha > 0)$ and Π_β^* $(\beta > 0)$ where

$$\Pi_\alpha = \bigcup_{\beta > 0} \left\{ F : \mu_F = \alpha\beta, \mathcal{E}e^{sX} \le \mathcal{E}e^{sY_{\alpha,\beta}} \text{ for } s \in \left(-\infty, \frac{1}{\beta}\right) \right\}$$

and

$$\Pi_\beta^* = \bigcup_{\alpha > 0} \left\{ F : \mu_F = \alpha\beta, \mathcal{E}e^{sX} \le \mathcal{E}e^{sY_{\alpha,\beta}} \text{ for } s \in \left(-\infty, \frac{1}{\beta}\right) \right\}.$$

Here, $Y_{\alpha,\beta}$ is a gamma random variable characterized by the probability density function

$$g_{\alpha,\beta}(x) = \frac{1}{\Gamma(\alpha)\beta^\alpha} x^{\alpha-1} e^{-\frac{x}{\beta}}, \quad x > 0,$$

where $\alpha > 0$, $\beta > 0$ are constants. Motivated by these results, he, therefore, posed the question as to whether the Theorem could be established even for the larger class $\Psi = \Pi$, where

$$\Pi \equiv \bigcup_{\alpha > 0} \Pi_\alpha = \bigcup_{\beta > 0} \Pi_\beta^*.$$

In a recent note, Basu and Bhattacharjee (1997) showed that the answer to this question is affirmative; they further dealt with Lin's query about the validity of results of the above type in respect of the class

$$\Theta_\alpha = \bigcup_{\beta > 0} \left\{ F : \mu_F = \alpha\beta, \mathcal{E}e^{-sX} \le \mathcal{E}e^{-sY_{\alpha,\beta}}, s \ge 0 \right\}, \quad \alpha > 0.$$

It may be noted that $\Theta_1 = \mathcal{L}$. As a partial response to the query raised by Lin, showed that the class Θ_α, $\alpha > 0$ is closed under weak convergence.

References

1. A. Alzaid, J.S. Kim, and F. Proschan (1991) Laplace ordering and its applications. *J. Appl. Prob.* 28, 116-130.
2. R.E. Barlow and F. Proschan (1981) *Statistical Theory of Reliability and Life testing: Probability Models.* To Begin With, Silver Spring, MD.

3. S.K. Basu and M.C. Bhattacharjee (1984) On weak convergence within the class of HNBUE life distributions. *J. Appl. Prob.* 21, 654-660.

4. S.K. Basu and M.C. Bhattacharjee (1997) Closure property under weak convergence of the class of life distributions that are Laplace-order dominated by gamma distributions. submitted.

5. S.K. Basu and M. Mitra (1997) Testing exponentiality against Laplace order dominance. submitted

6. S.K. Basu and G. Simons (1983) Moment spaces of IFR distributions, applications and related materials. In *Essays in Honour of Norman L. Johnson,* ed. P.K. Sen. North-Holland, Amsterdam.

7. A. Bhattacharjee and D. Sengupta (1996) On the coefficient of variation for the $\mathcal{L}$ and $\overline{\mathcal{L}}$ classes of life distributions. *Stat. and Prob. Letters,* 27, 177-180.

8. M.C. Bhattacharjee (1994) Ageing influenced by repairs, repair efficiency and realizability of renewal related distribution. *Internat. J. Reliability, Quality and Safety Engg.,* 1 (2), 147-159.

9. M.C. Bhattacharjee and S.K. Basu (1994) Aging with Laplace order conserving survival under perfect repairs. *Research Report No. CAMS-030, Department of Mathematics, New Jersey Institute of Technology.*

10. H.W. Block and T.H. Savits (1978) Shock models with NBUE survival. *J. Appl. Prob.* 15, 621-628.

11. L. Bondesson (1981) On preservation of classes of life distributions under reliability operations; some complementary results. *Research Report 1981-3,* Department of Mathematical Statistics, University of Umea.

12. G. Chaudhuri (1995) A note on the $\mathcal{L}$ class of life distributions. *Sankhya Ser. A,* 57 (1), 158-160.

13. G. Chaudhuri, J.V. Deshpande, and A.D. Dharmadhikari (1996) A lower bound on the $\mathcal{L}$-class of life distributions and its applications. *Calcutta Stat. Assn. Bulletin,* 46, 269-274.

14. J.D. Esary, A.W. Marshall, and F. Proschan (1973) Shock models and wear processes. *Ann. Prob.* 1, 627-649.

15. B. Klefsjö (1981) HNBUE survival under some shock models. *Scand. J. Statist.* 8, 39-47.

16. B. Klefsjö (1982) The HNBUE and HNWUE classes of life distributions. *Naval Res. Logist. Quart.* 29, 331-344.

17. B. Klefsjö (1983) A useful aging property based on the Laplace transform. *J. Appl. Prob.* 20, 615 - 626.

18. G.D. Lin (1997) On weak convergence within a family of life distributions. To appear in *J. Appl. Prob.*

19. M. Mitra, S.K. Basu, and M.C. Bhattacharjee (1995) Characterizing the exponential law under Laplace order domination. *Calcutta Stat. Assn. Bulletin,* 45, 171-180.

20. T. Rolski (1975) Mean residual life. *Bull. Internat. Statist. Inst.* 46, 266-270.

RECENT APPLICATIONS OF THE TTT-PLOTTING TECHNIQUE

BO BERGMAN

Division of Quality Technology, Linköping University, Linköping, Sweden.
E-mail: bober@ikp.liu.se

BENGT KLEFSJÖ

Division of Quality Technology & Statistics, Luleå University of Technology, Luleå, Sweden.
E-mail: bengt.klefsjo@ies.luth.se

The TTT-plot (TTT = Total Time on Test), an empirical and scale independent plot based on failure data, and the corresponding asymptotic curve, the scaled TTT-transform, were introduced by Barlow and Campo in 1975 and used for model identification purposes. Since then these tools have proven to be very useful in several practical and theoretical applications within reliability. This paper presents some recent applications of TTT-plotting. One is how to handle censored data and another one is an application for maintenance optimization when analysing data using the Proportional Hazards Model. We will also discuss how the TTT-plotting technique can be used to illustrate and study some test statistics for testing exponentiality. Furthermore, an application to design of experiments will be discussed.

1 Introduction

The TTT-plot (TTT = Total Time on Test), an empirical and scale independent plot based on failure data, and the corresponding asymptotic curve, named the scaled TTT-transform, were introduced by Barlow and Campo in 1975 and used for model identification purposes. Since then these tools have proven to be very useful in several applications within reliability. Examples of practical applications are analysis of aging properties, maintenance optimization, burn-in optimization and for power process analysis. The TTT-plot and the TTT-transform have also been sources of inspiration for more theoretical applications such as looking for test statistics for particular purposes and to study their power.

This paper gives an overview of some recent applications, both practical and theoretical. We will discuss TTT-plotting for censored data and indicate an application for maintenance optimization when analysing data using the Proportional Hazards Model. We will also discuss how the TTT-plotting technique can be used to illustrate Gnedenko's F-test for testing exponentiality and to study the Pitman efficiency. Furthermore, an application to design of experiments will be presented.

2 The TTT-plot and the scaled TTT-transform

Suppose that we have a complete ordered sample $0 = t(0) \leq t(1) \leq ... \leq t(n)$ of times to failure and the corresponding TTT-statistics $S_j = nt(1) + (n-1)(t(2)-t(1)) + ... + (n-j+1)(t(j)-t(j-1))$, $j = 1, 2, ..., n$, (for convenience we set $S_0 = 0$). Then the piecewise linear graph joining the points $(j/n, u_j)$, where $u_j = S_j/S_n$, $i = 0, 1, ..., n$, is called the *TTT-plot* based on these observations.

Accordingly, a TTT-plot lies within the unit square and consists of line segments starting at $(0, 0)$ and ending at $(1, 1)$. Figures 1 and 2 show examples of some TTT-plots.

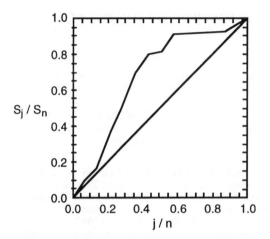

Figure 1   A TTT-plot based on times to failure of the engine in a laud haul dump machine working in a
              Swedish mine. (From Kumar et al., 1989.)

The reason for the acronym "TTT", which means "Total Time on Test", is that if all the
units are put into test at the same time then S_j is the total cumulated test time for all the
units at time t(j).

Since we can write

$$u_j = \frac{S_j}{S_n} = \frac{\displaystyle\int_0^{t_j} R_n(t)dt}{\displaystyle\int_0^{t_n} R_n(t)dt} = \frac{\displaystyle\int_0^{F_n^{-1}(j/n)} R_n(t)dt}{\displaystyle\int_0^{F_n^{-1}(n/n)} R_n(t)dt} \qquad (1)$$

where $R_n = 1 - F_n$ is the empirical survival function, it follows that when the sample size
n increases the TTT-plot converges (with probability one and uniformly; see Langberg et
al., 1980) to a continuous curve called the *scaled TTT-transform* of the life distribution
F(t) from which our sample has come. This is illustrated in Figure 2. Mathematically the
scaled TTT-transform is defined as

$$\varphi(u) = \int_0^{F^{-1}(u)} R(t)dt \qquad \text{for} \quad 0 \le u \le 1$$

where $R(t) = 1 - F(t)$ is the survival function and $\mu < \infty$ is the mean. Examples of scaled
TTT-transforms can be found in Figures 2 and 3.

Note that this transform is independent of scale. For instance, for a Weibull distribution with survival function $R(t) = 1 - F(t) = \exp(-(t/\alpha)^\beta)$, $t \ge 0$, we get the same transform
for a certain value of β, independently of the value of α, and every exponential
distribution $F(t) = 1 - \exp(-\lambda t)$, $t \ge 0$, is transformed to the diagonal in the unit square
independently of the value of the failure rate λ; see Figure 3.

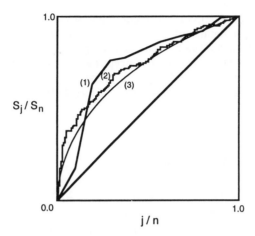

Figure 2   TTT-plots based on simulated Weibull data with β = 2.0 and n = 10 in (1), β = 2.0 and n = 100 in (2) and the scaled TTT-transform of a Weibull distribution with β = 2.0 in (3). (From Bergman & Klefsjö, 1984.)

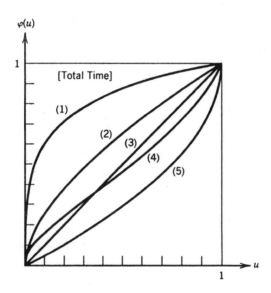

Figure 3   Scaled TTT-transforms of five different life distibutions; (1) normal with μ = 1 and σ = 0.3; (2) gamma distribution with shape parameter 2.0; (3) exponential distribution; (4) lognormal distribution with μ = 0 and σ = 1; (5) Pareto distribution with R(t) = (1+t)^{-2}, t ≥ 0. (From Bergman & Klefsjö, 1984.)

The scaled TTT-transform and the TTT-plot were first presented in a paper by Barlow & Campo (1975). They used these concepts for model identification purposes by comparing the TTT-plot to scaled TTT-transforms of different life distributions; see Figure 2. Since then a lot of other applications, both theoretical and practical, have appeared. Examples of such applications are age replacement problems with and without discounted costs, burn-in optimization, characterization of different aging properties and analysis of data from repairable systems. For details of these and other applications we refer to Bergman (1977), Bergman & Klefsjö (1982, 1984), Klefsjö (1991), Klefsjö & Kumar (1992) and Klefsjö & Westberg (1996a). The TTT-concept has also been useful when looking for suitable test statistics and when studying their properties. Such applications can be found in Klefsjö (1983), Bergman & Klefsjö (1985), Klefsjö (1989) and Bergman & Klefsjö (1989).

3 TTT-plotting with censored data

From (1) it is seen that the plotting positions (i/n, S_i/S_n) in a TTT-plot based on a complete sample can be interpreted as functions of the empirical distribution function F_n. If we have a censored sample it is therefore natural to change the empirical distribution function to another estimator, F_n^c say, of the true distribution function and then plot $(F_n^c(t(i)), S_i^c/S_n^c)$ for uncensored failure times t(i) and connect these points. One possibility then is to use the Kaplan-Meier estimator (see Kaplan & Meier, 1958). This idea was indicated e.g. in Bergman & Klefsjö (1984). Another possibility is to use the Piecewise Exponential Estimator (PEXE) introduced by Kitchin (1980) and also discussed by Kim & Proschan (1991). This approach is discussed and analysed in Westberg & Klefsjö (1994) and Westberg (1994).

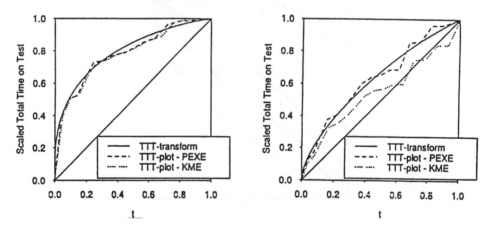

Figure 4   TTT plots based on a censored sample from the Weibull distribution with shape parameter 3.0 and 1.5, respectively. The sample size is 30 and the amount of censoring is 50%. One of the plots is based on the Kaplan-Meier estimator (KME) and the other one on the Piecewise Exponential Estimator (PEXE). The curve is the scaled TTT-transform of the corresponding Weibull distribution. (From Westberg & Klefsjö, 1994.)

The idea with PEXE is to estimate the unknown distribution function with a continuous curve between the uncensored observations consisting of pieces from different exponential distributions. The parameters in the different exponential pieces are estimators of the failure rate based on the MLE taking care of both uncensored and censored data points. For details see e.g. Kitchin (1980) or Westberg & Klefsjö (1994).

In Westberg (1994) some different extensions of the basic idea in Kitchin (1980) is presented. In that paper also some simulations are presented illustrating that the PEXE technique works well and that the yield increases when the amount of censoring increases. Figure 4 shows TTT-plots based on the PEXE and the Kaplan-Meier estimator.

4 TTT-plotting and Gnedenko's F-test

A lot of test statistics have been proposed for testing whether the exponential distribution is a suitable model (a summary can be found in d'Agostino & Stephens, 1986). One of these is Gnedenko's F-test, named after the Russian statistician Boris Vladimirovich Gnedenko (1912-1995). Gnedenko et al. (1969, p. 236) discussed a test statistic G(r) based on the normalized TTT-statistics u_j, $j = 1, 2, \dots, n$, and defined as

$$G(r) = \frac{\dfrac{1}{r} u_r}{\dfrac{1}{n-r}(1 - u_r)}$$

This statistic was suggested as a suitable statistic for testing exponentiality, i.e. constant failure rate, against increasing or decreasing failure rate. The same statistic was in fact also discussed already by Epstein (1960) as an omnibus test statistic for testing exponentiality. The reason for the name "F-test" is that the statistic is F-distributed under exponentiality (see e.g. Klefsjö & Westberg, 1996b).

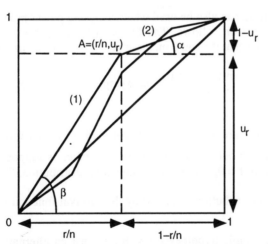

Figure 5   *A graphical illustration of Gnedenko's F-test G(r) directly in the TTT-plot. Since tan γ = (1−u_r)/(1−(r/n)) and tan δ = u_r/(r/n), G(r) can be interpreted as the ratio between the slopes of the lines marked (1) and (2), respectively.*

With the notations γ and δ from Figure 5, we get that

$$\tan \gamma = \frac{1 - u_r}{1 - \frac{r}{n}} \quad \text{and} \quad \tan \delta = \frac{u_r}{\frac{r}{n}}$$

Accordingly, we get that

$$G(r) = \frac{\tan \delta}{\tan \gamma}$$

i.e. $G(r)$ is the ratio between the slopes of the lines marked (1) and (2), respectively, in Figure 5.

If our sample is from an exponential distribution $G(r)$ is expected to be close to one since the corresponding TTT-plot is expected to wriggle around the diagonal. If $G(r)$ deviates too much from one, this indicates that our sample is from another life distribution than the exponential. From Figure 5 and the fact that the TTT-plot based on exponential observations converges to the diagonal we realize that in order for $G(r)$ to be as efficient as possible to detect the alternative distribution the point A in Figure 5 should be as far away as possible from the diagonal. Accordingly, the statistic $G(r)$ can not be very useful when testing exponentiality against a distribution the scaled TTT-transform of which crosses the diagonal near the value of r/n. One example of this is the lognormal distribution (see Figure 3). If we choose $r/n \approx 0.40$ it will be almost impossible to detect that our sample is from the lognormal distribution in Figure 3. This observation was made by Harris (1976) from another point of view and was used as an argument for proposing a generalization of Gnedenko's statistic; see Harris (1976) and Klefsjö & Westberg (1996b).

The TTT-transform can also be used when analysing power properties of Gnedenko's F-test. A brief discussion follows. The test statistic $G(r)$ is asymptotically normally distributed in the sense that if r and n increase in such a way that $r/n \to u$, $0 < u < 1$, then

$$\frac{\sqrt{n}(G(r) - \mu(\theta))}{\sigma(\theta)} \quad \text{is asymptotically } N(0,1)$$

where $\mu(\theta)$ and $\sigma(\theta)$ are functions of the scaled TTT-transform $\varphi_\theta(u)$ of the life distribution from which our sample is coming and which is supposed to depend on θ in such a way that when $\theta = \theta_0$ we get the exponential distribution. One example of this is the family of Pareto distributions, defined by the survival function $R(t) = (1+\theta t)^{-1/\theta}$, $t \geq 0$; here $\theta_0 = 0$.

When testing a simple hypothesis $\theta = \theta_0$ against an alternative $\theta > \theta_0$, different measures of asymptotic efficiencies can be applied, see e.g. Rao (1972). The most frequently used measure is probably the Pitman efficiency value, defined as

$$P = \left(\frac{\mu'(\theta_0)}{\sigma(\theta_0)}\right)^2$$

Calculations give (see Klefsjö & Westberg, 1996b) that the Pitman value, which here is a function of u and therefore is denoted P(u), is given by

$$P(u) = \frac{\left(\left(\frac{d\varphi(u)}{d\theta}\right)_{\theta=\theta_0}\right)^2}{u(1-u)}$$

For instance, for a Pareto distribution with survival function $R(t) = (1+\theta t)^{-1/\theta}$, $t \geq 0$, we get

$$\frac{d\varphi(u)}{d\theta} = (1-u)^{1-\theta}\ln(1-u)$$

and the Pitman value is therefore

$$P(u) = \frac{(1-u)(\ln(1-u))^2}{u}$$

By studying P(u) it is found that the function reaches its largest value for u = 0.7968121 ≈ 0.80. The corresponding maximum value of P(u) is 0.6475 ≈ 0.65. If we expect the alternative distribution to be a Pareto distribution we should therefore choose r/n ≈ 0.80 if we use the Gnedenko's F-test in order to have maximal possibility to detect the alternative; see also Figure 6. We can also see from Figure 6 that a value of r/n which is somewhat lower than 0.80 is better than a value higher than 0.80.

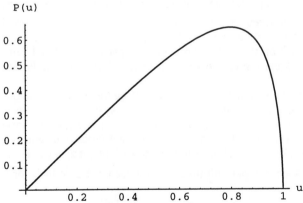

Figure 6   The curve illustrates the Pitman efficiency P(u) as a function of u, the limit of r/n, for Gnedenko's F-test when the sample is from a Pareto distribution.

In Klefsjö & Westberg (1996b) a further discussion can be found of how to use the TTT-plot and the TTT-transform as tools when analysing Gnedenko's F-test and some other related test statistics.

5 TTT-plotting and maintenance optimization using PHM

Maintenance scheduling has been widely investigated considering failure times only. However, information obtained using condition monitoring devices, whenever possible, is used more and more in industries for maintenance scheduling. This trend is accelerated by the availability of reliable sensors and the rapid development of information technology.

Under the age replacement policy, a system is assumed to be renewed by planned and unplanned maintenance either through replacements or thorough repairs. A maintenance cost function is formed considering the average long-run cost per unit time (see Barlow & Proschan, 1965) and the optimal replacement age can be found analytically or by using TTT-transforms and TTT-plots. The reliability function is supposed to depend on time alone.

In Kumar & Westberg (1997) the age replacement policy is extended to a reliability function based both on time and on the values of monitored variables assuming these to be constant within the life span of a single unit under study. There are several models for estimating the reliability function based on variables, but the Proportional Hazards Model, PHM, introduced by Cox (1972), appears very robust and requires few assumptions (Newby, 1993). Various applications of the PHM have been presented within biostatistics and recently also in reliability analysis. A general review of the PHM is given in Kumar & Klefsjö (1994). The suggestion in Kumar & Westberg (1997) is to use the estimate of the reliability function obtained in the PHM.

The basic approach in the PHM modelling is to assume that the hazard rate of a system consists of two multiplicative factors, the baseline hazard rate, $h_0(t)$, and generally an exponential function including the effects of the monitored variables. For example, these monitored variables can be lubricant pressure, temperature, particle contents in hydraulics or vibration levels. Hence, the hazard rate of a system can be written as (Cox, 1972) $h(t,z) = h_0(t) \exp(z\beta)$, where z is a row vector consisting of covariates, explanatory variables or any monitored variable, and β is a column vector consisting of the corresponding regression parameters. The unknown parameter β defines the influence of the monitored variables on the failure process. The partial likelihood function is generally used for estimating the regression vector (see e.g. Kalbfleisch & Prentice, 1980; Cox & Oakes, 1984).

Let c be the planned maintenance cost and K be the unplanned maintenance cost, i.e., K is the additional cost due to failure of the system during operation. The total maintenance cost $C(t,T,z)$ over the time interval $(0,t]$ is a random variable and T is the time interval between two planned maintenance activities. Our aim is to find either the value of T, say T_0, or the threshold value of the variable of z, say z:th, so that the long run average maintenance cost per unit time is minimised, i. e., we look for either T or z such that

$$C(T,z) = \frac{c + KF(T,z)}{\int_0^T R(t,z)dt}$$

is minimised.

The simple graphical approach to estimate the optimum maintenance time interval under the age replacement policy from a TTT-plot, first suggested by Bergman (1977), is extended by Kumar & Westberg (1997). The approach discussed in that paper is based upon the assumption that the values of monitored variables (such as level of metal particles in engine oil) are available for each installed unit. The idea is to estimate the survival function (and the distribution function) by using ideas from analysis using PHM. In the paper the ideas are illustrated using an example from Love & Guo (1991) in which times to failure of pressure gauges are studied.

To illustrate the concepts we consider an example from Love & Guo (1991). The example includes the times between replacements of pressure gauges which are replaced on failure or as part of a planned maintenance, see Table 1. The observations have been rearranged as per increasing magnitudes of the time intervals. The time points at which replacements occurred as part of planned maintenance are marked with asterisks.

Table 1:   The time to failure of pressure gauges which were replaced on failure or as part of planned maintenance (denoted by asterisks). (From Love & Guo, 1991).

Obs nr	1	2	3	4	5	6	7	8	9	10	11	12	13	14	15
Time to failure	32	42	44*	47	51	53	60*	61*	66	70	70	77	95	101	198
Monit. variable (Pressure)	5	4	5	5	5	4	3	4	3	4	5	4	3	3	3

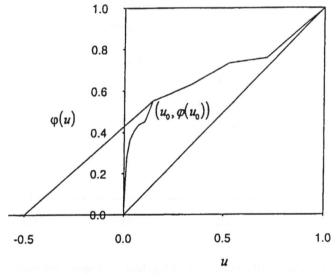

Figure 7a   A TTT-plot when the pressure is at the low level. The optimal replacement age is 66 hours. We have c/K = 0.5. (Data from Love & Guo, 1991, and the figure from Kumar & Westberg, 1997.)

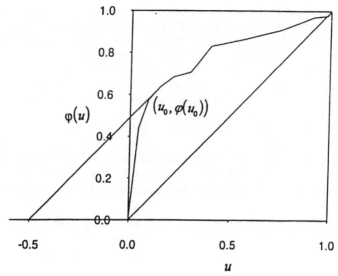

Figure 7b   A TTT-plot when the pressure is at the middle level. The optimal replacement age is 47 hours. We have c/K = 0.5. (Data from Love & Guo, 1991, and the figure from Kumar & Westberg, 1997.)

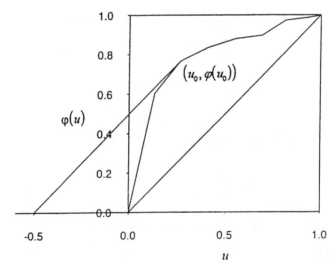

Figure 7c   A TTT-plot when the pressure is at the high level. . The optimal replacement age is 42 hours. We have c/K = 0.5. (Data from Love & Guo, 1991, and the figure from Kumar & Westberg, 1997.)

Analysis using the PHM resulted in the model

$$h(t,z) = h_0(t)\exp(1.188z)$$

where z is the pressure gauge. In order to estimate the optimal replacement age the TTT-plots corresponding to the different values of the pressure were drawn.The optimal replacement ages were obtained to 66, 47 and 42, hours when the pressure is 3, 4 and 5 units, respectively. See Figure 7 a-c. For more details, see Kumar & Westberg (1997).

An alternative approach to include covariates was indicated by Bergman (1977) and further studied by Al-Najjar (1990).

6 TTT-plotting and Design of Experiments

In industry, Design of Experiments in general and in particular fractional factorial experiments has gained increasing interest. One crucial point in the analysis of such experiments is the identification of active factors. It is important to have illustrative graphical procedures for this task, but also formal ones closely related to the illustrative ones. Already Daniel (1959) suggested the use of the half-normal plot of contrasts from the fractional factorial experiment as a tool for the identification procedure. Later a number of formal approaches have been suggested, see e.g. Dong (1993) and Venter & Steel (1996). However, the graphical procedures and the formal procedures are weakly related to each other. In Bergman & Sandvik Wiklund (1997) the TTT plotting technique is used to fill this gap. The idea is briefly described below. For more details, see Bergman & Sandvik Wiklund (1997).

Let y_i, i = 1, ..., m+1, be the observed values from an unreplicated fractional factorial experiment with independent, identically normally distributed errors with mean zero. Then the contrasts c_i, i = 1, ..., m, corresponding to non-active factors, are observations from a $N(0, \sigma^2)$ random variable, where σ^2 is the variance of a contrast. Thus, the corresponding squared contrasts $z_i = c_i^2$, i = 1, 2, ..., m, are observations from a scaled chi-square distribution with one degree of freedom, or more precisely, z_i/σ^2 is $\chi^2(1)$-distributed. If all squared contrasts correspond to non-active factors therefore the TTT-plot should be close to the scaled TTT-transform of a $\chi^2(1)$-distribution, since the TTT-plot is independent of scale. If there are some contrasts corresponding to active factors (main factors or interactions) then the empirical cumulative distribution function of all squared contrasts should be more "heavy-tailed" than that of a $\chi^2(1)$-distribution. Interpreting "heavy-tailed-ness" as DFR-ness (see Barlow, 1979) we get a visual support for our judgement concerning active factors (and interactions) from the TTT-plot since a TTT-plot based on data from a DFR-distribution should behave concavely (see Barlow & Campo, 1975). If there are active factors the TTT-plot accordingly should be more concave than a TTT-plot based on observations from a $\chi^2(1)$-distribution. To make the comparison more simple Bergman & Sandvik Wiklund (1997) suggest to plot $\varphi_n(i/n)$ against $\varphi(i/n)$, where $\varphi(u)$ is the scaled TTT-transform from a $\chi^2(1)$-distribution. In that way the TTT-transform from a $\chi^2(1)$-distribution is transformed to the diagonal of the unit square. This means that under the hypothesis of no active factors the transformed TTT-plot should be close to the diagonal.

The procedure is briefly illustrated on an experiment with the purpose to identify factors influencial on the presence of pores in a brake cylinder. Here just one of the response variables, weight, is shown. The design for the foundry experiment is a 2^{8-4} fractional factorial with resolution III, see Table 2.

B. Bergman and B. Klefsjö

Table 2   Design matrix, responses and confounding pattern for a foundry experiment conducted by the Swedish Foundry Association in a research project together with the Division of Quality Technology and Management at Linköping University. The letters correspond to the following factors: A-furnace temperature; B-casting pressure; D-air cooling and cleaning; E-vacuum; F-starting point for the second phase; H-plunger velocity in the first phase; L-plunger velocity in the second phase; O-cycle time. The two-factor interactions considered important a priori are shown in bold type.

| | | **EF** | BF | | | **AF** | | **EL** | **EO** | | | **BO** | BL | | |
| DE | DF | **AB** | AE | AD | BD | **BE** | DL | **AH** | **BH** | **DO** | DH | **AL** | **AO** | | |
No. A	B	**D**	E	F		**H**					**L**			O	y
1 −	−	+	−	+	+	−	−	+	+	−	+	−	−	+	262.4
2 −	−	+	−	+	+	−	+	−	−	+	−	+	+	−	262.7
3 −	−	+	+	−	−	+	−	+	+	−	+	+	−	−	259.4
4 −	−	+	+	−	−	+	+	−	−	+	+	−	−	+	261.8
5 −	+	−	−	+	−	+	−	+	−	+	+	−	+	−	265.6
6 −	+	−	−	+	−	+	+	−	+	−	−	+	−	+	264.6
7 −	+	−	+	−	+	−	−	+	−	+	−	+	−	+	263.3
8 −	+	−	+	−	+	−	+	−	+	−	+	−	+	−	263.5
9 +	−	−	−	−	+	+	−	−	+	+	+	+	−	−	260.4
10 +	−	−	−	−	+	+	+	+	−	−	−	−	+	+	260.6
11 +	−	−	+	+	−	−	−	−	+	+	−	−	+	+	261.1
12 +	−	−	+	+	−	−	+	+	−	−	+	+	−	−	262.6
13 +	+	+	−	−	−	−	−	−	−	−	+	+	+	+	264.6
14 +	+	+	−	−	−	−	+	+	+	+	−	−	−	−	264.7
15 +	+	+	+	+	+	+	−	−	−	−	−	−	−	−	265.4
16 +	+	+	+	+	+	+	+	+	+	+	+	+	+	+	265.2

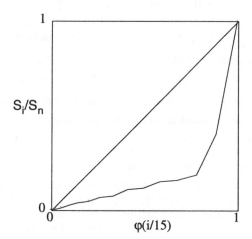

Figure 8   The transformed TTT-plot for foundry data obtained in an experiment in a foundry industry. The concavity of the plot indicates that there are active factors. This is confirmed by analytical test procedures. See Bergman & Sandvik Wiklund (1997).

The TTT-plot based on the squared contrasts from this experiment is illustrated in Figure 8. Since this transformed TTT-plot should be close to the diagonal under the hypothesis of no active factors, this plot clearly indicates actice factors. That is also confirmed from other formal test procedures; see Bergman & Sandvik Wiklund (1997).

To get a formal procedure close to the graphical one we might consider the cumulative TTT-statistic defined from the normalized TTT-statistics, u_j, $j = 1, 2, \ldots, n$, as

$$V_n = \sum_{j=1}^{n-1} u_j$$

If that statistic is small a DFR-ness with respect to the $\chi^2(1)$-distribution is indicated. Percentiles of the cumulative TTT-statistic under the $\chi^2(1)$-distribution is presented in Bergman & Sandvik Wiklund (1997). In that paper also a sequential procedure for selecting active factors is discussed, see also Bergman et al. (1997).

Conclusions

The TTT-plot and the TTT-transform introduced by Barlow and Proschan in 1975 are very powerful instruments in many different situations in reliability analysis. The power is mainly due to the fact that the tools are independent of scale and that they give, as many other graphical tools do, good possibilities for sensitivity analysis.

The application to design of experiments shows that the TTT-concept is applicable also in situations other than traditional reliability analysis.

We are convinced that many other applications and generalisations in different areas will be presented later on.

References

B. Al-Najjar. Licentiate Thesis No 248. Division of Quality Technology, Linköping University, Sweden (1990).

R.B. d'Agostino & M.A. Stephens. *Goodness-of-fit Techniques.* Marcel Dekker, New York (1986).

R.E. Barlow. *Naval Research Logistics Quarterly,* 26, 393-402 (1979).

R.E. Barlow & R. Campo. In *Reliability and Fault Tree Analysis*, ed. R.E. Barlow, J. Fussell and N.D. Singpurwalla, SIAM, Philadelphia, pp. 451-481 (1975).

R.E. Barlow & F. Proschan. *Mathematical Theory of Reliability and Life Testing. Probability Models.* John Wiley & Sons, New York (1965).

B. Bergman. In *Annual Reliability and Maintainability Symposium*, 467-471 (1977).

B. Bergman & B. Klefsjö. *IEEE Transactions on Reliability,* R-31, 478-481 (1982).

B. Bergman & B. Klefsjö. *Operations Research,* 32, 596-606 (1984).

B. Bergman & B. Klefsjö. In *Contributions to Probability and Statistics in honour of Gunnar Blom*, University of Lund, 49-72 (1985).

B. Bergman & B. Klefsjö. *Journal of Statistical Planning and Inference,* 21, 161-178 (1989).

B. Bergman & P. Sandvik Wiklund. Finding active factors from unreplicated fractional factorials utilising the Total Time on Test technique. Forthcoming paper (1997).

B. Bergman, F. Ekdahl, F. & P. Sandvik Wiklund. A comparison of selection procedures for finding active contrasts in unreplicated fractional factorials. The Second World Conference of the IASC, Pasadena, California (1997).

D.R. Cox. *Journal of Royal Statistical Society,* **B34**, 187-220 (1972).

D.R. Cox & D. Oakes. *Analysis of Survival Data.* Chapman and Hall, London (1984).

C. Daniel. *Technometrics,* **28**, 11-18 (1959).

F. Dong. *Statistica Sinica,* **3**, 209-217 (1993).

B. Epstein. *Technometrics,* **2**, 83-101 (1960).

B.V. Gnedenko, Y.K. Belyayev & A.D. Solovyev. *Mathematical Methods of Reliability Theory.* Academic Press, New York (1969).

C.M Harris. *Naval Research Logistics Quarterly,* **23**, 169-175 (1976).

J.D. Kalbfleisch & R.L. Prentice *The Statistical Analysis of Failure Time Data.* John Wiley & Sons, New York (1980).

E.L. Kaplan & P. Meier. *Journal of American Statistical Association,* **53**, 457-481 (1958).

J.S. Kim & F. Proschan. *IEEE Transactions on Reliability,* **R-40**, 134-139 (1991).

J. Kitchin. A new method for estimating life distributions from incomplete data. Unpublished PhD Dissertation. Florida State University (1980).

B. Klefsjö. *Scandinavian Journal of Statistics*, **10** , 65-75 (1983).

B. Klefsjö. *Microelectronics and Reliability,* **29**, 559-570 (1989).

B. Klefsjö. *Journal of Statistical Planning and Inference,* **29**, 99-110 (1991).

B. Klefsjö. & U. Kumar. *IEEE Transactions on Reliability,* **41**, 593-598 (1992).

B. Klefsjö. & U.Westberg. *Quality Engineering,* **9(2)**, 229-235 (1996a).

B. Klefsjö. & U. Westberg. *Arab Journal of Mathematical Sciences,* **2**, 129-149 (1996b).

D. Kumar & B. Klefsjö. *Reliability Engineering and System Safety,* **44**, 177-188 (1994).

U. Kumar, B. Klefsjö & S. Granholm. *Reliability Engineering and System Safety,* **26**, 341-361 (1989).

D. Kumar & U. Westberg. Maintenance scheduling under age replacement policy using Proportional Hazards Model and TTT-plotting. To appear in *European Journal of Operations Research* (1997).

N.A. Langberg, R.V. Leone & F. Proschan. *Annals of Probability,* **8,** 1163-1170 (1980).

C.E. Love & R. Guo. *Quality and Reliability Engineering International,* **7,** 7-17 (1991).

M. Newby. *IMA Journal of Mathematics Applied in Business and Industry,* **4,** 375-394 (1993).

R. Rao. *Linear Statistical Inference and Their Applications.* John Wiley & Sons, New York (1972).

J.H.Venter & S.J. Steel. *Technometrics,* **38,** 161-169 (1996).

U. Westberg. Licentiate Thesis 1994:24L. Division of Quality Technology & Statistics, Luleå University (1994).

U. Westberg & B. Klefsjö. *International Journal of Reliability, Quality and Safety,* **1,** 1-13 (1994).

DYNAMIC PROGRAMMING, RENEWAL FUNCTIONS, AND PERFECT VS. MINIMAL REPAIR COMPARISONS

M. C. Bhattacharjee

Center for Applied Mathematics & Statistics
Department of Mathematics
New Jersey Institute of Technology, Newark, NJ 07102

The average number of failures of a maintained equipment in any time interval $[0, t)$ is an indicator of how effective repairs are. Comparing replacement (perfect repair) vs. minimal repairs by this criterion requires estimating the ratio of the renewal and hazard functions of a survival distribution. Since the renewal function is rarely computable in explicit form; a general method for finding upper bounds on the renewal function is indicated, using ideas from dynamic programming. It is shown that not all classical aging properties are strong enough to ensure uniform superiority of perfect repairs over minimal repairs, a result which is counterintuitive.

1 Introduction and Summary

Minimal repairs and replacement (perfect repairs) are among the most popular and well understood strategies used for maintained systems. To compare their relative effectiveness via the ratio of corresponding expected number of failures in a arbitrary time interval; we use dynamic programming ideas to provide a method of checking for upper bounds on the renewal function $M(t)$ of the survival distribution F of the equipment. In section 2, we show that contrary to naive intuition, not all aging properties guarantee perfect repairs to be more efficient than minimal ones in the sense of having smaller expected number of failures. Among the classical nonparametric aging classes, NBUE and weaker aging properties are inadequate to ensure lesser average number of failures under perfect repairs (PR) than under minimal repairs (MR). In section 3, where our approach for bounding the renewal function is indicated (Theorem 3.1), we show as an application of our methods that under NBU hypothesis, the renewal function is bounded above by the hazard function - which is not the case when F is NBUE, as in the counterexamplein section 2 demonstrates that minimal repairs can outperform minimal repairs.

2 Are perfect repairs always better?

For an equipment with survival (life) distribution F, let $\bar{F}(t) := 1 - F(t)$ denote its reliability function and $\Lambda(t) := -\ln \bar{F}(t)$ its hazard function, $t \geq 0$. Let $N(t)$ be the counting process of failures, under any repair scheme. When repairs are effectively instantaneous, $N(t)$ is also the number of repairs in $[0, t)$. The *relative efficiency* of renewals (PR) compared to minimal repairs (MR) can be measured (Bhattacharjee (1993), (1994)) by the ratio

$$R_{PR:MR} = \frac{E_{MR} N(t)}{E_{PR} N(t)} = \frac{\Lambda(t)}{M(t)}, \tag{1}$$

since under perfect repairs $N(t)$ is a renewal process driven by F, with the renewal function $E_{PR}N(t) := M(t)$; while under minimal repairs, $N(t)$ is a nonhomogenious Poisson Process (NHPP) with mean value function $E_{MR}N(t) = \Lambda(t)$, the hazard function of the new equipment's life distribution. Clearly perfect repairs are preferable to minimal repairs, if and only if

$$R_{PR:MR} \geq 1 \Longleftrightarrow M(t) \leq \Lambda(t), \text{ all } t \geq 0. \tag{2}$$

The question then is : Is (2) free for all reasonable aging distributions such as those posssesing any of the classical aging properties?

Since a perfect repair (replacement/ renewal) always returns an equipment to "*good as new*" condition, while a minimal repair restores the equipment's level of performance to its condition just prior to failure; it would appear reasonable to believe that perfect repairs (PR) always outperform the minimal repairs (MR) in the sense that the former would result in a smaller number of failures on an average than under under minimal repairs (MR). Such belief however, tacitly assumes that the equipment is suffering some form of degradation over time, i.e., ,its survival distribution is aging in some sense. It is easy to construct examples of survival distributions F for which minimal repairs are vastly superior (see Bhattacharjee (1994)) under strong *anti-aging* assumption such as the DFR (decreasing failure rate) property.

What is surprizing and counterintuitive is that even assuming F is *aging* (i.e., assuming F belongs to any one of the standard nonparametric aging classes) is *not* enough to guarantee

$$E_{PR}N(t) \leq E_{MR}N(t), \text{ all } t > 0. \tag{3}$$

The example below shows that for (3) to hold, F must be fairly strongly aging, and aging profiles which are relatively weak won't do to justify superiority of PR over MR. The folllowing counterexample shows that the NBU (new better than used) property is sufficient to imply (3), but the weaker NBUE (new better than used in expectation) or the DMRL (decreasing mean residual life, which is stronger than NBUE but does not imply NBU) property, is not.

Counterexample. *Perfect repairs can be worse than minimal ones.* Consider the life d.f. F of a variable X such that $P(X = 1) = P(X = 3) = \frac{1}{2}$. Its survival function is :

$$\bar{F}(t) = 1_{(0,1]}(t) + \frac{1}{2}1_{(1,3]},$$

which then has the hazard function

$$\Lambda(t) = \begin{array}{ll} 0, & \text{if } 0 \leq t < 1 \\ \ln 2, & \text{if } 1 \leq t < 3 \\ \infty, & \text{if } 3 \leq t. \end{array}$$

Let F^{*n} denote the n-fold self convolution of F with itself. It is then easily checked that,

$$2 \leq t < 3 \Longrightarrow M(t) = F(2) + F^{*2}(2) = \frac{3}{4} > \ln 2 = \Lambda(t).$$

By computation, check F is DMRL and hence F is NBUE, but is *not* NBU. Our result (Corollary 3.2) in the next section shows that the stronger NBU property implies $R_{PR:MR} \geq 1$, which of course is not the case here.

3 Bounding the Renewal Function.

When the d.f. F driving the renewal process is known to possess specific nonparametric properties, it may be possible to exploit such specifications to find interesting upper bounds on the renewal function $M(t)$. Some bounds of this type are known in the literature. For example, Barlow and Proschan (1975) show that,

$$F \text{ is NBUE} \implies M(t) \leq \frac{t}{\mu}. \tag{4}$$

where μ is the mean lifetime.

Here is a general method to find upper bounds on the renewal function, using ideas from stochastic dynamic programming. Let,

$$\mathcal{Q} = \{Q \mid Q : R^+ \mapsto R^+, \ Q(0) = 0, \ Q \text{ is } \uparrow, \ Q(\infty) = \infty\}, \tag{5}$$

be the set of real valued, nonnegative, nondecreasing functions mapping $R^+ := [0, \infty)$ into itself such that the conditions stated in (5) are satisfied. Given a (survival) distribution F on R^+, define an *operator L* mapping $\mathcal{Q}$ into itself, by

$$LQ(t) := F(t) + \int_0^t Q(t - x)\, F(dx), \qquad t \geq 0. \tag{6}$$

The observation that $Q \in \mathcal{Q}$ implies $LQ \in \mathcal{Q}$ needs to be argued. To this end, first note that for any $s \geq 0$,

$$
\begin{aligned}
LQ(t + s) \quad &- \quad LQ(t) \\
&\geq \quad \{F(t + s) - F(t)\} + \int_0^t \{Q(t + s - x) - Q(t - x)\} F(dx) \\
&\geq \quad 0,
\end{aligned}
$$

since, F is a d.f., and Q is $\uparrow$ implies each of the terms within braces is nonnegative.

Again, $Q \uparrow$ and the definition of LQ implies $0 \leq LQ(t) \leq F(t)\{1 + Q(t)\}$. Since, $F(0) = 0$, we have $LQ(0) = 0$. Finally, choosing any ϵ such that $0 < \epsilon < t$ and $F(\epsilon) > 0$, we get,

$$
\begin{aligned}
LQ(t) \quad &\geq \quad F(t) + \int_0^\epsilon Q(t - x)\, F(dx) \\
&\geq \quad F(t) + Q(t - \epsilon) F(\epsilon) \\
&\to \quad 1 + Q(\infty) F(\epsilon) = \infty,
\end{aligned}
$$

as $t \to \infty$.

Let $\mathcal{Q}^+$ be the set of functions on R^+ into itself satisfying the first two conditions $Q(0) = 0, Q \uparrow$ in (5). Although the class of functions $\mathcal{Q}$ will be the focus of our interest, it may be noted that the operator L also maps the wider class Q^+ into itself. In particular the function 0 which identically vanishes on R^+ and its repeated iterates $L^n 0$ all belong to Q^+.

Definition. A function $Q \in Q$ is *excessive* if $LQ \leq Q$ pointwise.

Then, we have,

Theorem 3.1 *Given a survival d.f. F with renewal function $M(t)$,*
(i) $L^n 0 \to M \in Q$, *pointwise, as $n \to \infty$.*
(ii) L *is monotone.*
(iii) $Q \in Q$ *is excessive.* $\implies M \leq Q$ *pointwise.*
(iv) M *is the minimal fixed point of L.*

Proof. Monotonicity of L (i.e., $u \leq v \implies Lu \leq Lv$, for every d.f. F on R^+) is clear from the definition and (6) verify (i), note that for every $t \geq 0$,

$$L0(t) = F(t),$$

$$L^2 0(t) = F(t) + \int_0^t L0(t-x)\,F(dx)$$

$$= F(t) + \int_0^t F(t-x)\,F(dx) = F^{*2}(t).$$

By induction, it easily follows that for all $n = 1, 2, \cdots$

$$L^n 0(t) = \sum_{j=1}^n F^{*n}(t),$$

and hence, as $n \to \infty$, we get $L^n 0(t) \to M(t)$, by an appeal to the monotone convergence theorem. Funally, since a renewal function M is nondecreasing, $M(0) = 0$ and for any underlying suvival d.f.F with support $[0, \infty)$, satisfies $M(t) \to \infty$ as $t \to \infty$; we have $M \in Q$.

To check (iii), suppose $Q \in Q$ and $LQ \leq Q$ on R^+, Then by monotonicity of L, the repeated iterates of Q satisfy

$$Q \geq LQ \geq L^2 Q \geq \cdots \geq L^n Q,$$

for all $n \geq 1$. Since Q is nonnegative, another appeal to the monotonicity of L, and then a reference to claim (i) above, yields

$$Q \geq L^n Q \geq L^n 0 \to M, \text{ as } n \to \infty. \tag{7}$$

Thus $Q(t)$ dominates the renewal function $M_F(t)$ pointwise on R^+.

To verify (iv), note that the classic renewal equation :

$$M(t) = F(t) + \int_0^t M(t-x)\,F(dx), \qquad t \geq 0,$$

is equivalent to the statement $LM(t) = M(t)$, all $t \geq 0$, which shows M is a fixed point of L. If $Q \in \mathcal{Q}$ is any other fixed point of L, then by argumements parallel to those in (7), we get

$$Q = LQ = L^n Q \geq L^n 0 \longrightarrow M, \text{ as } n \to \infty.$$

Hence, $M(t) = \inf\{Q(t) : Q \in \mathcal{Q}, LQ = Q\}$. Thus the renewal function is the minimal fixed point of L. $\square$

While the technical content of Theorem 3.1 is fairly straightforward, it is of some interest to outline the dynamic programming analogy which motivates it. The idea of using such analogies applied to optimization problems in stochastic processes was first described by Blackwell (1964) in the context of some random walk type problems. For us, formulating the renewal function as the total "return" (expected payoff) in a game against nature can be described as follows.

Given a d.f. F on R^+, consider the following 'game against nature'. At any position $t > 0$; we observe a value of a random variable X distributed as F, next move to $\max(t - X, 0)$ and get an unit amount of payoff ("return") if $X \leq t$. If $X > t$, we have reached the origin and the game is finished. Otherwise, we proceed to the next move using the same rule. On the (x, y)-plane, the line $y = 0$ is thus an absorbing barrier. With an initial starting position $t > 0$, and under the strategy defined by the sequence of random variables $X_1, X_2, \cdots$ independently chosen according to the d.f F; the total expected return (= the total expected payoff) till absorption occurs, is then,

$$E(\sum_{n=1}^{\infty} 1_{\{\sum_{i=1}^n X_i \leq t\}}, \quad = \quad E \max\{n : \sum_{i=1}^n X_i \leq t\}$$
$$= \quad M(t),$$

the renewal function of F.

Suppose now that at any stage of the game with our current position at $t \geq 0$, we are given the option to quit playing and accept a terminal payoff $Q(t)$. This terminal paoff is given as a surrogate reward in lieu of potential payoffs we might accumalate in future moves if we continued play until absorption. Since i) there are no more payoffs once we hit the origin, and ii) further from the origin we start, greater we expect our total reward to be; reasonable conditions on the surrogate payoff functions Q are precisely those described by the class of functions $\mathcal{Q}$ in (5).

In the original game, the expected return of playing upto n-moves or absorption - whichever is sooner, is then clearly $L^n 0 \in \mathcal{Q}^+$ and converges to the the total expected return, the renewal function $M \in \mathcal{Q}$ as $n \to \infty$. In the modified game with options to quit, the expected return of playing for one move, starting at $t > 0$, and then quitting with the surrogate payoff $Q \in \mathcal{Q}$, is

$$E\{1_{X<t} + Q(\max(t - X, 0))\} = \int_0^t \{1 + Q(t - x)\} F(dx) + Q(0)\bar{F}(t)$$
$$= \quad LQ(t),$$

since $Q(0) = 0$. Theorem 3.1(iii) then says that, *if* it is better not to play at all by quitting immidiately with the surrogate payoff than playing for one move in the modified game, *then* such surrogate payoff functions Q majorize the the total expected payoff in the original game.

The next two results illustrate applications of the bounding technique in Theorem 3.1, by finding sharp surrogate functions under suitable aging assumptions.

The following result was fiven in Bhattacharjee (1993). Here we prove the upper bound in (8) below, by demonstrating that it is excessive.

Corollary 3.2 *Suppose F is NBU. Then,*

$$M(t) \leq -\ln\{1 - F(t)\}. \tag{8}$$

The bound is sharpi, and is attained if and only if F is exponential.

Proof. Recall, F is NBU if and only if its hazard function $\Lambda := -\ln \bar{F}$ is superadditive, i.e, iff

$$\Lambda(x + y) \geq \Lambda(x) + \Lambda(y), \quad \text{all } x, y \geq 0.$$

By Theorem 3.1, it is enough if $L\Lambda \leq \Lambda$ on R^+. Now, using the superadditivity of Λ,

$$
\begin{aligned}
L\Lambda(t) &= F(t) + \int_0^t \Lambda(t - x) \, F(dx) \\
&\leq F(t) + \int_0^t \{\Lambda(t) - \Lambda(x)\} \, F(dx) \\
&= \{1 + \Lambda(t)\} F(t) - \int_0^t \Lambda(x) F(dx). \tag{9}
\end{aligned}
$$

The last integral equals

$$
\begin{aligned}
\int_0^t \Lambda(x) \, F(dx) = \int_0^{\Lambda(t)} y e^{-y} \, dy &= 1 - \{1 + \Lambda(t)\} e^{-\Lambda(t)} \\
&= F(t) - \Lambda(t) \bar{F}(t),
\end{aligned}
$$

using the identity $\bar{F}(t) = e^{-\Lambda(t)}$. Substituting this in (9), we get,

$$
\begin{aligned}
L\Lambda(t) &\leq \{1 + \Lambda(t)\} F(t) + \Lambda(t) \bar{F}(t) - F(t) \\
&= \Lambda(t), \quad t \geq 0.
\end{aligned}
$$

Thus $\Lambda(t)$ is "excessive" for L, ,and hence, by Theorem 3.1, $M(t) \leq \Lambda(t)$ on $[0, \infty)$, whenever F is NBU. The bound is sharp, since it is attained when F is exponential. For the argument of the '*only if*' part, see Bhattacharjee (1993). □

The method of Theorem 3.1 can also be used to verify the well known bound in (4), when F is NBUE. The point of the derivation below is to underscore the fact that since the renewal function, being a fixed point of L, is necessarily excessive;

any sharp upper bound must be excessive, since it is necessarily a fixed point for the operator L corresponding to the distribution F for which it is the renewal function.

Corollary 3.3 *If F is NBUE, the upper bound in (4) on its renewal function is excessive. It is sharp, and is attained if and only if F is exponential.*

Proof. Suppose F is NBUE with mean μ. The function $Q(t) := t/\mu \in \mathcal{Q}$. We have,

$$
\begin{aligned}
LQ(t) &= F(t) + \int_0^t \left(\frac{t-x}{\mu}\right) F(dx) \\
&= (1+\frac{t}{\mu})F(t) - \frac{1}{\mu}\int_0^t x\, F(dx) \\
&= (1+\frac{t}{\mu})F(t) + \frac{1}{\mu}\{t\bar{F}(t) - \int_0^t \bar{F}(x)\, dx\} \\
&= (1+\frac{t}{\mu})F(t) + \frac{1}{\mu}\{t\bar{F}(t) + \int_t^\infty \bar{F}(x)\, dx - \mu\}.
\end{aligned}
$$

Using the NBUE property, for the integral on the right hand side, we have,

$$
\begin{aligned}
LQ(t) &\leq (1+\frac{t}{\mu})F(t) + \frac{1}{\mu}\{t\bar{F}(t) + \mu\bar{F}(t) - \mu\} \\
&= \frac{t}{\mu} \equiv Q(t),
\end{aligned}
$$

which completes the proof, by reference to Theorem 3.1 again. Sharpness of course is trivial, and follows by choosing F exponential.

To prove that the bound is attained *only if* F is exponential; let $F^*(s) = \int_0^\infty e^{-sx} F(dx)$ be the Laplace transform (L.T.) of a survival distribution F whose rebewal function $M(t)$ attains the bound in (4). Then, taking Laplace transform of both sides in the identity $M(t) = t/\mu$, we get,

$$
\frac{F^*(s)}{1-F^*(s)} = \int_0^\infty e^{-st} M(dt) = \mu^{-1}\int_0^\infty e^{-st}\, dt = \frac{1}{s\mu}, \qquad s > 0. \qquad (10)
$$

Clearly, the exponential Laplace transform $F^*(s) = \frac{1}{1+s\mu}$ is a solution of (10). There is no other 'completely monotone' (Feller (1967)) solution; since if $G^*(s)$, $s > 0$ is another such solution, then (10) implies

$$
\frac{F^*(s)}{1-F^*(s)} = \frac{G^*(s)}{1-G^*(s)},
$$

or, $F^*(s) = G^*(s)$, $s > 0$. By the uniqueness of Laplace transforms, G must be exponential.   □

4   Conclusion.

In this article, our concern has been to compare the expected number of failures, under perfect and minimal repairs, in any time interval $[0, t)$. Our formulation of the renewal function as the total expected payoff in a game against nature is useful in finding upper bounds on the renewal function and hence upper bounds on the efficiency of minimal repairs relative to replacements.

If there is a cost of repair; it may be more appropriate to compare the expected cost of the repair policies instead. Clearly, comparing the expected number of failures under either policy is equivalent to cost comparison if the cost of repair is fixed irrespective of whether we undertake either replacements or minial repairs. In other scenarios; where the cost of repair is allowed to depend on the nature of repair, with minimal repairs typically costing less than perfect ones; it would be of interest to ask : under what circumstances are perfect repairs preferable than minimal repairs? Such comparisons will be considered in a future communication.

References

1. R.E. Barlow and F. Proschan (1975). *Statistical Theory of Reliability, Probability Models.* Holt, Rinehart and Winston.
2. M.C. Bhattacharjee (1993). Aging renewal process characterizations of exponential distributions. *Microelectronics and Reliability.* **33**(14), 2143-2147.
3. M.C. Bhattacharjee (1994). Repair relative aging and efficiency of repairs. *Reliability and Quality in Design*, Proceedings of the ISSAT International Conference, H. Pham (ed.), International Society of Science and Applied Technologies, 52-58.
4. M.C. Bhattacharjee (1994). Aging influenced by repair, repair efficiency and realizability of a renewal related distribution. *Int. J. Reliab. Quality and Safety Engg.*, **1**(2), 147-159.
5. D. Blackwell (1964). Probability bounds via dynamic programming. *Proc. Symp. Appl. Math* Vol XVI, Amer. Math. Soc., Providence, R.I., 277-280.
6. W. Feller (1971). *An Introduction to Probability Theory and Its Applications.* Vol 2, 2nd Ed., John Wiley, New York.

TTT-TRANSFORM CHARACTERIZATION OF THE NBRUE PROPERTY AND TESTS OF EXPONENTIALITY [a]

MANISH C. BHATTACHARJEE

Center for Applied Mathematics Statistics
New Jersey Institute of Technology
Newark, NJ
Email: mabhat@chaos.njit.edu

PRANAB K. SEN

University of North Carolina
Chapel Hill, N.C.
Email: pksen@bios.unc.edu

Along with a characterization of the NBRUE class of life distributions, based on the 'total time on test' transform, the problem of testing for exponentiality against strictly NBRUE alternatives is appraised here with due emphasis on the basic role of the TTT-transformation in this perspective. The proposed test is related to some existing ones, and is more amenable to Type II censoring schemes that commonly arise in practice.

1 Introduction

Let $X_i, i \geq 1$ be the successive life times, starting with a new equipment, which is renewed at each failure point instantaneously by a statistically identical copy. Thus, we assume that the X_i are independent and identically distributed (i.i.d.) nonnegative random variables (r.v.) with a continuous distribution function (d.f.) F defined on $R^+ = (0, \infty)$. We denote the corresponding *survival function* by $\bar{F}(x) = 1 - F(x)$, and assume that the mean $\mu = \int_0^\infty \bar{F}(x)dx$ is finite and positive. Finiteness of the second moment, usually invoked as a regularity assumption to simplify considerations of large sample asymptotics, turns out to be guaranteed for the class of nonparametric survival distributions considered here.

Let $\{N(t), t \geq 0\}$ be the *counting process* relating to the number of renewals, under instantaneous replacement (perfect repair) policy, up to the time-point $t (\geq 0)$, so that $N(t)$ can only assume nonnegative integer values and is nondecreasing in t. Under such repeated renewals, it is well known that the *remaining life* $L(t) = \sum_{\{i \leq N(t)+1\}} X_i - t$ at time t converges in law (as $t \to \infty$) to a nondegenerate r.v. X_o that has the d.f.

$$TF(y) = \mu^{-1} \int_0^y \bar{F}(x)dx, \quad y \in R^+, \tag{1}$$

so that, writing $\overline{TF}(t) = 1 - TF(t) = \mu^{-1} \int_t^\infty \bar{F}(x)dx$, we can express the *mean remaining lifetime* (MRL) function $E(X_o - t|X_o > t)$ as

$$e_{TF}(t) = \{\int_t^\infty \overline{TF}(y)dy\}/\overline{TF}(t), \quad t \in R^+. \tag{2}$$

---

[a] AMS 1990 *subject classifications* Primary 62N05, Secondary 60K10, 62G20

Note that the mean life time of a new item is $e_F(0) = \mu$. Then the class $\mathcal{F}$ of d.f.'s for which

$$e_F(0) = \mu \geq e_{TF}(t), \quad \text{for all } t \in R^+ \tag{3}$$

is referred to as the *New Better than Renewal Used in Expectation* (NBRUE) class (Abouammoh, Ahmed and Barry (1993)). Sen and Bhattacherjee (1996) considered some tests, for F exponential versus it belongs to the NBRUE class, based on the familiar *total time on test* (TTT) statistics. In the present study, we exploit a new characterization of the NBRUE class that leads us to formulate some alternative test procedures which are adoptable even for censored data. The tests proposed by Sen and Bhattacharjee (1996) encounter some technical problems in this respect. Thus, our current proposal leads to more flexible tests procedures.

Along with the preliminary notions, the basic characterization result on the NBRUE class is presented in Section 2. Section 3 is devoted to the formulation of TTT-transformation based test statistics for testing exponentiality versus (strict) NBRUE alternatives. Section 4 deals with the censored data models with particular emphasis on Type II censoring. The concluding section relates to some general observations.

2   TTT-Transfomation based NBRUE Characterizations

Following Stoyan (1983), we define a convex ordering $\prec_c$ by

$$F \prec_c G \text{ if } \int_t^\infty \bar{F}(x)dx \leq \int_t^\infty \bar{G}(x)\,dx, \quad \text{for all } t \in R^+. \tag{4}$$

Also we define the *first derived distribution* $TF(x)$ as in (1). Clearly, the definition (3) of the NBRUE-aging property and (4) then shows that the NBRUE property is equivalent to the 'first derived distribution' being convex ordered relative to the survival distribution F, viz.,

$$F \text{ is NBRUE } \iff TF \prec_c F. \tag{5}$$

Motivated by the partial ordering characterization (5), Sen and Bhattacharjee (1996) considered the functional

$$J_F(t) = \int_t^\infty \{\bar{F}(x) - \bar{T}F(x)\}dx, \quad t \in R^+, \tag{6}$$

which is obviously nonnegative for the NBRUE class, is strictly positive for some $t > 0$ if F is strictly NBRUE, and vanishes identically ($J_F(t) \equiv 0$) when F is exponential. They used the sample counterparts of this functional to test for exponentiality against strict NBRUE alternatives, and incorporated both linear and Kolomogorov-Smirnov type statistics based on the TTT stochastic process. In this context, it may be noted that the above characterization involves the entire tail for both the distributions F and TF. For this reason, there is a technical problem in dealing with right-censored (Type I or II) data models where the right tails are unobservable. To eliminate this problem, we consider an alternative characterization that is amenable for right-censoring.

Note that, if F is NBRUE, then the convex ordering criterion (5) implies $E_{TF}(X^p) \leq E_F(X^p)$, for all $p \geq 0$. Also. for any $r \geq 1$, the moments of F and TF are related by, $E_{TF}(X^{r-1}) = (r\mu)^{-1} E_F(X^r)$. Hence, combining the two observations, we arrive at the conclusion :

$$F \text{ is NBRUE} \implies E_F(X^{r+1}) \leq (r+1)\mu E_F(X^r), \quad \text{for all } r > 0.$$

In particular, setting $r = 1$, it is easily seen that under the NBRUE hypothesis, the usual regularity assumption of finite variance is *free*, viz.,

$$F \text{ is NBRUE} \implies \mu_2 \leq 2\mu^2 < \infty \iff \eta \leq 1,$$

where σ is the population standard deviation, and $\eta = \sigma/\mu$ is the coefficient of variation (c.v.) of F.

Corresponding to the survival distribution F, define the function

$$\Delta_F(u) := \frac{1}{2}(1 - \eta^2) + \int_0^u \frac{\phi^{-1}(z) - z}{1 - \phi^{-1}(z)}\, dz, \quad u \in (0, 1), \tag{7}$$

where $\phi(u) := \psi_F(u)$ is the scaled *total time on test* (TTT)-transform of F, defined as,

$$\phi(u) = \mu^{-1} \int_0^{F^{-1}(u)} \bar{F}(x)\,dx, \quad 0 < u < 1, \tag{8}$$

and $F^{-1}(u) = \inf\{x > 0 : F(x) \geq u\}$. Then we have the following TTT-transform criterion for the NBRUE property.

Theorem 2.1 *Let F be a survival distribution. Then,*
(i) F is NBRUE if and only iff $\Delta_F(u) \geq 0$, for all $u \in (0, 1)$.
(ii) F is exponential $\iff \Delta_F(u) \equiv 0$, on $(0, 1)$.

Proof. (i) To prove the claim, we first derive the TTT-transform of the distribution TF as a functional of the TTT-transform of F. Since $\phi(F(t)) = TF(t)$, by (8); setting $z = \phi(F(t)) \in (0, 1)$ for $0 < t < \infty$, we get,

$$\frac{dz}{dt} = \bar{F}(t)/\mu = \{1 - \phi^{-1}(z)\}/\mu,$$

where $\phi^{-1}(z) = \inf\{u > 0 : \phi(u) \geq z\}$ is the usual inverse function. Hence, with μ_{TF} denoting the mean of TF, the scaled TTT-transform $\psi_{TF}(u)$ of TF can be expressed as

$$
\begin{aligned}
L(u) := \psi_{TF}(u) &= \frac{1}{\mu_{TF}} \int_0^T F^{-1}(t) \overline{TF}(t)\, dt \\
&= \frac{\mu}{\mu_{TF}} \int_0^u \frac{1 - z}{1 - \phi^{-1}(z)}\, dz, \\
&= \left\{\frac{1}{2}(1 + \eta^2)\right\}^{-1} \int_0^u \frac{1 - z}{1 - \phi^{-1}(z)}\, dz, \tag{9}
\end{aligned}
$$

since, $\mu_{TF} = \frac{\mu_2}{2\mu}$. Now, by the convex ordering characterization (5) of the NBRUE property and the remark following (6), we have,

$$F \text{ is NBRUE} \quad \Longleftrightarrow \quad J_F(t) \geq 0, \quad \text{all } t > 0$$

$$\Longleftrightarrow \quad \mu_{TF} - \int_0^t \overline{TF}(x)\, dx \leq \mu - \int_0^t \bar{F}(x)\, dx$$

$$\Longleftrightarrow \quad \mu_{TF}\{1 - \psi_{TF}(TF(t))\} \leq \mu\{1 - TF(t)\}$$

$$\Longleftrightarrow \quad \frac{\mu_{TF}}{\mu}\{1 - \psi_{TF}(TF(t))\} \leq \overline{TF}(t). \tag{10}$$

Setting $u = TF(t)$, $0 < u < 1$ in (10) then shows that, F is NBRUE *if and only if*:

$$1 - u = \overline{TF}(t) \quad \geq \quad \frac{\mu_{TF}}{\mu}\{1 - L(u)\} = \frac{1}{2}(1 + \eta^2) - \int_0^u \frac{1-z}{1-\phi^{-1}(z)}\, dz$$

$$\Longleftrightarrow \quad \frac{1}{2}(1 - \eta^2) + \int_0^u \frac{1-z}{1-\phi^{-1}(z)}\, dz - u \geq 0$$

$$\Longleftrightarrow \quad \Delta_F(u) \geq 0, \quad \text{all } 0 < u < 1, \tag{11}$$

where $\Delta_F(u)$ is the functional defined in (7).

(ii) If F is exponential, then its c.v $\eta = 1$ and its TTT-transform $\phi(u) \equiv u$, so that $\Delta_F(u) \equiv 0$, all $0 < u < 1$. Conversely, if $\Delta_F(u)$ vanishes on $(0,1)$; then,

$$\frac{1}{2}(1 - \eta^2) - I_F(u) = \Delta_F(u) \equiv 0,$$

where,

$$I_F(u) := u - \int_0^u \frac{1-z}{1-\phi^{-1}(z)}\, dz = \int_0^u \frac{\phi^{-1}(z) - z}{1-\phi^{-1}(z)}\, dz,$$

so that $I_F(u)$ is a constant (C), viz.,

$$I_F(u) = C \equiv \frac{1}{2}(1 - \eta^2), \quad \forall u \in (0,1).$$

This, of course, implies

$$\frac{\phi^{-1}(u) - u}{1 - \phi^{-1}(u)} = I_F'(u) = 0, \quad \forall u \in (0,1),$$

or, $\phi(u) = u$, on $(0,1)$; since the dinominator cannot identically vanish, in virtue of $\phi(u)$ being a nondegenerate d.f. on $(0,1)$. □

Note that the equivalence of the hypothesis of exponentiality to the criterion of identically vanishing $\Delta_F(u)$, in Theorem 2.1(ii), is true for *all* survival distributions, and does *not* require an appeal to any other nonparametric property. Together with the non-negativity of $\Delta_F(u)$ as a characterization of NBRUE property (Theorem 2.1(i)), this paves the way for constructing a suitable test statistic. In the next section, we will specifically incorporate the NBRUE characterization via TTT-transform in Theorem 2.1, for formulating suitable tests of exponentiality against strict NBRUE alternatives.

3   TTT Based Test Statistics

Recall, from Theorem 2.1, that F is NBRUE if and only if

$$\Delta_F(u) \equiv \frac{1}{2}(1 - \eta^2) + I_F(u) \geq 0, \quad \forall u \in (0,1), \tag{12}$$

where $\eta = \sigma/\mu =$ c.v. for the d.f. F, and the d.f. F is assumed to be absolutely continuous;

$$I_F(u) = \int_0^u \frac{\phi^{-1}(z) - z}{1 - \phi^{-1}(z)} \, dz, \quad 0 < u < 1, \tag{13}$$

and ϕ is the scaled TTT-transform of F, defined in (8). Based on a sample $(X_1, \cdots, X_n)$ of n i.i.d. r.v.s from the d.f F, let

$$0 \equiv X_{n:0} < X_{n:1} < \cdots < X_{n:n} < \infty \equiv X_{n:n+1}, \tag{14}$$

be the associated order statistics, and let

$$\begin{aligned}
d_{n0} &= 0 = d_{n,n+1}, \\
d_{nj} &= (n - j + 1)(Xn : j - X_{n:j-1}), \quad 1 \leq j \leq n; \tag{15}
\end{aligned}$$

and, further set,

$$\begin{aligned}
D_{nk} &= \sum_{j \leq k} d_{nj}, \quad u_{nk} = \frac{d_{nk}}{D_{nn}}, \text{ and} \\
U_{nk} &= \frac{D_{nk}}{D_{nn}} = \sum_{j \leq k} u_{nj}, \quad 0 \leq k \leq n. \tag{16}
\end{aligned}$$

Then, for the sample d.f. F_n, we have,

$$\bar{F}_n(t) = \frac{n - k + 1}{n}, \quad \text{for } X_{n:k-1} \leq t < X_{n:k}, \ k = 1, \cdots, n+1. \tag{17}$$

so that $\bar{F}_n(t)$ is nonincreasing, nonnegative and right continuous in $t(\geq 0)$ having n steps down of equal magnitude (n^{-1}) at the points $X_{n:k}; k = 1, \cdots, n$. The sample quantile function is

$$\begin{aligned}
F_n^{-1}(u) &= \inf\{t \geq 0 : F_n(t) \geq u\} \\
&= X_{n:k}, \quad \text{if } \frac{k-1}{n} < u \leq \frac{k}{n}, \quad k = 1, \cdots, n. \tag{18}
\end{aligned}$$

Note that $F_n^{-1}(u)$ is nondecreasing, nonnegative and left-continuous in $u \in (0,1)$ with jumps $(n - k + 1)^{-1} d_{nk}$ at $(k-1)/n$, $k = 1, \cdots, n$. Therefore, the 'plug-in' estimator of $\phi(.)$ is given by

$$\begin{aligned}
\phi_n(u) &= \hat{\mu}_n^{-1} \int_0^{F_n^{-1}(u)} {}^{'} \bar{F}_n(t) \, dt \\
&= \frac{D_{nk}}{D_{nn}} = U_{nk}, \quad \text{for } \frac{k-1}{n} < u < \frac{k}{n}; \ 1 \leq k \leq n. \tag{19}
\end{aligned}$$

Thus, $Z = \phi_n(u)$ is nonnegative, nondecreasing and left-continuous in $u \in (0,1)$ with jumps u_{nk} at $(k-1)/n$, , $k = 1, \cdots, n$. Since, for $(k-1)/n < u \le k/n$, and $\phi_n(u) = U_{nk} \ge Z > U_{n,k-1}$, we have,

$$\phi_n^{-1}(U_{nk}) \ge \phi_n^{-1}(z) > \phi_n^{-1}(U_{n,k-1}) = \frac{k-1}{n}, \text{ for all } U_{n,k-1} < z \le U_{nk}.$$

Hence,

$$\phi_n^{-1}(z) = \inf\{u \ge 0 \; : \; \phi_n(u) \ge z\} = \left(\frac{k-1}{n}\right)^+, \tag{20}$$

for $U_{n,k-1} < z \le U_{nk}$, $k = 1, \cdots, n$. Therefore, $\phi_n^{-1}(z)$ is nonnegative, nondecreasing and left-continuous in $z \in (0,1)$. Using these sample measures, and noting that the jump-points for $\phi_n(.)$ are the U_{nk} (as seen from (20)), the "plug-in" estimator of $I_F(u)$ at $u = U_{nk}$ is given by,

$$
\begin{aligned}
\hat{I}_{nk} &:= \hat{I}_n(U_{nk}) \\
&= \int_0^{U_{nk}} \frac{\phi_n^{-1}(z) - z}{1 - \phi_n^{-1}(z)} dz \\
&= \sum_{j \le k} \int_{U_{n,j-1}}^{U_{n,j}} \frac{j - nz - 1}{n - j + 1} dz \\
&= \sum_{j \le kl} \frac{n}{n - j + 1} u_{nj}\{\frac{j-1}{n} - U_{n,j-1}\} - \frac{n}{2} \sum_{j \le k} \frac{u_{nj}^2}{n - j + 1}, \tag{21}
\end{aligned}
$$

for $k = 0, 1, \cdots, n$. At this stage, we refer to the functional $J_F(t)$ defined in Section 2, and following Sen and Bhattacharjee (1996), we consider the "plug-in" estimator

$$\hat{J}_{nk} = \sum_{j=k+1}^{n} u_{nj}\{\frac{n}{n-j+1}[U_{n,j-1} - \frac{j-1}{n}]\} + \frac{n}{2} \sum_{j=k+1}^{n} \frac{u_{nj}^2}{n-j+1} \tag{22}$$

for $k = 0, 1, \cdots, n$. It follows readily that

$$\hat{J}_{nk} - \hat{J}_{n0} = \hat{I}_{nk}, \text{ for } k = 0, 1, \cdots, n. \tag{23}$$

Now to complete the definition of a 'plug-in' estimator of $\Delta_F(u)$, $u \in (0,1)$, we need an appropriate estimator of η^2 that would be sensitive to the NBRUE alternatives. Recall that,

$$
\begin{aligned}
\frac{1}{2}(1 - \eta^2) &= \frac{1}{2}\{1 - (\sigma^2/\mu^2)\} \\
&= (1 - \{E_F X^2 / 2(E_F X)^2\}) \ge 0, \text{ for } F \text{ NBRUE.} \tag{24}
\end{aligned}
$$

Moreover, the "plug-in" estimator of $\mu = E_F X$ is $\hat{\mu}_n = \frac{1}{n} D_{nn}$. We consider the following estimator of $E_F X^2$:

$$\hat{\xi}_n = \frac{1}{n} \sum_{k \le n} d_{nk}^2. \tag{25}$$

When F is exponential with mean μ, $E_F X^2 = 2\mu^2$ and $\hat{\xi}_n$ is an optimal unbiased estimator of $E_F X^2$. To estimate the term $\frac{1}{2}(1 - \eta^2)$, we consider the estimator :

$$
\begin{aligned}
\nu_n &= (1 - \hat{\xi}_n / 2\hat{\mu}_n^2) \\
&= \{1 - (n \sum_{k \leq n} d_{nk}^2)\}/2D_{nn}^2 \\
&= (1 - \frac{n}{2} \sum_{k \leq n} u_{nk}^2).
\end{aligned}
\tag{26}
$$

Combining (21) and (26), we get the estimators :

$$
\begin{aligned}
\hat{\Delta}_{nk} &= \nu_n + \hat{I}_{nk} \\
&= 1 - \frac{n}{2} \sum_{k \leq n} u_{nk}^2 - \frac{n}{2} \sum_{j \leq k} \frac{u_{nj}^2}{n - j + 1} \\
&\quad + \sum_{j \leq k} \frac{n}{n - j + 1} u_{nj} \{\frac{j - 1}{n} - U_{n,j-1}\}, \quad k = 0, 1, \cdots, n.
\end{aligned}
\tag{27}
$$

At this stage, we may proceed as in Sen and Bhattacharjee (1996), and consider the following test-statistics :

$$
(i) \quad L_n = \sum_{k \leq n} a_{nk} \hat{\Delta}_{nk},
\tag{28}
$$

$$
(ii) \quad K_n^+ = \max\{\hat{\Delta}_{nk} : 0 \leq k \leq n\},
\tag{29}
$$

where the a_{nk} are suitable normalizing constants (viz., $a_{nk} = (n - k + 1)/n$, $0 \leq k \leq n$).

For the linear statistic L_n, note that the Lemmas 4.3-4.9 of Sen and Bhattacharjee (1996) remain intact here. Moreover, $\sqrt{n}\hat{\nu}_n$ is aymptotically normal, with mean zero, when F is exponential. The third term on the right hand side of (27) is $O_p(n^{-1} \log n)$, uniformly in $k(\leq n)$, and hence, drops out in this asymptotic setup. On the rest, L_n can be expressed as a martingale sequence (with zero mean), and hence, its asymptotic normality follows by reference to the usual martingale central limit theorems. Now, by virtue of (23),

$$
\hat{\Delta}_{nk} = \hat{\nu}_{nk} + \hat{J}_{nk} - \hat{J}_{n0} = \hat{\nu}_n^* + \hat{J}_{nk}, \text{ say },
\tag{30}
$$

for $k = 0, 1, \cdots, n$. As in Sen and Bhattacharjee (1996), we define,

$$
Z_{nk} = \frac{1}{nk} \{U_{nk} - \frac{k}{n}\}; \quad k = 1, \cdots, n - 1; \quad Z_{nn} = 0.
\tag{31}
$$

Then, by their equation (4.21), we have,

$$
\max\{| \hat{J}_{nk} - \sum_{j=k}^{n} Z_{nj} | : 1 \leq k \leq n\} = O_p\left(\frac{\log n}{n}\right),
\tag{32}
$$

so that,

$$\max\{|\hat{J}_{nk} - \hat{J}_{n0} + \sum_{j<k} Z_{nj}|: 1 \leq k \leq n\} = O_p\left(\frac{\log n}{n}\right). \qquad (33)$$

Therefore, by (30) and (33), we have

$$\max\{|\hat{\Delta}_{nk} + \frac{n}{2}\sum_{k\leq n}\{u_{nk}^2 - \frac{2}{n(n+1)}\} - \sum_{j<k} Z_{nj}|: 0 \leq k \leq n\} = O_p\left(\frac{\log n}{n}\right). \quad (34)$$

Let us define the σ-field $\{\mathcal{B}_{nk}, 0 \leq k \leq n\}$ as in Sen and Bhattacharjee (1996), and note that by their Lemma 4.1, $\{Z_{nk}, \mathcal{B}_{nk}; 0 \leq k \leq n\}$ is a zero-mean martingale (array). Recall that the Z_{nk} can be rewritten as

$$Z_{nk} = (n-k)^{-1}\sum_{j\leq k}(u_{nj} - \frac{1}{n}), \quad 1 \leq k \leq n. \qquad (35)$$

Therefore, (34) can be incorporated to verify that when F is exponential,

$$n^{\frac{1}{2}}\hat{\Delta}_{n[np]} \xrightarrow{D} N(0, \sigma_p{}^2), \quad \text{as } n \to \infty, \qquad (36)$$

and, in fact,

$$\{n^{\frac{1}{2}}\hat{\Delta}_{n[np]}, 0 \leq p \leq 1\} \xrightarrow{D} \mathbf{W} = \{W(p); 0 \leq p \leq 1\}, \qquad (37)$$

where $\mathbf{W}$ is Gaussian, and $\sigma_p{}^2$ is a known function of $p\,(0 \leq p \leq 1)$. Hence, the limiting distribution of the test statistics can be handled as in Sen and Bhattacharjee (Theorems 4.1 and 4.3, (1996)). In the remaining of this section, we elaborate the covariance kernel of this Gaussian process, as it would be needed in turn to provide suitable approximations for the critical level of test statistics based on the stochastic process in (36)-(37).

Note that by virtue of (34) and (35), we can write

$$\begin{aligned}
\hat{\Delta}_{nk} &= \sum_{j<k} Z_{nj} - \frac{n}{2}\sum_{k\leq n}\{u_{nk}^2 - \frac{2}{n(n+1)}\} + O_p(\frac{\log n}{n}), \\
&= \sum_{j<k} a_{nj}^{(k)}\{u_{nj} - \frac{1}{n}\} - \frac{n}{2}\sum_{j\leq n}\{u_{nj}^2 - \frac{2}{n(n+1)}\} + O_p(\frac{\log n}{n}), \quad (38)
\end{aligned}$$

uniformly in $k(= 1, \ldots, n)$, where

$$a_{nj}^{(k)} = \sum_{r=1}^{k-1}(n-r)^{-1}, \text{ for } j < k, \text{ and } 0, \text{ otherwise.} \qquad (39)$$

We may recall that under the exponential model (i.e., when H_0 holds), the u_{nj} have the dirichlet distribution, for which moments are easy to compute. In particular, we have the following:

$$E(u_{nj}^2 - \frac{2}{n(n+1)})^2 = \frac{4}{n(n+1)}[\frac{6}{(n+2)(n+3)} - \frac{1}{n(n+1)}], \forall j(\leq n); \qquad (40)$$

and, for any pair $j \neq r$,

$$E(u_{nj}^2 - \frac{2}{n(n+1)})(u_{nr}^2 - \frac{2}{n(n+1)}) = \frac{4}{n(n+1)}[\frac{1}{(n+2)(n+3)} - \frac{1}{n(n+1)}]. \quad (41)$$

Therefore, some routine computations lead us to the following:

$$Var(\frac{n}{2}\sum_{j\leq n}[u_{nj}^2 - \frac{2}{n(n+1)}]) = \frac{n^2(n-1)}{(n+1)^2(n+2)(n+3)} = n^{-1} + O(n^{-2}). \quad (42)$$

In the same way, we can show that under the exponentiality,

$$Var(\sum_{j<k} a_{nj}^{(k)}\{u_{nj} - \frac{1}{n}\}) = \frac{1}{n(n+1)}[\sum_{j<k}(a_{nj}^{(k)})^2 - \frac{(k-1)^2}{n}(\frac{1}{k-1}\sum_{j<k}a_{nj}^{(k)})^2]. \quad (43)$$

Likewise, under the exponentiality,

$$Cov(\sum_{j<k} a_{nj}^{(k)}\{u_{nj} - \frac{1}{n}\}, \frac{n}{2}\sum_{i\leq n}\{u_{nj}^2 - \frac{2}{n(n+1)}\})$$

$$= \sum_{j<k} a_{nj}^{(k)}\frac{2}{(n+1)(n+2)} - \frac{2}{(n+1)(n+2)}(\sum_{j<k}a_{nj}^{(k)})$$

$$= 0. \quad (44)$$

Therefore, letting $k = [np] + 1$, the variance σ_p^2 can be computed directly from (42) and (43) after noting that the $a_{nj}^{(k)}$ can be well approximated by

$$a_{nj}^{(k)} \sim log(n-j+1) - log(n-k+1) = log\{\frac{1-(j-1)/n}{1-(k-1)/n}\}, \quad j < k(\leq n), \quad (45)$$

which insures the existence of the limits for the variance factors. A similar case holds for the covariance terms.

4 Type II Censoring Case

When the life tests on n units are curtailed at the occurence of K-th failure (for some $K \leq n$), the sample is Type II censored, with the observed ordered failure times $X_{n:1} < X_{n:2} < \cdots < X_{n:K}$, while $X_{n:K+1} < \cdots < X_{n:n}$ are unobserved ($K \leq n$). At the observed time $X_{n:K}$ of the K-th failure, we have the modified estimators,

$$\hat{\mu}_{nK} = \frac{1}{K}\sum_{j\leq K} d_{nj} = \frac{1}{K}D_{nK}; \quad (46)$$

$$\hat{\xi}_{nk} = \frac{1}{K}\sum_{j\leq K} d_{nj}^2, \quad (47)$$

of μ and $E_F X^2$ respectively, using only the observed data points up to the K-th failure. Compared to (20), the "plug-in" estimator that we use in this censored case, is

$$\phi_{nK}^{-1}(z) = \left(\frac{j-1}{n}\right)^+, \text{ for } a_n(j-1,K) < z \le a_n(j,K), \ j \le K. \quad (48)$$

Then, in contrast to (21), we have here,

$$\begin{aligned}
\hat{I}_n(j,K) &:= \hat{I}_n(a_n(j,K)) \\
&= \sum_{k \le j} \int_{a_n(k-1,K)}^{a_n(k,K)} \frac{k-nz+1}{n-k+1} \, dz \\
&= \frac{K^2}{D_{nK}^2} \sum_{k \le j} \frac{d_{nk}}{n-k+1} \{ \frac{k-1}{n} \frac{D_{nK}}{K} - \frac{1}{2n}(D_{n,k-1} + D_{nk}) \}, \\
&\hspace{6cm} j = 1, \cdots, K. \quad (49)
\end{aligned}$$

for $j = 1, \cdots, K$. Thus, for $j \le K$, we use the censored sample estimators

$$\begin{aligned}
\hat{\Delta}_{nj}^{(K)} &= \hat{\nu}_{nK} + \hat{I}_n(j,K) \\
&= 1 - \frac{K}{2} \sum_{k \le K} \left(\frac{d_{nk}}{D_{nK}}\right)^2 \\
&\quad + \frac{K^2}{D_{nK}^2} \sum_{k \le j} \frac{d_{nk}}{n-k+1} \{ \frac{k-1}{n} \frac{D_{nK}}{K} - \frac{1}{2n}(D_{n,k-1} + D_{nk}) \}, \\
&\hspace{6cm} \text{for } j \le K, \quad (50)
\end{aligned}$$

in lieu of (27). Hence, defining

$$W_{nK}(u) = n^{\frac{1}{2}} \{ D_{nK}^{-1} D_{n[Ku]} - u \}, \quad 0 \le u \le 1, \quad (51)$$

we may use the weak convergence properties of $W_{nK}(.)$ and proceed as in the uncensored case. The necessity of resampling plans is even more in this case.

Note that for F exponential, $W_{nK}(u)$ has the same distribution as $W_K(u)$, $0 \le u \le 1$, whenever $K \ge 1$. This enables us to recapture the same asymptotic distribution theory for $W_{nK}(.)$, and hence, the test statistics in the censored case. While this works out well for type-II censoring (where K is prefixed); for type-I censoring, K is itself a random variable (having a binomial distribution with parameters $(n, F(T))$, where T is the prefixed censoring time. This introduces some additional complications. However since, for large n, as $k/n \to F(T)$ a.s. , and hence, the results for type-II censoring also pertain to censorings of type-I.

5   Conclusion

In many applied contexts, the use of censored sampling schemes for life-tests are often dictated by pragmatic reasons. Such reasons may due to various factors, such as (i) the need to economize when the equipments are costly and we terminate life tests as soon as a predetermined affordable number of units have been sacrificed (Type II censoring, or due to (ii) a maximum time limit available for life-tests (Type I censoring), or (iii) for other logistic reasons that give rise to many other censoring schemes such as *random censoring*.

Both type-I and type-II censoring are cases of *right censoring*, where the observations are truncated from above. This truncation corresponds to the unobservable data points as a consequence of curtailing the life tests at a predetermined time or, after a preassigned number of failures have been observed. While our analyses and discussions in Section 4 outline the relevant asymptotics and large sample distribution theory of the test statistics developed here for the two 'right censoring' cases; it should be remarked that the impact of various censoring schemes on the statistical functionals used to develop appropriate test statistics, often result in novel nonstandard situations from a technical viewpoint.

As remarked in Section 2, the TTT-transform based approach used here in formulating our test statistics is more amenable to right censoring than the approach we had used earlier in Sen and Bhattacharjee (1996), based on the characterization of the NBRUE property via the functional $J_F(t)$, which depends on the entire right tail of F and TF. The case of random censoring, which is also important in practice, remains to be explored for our problem (of testing exponentiality against strict NBRUE alternatives), and will be considered in a future communication.

References

1. A.M. Abouammoh, A.N. Ahmed and A.M. Barry (1993). Shock models and testing for the Renewal Mean Remaining Life. *Microelectronics and Reliability*, **33**, 729-740.
2. P.K. Sen and M.C. Bhattacharjee (1996). Tests for a property of Aging under Renewals : Rationality and general Asymptotics. *Frontiers in Probability & Statistics*, Mukherjee, S.P., Basu, S., and Sinha, B. K. (eds.), Narosa Publishing House, New Delhi. pp. 329-341.
3. D. Stoyan (1983). *Comparison Methods for Queues and other Stochastic Systems*. John Wiley, New York.

ON SMOOTHED FUNCTIONAL ESTIMATION UNDER RANDOM CENSORING

YOGENDRA P. CHAUBEY
Department of Mathematics and Statistics,
Concordia University,
Montréal, PQ H4B 1R6, CANADA
E-mail: chaubey@vax2.concordia.ca

PRANAB K. SEN
Department of Statistics and Biostatistics,
University of North Carolina at Chapel Hill,
Chapel Hill, NC 27599-7400 USA
E-mail: pksen@bios.unc.edu

A smoothing technique based on Hille's theorem adopted by Chaubey and Sen [7,8] for the estimation of survival, density, and hazard functions for complete data is incorporated here for random censoring case. In that way, Stute and Wang's [25] results are generailzed for such smoothed versions of the estimated functionals.

1 Introduction

For a nonnegative random variable $(r.v.)$ X, with a distribution function $(d.f.)$ F, we denote the survival function $(s.f)$ by $S(t) = 1 - F(t)$, $t \geq 0$. Whenever F is absolutely continuous, the density f is defined by $f(t) = F'(t)$, $t \geq 0$. In the latter case we may also define the hazard function $(h.f.)$ h by $h(t) = f(t)/S(t) = -(\partial/\partial t) \log S(t)$. Finally, the cumulative hazard function $(c.h.f.)$ H is defined by $H(t) = \int_0^t h(u)du$ $t \geq 0$. Note that by definition

$$S(t) = \exp\{-H(t)\}, \ t \geq 0,$$

so that $H(0) = 0$, $H(\infty) = \infty$, and $H(t)$ is $\nearrow$ in $t \geq 0$. We are basically interested in the estimation of these functionals, namely, $S(\cdot)$, $f(\cdot)$, $h(\cdot)$, and $H(\cdot)$.

In a complete sample case, we have n independent and identically distributed $(i..i.d.)$ $r.v.$'s X_1, $X_2, ...$ with the common $d.f.$ F(and $s.f.$ S), and an optimum estimator of $S(\cdot)$ is the sample counterpart

$$S_n(t) = \frac{1}{n}\sum_{i=1}^n I(X_i \geq t), \ t \geq 0. \tag{1}$$

However, $S_n(\cdot)$ being a step function, may not be that suitable for estimation of the hazard or density function. A host of *smoothing techniques* have been therefore proposed and studied in the literature. We refer to Devroye[9] and Wertz[29] for density estimation, Nadaraya [15,16] for the estimation of distribution function, Watson and Leadbetter [26,27], Rice and Rosenblatt [19], Singpurwalla and Wong [22,23], and Patil [17] for that of hazard and related functionals (see also Révész [18] for a comparative disposition of some common approaches to density estimation, which is the basis behind estimation of other functionals).

The current popular *kernel smoothing* may attach a positive value of probability for negative values of t near zero, and hence may not be appropriate for smooth estimation in survival analysis. Chaubey and Sen[7] formulated an alternative approach based on the classical Hille's[13] theorem (on uniform smoothing in real analysis), and obtained some smooth estimators of $S(\cdot)$ and $f(\cdot)$, which may have some advantage over their counterparts based on the usual *kernel* method of smoothing. They further studied the use of the resulting estimator in smooth estimation of the hazard and cumulative hazard functions (Chaubey and Sen[8]).

The purpose of the present paper is to adapt the above novel method in proposing smooth estimation of the survival distribution under random censorship model. In this case, we do not have the complete sample, rather we observe the censored life times $Z_i = \min(X_i, Y_i)$ and the indicator variable $\delta_i = I(X_i \leq Y_i)$, where $Y_1, Y_2, \ldots$ is an independent sequence of non-negative independent random variables representing random censoring times. For notational convenience, assume that all the random variables are defined on a probability space $(\Omega, \mathcal{B}, P)$. The role of $S_n = 1 - F_n$ is played by the product-limit estimator (PLE) as defined by

$$\hat{S}_n(t) = 1 - \hat{F}_n(t) = \prod_{i=1}^{n} \left(1 - \frac{\delta_{[i:n]}}{n - i + 1}\right)^{I_{[Z_{i:n} \leq t]}} \tag{2}$$

where $(Z_{n:0} = 0) < Z_{n:1} < \ldots < Z_{n:n} \; (< Z_{n:n+1} < +\infty)$ denote the ordered statistics corresponding to the observations Z_i, $i \leq n$, and $\delta_{[i:n]}$ is the value of δ correponding to $Z_{i:n}$.

This estimator was derived by Kaplan and Meier[14] as the nonparametric maximum likelihood estimator of the distribution function in the above context and it has been heralded as a breakthrough in Statistics (see Breslow[5]). It has received thorough scrutiny and attention to its theoretical and applied aspects. We cite here a few references which show the progress of the study of the properties of this estimator. Breslow and Crowley[6] studied its large sample behavior in comparing it with life table estimates, Gill[12] studied its weak convergence, and Wellner[28] studied the strong consistency. We also refer to Shorack and Wellner[21] for the state of the knowledge of the properties of this estimator until 1986. Recently, Stute[24] and Stute and Wang[25] have established strong consistency and asymptotic normality of functions of the type $\int \phi dF$ for certain functions ϕ which are generalizations of some of the previously obtained results. We would also like to refer to the bibliographical note in Anderson *et al.*[3] which discusses the smoothed version of Nelson-Aalen estimator of the hazard function. A smoothed version of the product limit estimator may be obtained from this. However, our approach is to start with a smooth estimator of the survival function, and exploiting the smoothness (differentiability) aspects derive other relevant functionals from it.

Section 2 presents the smooth estimator of the survival, density, and hazard functions, and section 3 presents the asymptotic properties of the survival function. The asymptotic properties of the derived smooth estimator of the density function are presented in section 4, and section 5 presents such properties for the hazard and cumulative hazard functions. The final section presents some remarks.

2   A Smooth Estimator of Survival Function

The common distribution of the observations $Z_1, Z_2, ..., Z_n$ is denoted by D, and the subdistributions corresponding to the events $\delta = 0$ and $\delta = 1$ are denoted by D_0 and D_1, respectively. We will also use the notation, $\bar{F} = 1 - F$, $\bar{G} = 1 - G$, etc. Further, for a distribution function F, τ_F is defined by $\tau_F = inf\{x : F(x) = 1\}$. The following representation of the PLE from Bhattacharjee and Sen [4] will be subsequently used for further analysis. Let m_n equal total number of failure points, and let these ordered failure points be denoted by $Z_{n:i}^*$, $i = 1, 2, ..., m_n$, then we have

$$\hat{S}_n(t) = \prod_{i \leq j}\left(\frac{n - k_i}{n - k_i + 1}\right), \ Z_{n:j}^* \leq t < Z_{n:j+1}^*, \ 0 \leq j \leq m_n, \tag{3}$$

where $Z_{n:j}^* = Z_{n:k_j}$ ($Z_{n:0}^* = 0, Z_{n:m_n+1}^* = \infty$). Consequently $\hat{S}_n(t)$ is also a step function with downward steps only at failure points $Z_{n:j}^*, 1 \leq j \leq m_n$. The smoothing algorithm, in the complete data case, is defined as a weighted average of a sequence of discrete points of the c.d.f., namely, $S_n(k/\lambda_n), 0 \leq k \leq n$ using an array of weights, $\{w_{nk}(t\lambda_n) : 0 \leq k \leq n, n \geq 1\}$, where,

$$w_{nk}(y) = \{\frac{y^k}{k!}\}/\{\sum_{j=0}^{n}\frac{y^j}{j!}\}, \ 0 \leq k \leq n, \tag{4}$$

and λ_n is is a sequence (possibly stochastic) which depends on whether the support of the underlying distribution is finite or infinite and on existence of a moment of a given order. Typically, we must have $\lambda_n \to \infty$ a.s. and $n^{-1}\lambda_n \to 0$ a.s. as $n \to \infty$. Assuming the existence of the first moment of F, Chaubey and Sen [7] suggest a data-adaptive choice $\lambda_n = n/X_{n:n}$. Such a choice is not valid in the present censored case as can be seen as follows. The parallel choice for λ_n in the complete data case will be $\lambda_n = m_n/Z_{n:m_n}^*$. Since, for this choice, we can write $\lambda_n = (m_n/n)(n^{-1}Z_{n:m_n}^*)^{-1}$ and $m_n/n \overset{a.s}{\to} \pi$ as $n \to \infty$, where $\pi = P(X < Y)$, we can claim that

$$|\lambda_n - n\pi Z_{n:n\pi}^{*-1}| \overset{a.s.}{\to} 0 \text{ as } n \to \infty.$$

Furthermore, since, $Z_{n:n\pi}^* \overset{a.s.}{\to} D_1^{-1}(\pi)$ as $n \to \infty$, where $D_1(t) = P(Z \leq t, \delta = 1)$ is a subdistribution of Z corresponding to $\delta = 1$. Hence, as $n \to \infty$, $(n/\lambda_n) \to 1/(\pi D_1^{-1}(\pi))$ a.s., which is finite for $0 < \pi < 1$; it equals ∞ or 0 accordingly as $\pi = 0$ or 1. Below, we show that the data-adaptive choice $\lambda_n = n/Z_{n:n}$ may still be appropriate if we make certain assumptions and consider a simple modification of $\hat{S}_n(t)$ as in Efron [10], namely,

$$\check{S}_n(t) = \begin{cases} \hat{S}_n(t) = \prod_{i \leq j}\left(\frac{n-k_i}{n-k_i+1}\right), \ Z_{n:j}^* \leq t < Z_{n:j+1}^*, \ 0 \leq j \leq m_n, \\ \\ 0 \text{ for } t > Z_{n:n} \end{cases} \tag{5}$$

where for $m_n < n$, we set $Z_{n:n} = Z_{n:m_n+1}^*$. The assumption we make here about the distributions F and G, is that they have infinite support, consequently, the distribution D also has the infinite support. For, if either one of F or G has a finte support $[0, T_0]$, then $\bar{D}(t) = 0$ for $t \geq T_0$. As is apparent from the corollary to the theorem of Stute [24], $\hat{S}_n(t)$ is strongly consistent in the interval $[0, T_0]$ for $S(t)$ if T_0 is not a jump-point of F. However, this convergence must be very slow, because the unmodified PL estimator will have a positive mass before $Z_{n:n}$ with probability π, hence the estimator given above is more appropriate for smoothing purposes. Moreover, with smooth estimator, in finite samples the flat mass between $Z_{n:m_n}^*$ and $Z_{n:n}$ will be smoothed out to give a decreasing trend, and therefore, rate of convergence in the tails with the smoothed estimator may be accelerated. The assumption of infnte support also leads us to show that $Z_{n:m_n}^* \to \infty$. This can be seen as follows. Defining $p(t_n) = D_1(t_n) = \int_0^{t_n} f(t)\bar{G}(t)dt$, we have $P[Z_{n:m_n}^* \leq t_n] = [p(t_n)]^n$. We can see that letting $p(t_n) = 1 - (a_n/n)$, where $a_n \to \infty$ but $n^{-1}a_n \to 0$, e.g., $a_n = c \log n$, $c > 0$, we have

$$P[Z_{n:m_n}^* \leq t_n] = (1 - \frac{a_n}{n})^n \sim e^{-a_n} \to 0 \text{ as } n \to \infty.$$

Hence, $Z_{n:m_n}^* \overset{a.s.}{\to} \infty$, as $n \to \infty$. Thus, the choice $\lambda_n = n/\lambda_n$ satisfies the conditions, as earlier required, and we are ready to define the smooth estimator of $S(t)$ similar to the one proposed in Chaubey and Sen [7] as given by:

$$\tilde{S}_n(t) = \sum_{k=0}^{m_n} w_{nk}(t\lambda_n)\hat{S}_n(k/\lambda_n), \ t \in \mathbf{R}^+, \tag{6}$$

where $\lambda_n = n/Z_{n:n}$ for now. A different choice may have to be made for density and hazard functions as we shall see later.

Note that by definition, $\tilde{S}_n(t)$ is smooth, bounded, and continuously differentiable with respect to $t \in \mathbf{R}^+$. Therefore, we have the following derived estimators for the density, cumulative hazard, and hazard functions given respectively as:

$$\begin{aligned}
\tilde{f}_n(t) &= -(d/dt)\tilde{S}_n(t) \\
&= \lambda_n\{\tilde{S}_n(t)[1 - w_{nn}(t\lambda_n)] \\
&\quad - \textstyle\sum_{k=0}^n w_{nk}(t\lambda_n)\hat{S}_n((k+1)/\lambda_n)\},
\end{aligned} \tag{7}$$

$$\tilde{H}_n(t) = -\log \tilde{S}_n(t), \text{ and} \tag{8}$$

$$\begin{aligned}
\tilde{h}_n(t) &= -(d/dt)\tilde{H}_n(t) = \tilde{f}_n(t)/\tilde{S}_n(t) \\
&= \lambda_n\{1 - w_{nn}(t\lambda_n)\} - \frac{\sum_{k=0}^n w_{nk}(t\lambda_n)\hat{S}_n(\frac{k+1}{\lambda_n})}{\sum_{k=0}^n w_{nk}(t\lambda_n)\hat{S}_n(\frac{k+1}{\lambda_n})}
\end{aligned} \tag{9}$$

for $t \in \mathbf{R}^+$. In this paper we will investigate the asymptotic properties of the smooth estimators of the survival, density, and hazard functions.

3   Asymptotic Properties of $\tilde{S}_n(\cdot)$

The motivation in Chaubey and Sen [7] comes from the following theorem due to Hille [13] (see Feller [11] p. 219).

Theorem 3.1. *Let $u(t)$ be a bounded, continuous function on* $\mathbf{R}^+$. *Then*

$$
e^{-\lambda t} \sum_{k \geq 0} u(k/\lambda)(\lambda t)^k/k! \rightarrow u(t), \quad \text{as } \lambda \rightarrow \infty, \tag{10}
$$

uniformly in any finite interval $\mathbf{J}$ *contained in* $\mathbf{R}^+$.

Remark 3.1. If in addition to being bounded and continuous, $u(t)$ is also monotone, then the convergence is uniform on $\mathbf{R}^+$.

Remark 3.2. As we shall see later, the smoothing technique proposed here involves a truncated version of the Poisson series. In view of that, for large t or when $t\lambda_n$ is large, we need to use a modified approach.

In order to make full use of this elegant theorem, first we use a version of $S(\cdot)$ which provides the intuitive basis for our $\tilde{S}(\cdot)$ and motivates the purpose. The main consistency result on $\hat{S}_n(t)$ is given below which uses the theorem in Stute and Wang [25] (see also Shorack and Wellner [21], Thoerem 1, p. 304) from which the uniform consistency of over $\mathbf{R}^+$ follows under the assumptions that F and G are absolutely continuous on $\mathbf{R}^+$.

Theorem 3.2. (Stute and Wang [25]) *Assume that F and G have no jumps in common, and denote by A the set of all atoms of E, then for any Borel-measurable, F-integrable function φ, with probability one and in mean*

$$
\lim_{n \rightarrow \infty} \int \varphi(x)\hat{F}_n(dx) = \int_{(x < \tau_D)} \varphi(x)F(dx) + 1_{\{\tau_D \in A\}}\varphi(\tau_D)F\{\tau_D\}, \tag{11}
$$

where $F\{a\} = F(a) - F(a^-)$ and $\tau_D = \inf\{x : D(x) = 1\}$.

From the above theorem we get the following corollaries.

Corollary 3.1. *Under the conditions in Theorem 3.2 we have*

$$
\sup_{t \leq \tau_D} |\hat{S}_n(t) - S_0(t)| \rightarrow 0 \text{ a.s. as } n \rightarrow \infty, \tag{12}
$$

where

$$
S_0(t) = \begin{cases} F'(t) \text{ if } t < \tau_D \\ S(\tau_D^-) - 1_{\{\tau_D \in A\}} F\{\tau_D\} \text{ if } t \geq \tau_D \end{cases}. \tag{13}
$$

Corollary 3.2. *Under the conditions in Theorem 3.1, $\hat{F}_n$ is uniformly consistent on $(-\infty, \tau_D)$ for F iff either $F\{\tau_D\} = 0$ or $F\{\tau_D\} > 0$ and $G(\tau_D^-) < 1$.*

Remark 3.3. Uniform consistency of $\hat{S}_n(t)$ defined in (5) is obtained if F and G are both continuous and $\tau_F = \tau_G = \infty$.

With the above exposition we are in a position to state the main consistency result for the smoothed estimator of the distribution function.

Theorem 3.3. *If $F(t)$ and $G(t)$ are continuous (a.e.), $\lambda_n \to \infty$, and $n^{-1}\lambda_n \to 0$ then*

$$\sup_{t \in \mathbf{R}+} |\tilde{S}_n(t) - S(t)| \to 0 \text{ a.s., as } n \to \infty. \tag{14}$$

Proof: We provide a Hille's analogue of $S(\cdot)$ for our susequent purpose. Since $S(t)$ is bounded, continuous, nonnegative, and nonincreasing (on $\mathbf{R}^+$), by (10), we can claim that

$$S^*(t; \lambda) = e^{-\lambda t} \sum_{k \geq 0} S(k/\lambda)(\lambda t)^k/k! \to S(t), \text{ as } \lambda \to \infty, \tag{15}$$

uniformly in any finite interval $\mathbf{J} \subseteq \mathbf{R}^+$. Since $\lambda_n = o(n)$, a.s., for every $t \in \mathbf{J}$, using the tail behavior of the exponential series, it follows that as $n \to \infty$,

$$S_n^*(t; \lambda_n) = \sum_{k=0}^{n} w_{nk}(t, \lambda_n) S(k/\lambda_n) \to S(t), \text{ uniformly in } t \in \mathbf{J}. \tag{16}$$

Our basic strategy is to to compare $\tilde{S}_n$ and S_n^*. Note that by Eqs. (6), (16), and Cor. 3.2,

$$
\begin{aligned}
\sup\{|\tilde{S}_n(t) - S_n^*(t, \lambda_n)| : t \in \mathbf{J}\} &\leq \max_{k \leq n} |\hat{S}_n(k/\lambda_n) - S(k/\lambda_n)| \\
&\leq \sup_{x \in \mathbf{R}+} |\hat{S}_n(x) - S(x)| \\
&\to 0 \text{ a.s., as } n \to \infty.
\end{aligned}
\tag{17}
$$

Therefore, by Eqs. (15), (16), and (17), as $n \to \infty$, we have for $\mathbf{J} \subset \mathbf{R}^+$

$$\sup_{t \in \mathbf{J}} |\tilde{S}_n(t) - S(t)| \to 0 \text{ a.s., as } n \to \infty, \tag{18}$$

whenever $\lambda_n \to \infty$ and $n^{-1}\lambda_n \to 0$ as $n \to \infty$. On the other hand, $S(t)$ and $\tilde{S}_n(t)$ are both nonnegative, nonincreasing, and bounded (on $\mathbf{R}^+$) and they converge to 0 as $t \to \infty$. Thus, for every $\eta > 0$, there exists a $t_\eta(< \infty)$, such that $S(t_\eta) < \eta$, so that letting $\mathbf{J} = [0, t_\eta]$, we have by Eq.(18), $\tilde{S}_n(t) < \eta$ a.s., as $n \to \infty$. As a result, $S(t) \leq S(t_\eta) < \eta$, $\forall t \geq t_\eta$ and $\tilde{S}_n(t) \leq \tilde{S}_n(t_\eta) < \eta$, $\forall t \geq t_\eta$ a.s., as $n \to \infty$, so that $\sup\{|\tilde{S}_n(t) - S(t)| : t \geq t_\eta\} \leq 2\eta$ a.s., as $n \to \infty$. Since $\eta(> 0)$ is arbitrary, the proof is completed.

The asymptotic normality of the smooth estimator follows from that of the PLE (see Gill[12], Shorack and Wellner[21], and Stute[24]) and Theorem 3.3.

Theorem 3.4. *Assume that F and G are absolutely continuous with infinite support, and let τ be such that $F(\tau) < 1$ and $G(\tau) < 1$*

$$n^{1/2}(\tilde{S}_n(\cdot) - S(\cdot)) \Rightarrow W(.) \tag{19}$$

in Skorohod-topology $D[0, \tau]$ as $n \to \infty$, where $W(.)$ is a Gaussian process with zero mean and covariance function

$$\sigma(s, t) = S(s)S(t) \int_0^{s \wedge t} \frac{1}{\overline{G}(v)[\overline{F}(v)]^2} dF(v). \tag{20}$$

Next, we establish a parallel result to the complete data case for studying the closeness of the smooth estimator to the product limit estimator. The asymptotic normality in the above theorem follows from the representation (see Stute [24])

$$\hat{S}_n(t) - S(t) = n^{-1} \sum_{i=1}^n U_i(t) + R_n(t), \tag{21}$$

where $U_1(t), U_2(t)...$ are *i.i.d.* with mean zero and $R_n(t) = o_P(1/\sqrt{n})$. However, this representation is not sufficient for studying the closeness of the smoothed estimator to the unsmoothed estimator. For this purpose, we use strong approximation results of Aly *et al.* [2] for the process

$$\beta_n(t) = n^{1/2}[S_n(t) - S(t)], \quad t \in \mathbf{R}^+, \tag{22}$$

and establish the following theorem.

Theorem 3.5. *Suppose that F and G are both abolutely continuous on $[0, \infty)$ with moments of order at least $r > 1/2$, F having bounded derivative, then we have for any $\eta \geq 0$ and $T < \infty$,*

$$\sup_{0 \leq t \leq T} |\tilde{S}_n(t) - \hat{S}_n(t)| = O\left(n^{-\frac{3}{4} + \frac{1}{8r}} (\log n)^{1+\eta}\right) \quad a.s. \text{ as } n \to \infty. \tag{23}$$

Proof: The proof is similar to that of a corresponding theorem in Chaubey and Sen [7]. The main result which is useful here is Corrolary 2.1 of Aly *et al.* [2]. Using this result we may assert that, for a sequence $a_n \to 0$ as $n \to \infty$, and $T < \infty$,

$$\begin{aligned}
\sup_{0 \leq t \leq T} \sup_{|s| \leq a_n} &n^{1/2}[\hat{S}_n(t + s) - \hat{S}_n(t) - S(t + s) + S(t)] \\
&= O\left(n^{-1/2}(\log n)^{5/2} \vee ((a_n \log(n/a_n))^{(1/2)}\right) \quad a.s. \text{ as } n \to \infty.
\end{aligned} \tag{24}$$

We start with the decomposition

$$\begin{aligned}
\tilde{S}_n(t) - \hat{S}_n(t) &= \sum_{k=0}^n w_{nk}(t, \lambda_n)\{\hat{S}_n(k/\lambda_n) - S(k/\lambda_n)\} \\
&\quad + \{S_n^*(t, \lambda_n) - \hat{S}_n(t)\}
\end{aligned} \tag{25}$$

$$\begin{aligned}
&= \sum_{k=0}^n w_{nk}(t, \lambda_n)\{\hat{S}_n(k/\lambda_n) - S(k/\lambda_n) - \hat{S}_n(t) + S(t)\} \\
&\quad + \{S_n^*(t, \lambda_n) - S(t)\}.
\end{aligned} \tag{26}$$

Let $\mathbf{N} = \{0, 1, \ldots, n\}$, and for every fixed $t \in \mathbf{R}^+$, let

$$\mathbf{N}_t = \{k \in \mathbf{N} : |k/\lambda_n - t| \leq (\lambda_n^{-1/2}(\log n)^{\frac{1}{2}(1+\delta)})\} and \tag{27}$$

$$\mathbf{N}_t^c = \mathbf{N} \backslash \mathbf{N}_t. \tag{28}$$

Note that under the assumed moment condition $Z = \min\{X, Y\}$ has moments of at least 2r, we can claim that $Z_{n:n} = o(n^{1/2r})$. Thus, $\lambda_n = n/Z_{n:n}$ is of maximum order $O(n^{1-\frac{1}{2r}})$. Taking $a_n = n^{-\frac{1}{2}+\frac{1}{4r}}(\log n)^{\frac{1}{2}(1+\delta)}$ in Eq. (24), we can claim for $n \to \infty$,

$$\sup_{0 \leq t \leq T < \infty} \{\max_{k \in \mathbf{N}_t} |\hat{S}_n(k/\lambda_n) - S(k/\lambda_n) - \hat{S}_n(t) + S(t)|\} = O\left(n^{-\frac{3}{4}+\frac{1}{8r}}(\log n)^{1+\eta}\right) \ a.s.,$$

$$\tag{29}$$

where $\eta \geq 0$.

Thus, the first term on the right hand side of the Eq. (26) is of order $O(n^{-\frac{3}{4}+\frac{1}{8r}}(\log n)^{1+\eta})$ $a.s.$ as $n \to \infty$. Furthermore using lemma 3.1 of Chaubey and Sen [7], which asserts that

$$\sum_{k \in \mathbf{N}_t^c} w_{nk}(t, \lambda_n) = o(n^{-1}), \ a.s., \ as \ n \to \infty, \tag{30}$$

we have

$$\sum_{k \in \mathbf{N}_t^c} \{(k/\lambda_n) - t\}w_{nk}(t, \lambda_n) = o(n^{-1}), \ a.s., \ as \ n \to \infty, \tag{31}$$

and

$$\sup\{|k/\lambda_n - t|^2 : k \in \mathbf{N}_t\} \leq \lambda_n^{-1}(\log n)^{1+\delta}. \tag{32}$$

Since $f'(\cdot)$ is assumed to be bounded a.e., we have for $k \in \mathbf{N}_t$,

$$S(k/\lambda_n) - S(t) = -f(t)(k/\lambda_n - t) + O((k/\lambda_n - t)^2), \tag{33}$$

so that by Eqs. (28), (30), (31), (32), and (33), we have

$$|S_n^*(t, \lambda_n) - S(t)| = O(n^{-1}(\log n)^{1+\delta}) \ a.s. \ as \ n \to \infty. \tag{34}$$

This completes the proof of Eq. (23).

Remark 3.4. In the above theorem, the *a.s.* convergence rate can not be guaranteed for the entire $\mathbf{R}^+$, as we were able to in the case of no-censoring in Chaubey and Sen [7]. Moreover, it is worth contrasting that the rate achieved here is lower than that in the complete data case, namely $O(n^{-3/4}(\log n)^{1+\eta})$. Such a rate can only be established if we assume the existence of the moment generating function of Z. In survival analysis, we often assume the existence of first moment, which provides a rate of $O(n^{-5/8}(\log n)^{(1+\eta)})$. The convergence rate in the tails depend on the tail behaviour of both the distributions F and G, though, we can assure for very large values of t that $|\tilde{S}_n(t) - \hat{S}_n(t)| \overset{a.s.}{\to} 0$ as $n \to \infty$. This shall be pursued in a future communication. We shall use a non-stochastic choice, $\lambda_n = O(n^\alpha)$ where $\alpha < (3/4)$ for studying the smoothness of other derived estimators.

In the next section we establish the asymptotic properties of the derived estimator of $f(t)$.

4 Asymptotic Properties of $\tilde{f}_n(\cdot)$

For $\tilde{f}_n(\cdot)$, defined by Eq. (7), first, consider the following.

Theorem 4.1. *Under the hypothesis of Theorem 3.3, if* $\lambda_n = O(n^\alpha)$ *for some* $\alpha < 3/4$,

$$\|\tilde{f}_n - f\| \to 0 \quad a.s. \text{ as } n \to \infty. \tag{35}$$

Proof: First, note that by Theorems 3.1 and 3.2, $\tilde{S}_n(t)$ is non–increasing in $t \in \mathbf{R}^+$, and $\tilde{f}_n(t)$ is continuous a.e. Thus, for every $\eta > 0$, there exists a $C(= C_\eta < \infty)$, such that

$$\tilde{S}_n(t) - \tilde{S}_n(t + y) < \eta \quad a.s., \ \forall\, t \geq C, \ y \geq 0. \tag{36}$$

Since, the left hand side of Eq. (36) is equal to $\int_t^{t+y} \tilde{f}_n(u)du$, a direct application of the first mean value theorem (of calculus) yields that by choosing y such that η/y is small, as $n \to \infty$,

$$\tilde{f}_n(t) \leq \eta' \quad a.s. \text{ for every } t \geq C, \tag{37}$$

where $\eta' \to 0$ as $\eta \to 0$. Also, repeating the same argument with $S(t)$, we have $f(t) < \eta'$, $\forall t \geq C$. Consequently, we have $\sup\{|\tilde{f}_n(t) - f(t)| : t \geq C\} \leq 2\eta'$, a.s. as $n \to \infty$. Thus to prove (35), it suffices to show that

$$\sup\{|\tilde{f}_n(t) - f(t)| : 0 \leq t \leq C\} \to 0 \quad a.s., \text{ as } n \to \infty. \tag{38}$$

To prove this we write $\tilde{f}_n(t)$ defined by Eq. (7)

$$\begin{aligned}
\tilde{f}_n(t) &= [-\lambda_n w_{nn}(t\lambda_n)\tilde{S}_n(t)] \\
&\quad + [\lambda_n \textstyle\sum_{k=0}^n w_{nk}(t, \lambda_n)[S(\tfrac{k}{\lambda_n}) - S(\tfrac{k+1}{\lambda_n})] \\
&\quad + [\lambda_n \textstyle\sum_{k=0}^n w_{nk}(t, \lambda_n)[\hat{S}_n(\tfrac{k}{\lambda_n}) - \hat{S}_n(\tfrac{k+1}{\lambda_n}) - S(\tfrac{k}{\lambda_n}) + S(\tfrac{k+1}{\lambda_n})] \\
&= T_{n1}(t) + T_{n2}(t) + T_{n3}(t), \text{ say.}
\end{aligned} \tag{39}$$

The proof is virtually the same as that of that of Theorem 4.1 in Chaubey and Sen [7] . The key steps are sketched here. First, similar to the proof of Eq. (30), for every (fixed) t, $nw_{nn}(t\lambda_n) \to 0$ a.s, as $n \to \infty$, and since by assumption $n^{-1}\lambda_n \to 0$ a.s., as $n \to \infty$, $\sup\{|T_{n1}(t)| : t \in [0, C]\} \to 0$, a.s. as $n \to \infty$. Making use of Eq. (30) and the fact that $\lambda_n = O(n^\alpha)$ for some $\alpha \leq (3/4)$, we conclude that $\sup|\{T_{n3}(t)| : t \in [0, C]\} \to 0$ a.s as $n \to \infty$. Using the boundedness of the density and its derivative, we may repeat the steps in Theorem 3.3, and conclude that

$$\sup\{|T_{n2}(t) - f(t)| : t \in [0, C]\} \to 0, \quad a.s. \text{ as } n \to \infty. \tag{40}$$

This completes the proof of the theorem.

Reamrk 4.1. As remarked in Chaubey and Sen [7] the existence and boundedness of $f'(\cdot)$ are not really needed for the theorem to be true. If we assume that $f(t)$ satisfies a Lipschitz condition of order α, for some $\alpha > 0$, i.e.,

$$|f(t) - f(s)| \leq K|t - s|^\alpha, \quad \text{forevery } t, s \in \mathbf{R}^+, \tag{41}$$

where $K(< \infty)$ does not depend on (s, t).

Next we give the asymptotic representation for $\tilde{f}_n(t)$ which would provide the asymptotic normality as well as the rate of convergence of the estimator. In this respect, it is customary (as in the kernel/nearest neighborhood methods) to assume the existence and boundedness of the second derivative $f''(t), t \in \mathbf{R}$. In our setup, we will appeal to a less stringent condition: $f'(t)$ satisfies a Lipschitz order α condition, for some $\alpha > 0$, i.e., there exists a positive $K(< \infty)$, such that

$$|f'(t) - f'(s)| \le K|t - s|^\alpha, \text{ for every } t, s \in \mathbf{R}^+. \tag{42}$$

Under the condition in Eq (42), Chaubey and Sen[7] (see their Eq. (4.13)) obtained the following representation for every(fixed) $t \in \mathbf{R}^+$:

$$\tilde{f}_n(t) = f(t) + \frac{1}{2\lambda_n} f'(t) + T_{n3}(t) + O(\lambda_n^{-1-\alpha}) \text{ a.s.}. \tag{43}$$

Before we proceed to incorporating Eq. (43) in the proof of asymptotic normality of the standardized version of $\tilde{f}_n(t)$, we may note that the variance of $T_{n3}(t)$ (as will be shown later on; see (47)) is $O(n^{-1}\lambda_n^{1/2})$, so that the mean squared error (MSE) of $\tilde{f}_n(t)$ is $O(n^{-1}\lambda_n^{1/2}) + O(\lambda_n^{-2})$, as $n \to \infty$. As such, we need to limit λ_n bounded from below by $cn^{2/5}$, for some $c > 0$. Choosing $\lambda_n = o(n^{2/5})$ enhances the bias term(to $O(n^{-2/5})$), while for $\lambda_n \sim n^{2/5}$, the normalizing factor $n^{2/5}$ and the bias term are of comparable order of magnitude. We shall make more comments on it later on.

Theorem 4.2. *Consider a nonstochastic, nondecreasing sequence $\{\lambda_n\}$ of positive numbers such that $\lambda_n = O(n^{2/5})$, and assume that (41) holds. Then, for every fixed $t \in \mathbf{R}^+$,*

$$n^{2/5}(\tilde{f}_n(t) - f(t)) - \frac{1}{2}\delta^{-2} f'(t) \xrightarrow{D} \mathcal{N}(0, \gamma_t^2), \text{ as } n \to \infty, \tag{44}$$

where

$$\gamma_t^2 = \frac{1}{2}(2\pi t)^{1/2} f(t)\delta; \quad \delta = \lim_{n \to \infty} \left(n^{-1/5}\lambda_n^{1/2}\right). \tag{45}$$

Proof: Again for the details of the proof we refer to Chaubey and Sen[7]. Basically, we need to prove that for every fixed $t > 0$,

$$n^{2/5} T_{n3}(t) \sim \mathcal{N}(0, \gamma_t^2), \text{ as } n \to \infty. \tag{46}$$

Note that the expression $T_{n3}(t)$ (see Eq. 39) involves a linear combination of the (centered) multinomial proportions $\{S_n(\frac{k}{\lambda_n}) - S_n(\frac{k+1}{\lambda_n}), 0 \le k \le n\}$, so that $E\, T_{n3}(t) = 0$ and the leading term of $Var(T_{n3}(t))$ is shown to be (see Eq. (4.19) of Chaubey and Sen[7])

$$\frac{1}{2}(2\pi t)^{-1/2} f(t)(\lambda_n^{1/2}/n), \tag{47}$$

which establishes the theorem.

The representation given in Eq. (43) extends to the entire $\mathbf{R}^+$ as such the following theorem as established in Chaubey and Sen[7] also holds in the present case (see the same for the details of the proof).

Theorem 4.3. *Under the conditions of Theorem 4.2, for a compact set $\mathcal{C} \subset \mathbf{R}^+$,*

$$\left\{ \left(n^{2/5}[\tilde{f}_n(t) - f(t)] - \frac{1}{2\delta_n^2} f'(t) \right), \, t \in \mathcal{C} \right\} \xrightarrow{\mathcal{D}} \text{Gaussian function,}$$

with the covariance function $\gamma_t^2 \delta_{st}$ where $\delta_{st} = 0$ for $s \neq t$ and 1 for $s = t$ and $\delta_n = \left(n^{-1/5} \lambda_n^{1/2} \right)$.

Remark 4.2. For the analysis of our estimator we do not need to assume the boundedness of the 2nd derivative of the density function as assumed in the kernel method. However, if we can assume f'' to be bounded, we can use the proposed method in this paper to estimate f' smoothly, choosing $\lambda_n \to \infty$ at the rate of $O(\lambda_n^{2/5})$.

Remark 4.3. Letting $\lambda_n = cn^h$ for some constant $c > 0$ and $0 < h < 1$, we find that

$$Bias^2(\tilde{f}_n) \approx c^{-2}(f'(t)/2)^2 \, n^{-2h} \tag{48}$$

and

$$MSE(\tilde{f}_n) \approx c^{-2}(f'(t)/2)^2 \, n^{-2h} + \frac{1}{2}\sqrt{c/(2\pi t)} f(t) \, n^{\frac{h}{2}-1}. \tag{49}$$

If we desire $\{bias\}^2$ going to zero at a faster rate than the variance, then we must have $2h > 1 - \frac{h}{2}$, *i.e.* $h > 2/5$. We note that for $h = 2/5$, bias^2 and variance go to zero at the same rate, namely $n^{-4/5}$, which is the same rate achieved by kernel method. Also, note that kernel method requires existence of 4 derivatives to achieve this rate.

Remark 4.4. The value of the constant c in the choice $\lambda_n = cn^{2/5}$ can be obtained by considering minimization of some criterion such as Mean Integrated Squared Error (MISE) or the Weighted Mean Integrated Squared Error (WMISE), given by:

$$MISE = \int_0^\infty MSE(\tilde{f}_n(t))dt \tag{50}$$

and

$$WMISE = \int_0^\infty MSE(\tilde{f}_n(t))f(t)dt, \tag{51}$$

as in the case of kernel method (see Chaubey and Sen[7]).

The next section studies the asymptotic properties of the smooth estimator of cumulative hazard and hazard functions.

5 Asymptotic Properties of $\tilde{H}_n(\cdot)$ and $\tilde{h}_n(\cdot)$

The strong consistency of the derived functionals $\tilde{H}_n(\cdot)$ and $\tilde{h}_n(\cdot)$ can be established only for compact intervals. First, we establish strong consistency of the estimators over a compact interval in the form of following theorem.

Theorem 5.1. *Let the pdf f be defined on $\mathbf{R}^+$ such that $E(X) < \infty$ and $H(X) < \infty$ for all $t \in \mathbf{R}^+$. Then for any compact interval $C \subset \mathbf{R}^+$, we have*

$$\| \tilde{H}_n - H \|_C = \sup_{t \in C} |\tilde{H}_n(t) - H(t)| \to 0 \quad \text{a.s. as } n \to \infty, \tag{52}$$

if $\lambda_n = o(n)$, and if in addition $\lambda_n = O(n^\alpha)$ for $\alpha < 3/4$,

$$\| \tilde{h}_n - h \|_C = \sup_{t \in C} |\tilde{h}_n(t) - h(t)| \to 0 \quad \text{a.s. as } n \to \infty. \tag{53}$$

Proof: Note that for t such that $0 < S(t) \le 1$

$$\begin{aligned}
\tilde{H}_n(t) &= -\log[\tilde{S}_n(t) - S(t) + S(t)] \\
&= -\log S(t) - \log(1 + x),
\end{aligned} \tag{54}$$

where

$$x = (\tilde{S}_n(t) - S(t))/S(t). \tag{55}$$

Using Theorem 3.3, for the *strong consistency* of $\tilde{S}_n(t)$ we conclude that $x \to 0$ a.s. as $n \to \infty$. Further using the Mann-Wald theorem (see Theorem 2.3.4 of Sen and Singer [20]) we conclude that for every t such that $S(t) > 0$,

$$|\tilde{H}_n(t) - H(t)| \to 0 \quad \text{a.s. as } n \to \infty. \tag{56}$$

Since $H(\cdot)$ and $\tilde{H}_n(\cdot)$ are continuous and monotone in t, pointwise convergence implies uniform convergence. Hence we have

$$\sup_{t \in C} |\tilde{H}_n(t) - H(t)| \to 0 \text{ a.s. as } n \to \infty, \tag{57}$$

which proves Eq. (52). Further using Theorems 3.3 and 4.1 on the almost sure convergence of $\tilde{S}_n$ and $\tilde{f}_n$, respectively, and a further use of Slutsky's theorem prove Eq. (53).

We may also establish the asymptotic normality of $\tilde{H}_n(t)$ by using Theorem 3.4 and the Mann-Wald theorem. Towards this end we compute the asymptotic bias and mean square error of $\tilde{H}_n(t)$. Expanding the logarithm in Eq. (54), we can easily conclude that for t such that $S(t) > 0$:

$$E(\tilde{H}_n(t) - H(t)) \sim \frac{1}{2n} \frac{V(t)}{S(t)} \tag{58}$$

and

$$E(\tilde{H}_n(t) - H(t))^2 \sim \frac{1}{n} \frac{V(t)}{S(t)}, \tag{59}$$

where

$$V(t) = \int_0^t \frac{1}{\bar{G}(v)[\bar{F}(v)]^2} dF(v). \tag{60}$$

Thus, we have the following theorem.

Theorem 5.2. *For every fixed* t, $\lambda_n \to \infty$, *and* $n^{-1}\lambda_n \to 0$ *as* $n \to \infty$, *we have*

$$\sqrt{n}[\tilde{H}_n(t) - H(t)] \xrightarrow{D} \mathcal{N}(0, \, V(t)/S(t)), \quad \text{as } n \to \infty. \tag{61}$$

For establishing a similar result for $\tilde{h}_n(t)$ we find that for t such that $S(t) > 0$

$$\tilde{h}_n(t) \sim h(t) + \frac{1}{S(t)}(\tilde{f}_n(t) - f(t)) + \frac{f(t)}{S^2(t)}(\tilde{S}_n(t) - S(t)) \quad \text{a.s.} \tag{62}$$

Using the asymptotic results established in Sec. 3 and 4, we obtain

$$E(\tilde{h}_n(t) - h(t)) \sim \frac{1}{2\lambda_n} \frac{f'(t)}{S(t)}. \tag{63}$$

Further, since, $Var(\tilde{f}_n(t)) = O(1/\lambda_n^2)$ and $Var(\tilde{S}_n(t)) = O(1/n^2)$, the order of $Cov(\tilde{f}_n(t), \tilde{S}_n(t))$ can not exceed $1/\sqrt{n}\lambda_n$, which can thus be ignored in computing the asymptotic variance of $\tilde{h}_n(t)$. Hence, we have the following theorem.

Theorem 5.3. *When* $\lambda_n/n^{2/5} \not\to 0$ *but* $\lambda_n < n^{3/4}$, *and for every fixed* t *such that* $S(t) > 0$ *we have*

$$n^{2/5}(\tilde{h}_n(t) - h(t) - \frac{1}{2\delta^2} \frac{f'(t)}{S(t)}) \xrightarrow{D} \mathcal{N}(0, \, (\gamma_t/S(t))^2), \quad \text{as } n \to \infty, \tag{64}$$

where

$$\gamma_t^2 = (\frac{\pi t}{2})^{1/2} f(t)\delta; \quad \delta = \lim_{n \to \infty} (n^{-1/5}\lambda_n^{1/2}), \tag{65}$$

and the bias term $f'(t)/\{2\delta^2 S(t)\}$ *drops out if* $n^{-2/5}\lambda_n \to \infty$.

Proof: The proof of the theorem follows from the discussion preceding the theorem and the asymptotic normality results as obtained in Sec. 3 and 4 with respect to $\tilde{S}_n(t)$ and $\tilde{f}_n(t)$ and the representation in Eq. (62).

6 Concluding Remarks

The basic problem with the original Product Limit Estimator (PLE) may not vanish in the tail unless the largest order statistic is a failure point. Efron's [10] modification forces it to zero beyond $Z_{n:n}$. Even with this modification, if the censoring distribution has lighter tails, then the PLE continues to be flat and may not be small for longer intervals with a high probability. The opposite event holds when the censoring distribution has heavier tails. In this respect, the proposed smoothing has a better performance in the tail. Also, the smooth estimator proposed here has merit in that it gives reasonable smoothing over the whole $\mathbf{R}^+$, where as the smooth estimator obtained from that of the Nelson-Aalen estimator is valid only for a finite range. The treatment of the functionals for large values of t requires more delicate consideration of the tail behavior of the individual distributions F and G. Some further modifiaction of the proposed smoothing may be necessary in this context. We intend to pursue it in a subsequent communication.

Acknowledgments

This research is partially supported from a research grant from NSERC of Canada.

References

1. M.G. Akritas and M.P. LaValley, Statistical analysis with incomplete data: A selective review, in *Handbook of Statistics 15*, pp. 551, eds., G.S. Maddala and C.R. Rao (North Holland, New York, 1997).
2. E.A.A. Aly, M. Csörgő and L. Horváth, Strong approximations of the quantile process of the product-limit estimator, J. Multivar. Anal. **16**, 185 (1985).
3. P.K. Andersen, O. Borgon, R.D. Gill and N. Keiding, *Statistical Models Based on Counting Processes*(Springer–Verlag, New York, 1993).
4. M.C. Bhattacharjee and P.K. Sen, Kolmogorov-Smirnov type tests for NB(W)UE alternatives under censoring schemes, in *Analysis of Censored Data*,eds., H.L. Koul and J.V. Deshpande, IMS Lecture-Monograph Series **27**, 25 (1995).
5. N.E. Breslow, Introduction to Kaplan and Meier (1958) Nonparametric estimation from incomplete observations, in *Breakthroughs in Statistics 3*, Vol. II, eds., Samuel Kotz and Norman L. Johnson(Springer– Verlag, New York, 1993).
6. N.E. Breslow and J.J. Crowley, A large sample study of the life table and-product limit estimates under random censorship, Ann. Stat. **2**, 437 (1974).
7. Y.P. Chaubey and P.K. Sen, On smooth estimation of survival and density functions. Statistics and Decisions **14/1**, 1 (1996).
8. Y.P. Chaubey and P.K. Sen, On smooth estimation of hazard and cumulative hazard functions. Proc. Golden Jubilee of Calcutta University, (*to appear*)(1997).
9. L. Devroye, *A Course in Density Estimation*. Birkhaüser, Boston, 1989
10. B. Efron, The two sample problem with censored data,Proc. Fifth Berkeley Sympos. Math. Statisti. Probab. **4**, 831 (1967).
11. W. Feller, *An Introduction to Probability Theory and its Applications, Vol. II.*(John Wiley and Sons, New York, 1965).
12. R.D. Gill, Large sample behavior of the product-limit estimator on the whole line,Ann. Stat. **11**, 49 (1983).

13. E. Hille. *Functional Analysis and Semigroups*, (Am. Math. Colloq. Pub. 31, New York, 1948).

14. E.L. Kaplan and P. Meier, Nonparametric estimation from incomplete observations, J. Amer Staist. Assoc. **53**, 457 (1958).

15. E. A. Nadaraya, Some new estimates for distribution functions, Theor. Prob. App. **9**, 497 (1964).

16. E. A. Nadaraya, On non–parametric estimates of density functions and regression curves, Theor. Prob. App. **10**, 196 (1965).

17. P.N. Patil, Bandwidth choice for nonparametric hazard rate estimaion, J. Statist. Plann. Inf. **35**, 15 (1993).

18. P. Révész, Density estimation, in *Handbook of Statistics 4*, pp. 531, eds., P.R. Krishnaiah and C.R. Rao (North Holland, New York, 1984).

19. J. Rice and M. Rosenblatt, Estimation of the log survivor function and hazard function,Sankhyā **38**, 60 (1976).

20. P.K.Sen and J.M. Singer, *Large Sample Methods in Statistics* (Chapman and Hall, New York, 1993).

21. G.R. Shorack and J.A. Wellner, *Empirical Processes with Applications in Statistics*(John Wiley, New York, 1986).

22. N.D. Singpurwalla and M. Wong, Estimation of the failure rate–A survey of nonparametric methods, Part I: Non–Bayesian Methods, Commun. Statist.–Theor. Meth. **12(5)**, 559 (1983).

23. N.D. Singpurwalla and M. Wong, Kernel estimators of the Failure–rate function and density estimation: An analogy, Jour. Amer. Statist. Assoc. **78**, 478 (1983).

24. W. Stute, The central limit theorem under random censorship, Ann. Stat. **23**, 422 (1995).

25. W. Stute and J.L. Wang, The strong law under random censordhip, Ann. Stat. **21**, 1591 (1993).

26. G.S. Watson and M.R. Leadbetter, Hazard analysis I, Biometrika **51**, 175 (1963).

27. G.S. Watson and M.R. Leadbetter, Hazard analysis II, Sankhyā **A, 26**, 110 (1964).

28. J.A. Wellner, A heavy censoring limit theorem for the product-limit estimator, Ann. Stat. **15**, 1313 (1987).

29. W. Wertz, *Statistical Density Estimation, A survey* (Vandenhoeck & Ruprecht, Göttingen, 1978).

APPLICATIONS OF HIGHER ORDER MOMENTS
AND THE EDGEWORTH APPROXIMATION
TO RELIABILITY AND LIFE TESTING

A. CHILDS

Department of Mathematics, Computing and Information Systems
St. Francis Xavier University, P.O. Box 5000,
Antigonish, Nova Scotia B2G 2W5, Canada
E-mail: achilds@juliet.stfx.ca

N. BALAKRISHNAN

Department of Mathematics and Statistics, McMaster University
1280 Main St. West, Hamilton, Ontario L8S 4K1, Canada
E-mail: bala@mcmail.cis.mcmaster.ca

In this chapter we first introduce some L-estimators and pivotal quantities that can be used for making inferences for the parameters of a distribution. We then describe how the Edgeworth approximation can be used to find the approximate distribution of these pivotal quantities. We will see that use of the Edgeworth approximation requires knowledge of the higher order moments of order statistics. Methods that have been used to obtain these higher order moments will then be reviewed for several distributions. We then use these higher order moments along with the Edgeworth approximation to derive approximate confidence intervals for the parameters of the exponential, Laplace, Pareto, and power function distributions. Comparisons will also be made with the corresponding results based on simulations.

1 Introduction

1.1 The L-estimators and Pivotal Quantities

Let $X_{r+1:n} \leq X_{r+2:n} \leq \cdots \leq X_{n-s:n}$ be a doubly Type-II censored sample available from a population with pdf

$$f(x; \mu, \sigma) = \frac{1}{\sigma} g\left(\frac{x - \mu}{\sigma}\right), \quad -\infty < x < \infty, \; -\infty < \mu < \infty, \; \sigma > 0. \qquad (1.1)$$

Here, out of n items placed on a life-testing experiment, the smallest r and the largest s items have been censored. The most common situation in a life-testing problem is, of course, a Type-II right-censored sample (with $r = 0$ and $s > 0$) as the experimenter will often terminate the experiment as soon as a certain number of items have failed instead of waiting for all the items to fail; see Mann, Schafer and Singpurwalla[1], Bain[2], Lawless[3], Cohen and Whitten[4], Bain and Engelhardt[5], and Balakrishnan and Cohen[6].

Inferences regarding the parameters of the underlying distribution can be based on estimators of μ and σ that are linear functions of order statistics, or L-estimators,

$$\mu^* = \sum_{i=r+1}^{n-s} a_i X_{i:n}, \qquad \sigma^* = \sum_{i=r+1}^{n-s} b_i X_{i:n}.$$

If $\sum_{i=r+1}^{n-s} a_i = 1$, and $\sum_{i=r+1}^{n-s} b_i = 0$ (which will be the case whenever μ^* and σ^* are unbiased) then the following statistics

$$P_1 = \frac{\mu^* - \mu}{\sigma\sqrt{V_1}}, \quad P_2 = \frac{\sigma^* - \sigma}{\sigma\sqrt{V_2}}, \text{ and } \quad P_3 = \frac{\mu^* - \mu}{\sigma^*} \tag{1.2}$$

are all pivotal quantities. P_1 can be used to draw inference for μ when σ is known, while P_3 can be used to draw inference for μ when σ is unknown. Similarly, P_2 can be used to draw inference for σ when μ is unknown.

Notice also that P_1 and P_2 in (1.1) can be written as

$$P_1 = \frac{\sum\limits_{i=r+1}^{n-s} a_i Z_{i:n}}{\sqrt{V_1}} = \frac{P_1^*}{\sqrt{V_1}} \text{ and } P_2 = \frac{\sum\limits_{i=r+1}^{n-s} b_i Z_{i:n} - 1}{\sqrt{V_2}} = \frac{P_2^* - 1}{\sqrt{V_2}}.$$

Thus, they are linear functions of the standardized order statistics $Z_{i:n} = \frac{X_{i:n} - \mu}{\sigma}$. In order to use P_1 or P_2 to draw inferences, their distribution must be known. In some instances, their distribution will be approximately normal for large samples. But for small samples, the normal approximation will not be adequate.

In this chapter, we review some techniques that have been used to approximate the distribution of a linear function order statistics for small samples. Many of these make use of the Edgeworth approximation. And in each case, the techniques require knowledge of the higher order moments of order statistics.

1.2 The Edgeworth Approximation and Higher Order Moments

Let us denote the single moments $E(Z_{i:n}^a)$ by $\mu_{i:n}^{(a)}$, the double moments $E(Z_{i:n}^a Z_{j:n}^b)$ by $\mu_{i,j:n}^{(a,b)}$, the triple moments $E(Z_{i:n}^a Z_{j:n}^b Z_{k:n}^c)$ by $\mu_{i,j,k:n}^{(a,b,c)}$, and the quadruple moments $E(Z_{i:n}^a Z_{j:n}^b Z_{k:n}^c Z_{l:n}^d)$ by $\mu_{i,j,k,l:n}^{(a,b,c,d)}$ for $1 \leq i < j < k < l \leq n$ and $a, b, c, d \geq 0$.

The Edgeworth approximation for the distribution of a standardized statistic T (with mean 0 and variance 1) is given by

$$F(t) \approx \Phi(t) - \phi(t)\left\{ \frac{\sqrt{\beta_1}}{6}(t^2 - 1) + \frac{(\beta_2 - 3)}{24}(t^3 - 3t) + \frac{\beta_1}{72}(t^5 - 10t^3 + 15t) \right\},$$

$$\tag{1.3}$$

where $\sqrt{\beta_1}$ and β_2 are the coefficients of skewness and kurtosis, respectively, of T, and $\Phi(t)$ is the cumulative distribution function of the standard normal distribution with corresponding pdf $\phi(t)$.

Since $\sqrt{\beta_1} = \frac{E[(T - E(T))^3]}{E[(T - E(T))^2]^{3/2}} = \frac{E(T^3)}{E(T^2)^{3/2}}$ and $\beta_2 = \frac{E(T^4)}{E(T^2)^2}$, we see that when T is a linear function of order statistics, $T = \sum\limits_{i=1}^{n} a_i Z_{i:n}$, use of the Edgeworth approximation requires knowledge of

$$E(T^4) = E\left[\left(\sum_{i=1}^{n} a_i Z_{i:n}\right)^4\right] = \sum \binom{4}{r_1, r_2, \ldots, r_n} a_1^{r_1} a_2^{r_2} \cdots a_n^{r_n} E(Z_{1:n}^{r_1} Z_{2:n}^{r_2} \cdots Z_{n:n}^{r_n}),$$

where the sum is over all nonnegative integer valued vectors $(r_1, r_2, \ldots, r_n)$ satisfying $r_1 + r_2 + \cdots + r_n = 4$. Therefore all of the quadruple moments $\mu_{i,j,k,l:n}^{(a,b,c,d)}$ for $a + b + c + d = 4$, the triple moments $\mu_{i,j,k:n}^{(a,b,c)}$ for $a + b + c = 4$, the double moments $\mu_{i,j:n}^{(a,b)}$, for $a + b = 4$, and the single moments $\mu_{i:n}^{(4)}$, are required. Similar expressions for $E(T^3)$ and $E(T^2)$ reveal that use of the Edgeworth approximation for a linear function of order statistics also requires knowledge of $\mu_{i,j,k:n}^{(a,b,c)}$, for $a + b + c = 3$, $\mu_{i,j:n}^{(a,b)}$ for $2 \leq a + b \leq 3$, and $\mu_{i:n}^{(a)}$ for $2 \leq a \leq 3$. The required higher order moments may be obtained using the appropriate density function of order statistics.

The probability density function of $X_{i:n}$ for $1 \leq i \leq n$ is (see David[7], pp. 9) given by

$$f_{i:n}(x) = \frac{n!}{(i-1)!\,(n-i)!} \{F(x)\}^{i-1} \{1 - F(x)\}^{n-i} f(x), \quad -\infty < x < \infty,$$

From this we may compute the single moments of $X_{i:n}$ as follows,

$$\mu_{i:n}^{(a)} = \int_{-\infty}^{\infty} x^a f_{i:n}(x)\,dx, \quad a = 1, 2, \ldots, \ 1 \leq i \leq n.$$

The joint density function of $X_{i:n}$ and $X_{j:n}$ for $1 \leq i < j \leq n$ is given by

$$
\begin{aligned}
f_{i,j:n}(x, y) = {} & \frac{n!}{(i-1)!\,(j-i-1)!\,(n-j)!} \\
& \times \{F(x)\}^{i-1} \{F(y) - F(x)\}^{j-i-1} \{1 - F(y)\}^{n-j} f(x) f(y) \\
& \qquad\qquad\qquad\qquad\qquad\qquad - \infty < x < y < \infty.
\end{aligned}
$$

and can be used to compute the double moments,

$$\mu_{i,j:n}^{(a,b)} = \iint\limits_{-\infty < x < y < \infty} x^a y^b f_{i,j:n}(x, y)\,dx\,dy, \quad 1 \leq i < j \leq n, \ a, b \geq 1.$$

The triple moments, $\mu_{i,j,k:n}^{(a,b,c)}$, may be computed using the joint density function of $X_{i:n}$, $X_{j:n}$, and $X_{k:n}$ for $1 \leq i < j < k \leq n$,

$$
\begin{aligned}
f_{i,j,k:n}(x, y, z) = {} & \frac{n!}{(i-1)!\,(j-i-1)!\,(k-j-1)!\,(n-k)!} \\
& \times \{F(x)\}^{i-1} \{F(y) - F(x)\}^{j-i-1} \{F(z) - F(y)\}^{k-j-1} \\
& \times \{1 - F(z)\}^{n-k} f(x) f(y) f(z) \\
& \qquad\qquad\qquad\qquad\qquad\qquad - \infty < x < y < z < \infty,
\end{aligned}
$$

as follows,

$$
\begin{aligned}
\mu_{i,j,k:n}^{(a,b,c)} = {} & \iiint\limits_{-\infty < x < y < z < \infty} x^a y^b z^c f_{i,j,k:n}(x, y, z)\,dx\,dy\,dz \\
& \qquad\qquad\qquad\qquad 1 \leq i < j < k \leq n, \ a, b, c \geq 1.
\end{aligned}
$$

And the joint density function of $X_{i:n}$, $X_{j:n}$, $X_{k:n}$, and $X_{l:n}$ for $1 \le i < j < k < l \le n$,

$$f_{i,j,k,l:n}(x,y,z,w) = \frac{n! \cdot}{(i-1)!\,(j-i-1)!\,(k-j-1)!\,(l-k-1)!\,(n-l)!}$$
$$\times \{F(x)\}^{i-1}\{F(y)-F(x)\}^{j-i-1}\{F(z)-F(y)\}^{k-j-1}$$
$$\times \{F(w)-F(z)\}^{l-k-1}\{1-F(w)\}^{n-l}f(x)f(y)f(z)f(w)$$
$$-\infty < x < y < z < w < \infty,$$

can be used to compute the quadruple moments,

$$\mu_{i,j,k,l:n}^{(a,b,c,d)} = \iiiint\limits_{-\infty<x<y<z<w<\infty} x^a y^b z^c w^d f_{i,j,k,l:n}(x,y,z,w)\,dx\,dy\,dz\,dw$$

$$1 \le i < j < k < l \le n,\ a,b,c,d \ge 1.$$

Several methods have been developed to evaluate the above integrals for a variety of distributions. For the exponential distribution, the above expressions have been used by Balakrishnan and Gupta[8] to derive recurrence relations for the required higher order moments. This work will be reviewed in the following section. In Section 3, we will show how the results of Balakrishnan and Gupta[8] for the exponential distribution have been used by Balakrishnan *et. al*[9] to obtain the required higher order moments for the Laplace distribution. Exact explicit expressions, as well as recurrence relations, for higher order moments have been derived for the Pareto and power function distribution by Childs *et. al*[10] and Sultan *et. al*[11], respectively. These developments are reviewed in Sections 4 and 5. In Section 6, we present the BLUE's of μ and σ. Then, in Section 7, we explain for each distribution the approximate inference results that have been obtained using the results presented in the previous sections. Finally, in Section 8, we give some numerical examples.

2 Results for the Exponential Distribution

For the exponential distribution

$$f(x;\sigma) = \frac{1}{\sigma}e^{-x/\sigma},\ x \ge 0,\ \sigma > 0, \tag{2.1}$$

$$F(x;\sigma) = 1 - e^{x/\sigma},\ x \ge 0,\ \sigma > 0.$$

The differential equation

$$f(x) = \frac{1}{\sigma}\{1 - F(x)\}$$

has been used by Joshi[12] to derive the following recurrence relations the single and double moments of order statistics.

Relation 2.1: For $n \ge 1$ and $a = 1, 2, \ldots,$

$$\mu_{1:n}^{(a)} = \frac{a}{n}\mu_{1:n}^{(a-1)}.$$

Relation 2.2: For $2 \leq i \leq n$ and $a = 1, 2, \ldots,$

$$\mu_{i:n}^{(a)} = \mu_{i-1:n}^{(a)} + \frac{a}{n-i+1}\mu_{i:n}^{(a-1)}.$$

Relation 2.3: For $1 \leq i \leq n - 1$ and $a, b = 1, 2, \ldots,$

$$\mu_{i,i+1:n}^{(a,b)} = \mu_{i:n}^{(a+b)} + \frac{b}{n-i}\mu_{i,i+1:n}^{(a,b-1)}.$$

Relation 2.4: For $1 \leq i < j \leq n$, $j - i \geq 2$, and $a, b = 1, 2, \ldots,$

$$\mu_{i,j:n}^{(a,b)} = \mu_{i,j-1:n}^{(a,b)} + \frac{b}{n-j+1}\mu_{i,j:n}^{(a,b-1)}.$$

More recently, for the purpose of examining approximations to the distribution of linear functions of order statistics, Balakrishnan and Gupta[8] have extended these relations to include triple and quadruple moments:

Relation 2.5: For $1 \leq i < j \leq n - 1$, $n \geq 3$ and $a, b, c = 1, 2, \ldots,$

$$\mu_{i,j,j+1:n}^{(a,b,c)} = \mu_{i,j:n}^{(a,b+c)} + \frac{c}{n-j}\mu_{i,j,j+1:n}^{(a,b,c-1)}.$$

Relation 2.6: For $1 \leq i < j < k \leq n$, $k - j \geq 2$ and $a, b, c = 1, 2, \ldots,$

$$\mu_{i,j,k:n}^{(a,b,c)} = \mu_{i,j,k-1:n}^{(a,b,c)} + \frac{c}{n-k+1}\mu_{i,j,k:n}^{(a,b,c-1)}.$$

Relation 2.7: For $1 \leq i < j < k \leq n - 1$, $n \geq 4$ and $a, b, c, d = 1, 2, \ldots,$

$$\mu_{i,j,k,k+1:n}^{(a,b,c,d)} = \mu_{i,j,k:n}^{(a,b,c+d)} + \frac{d}{n-k}\mu_{i,j,k,k+1:n}^{(a,b,c,d-1)}.$$

Relation 2.8: For $1 \leq i < j < k < l \leq n$, $l - k \geq 2$ and $a, b, c, d = 1, 2, \ldots,$

$$\mu_{i,j,k,l:n}^{(a,b,c,d)} = \mu_{i,j,k,l-1:n}^{(a,b,c,d)} + \frac{d}{n-l+1}\mu_{i,j,k,l:n}^{(a,b,c,d-1)}.$$

3 The Laplace Distribution

Moments of order statistics for the Laplace distribution,

$$f(x; \mu, \sigma) = \frac{1}{2\sigma}e^{-|x-\mu|/\sigma}, \quad -\infty < x < \infty, \; -\infty < \mu < \infty, \; \sigma > 0, \qquad (3.1)$$

can be obtained by exploiting its special relationship to the exponential distribution. Namely if $f^*(x)$ is the exponential pdf in (3.1) and $f(x)$ is the Laplace pdf in (4.1) with $\mu = 0$ then for $x \geq 0$,

$$F^*(x) = 2F(x) - 1 \quad \text{and} \quad f^*(x) = 2f(x).$$

In general, let $Z_1, Z_2, \ldots, Z_n$ be I.I.D. random variables with probability density function $f(z)$ symmetric about 0 (without loss of generality), and cumulative distribution function $F(z)$. Then, for $z \geq 0$ let

$$F^*(z) = 2F(z) - 1 \quad \text{and} \quad f^*(z) = 2f(z). \tag{3.2}$$

That is, the density function $f^*(z)$ is obtained by folding the density function $f(z)$ at zero (the point of symmetry).

Let $Y_{1:n} \leq Y_{2:n} \leq \cdots \leq Y_{n:n}$ denote the order statistics obtained from n I.I.D. random variables $Y_1, Y_2, \ldots, Y_n$ having probability density function $f^*(z)$ and cumulative distribution function $F^*(z)$ as given in (3.2), and let $Z_{1:n} \leq Z_{2:n} \leq \cdots \leq Z_{n:n}$ denote the order statistics obtained from the random variables $Z_1, Z_2, \ldots, Z_n$.

Let us now denote the single moments $E(Z_{i:n}^a)$ by $\mu_{i:n}^{(a)}$, the double moments $E(Z_{i:n}^a Z_{j:n}^b)$ by $\mu_{i,j:n}^{(a,b)}$, the triple moments $E(Z_{i:n}^a Z_{j:n}^b Z_{k:n}^c)$ by $\mu_{i,j,k:n}^{(a,b,c)}$, and the quadruple moments $E(Z_{i:n}^a Z_{j:n}^b Z_{k:n}^c Z_{l:n}^d)$ by $\mu_{i,j,k,l:n}^{(a,b,c,d)}$ for $1 \leq i < j < k < l \leq n$ and $a, b, c, d \geq 0$. Similarly, let us denote the corresponding moments of order statistics $Y_{i:n}$ by $\nu_{i:n}^{(a)}$, $\nu_{i,j:n}^{(a,b)}$, $\nu_{i,j,k:n}^{(a,b,c)}$ and $\nu_{i,j,k,l:n}^{(a,b,c,d)}$ for $1 \leq i < j < k < l \leq n$ and $a, b, c, d \geq 0$.

Then by making use of the relations in (3.1), Govindarajulu[13] established the following two recurrence relations.

Relation 3.1: For $i = 1, 2, \ldots, n$ and $a \geq 1$,

$$\mu_{i:n}^{(a)} = \frac{1}{2^n} \left\{ \sum_{t=0}^{i-1} \binom{n}{t} \nu_{i-t:n-t}^{(a)} + (-1)^a \sum_{t=i}^{n} \binom{n}{t} \nu_{t-i+1:t}^{(a)} \right\}.$$

Relation 3.2: For $1 \leq i < j \leq n$ and $a, b \geq 1$,

$$\mu_{i,j:n}^{(a,b)} = \frac{1}{2^n} \left\{ \sum_{t=0}^{i-1} \binom{n}{t} \nu_{i-t,j-t:n-t}^{(a,b)} + (-1)^a \sum_{t=i}^{j-1} \binom{n}{t} \nu_{t-i+1:t}^{(a)} \nu_{j-t:n-t}^{(b)} \right.$$

$$\left. + (-1)^{a+b} \sum_{t=j}^{n} \binom{n}{t} \nu_{t-j+1,t-i+1:t}^{(b,a)} \right\}.$$

More recently, for the purpose of examining inference problems using the Edgeworth approximation, Balakrishnan et. al[9] extended these results to the triple and quadruple moments of order statistics. Their results are as follows.

Relation 3.3: For $1 \leq i < j < k \leq n$ and $a, b, c \geq 1$,

$$\mu_{i,j,k:n}^{(a,b,c)} = \frac{1}{2^n} \left\{ \sum_{t=0}^{i-1} \binom{n}{t} \nu_{i-t,j-t,k-t:n-t}^{(a,b,c)} + (-1)^a \sum_{t=i}^{j-1} \binom{n}{t} \nu_{t-i+1:t}^{(a)} \nu_{j-t,k-t:n-t}^{(b,c)} \right.$$

$$+ (-1)^{a+b} \sum_{t=j}^{k-1} \binom{n}{t} \nu_{t-j+1,t-i+1:t}^{(b,a)} \nu_{k-t:n-t}^{(c)}$$

$$\left. + (-1)^{a+b+c} \sum_{t=k}^{n} \binom{n}{t} \nu_{t-k+1,t-j+1,t-i+1:t}^{(c,b,a)} \right\}.$$

Relation 3.4: For $1 \leq i < j < k < l \leq n$ and $a, b, c, d \geq 1$,

$$
\mu_{i,j,k,l:n}^{(a,b,c,d)} = \frac{1}{2^n} \Bigg\{ \sum_{t=0}^{i-1} \binom{n}{t} \nu_{i-t,j-t,k-t,l-t:n-t}^{(a,b,c,d)}
$$

$$
+ (-1)^a \sum_{t=i}^{j-1} \binom{n}{t} \nu_{t-i+1:t}^{(a)} \nu_{j-t,k-t,l-t:n-t}^{(b,c,d)}
$$

$$
+ (-1)^{a+b} \sum_{t=j}^{k-1} \binom{n}{t} \nu_{t-j+1,t-i+1:t}^{(b,a)} \nu_{k-t,l-t:n-t}^{(c,d)}
$$

$$
+ (-1)^{a+b+c} \sum_{t=k}^{l-1} \binom{n}{t} \nu_{t-k+1,t-j+1,t-i+1:t}^{(c,b,a)} \nu_{l-t:n-t}^{(d)}
$$

$$
+ (-1)^{a+b+c+d} \sum_{t=l}^{n} \binom{n}{t} \nu_{t-l+1,t-k+1,t-j+1,t-i+1:t}^{(d,c,b,a)} \Bigg\}.
$$

Relations 3.1-3.4 can be used recursively to compute the single, double, and quadruple moments of order statistics from a symmetric distribution by making use of the corresponding quantities from its folded distribution. In particular, after computing all these moments of order up to 4, one can determine the mean, the variance, and the coefficients of skewness and kurtosis of any linear function of order statistics from that symmetric distribution. These measures can then be utilized to develop Edgeworth approximations for distributions of linear functions of order statistics as we will illustrate in Section 7.

For the case when the order statistics $Z_{i:n}$ arise from the standard Laplace distribution [with $\mu = 0$ and $\sigma = 1$ in (1.1)], the computation of the single, double, triple and quadruple moments by means of Relations 1-4 require the knowledge of these moments from the standard exponential distribution, with pdf $f^*(z) = e^{-z}$, $z \geq 0$. In this case, the additive Markov chain representation of standard exponential order statistics [Sukhatme[14], Rényi[15]] given by,

$$
Y_{i:n} = \sum_{t=1}^{i} \frac{E_t}{n-t+1}, \quad i = 1, 2, \dots, n,
$$

where E_t's are I.I.D. standard exponential random variables, makes it possible to write down the necessary moments $\nu_{i:n}^{(a)}$, $\nu_{i,j:n}^{(a,b)}$, $\nu_{i,j,k:n}^{(a,b,c)}$ and $\nu_{i,j,k,l:n}^{(a,b,c,d)}$ in simple explicit algebraic forms. For example,

$$
\nu_{i:n} = \sum_{t=1}^{i} \frac{1}{n-t+1}, \tag{3.3}
$$

$$
\nu_{i:n}^{(2)} = \sum_{t=1}^{i} \frac{1}{(n-t+1)^2} + \left(\sum_{t=1}^{i} \frac{1}{n-t+1} \right)^2, \tag{3.4}
$$

and

$$\nu_{i,j:n} = \sum_{t=1}^{i} \frac{1}{(n-t+1)^2} + \left(\sum_{t=1}^{i} \frac{1}{n-t+1}\right)\left(\sum_{t=1}^{j} \frac{1}{n-t+1}\right). \tag{3.5}$$

Alternatively, one may use the recurrence relations for the moments of exponential order statistics presented in the previous section in order to compute the necessary moments $\nu_{i:n}^{(a)}$, $\nu_{i,j:n}^{(a,b)}$, $\nu_{i,j,k:n}^{(a,b,c)}$ and $\nu_{i,j,k,l:n}^{(a,b,c,d)}$ in a simple recursive manner.

4 The Pareto Distribution

Let $X_1, X_2, \ldots, X_n$ be a random sample of size n from a two-parameter Pareto population with probability density function

$$f(x; \mu, \sigma) = \nu \sigma^\nu (x - \mu)^{-(\nu+1)}, \qquad x \geq \mu + \sigma, \nu > 0, \sigma > 0 \tag{4.1}$$

and cumulative distribution function

$$F(x; \mu, \sigma) = 1 - \sigma^\nu (x - \mu)^{-\nu} \qquad x \geq \mu + \sigma, \nu > 0, \sigma > 0. \tag{4.2}$$

Let $X_{1:n} \leq X_{2:n} \cdots \leq X_{n:n}$ be the order statistics obtained by arranging the above sample in increasing order of magnitude. Notice from (1.1) and (1.2) that

$$f(x) = \frac{\nu}{x - \mu}\{1 - F(x)\}. \tag{4.3}$$

Balakrishnan and Joshi[16] used the above differential equation to derive the following recurrence relations satisfied by the single and the double moments of order statistics from the Pareto model.

Relation 4.1: For $n \geq 1$ and $a = 1, 2, \ldots$,

$$\mu_{1:n}^{(a)} = \frac{n}{n - a/\nu}.$$

Relation 4.2: For $2 \leq i \leq n$ and $a = 1, 2, \ldots$,

$$\mu_{i:n}^{(a)} = \frac{n - i + 1}{n - i + 1 - a/\nu}\mu_{i-1:n}^{(a)}.$$

Relation 4.3: For $1 \leq i \leq n - 1$ and $a, b = 1, 2, \ldots$,

$$\mu_{i,i+1:n}^{(a,b)} = \frac{n - i}{n - i - b/\nu}\mu_{i:n}^{(a+b)}.$$

Relation 4.4: For $1 \leq i < j \leq n$, $j - i \geq 2$, and $a, b = 1, 2, \ldots$,

$$\mu_{i,j:n}^{(a,b)} = \frac{n - j + 1}{n - j + 1 - b/\nu}\mu_{i,j-1:n}^{(a,b)}.$$

Equivalent exact explicit expressions for the single and double moments of order statistics have been derived by Huang[17]:

$$\mu_{i:n}^{(a)} = \frac{\Gamma(n+1)}{\Gamma(n-i+1)} \frac{\Gamma(n-i+1-a/\nu)}{\Gamma(n+1-a/\nu)}$$

and

$$\mu_{i,j:n}^{(a,b)} = \frac{\Gamma(n+1)}{\Gamma(n-j+1)} \frac{\Gamma(n-j+1-b/\nu)}{\Gamma(n-i+1-b/\nu)} \frac{\Gamma(n-i+1-(a+b)/\nu)}{\Gamma(n+1-(a+b)/\nu)}.$$

More recently, for the purpose of examining inference problems using the Edgeworth approximation, Childs *et. al*[10] extended these results to the triple and quadruple moments of order statistics. Their recurrence relations are as follows.

Relation 4.5: For $1 \le i < j \le n-1$, $n \ge 3$ and $a, b, c = 1, 2, \ldots$,

$$\mu_{i,j,j+1:n}^{(a,b,c)} = \frac{n-j}{n-j-c/\nu} \mu_{i,j:n}^{(a,b+c)}.$$

Relation 4.6: For $1 \le i < j < k \le n$, $k-j \ge 2$ and $a, b, c = 1, 2, \ldots$,

$$\mu_{i,j,k:n}^{(a,b,c)} = \frac{n-k+1}{n-k+1-c/\nu} \mu_{i,j,k-1:n}^{(a,b,c)}.$$

Relation 4.7: For $1 \le i < j < k \le n-1$, $n \ge 4$ and $a, b, c, d = 1, 2, \ldots$,

$$\mu_{i,j,k,k+1:n}^{(a,b,c,d)} = \frac{n-k}{n-k-d/\nu} \mu_{i,j,k:n}^{(a,b,c+d)}.$$

Relation 4.8: For $1 \le i < j < k < l \le n$, $l-k \ge 2$ and $a, b, c, d = 1, 2, \ldots$,

$$\mu_{i,j,k,l:n}^{(a,b,c,d)} = \frac{n-l+1}{n-l+1-d/\nu} \mu_{i,j,k,l-1:n}^{(a,b,c,d)}.$$

The exact expressions are given by

$$\mu_{i,j,k:n}^{(a,b,c)} = \frac{\Gamma(n+1)}{\Gamma(n-k+1)} \frac{\Gamma(n-k+1-c/\nu)}{\Gamma(n-j+1-c/\nu)} \frac{\Gamma(n-j+1-(b+c)/\nu)}{\Gamma(n-i+1-(b+c)/\nu)}$$
$$\times \frac{\Gamma(n-i+1-(a+b+c)/\nu)}{\Gamma(n+1-(a+b+c)/\nu)}$$

and

$$\mu_{i,j,k,l:n}^{(a,b,c,d)} = \frac{\Gamma(n+1)}{\Gamma(n-l+1)} \frac{\Gamma(n-l+1-d/\nu)}{\Gamma(n-k+1-d/\nu)} \frac{\Gamma(n-k+1-(c+d)/\nu)}{\Gamma(n-j+1-(c+d)/\nu)}$$
$$\times \frac{\Gamma(n-j+1-(b+c+d)/\nu)}{\Gamma(n-i+1-(b+c+d)/\nu)} \frac{\Gamma(n-i+1-(a+b+c+d)/\nu)}{\Gamma(n+1-(a+b+c+d)/\nu)}.$$

5 The Power Function Distribution

Let $X_1, X_2, \ldots, X_n$ be a random sample of size n from a two-parameter power function population with probability density function

$$f(x; \mu, \sigma) = \nu \sigma^{-\nu}(x - \mu)^{\nu - 1}, \qquad \mu \le x \le \mu + \sigma, \nu > 0, \sigma > 0 \tag{5.1}$$

and cumulative distribution function

$$F(x; \mu, \sigma) = \sigma^{-\nu}(x - \mu)^{\nu} \qquad \mu \le x \le \mu + \sigma, \nu > 0, \sigma > 0. \tag{5.2}$$

Let $X_{1:n} \le X_{2:n} \cdots \le X_{n:n}$ be the order statistics obtained by arranging the above sample in increasing order of magnitude. Notice from (1.1) and (1.2) that

$$f(x) = \frac{\nu}{x - \mu} F(x). \tag{5.3}$$

Balakrishnan and Joshi[18] used the above differential equation to derive the following recurrence relations satisfied by the single and the double moments of order statistics from the power function model.

Relation 5.1: For $n \ge 1$ and $a = 1, 2, \ldots$,

$$\mu_{n:n}^{(a)} = \frac{n}{n + a/\nu}.$$

Relation 5.2: For $1 \le i \le n - 1$ and $a = 1, 2, \ldots$,

$$\mu_{i:n}^{(a)} = \frac{i}{i + a/\nu} \mu_{i+1:n}^{(a)}.$$

Relation 5.3: For $1 \le i \le n - 1$ and $a, b = 1, 2, \ldots$,

$$\mu_{i,i+1:n}^{(a,b)} = \frac{i}{i + a/\nu} \mu_{i+1:n}^{(a+b)}.$$

Relation 5.4: For $1 \le i < j \le n$, $j - i \ge 2$, and $a, b = 1, 2, \ldots$,

$$\mu_{i,j:n}^{(a,b)} = \frac{i}{i + a/\nu} \mu_{i+1,j:n}^{(a,b)}.$$

Equivalent exact explicit expressions for the single and double moments of order statistics have been derived by Malik[19]:

$$\mu_{i:n}^{(a)} = \frac{\Gamma(n+1)}{\Gamma(i)} \frac{\Gamma(i + a/\nu)}{\Gamma(n + 1 + a/\nu)}$$

and

$$\mu_{i,j:n}^{(a,b)} = \frac{\Gamma(n+1)}{\Gamma(i)} \frac{\Gamma(i + a/\nu)}{\Gamma(j + a/\nu)} \frac{\Gamma(j + (a+b)/\nu)}{\Gamma(n + 1 + (a+b)/\nu)}.$$

More recently, for the purpose of examining inference problems using the Edgeworth approximation, Sultan et. al[11] extended these results to the triple and quadruple moments of order statistics. Their recurrence relations are as follows.

Relation 5.5: For $1 \leq i < k \leq n, k - i \geq 2, n \geq 3$ and $a, b, c = 1, 2, \ldots$,

$$\mu_{i,i+1,k:n}^{(a,b,c)} = \frac{i}{i + a/\nu} \mu_{i+1,k:n}^{(a+b,c)}.$$

Relation 5.6: For $1 \leq i < j < k \leq n, j - i \geq 2$ and $a, b, c = 1, 2, \ldots$,

$$\mu_{i,j,k:n}^{(a,b,c)} = \frac{i}{i + a/\nu} \mu_{i+1,j,k:n}^{(a,b,c)}.$$

Relation 5.7: For $1 \leq i < k < l \leq n, k - i \geq 2, n \geq 4$ and $a, b, c, d = 1, 2, \ldots$,

$$\mu_{i,i+1,k,l:n}^{(a,b,c,d)} = \frac{i}{i + a/\nu} \mu_{i+1,k,l:n}^{(a+b,c,d)}.$$

Relation 5.8: For $1 \leq i < j < k < l \leq n, j - i \geq 2$ and $a, b, c, d = 1, 2, \ldots$,

$$\mu_{i,j,k,l:n}^{(a,b,c,d)} = \frac{i}{i + a/\nu} \mu_{i+1,j,k,l:n}^{(a,b,c,d)}.$$

The exact expressions are given by

$$\mu_{i,j,k:n}^{(a,b,c)} = \frac{\Gamma(n+1)}{\Gamma(i)} \frac{\Gamma(i+a/\nu)}{\Gamma(j+a/\nu)} \frac{\Gamma(j+(a+b)/\nu)}{\Gamma(k+(a+b)/\nu)} \frac{\Gamma(k+(a+b+c)/\nu)}{\Gamma(n+1+(a+b+c)/\nu)}$$

and

$$\mu_{i,j,k,l:n}^{(a,b,c,d)} = \frac{\Gamma(n+1)}{\Gamma(i)} \frac{\Gamma(i+a/\nu)}{\Gamma(j+a/\nu)} \frac{\Gamma(j+(a+b)/\nu)}{\Gamma(k+(a+b)/\nu)} \frac{\Gamma(k+(a+b+c)/\nu)}{\Gamma(l+(a+b+c)/\nu)}$$
$$\times \frac{\Gamma(l+(a+b+c+d)/\nu)}{\Gamma(n+1+(a+b+c+d)/\nu)}.$$

6 BLUE'S of μ and σ

In this section, we will first present general expressions for the BLUE's of μ and σ in (1.1), and then we will show how these expressions simplify to give exact explicit expressions for the BLUE's in the case of the exponential, Pareto, and power function distributions. For the Laplace distribution, exact explicit expressions are not available. However we will present exact explicit expressions for the MLE's (modified to make them unbiased).

Let

$$X_{r+1:n} \leq X_{r+2:n} \leq \cdots \leq X_{n-s:n}$$

denote a general doubly Type-II censored sample from the distribution in (1.1), and let $Z_{i:n} = (X_{i:n} - \mu)/\sigma$, $i = r + 1, r + 2, \ldots, n - s$, be the corresponding order statistics from the standardized distribution. Let us denote $\mathrm{E}(Z_{i:n})$ by $\mu_{i:n}$, $\mathrm{Var}(Z_{i:n})$ by $\sigma_{i,i:n}$, and $\mathrm{Cov}(Z_{i:n}, Z_{j:n})$ by $\sigma_{i,j:n}$; further, let

$$\mathbf{X} = (X_{r+1:n}, X_{r+2:n}, \ldots, X_{n-s:n})^T,$$

$$\mu = (\mu_{r+1:n}, \mu_{r+2:n}, \ldots, \mu_{n-s:n})^T,$$

$$1 = \underbrace{(1, 1, \ldots, 1)}_{n-r-s}^T,$$

and

$$\Sigma = ((\sigma_{i,j:n})), r+1 \leq i, j \leq n-s.$$

Then, the BLUE's of μ and σ are given by

$$\mu^* = \left\{ \frac{\mu^T \Sigma^{-1} \mu 1^T \Sigma^{-1} - \mu^T \Sigma^{-1} 1 \mu^T \Sigma^{-1}}{(\mu^T \Sigma^{-1} \mu)(1^T \Sigma^{-1} 1) - (\mu^T \Sigma^{-1} 1)^2} \right\} X = \sum_{i=r+1}^{n-s} a_i X_{i:n} \tag{6.1}$$

and

$$\sigma^* = \left\{ \frac{1^T \Sigma^{-1} 1 \mu^T \Sigma^{-1} - 1^T \Sigma^{-1} \mu 1^T \Sigma^{-1}}{(\mu^T \Sigma^{-1} \mu)(1^T \Sigma^{-1} 1) - (\mu^T \Sigma^{-1} 1)^2} \right\} X = \sum_{i=r+1}^{n-s} b_i X_{i:n}. \tag{6.2}$$

Furthermore, the variances and covariances of these BLUE's are given by

$$\mathrm{Var}(\mu^*) = \sigma^2 \left\{ \frac{\mu^T \Sigma^{-1} \mu}{(\mu^T \Sigma^{-1} \mu)(1^T \Sigma^{-1} 1) - (\mu^T \Sigma^{-1} 1)^2} \right\} = \sigma^2 V_1, \tag{6.3}$$

$$\mathrm{Var}(\sigma^*) = \sigma^2 \left\{ \frac{1^T \Sigma^{-1} 1}{(\mu^T \Sigma^{-1} \mu)(1^T \Sigma^{-1} 1) - (\mu^T \Sigma^{-1} 1)^2} \right\} = \sigma^2 V_2, \tag{6.4}$$

and

$$\mathrm{Cov}(\mu^*, \sigma^*) = -\sigma^2 \left\{ \frac{\mu^T \Sigma^{-1} 1}{(\mu^T \Sigma^{-1} \mu)(1^T \Sigma^{-1} 1) - (\mu^T \Sigma^{-1} 1)^2} \right\} = \sigma^2 V_3. \tag{6.5}$$

Similar expressions hold for BLUE's of the scale parameter σ when μ is known; for details, refer to David[7], Balakrishnan and Cohen[6], and Arnold, Balakrishnan and Nagaraja[20].

6.1 BLUE's for the Exponential Distribution

For the exponential distribution, an exact explicit expression for the BLUE of σ is given in Balakrishnan and Cohen[6] (pp. 76-80), and is as follows.

$$\sigma^* = \sum_{i=r+1}^{n-s} b_i X_{i:n}, \tag{6.6}$$

where

$$b_i = \frac{b_i^*}{K}, \quad K = (n - r - s - 1) + \frac{\left\{ \sum\limits_{j=n-r}^{n} \frac{1}{j} \right\}^2}{\left\{ \sum\limits_{j=n-r}^{n} \frac{1}{j^2} \right\}}$$

and

$$b_i^* = \begin{cases} \left\{ \left[\sum\limits_{j=n-r}^{n} \frac{1}{j} \right] \Big/ \left[\sum\limits_{j=n-r}^{n} \frac{1}{j^2} \right] \right\} - (n - r - 1) & \text{for } i = r + 1 \\ 1 & \text{for } r + 2 \leq i \leq n - s - 1 \\ s + 1 & \text{for } i = n - s \end{cases}$$

The variance of this estimator is

$$Var(\sigma^*) = \frac{\sigma^2}{K}.$$

6.2 BLUE's and MLE's for the Laplace Distribution

From Equations (6.1) and (6.2), Govindarajulu[21] computed the coefficients a_i and b_i for symmetrically Type-II censored samples (with $r = s$) for sample sizes n up to 20. He also presented the values of V_1 and V_2 in Equations (6.3) and (6.4). Similar tables have been prepared recently by Balakrishnan, Chandramouleeswaran, and Ambagaspitiya[22] for Type-II right-censored samples (with $r = 0$ and $s = 0(1)n - 2$) for sample sizes $n = 3(1)20$. They also presented some percentage points of three pivotal quantities, based on BLUE's μ^* and σ^*, which will enable one to construct confidence intervals and carry out tests of hypotheses for the parameters μ and σ.

Unlike the BLUE's, special tables are not needed for the MLE's of μ and σ. Recently, Childs and Balakrishnan[23] derived the MLE's of μ and σ based on the general doubly Type-II censored sample,

$$X_{r+1:n} \leq X_{r+2:n} \leq \cdots \leq X_{n-s:n},$$

from the Laplace distribution (3.1). They found that when $r + 1 < n - s < n/2$,

$$\hat{\mu} = X_{n-s:n} + \hat{\sigma} \ln \frac{n/2}{n - s} \tag{6.7}$$

and

$$\hat{\sigma} = \frac{1}{n - s - r} \left[(n - s) X_{n-s:n} - \sum_{i=r+1}^{n-s} X_{i:n} - r X_{r+1:n} \right]; \tag{6.8}$$

when $\frac{n}{2} + 1 \leq r + 1 < n - s$,

$$\widehat{\mu} = X_{r+1:n} - \widehat{\sigma} \ln \frac{n/2}{n - r} \tag{6.9}$$

and

$$\widehat{\sigma} = \frac{1}{n - s - r}\left[sX_{n-s:n} + \sum_{i=r+1}^{n-s} X_{i:n} - (n - r)X_{r+1:n}\right]; \tag{6.10}$$

and when $r + 1 \leq n/2 < n - s$,

$$\widehat{\mu} = \begin{cases} \frac{1}{2}(X_{m:2m} + X_{m+1:2m}) & \text{if } n \text{ is even, } n = 2m \\ X_{m+1:2m+1} & \text{if } n \text{ is odd, } n = 2m + 1 \end{cases} \tag{6.11}$$

(actually, when n is even any value in $[X_{m:2m}, X_{m+1:2m}]$ is an MLE for μ, but we will use the one given since it is unbiased) and

$$\widehat{\sigma} = \begin{cases} \dfrac{1}{n-s-r}\left[sX_{n-s:n} + \displaystyle\sum_{i=m+1}^{n-s} X_{i:n} - \sum_{i=r+1}^{m} X_{i:n} - rX_{r+1:n}\right] & \text{if } n \text{ is even, } n = 2m \\ \dfrac{1}{n-s-r}\left[sX_{n-s:n} + \displaystyle\sum_{i=m+2}^{n-s} X_{i:n} - \sum_{i=r+1}^{m} X_{i:n} - rX_{r+1:n}\right] & \text{if } n \text{ is odd, } n = 2m + 1 \end{cases}$$

$$\tag{6.12}$$

Unlike the BLUE's presented in the previous section, the MLE's are explicit linear functions of order statistics,

$$\widehat{\mu} = \sum_{i=r+1}^{n-s} \widehat{a}_i X_{i:n} \quad \text{and} \quad \widehat{\sigma} = \sum_{i=r+1}^{n-s} \widehat{b}_i X_{i:n},$$

where special tables are not required for the computation of the $\widehat{a}_i$'s or $\widehat{b}_i$'s [they are given explicitly in (6.6)-(6.11)]. Since the MLE for σ is not unbiased, we obtain an unbiased estimator of σ, $\widetilde{\sigma}$, based on the MLE (to be used in the following section) by dividing by its expected value when the underlying random variables come from the standard Laplace distribution ($\mu = 0$, $\sigma = 1$),

$$\widetilde{\sigma} = \frac{\displaystyle\sum_{i=r+1}^{n-s} \widehat{b}_i X_{i:n}}{\displaystyle\sum_{i=r+1}^{n-s} \widehat{b}_i \mu_{i:n}} = \sum_{i=r+1}^{n-s} \widetilde{b}_i X_{i:n}, \tag{6.13}$$

where $\mu_{i:n}$ is given in Relation 3.1, and the corresponding $\nu_{i:n}$ is given in (3.3). Furthermore, a closed form expression for its variance may be obtained using Relations 3.1 and 3.2 in conjunction with Equations (3.3)-(3.5). We have,

$$\text{Var}(\widetilde{\sigma}) = \sigma^2\left\{\sum_{i=r+1}^{n-s} \widetilde{b}_i^2 \mu_{i:n}^2 + 2\sum_{i=r+1}^{n-s-1} \sum_{j=i+1}^{n-s} \widetilde{b}_i \widetilde{b}_j \mu_{i,j:n} - \left(\sum_{i=r+1}^{n-s} \widetilde{b}_i \mu_{i:n}\right)^2\right\} = \sigma^2 \widetilde{V}_2,$$

where exact explicit expressions for $\mu_{i:n}$, $\mu_{i:n}^{(2)}$, and $\mu_{i,j:n}$ are given in Relations 3.1 and

3.2, with the corresponding $\nu_{i:n}$, $\nu_{i:n}^{(2)}$, and $\nu_{i,j:n}$ given in Equations (3.3)-(3.5). An analogous equation holds for the variance $\widehat{V}_1$ of $\widehat{\mu}$.

It is also shown in Childs and Balakrishnan[23] that the performance of the unbiased MLE for σ is very similar to that of the BLUE in terms of variance, $\sqrt{\beta_1}$, and β_2.

6.3 BLUE's for the Pareto Distribution

Childs et. al[10] have used the formulas in (6.1)-(6.5) to derive exact explicit expressions for the BLUE's of μ and σ, as well as the variances, and the covariance for the Pareto distribution (4.1). They use the fact that the covariance matrix of the standardized Pareto order statistics $(\sigma_{i,j:n})$ is of the form $(a_i b_j)$ where

$$a_i = \frac{\Gamma(n-i+1-2/\nu)}{\Gamma(n-i+1-1/\nu)\Gamma(n+1-2/\nu)} - \frac{\Gamma(n-i+1-1/\nu)\Gamma(n+1)}{\Gamma(n-i+1)\Gamma(n+1-1/\nu)^2} \quad (6.14)$$

and

$$b_j = \frac{\Gamma(n+1)\Gamma(n-j+1-1/\nu)}{\Gamma(n-j+1)}. \quad (6.15)$$

They are then able to invert the covariance matrix $(\sigma_{i,j:n})$ and obtain explicit expressions for the censored BLUE's μ^* and σ^* of μ and σ respectively, the variances, $\mathrm{Var}(\mu^*)$ and $\mathrm{Var}(\sigma^*)$, and the covariance $\mathrm{Cov}(\mu^*, \sigma^*)$, as described, for example, in Arnold, Balakrishnan, and Nagaraja (1992). We have,

$$\mu^* = \left(\left\{ \frac{a_{r+2} - a_{r+1}}{a_{r+1}(a_{r+2}b_{r+1} - a_{r+1}b_{r+2})} - \frac{1}{a_{r+1}b_{r+1}} \right\} X_{r+1:n} \right.$$

$$+ \sum_{i=r+1}^{n-s-2} \frac{a_i(b_{i+1} - b_{i+2}) + a_{i+1}(b_{i+2} - b_i) + a_{i+2}(b_i - b_{i+1})}{(a_{i+1}b_i - a_i b_{i+1})(a_{i+2}b_{i+1} - a_{i+1}b_{i+2})} X_{i+1:n}$$

$$\left. + \frac{b_{n-s-1} - b_{n-s}}{b_{n-s}(a_{n-s}b_{n-s-1} - a_{n-s-1}b_{n-s})} X_{n-s:n} \right) \bigg/ \left(\sum_{i,j} \sigma^{ij} - \frac{1}{a_{r+1}b_{r+1}} \right),$$

$$\sigma^* = \left(\left\{ \sum_{i,j} \sigma^{ij} - \frac{a_{r+2} - a_{r+1}}{a_{r+1}(a_{r+2}b_{r+1} - a_{r+1}b_{r+2})} \right\} X_{r+1:n} \right.$$

$$- \sum_{i=r+1}^{n-s-2} \frac{a_i(b_{i+1} - b_{i+2}) + a_{i+1}(b_{i+2} - b_i) + a_{i+2}(b_i - b_{i+1})}{(a_{i+1}b_i - a_i b_{i+1})(a_{i+2}b_{i+1} - a_{i+1}b_{i+2})} X_{i+1:n}$$

$$\left. - \frac{b_{n-s-1} - b_{n-s}}{b_{n-s}(a_{n-s}b_{n-s-1} - a_{n-s-1}b_{n-s})} X_{n-s:n} \right) \bigg/ \left(\frac{b_{r+1}}{c} \sum_{i,j} \sigma^{ij} - \frac{1}{c a_{r+1}} \right),$$

$$Var(\mu^*) = \sigma^2 \Big/ \left(\sum_{i,j} \sigma^{ij} - \frac{1}{a_{r+1}b_{r+1}} \right), \qquad Var(\sigma^*) = \sigma^2 \Big/ \left(\frac{b_{r+1}}{a_{r+1}c^2} - \frac{1}{c^2 a_{r+1}^2 \sum_{i,j} \sigma^{ij}} \right),$$

and

$$\text{Cov}(\mu^*, \sigma^*) = -\sigma^2 \Big/ \left(\frac{b_{r+1}}{c} \sum_{i,j} \sigma^{ij} - \frac{1}{ca_{r+1}} \right),$$

where a_i and b_i are given in (6.14) and (6.15) respectively,

$$c = \Gamma(n + 1 - 1/\nu),$$

and $\sum_{i,j} \sigma^{ij}$ is the sum of all of the elements of the inverse matrix of the covariance matrix $(\sigma_{i,j:n})$, which is given by

$$\sum_{i,j} \sigma^{ij} = \sum_{i=r+2}^{n-s-1} \frac{a_{i-1}(2b_i - b_{i+1}) - 2a_i b_{i-1} + a_{i+1}b_{i-1}}{(a_i b_{i-1} - a_{i-1}b_i)(a_{i+1}b_i - a_i b_{i+1})}$$

$$+ \frac{a_{r+2} - 2a_{r+1}}{a_{r+1}(a_{r+2}b_{r+1} - a_{r+1}b_{r+2})} + \frac{b_{n-s-1}}{b_{n-s}(a_{n-s}b_{n-s-1} - a_{n-s-1}b_{n-s})}.$$

In the full sample case ($r = s = 0$), exact explicit expressions for the BLUE's may be obtained from the results of Kuldorff and Vännman[24], while exact explicit expressions for the right-censored case ($r = 0$) have been derived recently by Childs and Balakrishnan[25].

6.4 BLUE's for the Power Function Distribution

Exact explicit expressions similar to those presented in the previous section have been derived by Sultan et. al[11] for the power function distribution. They again used the fact that the covariance matrix of the standardized power function order statistics $(\sigma_{i,j:n})$ is of the form $(a_i b_j)$, where

$$a_i = \frac{\Gamma(n + 1)\Gamma(i + 1/\nu)}{\Gamma(i)} \tag{6.16}$$

and

$$b_j = \frac{\Gamma(j + 2/\nu)}{\Gamma(j + 1/\nu)\Gamma(n + 1 + 2/\nu)} - \frac{\Gamma(j + 1/\nu)\Gamma(n + 1)}{\Gamma(j)\Gamma(n + 1 + 1/\nu)^2}, \tag{6.17}$$

to invert the covariance matrix $(\sigma_{i,j:n})$ and obtain explicit expressions for the censored BLUE's μ^* and σ^* of μ and σ respectively, the variances $Var(\mu^*)$ and $Var(\sigma^*)$, and the covariance $\text{Cov}(\mu^*, \sigma^*)$, as described, for example, in Arnold, Balakrishnan, and Nagaraja[20]. We have,

$$\mu^* = \left(\frac{a_{r+2} - a_{r+1}}{a_{r+1}(a_{r+2}b_{r+1} - a_{r+1}b_{r+2})} X_{r+1:n} \right.$$

$$+ \sum_{i=r+1}^{n-s-2} \frac{a_i(b_{i+1} - b_{i+2}) + a_{i+1}(b_{i+2} - b_i) + a_{i+2}(b_i - b_{i+1})}{(a_{i+1}b_i - a_ib_{i+1})(a_{i+2}b_{i+1} - a_{i+1}b_{i+2})} X_{i+1:n}$$

$$+ \left\{ \frac{b_{n-s-1} - b_{n-s}}{b_{n-s}(a_{n-s}b_{n-s-1} - a_{n-s-1}b_{n-s})} - \frac{1}{b_{n-s}a_{n-s}} \right\} X_{n-s:n} \right) \bigg/ \left(\sum_{i,j} \sigma^{ij} - \frac{1}{a_{n-s}b_{n-s}} \right),$$

$$\sigma^* = \left(\frac{a_{r+1} - a_{r+2}}{a_{r+1}(a_{r+2}b_{r+1} - a_{r+1}b_{r+2})} X_{r+1:n} \right.$$

$$- \sum_{i=r+1}^{n-s-2} \frac{a_i(b_{i+1} - b_{i+2}) + a_{i+1}(b_{i+2} - b_i) + a_{i+2}(b_i - b_{i+1})}{(a_{i+1}b_i - a_ib_{i+1})(a_{i+2}b_{i+1} - a_{i+1}b_{i+2})} X_{i+1:n}$$

$$+ \left\{ \sum_{i,j} \sigma^{ij} - \frac{b_{n-s-1} - b_{n-s}}{b_{n-s}(a_{n-s}b_{n-s-1} - a_{n-s-1}b_{n-s})} \right\} X_{n-s:n} \right) \bigg/ \left(\frac{a_{n-s}}{c} \sum_{i,j} \sigma^{ij} - \frac{1}{cb_{n-s}} \right),$$

$$Var(\mu^*) = \sigma^2 \bigg/ \left(\sum_{i,j} \sigma^{ij} - \frac{1}{a_{n-s}b_{n-s}} \right), \qquad Var(\sigma^*) = \sigma^2 \bigg/ \left(\frac{a_{n-s}}{b_{n-s}c^2} - \frac{1}{c^2 b_{n-s}^2 \sum_{i,j} \sigma^{ij}} \right),$$

and

$$Cov(\mu^*, \sigma^*) = -\sigma^2 \bigg/ \left(\frac{a_{n-s}}{c} \sum_{i,j} \sigma^{ij} - \frac{1}{cb_{n-s}} \right),$$

where a_i and b_i are given in (6.16) and (6.17) respectively,

$$c = \Gamma(n + 1 + 1/\nu),$$

and $\sum_{i,j} \sigma^{ij}$ is the sum of all of the elements of the inverse matrix of the covariance matrix $(\sigma_{i,j:n})$, which is given in the previous section [with the corresponding a_i and b_i given in (6.16) and (6.17)].

7 Approximate Inference

7.1 The Exponential Distribution

The BLUE of σ for the exponential distribution (6.6) has the property that when $r = 0$, that is, for a right censored sample, $2(n - s)\frac{\sigma^*}{\sigma}$ has a chi-square distribution with $2(n - s)$ degrees of freedom. This result can be used to obtain confidence intervals for σ and carry out tests of hypotheses concerning σ. However, such a result does not hold for the BLUE in (6.6) based on a doubly Type-II censored sample.

With this in mind, Balakrishnan and Gupta[8] looked at the higher moments of order statistics from the exponential distribution to see whether or not the chi-square distributional result holds approximately for doubly Type-II censored samples. They observed that if $\gamma = \beta_2 - 1.5\beta_1 - 3$, then $\gamma = 0$ represents the chi-square line in the Pearson (β_1, β_2)-plane. Therefore, for the case of a right censored sample from the exponential distribution, γ will be equal to 0 when $\sqrt{\beta_1}$ and β_2 are the coefficients of skewness and kurtosis of the BLUE σ^*. By calculating $\sqrt{\beta_1}$ and β_2 of σ^* for doubly censored samples, they presented tables that examined the value of γ for $n = 8(1)12$ for various levels of censoring.

It can be seen from the tables of Balakrishnan and Gupta[8], that indeed when $r = 0$, γ is always 0. But when $r \neq 0$, γ is very close to zero at all levels of censoring, even for n as small as 8. Furthermore, as the sample size increases, the approximation $\gamma \approx 0$ becomes even better. We can therefore conclude that the distribution of $\frac{\sigma^*}{\sigma}$, in the case of doubly Type-II censored samples, may be very closely approximated by a chi-square distribution with degrees of freedom determined approximately (for example, by looking at the variance). An example will be given in Section 8.

7.2 The Laplace Distribution

By making use of the relations in Section 2, the values of a_i, b_i, V_1, and V_2 [tabulated by Balakrishnan, Chandramouleeswaran, and Ambagaspitiya[22]], and the exact values of $\widehat{a}_i$, $\widetilde{b}_i$ [given in (6.11) and obtained from (6.12), respectively], $\widehat{V}_1$, and $\widetilde{V}_2$ (from the exact explicit expressions described in Section 6) for the case of Type-II right-censored samples,

$$X_{1:n} \leq X_{2:n} \leq \cdots \leq X_{n-s:n},$$

Balakrishnan et. al[9] determined the values of the mean, the variance, and the coefficients of skewness and kurtosis ($\sqrt{\beta_1}$ and β_2) of P_1^*, P_2^*, $\widehat{P}_1$ and $\widetilde{P}_2$ (the analogous quantities using the unbiased MLE's) for $n = 5(1)10(5)20$ and $s = 0(1)(\frac{n}{2} - 1)$ for n even and $s = 0(1)[\frac{n}{2}]$ for n odd. However, they noted that the values obtained using the BLUE for σ are very similar to those obtained using the MLE. In fact the agreement is often to two decimals for small samples, and three or four decimals for larger samples. It is for this reason, and the fact that the MLE's are explicit linear functions of the order statistics, that we only discuss the results relating to the MLE's.

An examination of the β_2 values for $\widehat{P}_1$ and $\widetilde{P}_2$ in the tables of Balakrishnan et. al[9] reveals that the distribution of $\widehat{P}_1$ (and hence of P_1) is slightly heavier tailed than normal and, therefore, an Edgeworth approximation in this case will be quite appropriate. An examination of the $(\sqrt{\beta_1}, \beta_2)$ values reveals that the distribution of $\widetilde{P}_2$ (and hence of P_2) is positively skewed and also heavier tailed than the normal, but lies in the range of an Edgeworth approximation; for details on the possible range for Edgeworth approximation, see Barton and Dennis[26] and Johnson, Kotz and Balakrishnan[27].

By making use of the $(\sqrt{\beta_1}, \beta_2)$ values, lower and upper 1%, 2.5%, 5% and 10% points of P_1 and P_2 may be obtained through the Edgeworth approximation in (1.3). These values, for the case of Type-II right-censored samples ($r = 0$) for $s = 0(1)(\frac{n}{2} - 1)$

for n even and $s = 0(1)[\frac{n}{2}]$ for n odd and sample size $n = 5(1)10(5)20$, have been calculated by Balakrishnan et. al[9]. For the purpose of comparison, these percentage points were also determined by simulations (based on 5000 runs).

From the tables of Balakrishnan et. al[9] we see that the Edgeworth approximation of the distributions of P_1 provides quite close agreement with the simulated percentage points. The largest discrepancy occurs at the extreme lower and upper tails of the distribution, but only for small sample sizes. As the sample size increases, the agreement becomes quite close at all levels of censoring, even at the extremes of the distribution.

The Edgeworth approximation of the distributions of P_2 also provide close agreement with the simulated percentage points. This time however, the discrepancy for small sample sizes only occurs at the upper tail. But again, as the sample size increases, the discrepancy becomes quite small at all levels of censoring, even at the extremes of the distribution.

In conclusion, we observe that the Edgeworth approximations of the distributions of P_1 and P_2 all work quite satisfactorily even in samples of size as small as 5, and they indeed improve in accuracy as the sample size n increases. It should also be pointed out here that a similar Edgeworth approximation can not be developed for the percentage points of the pivotal quantities P_3 or P_3' in (5.1) and (5.2) since it is not a linear function of order statistics. However, approximate inference based on P_1 with σ replaced by $\tilde{\sigma}$ provides quite close results to those based on P_3. For this purpose, Balakrishnan et. al[9] have also presented some selected percentage points of P_3 determined by simulations (based on 5000 runs).

7.3 The Pareto Distribution

By making use of the recurrence relations in Section 4, and the values of a_i and b_i obtained from the exact explicit expressions for the BLUE's given in Section 6 for the case of Type-II right-censored samples,

$$X_{1:n} \leq X_{2:n} \leq \cdots \leq X_{n-s:n},$$

Childs et. al[10] determined the values of the mean, variance and the coefficients of skewness and kurtosis ($\sqrt{\beta_1}$ and β_2) of P_1^*, and P_2^*, for $n = 5(1)10(5)20$, $\nu = 5(10)25$, and $s = 0(1)(\frac{n}{2} - 1)$ for n even and $s = 0(1)[\frac{n}{2}]$ for n odd. By examining the ($\sqrt{\beta_1}, \beta_2$), Childs et. al[10] observed that the distribution of P_1^* (and hence of P_1) is negatively skewed and heavier tailed than normal. For small values of n and ν (most notably $n = 5, \ldots, 8$ when $\nu = 5$) the values of ($\sqrt{\beta_1}, \beta_2$) are not within the range of the Edgeworth approximation; for details on the possible range for Edgeworth approximation, see Barton and Dennis[26] and Johnson, Kotz, and Balakrishnan[27]. However, for larger values of n when $\nu = 5$, and even for some of the smaller values of n for $\nu = 15$ and 25, the ($\sqrt{\beta_1}, \beta_2$) values lie well within the range of an Edgeworth approximation.

By examining the ($\sqrt{\beta_1}, \beta_2$) values for P_2^*, Childs et. al[10] observed that they are very similar to those of P_1^*. The β_2 values are almost identical, while the $\sqrt{\beta_1}$ values are also almost identical, except for a negative sign. This similarity can be explained by a careful examination of the exact explicit expressions for the BLUE's given in Section 6. These expressions reveal that $a_i = -cb_i$ for $i = r + 1, 3, \ldots n - s - 1$ while a_1 is very similar to $-cb_1$, and a_{n-s} is very similar to $-cb_{n-s}$.

Childs *et. al*[10] also determined the lower and upper 1%, 2.5%, 5%, and 10% points of P_1 and P_2 through the Edgeworth approximation in (1.3) for $\nu = 5(10)25$. They calculated these values for the case of Type-II right-censored samples $(r = 0)$ for $s = 0(1)(\frac{n}{2} - 1)$ for n even and $s = 0(1)[\![\frac{n}{2}]\!]$ for n odd and sample size $n = 5(1)10$. For the purpose of comparison, these percentage points were also determined by simulations (based on 5000 runs).

From their tables Childs *et. al*[10] found that the Edgeworth approximation of the distribution of P_1 provides quite close agreement with the simulated percentage points. The largest discrepancy occurs at the extreme lower tails of the distribution, but only for small sample sizes and small values of ν. As the sample size and/or ν increases, the agreement becomes quite close at all levels of censoring, even at the lower extremes of the distribution.

Childs *et. al*[10] also observed that the Edgeworth approximation of the distribution of P_2 also provides close agreement with the simulated percentage points, even a bit closer than P_1. This time however, the largest discrepancies occur at the upper tails of the distribution for small sample sizes, but the discrepancies are not as great as those of P_1. Again, as the sample size and/or ν increases, the discrepancy becomes quite small at all levels of censoring, even at the extremes of the distribution.

In conclusion, we observe that the Edgeworth approximations of the distributions of P_1 and P_2 both work quite satisfactorily even in samples of size as small as 5 (for large ν), and they improve in accuracy as the sample size and/or ν increases. It should also be pointed out here that a similar Edgeworth approximation can not be developed for the percentage points of the pivotal quantities P_3 since it is not a linear function of order statistics. However, as with the Laplace distribution, approximate inference based on P_1 with σ replaced by σ^* provides quite close results to those based on P_3. For this purpose, Childs *et. al*[10] have presented some selected percentage points of P_3 (for $\nu = 5$) determined by simulations (based on 5000 runs).

7.4 The Power Function Distribution

By making use of the recurrence relations in Section 4, and the values of a_i and b_i obtained from the exact explicit expressions for the BLUE's given in Section 6 for the case of Type-II right-censored samples,

$$X_{1:n} \leq X_{2:n} \leq \cdots \leq X_{n-s:n},$$

Sultan *et. al*[11] determined the values of the mean, the variance, and the coefficients of skewness and kurtosis ($\sqrt{\beta_1}$ and β_2) of P_1^*, and P_2^*, for $n = 5(1)10(5)20$, $\nu = 3/2, 3$ and 6, and $s = 0(1)(\frac{n}{2} - 1)$ for n even and $s = 0(1)[\![\frac{n}{2}]\!]$ for n odd. They observed that when $\nu = 3/2$ the distribution of P_1^* (and hence of P_1) is positively skewed and heavier tailed than normal except for small sample sizes where it is positively skewed and lighter tailed than normal. When $\nu = 3$, the distribution of P_1^* is lighter tailed than normal for all sample sizes and negatively skewed for small sample sizes. When $\nu = 6$, the distribution of P_1^* is negatively skewed for all sample sizes and slightly heavier tailed than normal for small sample sizes. In each case, the values of ($\sqrt{\beta_1}, \beta_2$) lie well within the range of the Edgeworth approximation.

Sultan *et. al*[11] made similar observations regarding the ($\sqrt{\beta_1}, \beta_2$) values for P_2^* except that the $\sqrt{\beta_1}$ values for P_2^* usually are opposite in sign to those of P_1^*. This similarity can be explained by a careful examination of the exact explicit expressions for the BLUE's given in Section 6. These expressions reveal that $a_i = -cb_i$ for $i = 2, 3, \ldots n - s - 1$, while a_1 is very similar to $-cb_1$, and a_{n-s} is very similar to $-cb_{n-s}$.

Sultan et. al[11] also determined the lower and upper 1%, 2.5%, 5%, and 10% points of P_1 and P_2 through the Edgeworth approximation in (1.3) for $\nu = 3/2, 3$ and 6 for the case of Type-II right-censored samples $(r = 0)$ for $s = 0(1)(\frac{n}{2} - 1)$ for n even and $s = 0(1)[\frac{n}{2}]$ for n odd and sample size $n = 5(1)10(5)20$. For the purpose of comparison, these percentage points were also determined by simulations (based on 5000 runs).

From their tables Sultan et. al[11] made observations very similar to the Pareto distribution discussed in the previous section. They concluded that the Edgeworth approximations of the distributions of P_1 and P_2 both work quite satisfactorily even in samples of size as small as 5 (for large ν), and they improve in accuracy as the sample size and/or ν increases. Sultan et. al[11] have also presented some selected percentage points of P_3 (for $\nu = 5$), determined by simulations (based on 5000 runs), for the purpose of making inferences about μ when σ is unknown.

8 Numerical Illustrations

Example 1: The Exponential Distribution

Balakrishnan and Gupta[8] presented the following example using the data from Nelson[28] (Table 2.1) arising from an experiment on insulating fluid breakdowns. They considered the data

$$2, 2, 3, 9, 13, 47, 50, 55, 71$$

to be a doubly Type-II censored sample with $n = 12$, $r = 3$, and $s = 0$. Using the expression in (6.6) they found that

$$\sigma^* = 21.2915 \text{ seconds.}$$

In this case, the value of γ is 0.001905, indicating that the distribution of $24\sigma^*/\sigma$ may be closely approximated by a chi-square distribution with 24 degrees of freedom. Consequently, the approximate 90% confidence interval for σ is

$$\left[\frac{24\sigma^*}{36.4151}, \frac{24\sigma^*}{13.8484} \right] = [14.0325, 36.8993].$$

As a check of the accuracy of this result, Balakrishnan and Gupta[8] pointed out that Viveros[29] obtained a 90% conditional confidence interval for σ in this case to be $[14.00, 36.79]$, which agrees quite closely with the approximate result.

Example 2: The Laplace Distribution

In order to illustrate the usefulness of the inference procedures discussed in the previous sections, we consider here a simulated data set of size $n = 20$ (with $\mu = 50$ and $\sigma = 5$):

32.007, 37.757, 43.848, 46.268, 46.907, 47.262, 47.290, 47.593, 48.065, 49.254,
50.278, 50.487, 50.662, 53.336, 53.493, 53.567, 53.981, 54.942, 55.695, 66.396.

Using this sample, the BLUE's and MLE's were calculated based on complete as well as Type-II right-censored samples ($r = 0$) by making use of the tables of Balakrishnan, Chandramouleeswaran and Ambagaspitiya[22], and the explicit expressions in (4.5) and (4.6), respectively. These estimates are presented in the following table.

n	s	μ^*	$\widehat{\mu}$	σ^*	$\widetilde{\sigma}$
20	0	49.561	49.766	4.947	4.964
	1	49.561	49.766	4.635	4.653
	2	49.561	49.766	4.814	4.834
	3	49.561	49.766	4.931	4.952

With these estimates and the use of the tables of Balakrishnan *et. al*[9], we determined the 90% confidence intervals for μ (when σ is known to be 5) based on the Edgeworth approximation and on the simulated percentage points using both P_1 and P_1'. These are presented in the table below.

		Edgeworth C.I.		Simulated C.I.	
n	s	P_1	P_1'	P_1	P_1'
20	0	$(47.501, 51.621)$	$(47.667, 51.865)$	$(47.454, 51.668)$	$(47.637, 51.895)$
	1	$(47.501, 51.621)$	$(47.667, 51.865)$	$(47.530, 51.656)$	$(47.637, 51.895)$
	2	$(47.501, 51.621)$	$(47.667, 51.865)$	$(47.517, 51.618)$	$(47.637, 51.895)$
	3	$(47.501, 51.621)$	$(47.667, 51.865)$	$(47.466, 51.631)$	$(47.637, 51.895)$

It is clear that the confidence interval based on the Edgeworth approximation is very close to the confidence interval determined by simulations, at all levels of censoring.

Similarly, we determined the 90% confidence intervals for σ, and they are presented below.

		Edgeworth C.I.		Simulated C.I.	
n	s	P_2	P_2'	P_2	P_2'
20	0	$(3.535, 7.522)$	$(3.547, 7.550)$	$(3.565, 7.471)$	$(3.576, 7.526)$
	1	$(3.285, 7.138)$	$(3.298, 7.168)$	$(3.332, 7.226)$	$(3.344, 7.257)$
	2	$(3.383, 7.518)$	$(3.396, 7.550)$	$(3.381, 7.483)$	$(3.395, 7.518)$
	3	$(3.433, 7.820)$	$(3.447, 7.854)$	$(3.415, 7.795)$	$(3.429, 7.832)$

Once again, we observe that the confidence intervals based on the Edgeworth approximation are very close to those based on simulations.

In the case when σ is unknown, the Edgeworth approximation method cannot be used to draw inference for μ using P_3 or P_3'. However, as pointed out in the last section, the Edgeworth approximation for the distribution of the pivotal quantity P_1 may be used in this case with σ replaced by σ^*, or P_1' may be used with σ replaced by $\widetilde{\sigma}$, in order to draw approximate inference for μ. By this method, we determined the 90% confidence intervals for μ, which are presented in the following table for the choices of $r = 0$, $s = 0(1)3$. Also presented in this table are the corresponding 90% confidence intervals for μ based on the simulated percentage points of the pivotal quantities P_3 and P_3' given in Balakrishnan *et. al*[9].

		Edgeworth C.I.		Simulated C.I.	
n	s	$P_1\,(\sigma = \sigma^*)$	$P_1'\,(\sigma = \tilde{\sigma})$	P_3	P_3'
20	0	$(47.526, 51.596)$	$(47.678, 51.854)$	$(47.384, 51.738)$	$(47.483, 52.000)$
	1	$(47.654, 51.468)$	$(47.809, 51.723)$	$(47.614, 51.647)$	$(47.719, 51.906)$
	2	$(47.581, 51.541)$	$(47.733, 51.799)$	$(47.539, 51.583)$	$(47.687, 51.845)$
	3	$(47.532, 51.590)$	$(47.683, 51.849)$	$(47.391, 51.681)$	$(47.538, 51.895)$

It is quite clear that the confidence intervals using the approximate Edgeworth method based on the pivotal quantities P_1 or P_1' are all very close to the confidence intervals determined by simulation of the distribution of the pivotal quantity P_3 or P_3' at all levels of censoring.

Acknowledgments

The authors would like to thank the Natural Sciences and Engineering Research Council of Canada for funding this research.

References

1. Mann, N.R., Schafer, R.E., and Singpurwalla, N.D. (1974). *Methods for Statistical Analysis of Reliability and Life Data*, John Wiley & Sons, New York.
2. Bain, L. J. (1978). *Statistical Analysis of Reliability and Life-Testing Models: Theory and Methods*, Marcel Dekker, New York.
3. Lawless, J.F. (1982). *Statistical Models and Methods for Lifetime Data*, John Wiley & Sons, New York.
4. Cohen, A.C. and Whitten, B.J. (1988). *Parameter Estimation in Reliability and Life Span Models*, Marcel Dekker, New York.
5. Bain, L.J. and Engelhardt, M. (1991). *Statistical Analysis of Reliability and Life-Testing Models: Theory and Methods*, Second edition, Marcel Dekker, New York.
6. Balakrishnan, N. and Cohen, A.C. (1991). *Order Statistics and Inference: Estimation Methods*, Academic Press, Boston.
7. David, H.A. (1981). *Order Statistics*, Second edition, John Wiley & Sons, New York.
8. Balakrishnan, N. and Gupta, S.S. (1997). Higher order moments of order statistics from exponential and right-truncated exponential distributions and applications to life-testing problems, In *Handbook of Statistics - 16: Order Statistics and Their Applications* (Eds., C.R. Rao and N. Balakrishnan), North-Holland, Amsterdam (to appear).
9. Balakrishnan, N., Childs, A., Chandramouleeswaran, M.P., and Govindarajulu, Z. (1997). Inference on parameters of the Laplace distribution based on Type-II censored samples using Edgeworth approximation, *Submitted for Publication*.
10. Childs, A., Sultan, K, and Balakrishnan, N. (1997). Higher order moments of order statistics from the Pareto distribution and Edgeworth approximation inference, *Submitted for Publication*.

11. Sultan, K., Childs, A., and Balakrishnan, N. (1997). Higher order moments of order statistics from the power function distribution and Edgeworth approximation inference, *Submitted for Publication*.

12. Joshi, P.C. (1978). Recurrence relations between moments of order statistics from exponential and truncated exponential distributions, *Sankhyā, Series B* **39**, 362-371.

13. Govindarajulu, Z. (1963). Relationships among moments of order statistics in samples from two related populations, *Technometrics* **5**, 514-518.

14. Sukhatme, P.V. (1937). Tests of significance for samples of the χ^2 population with two degrees of freedom, *Ann. Eugen.* **8**, 52-56.

15. Rényi, A. (1953). On the theory of order statistics, *Acta Math. Acad. Sci. Hung.* **4**, 191-231.

16. Balakrishnan, N. and Joshi, P.C. (1982). Moments of order statistics from doubly truncated Pareto distribution, *Journal of the Indian Statistical Association* **20**, 109-117.

17. Huang, J. S. (1975). A note on order statistics from the Pareto distribution, *Scandinavian Actuarial Journal*, **2**, 187-190.

18. Balakrishnan, N. and Joshi, P.C. (1981). Moments of order statistics from doubly truncated power function distribution, *Aligarh J. Statist.* **1**, 98-105.

19. Malik, H.J. (1967). Exact moments of order statistics from a power-function distribution, *Skand. Aktuarietidskr.* 64-69.

20. Arnold, B.C., Balakrishnan, N. and Nagaraja, H.N. (1992). *A First Course in Order Statistics*, John Wiley & Sons, New York.

21. Govindarajulu, Z. (1966). Best linear estimates under symmetric censoring of the parameters of a double exponential population, *Journal of the American Statistical Association* **61**, 248-258.

22. Balakrishnan, N., Chandramouleeswaran, M.P., and Ambagaspitiya, R.S. (1996). BLUE's of location and scale parameters of Laplace distribution based on Type-II censored samples and associated inference, *Microelectronics and Reliability* **36**, 371-374.

23. Childs, A. and Balakrishnan, N. (1997). Maximum likelihood estimation of Laplace parameters based on general Type-II censored samples, *Statistische Hefte* (to appear).

24. Kulldorff, G., and Vännman, K., (1973). Estimation of the location and scale parameters of a Pareto distribution by linear functions of order statistics, *Journal of the American Statistical Association* **68**, 218-227.

25. Childs, A. and Balakrishnan, N. (1997). Generalized recurrence relations for moments of order statistics from non-identical Pareto and truncated Pareto random variables with applications to robustness, In *Handbook of Statistics - 16: Order Statistics and Their Applications* (Eds., C.R. Rao and N. Balakrishnan), North-Holland, Amsterdam (to appear).

26. Barton, D.E. and Dennis, K.E.R. (1952). The conditions under which Gram-Charlier and Edgeworth curves are positive definite and unimodal, *Biometrika* **39**, 425-427.

27. Johnson N.L., Kotz S., and Balakrishnan N., (1994). *Continuous Univariate Distributions*, Vol. 1, Second edition, John Wiley & Sons, New York.

28. Nelson, W. (1982). *Applied Life Data Analysis*, John Wiley & Sons, New York.

29. Viveros, R. (1995). Conditional inference and applications, In *The Exponential Distribution: Theory, Methods and Applications* (Eds., N. Balakrishnan and A.P. Basu), Gordon and Breach, Langhorne, PA.

RECENT DEVELOPMENTS IN INFORMATION THEORY AND RELIABILITY ANALYSIS

NADER EBRAHIMI
Division of Statistics
Northern Illinois University
DeKalb, IL 60115 USA
E-mail: nader@math.niu.edu

EHSAN S. SOOFI
School of Business Administration
University of Wisconsin-Milwaukee
P.O. Box 742, Milwaukee, WI 53201 USA
E-mail: esoofi@csd.uwm.edu

Information theory provides a convenient framework for many common problems we face in the field of reliability. This paper reviews some aspects of information-theoretic statistics for reliability analysts. In particular, measuring information and information diagnostics recently developed for reliability modeling are reviewed.

1 Introduction

The logical foundation, elegance, and versatility of the information theoretic approach has been increasingly attracting the attention of researchers in many fields including reliability. Although, the available entropy based methods are not yet commonly used in the mainstream reliability analysis, because of the simplicity of information theoretic methods, research in this area has continued to be active and expanding. Ebrahimi, Habibullah and Soofi (1993) provided an overview of the developments in applications of information theory to reliability problems. See also Singpurwalla (1996). In this paper, we update those papers by summarizing the new developments in connection with reliability analysis.

Section 2 of this paper presents the basic information functions and their modifications for measuring uncertainty associated with the residual life time distribution, introduced by Ebrahimi (1996), Ebrahimi and Pellerey (1995), Ebrahimi and Kirmani (1996 a,b,c).

Section 3 describes various entropy-based diagnostics and tests of distributional hypotheses developed during recent years, which are useful for reliability model building.

Another information-theoretic line of research which has wide applications in reliability is development of measures that quantify the amount of information about the immediate future contained in the past. Section 4 is devoted to describing such measures.

2 Measures of Information for Reliability Analysis

In this section we define a quantity called the dynamic version of the entropy. We also define the dynamic version of Kullback-Leibler discrimination information.

2.1 Dynamic version of the entropy

The entropy of a random variable X with a probability function F is defined by

$$H(X) = H(F) = -\int_0^\infty (\log f(x))dF(x), \tag{1}$$

where $f(x) = dF(x)$ is the probability density (mass) function for the absolutely continuous (discrete) distribution F.

The entropy measures the "uniformity" of a distribution. As $H(F)$ increases, $f(x)$ approaches to a uniform. Consequently, the concentration of probabilities decreases and it becomes more difficult to predict an outcome of a draw from $f(x)$. In this sense $H(F)$ is a measure of uncertainty associated with F. The quantity $-H(F)$ is used as a measure of statistical information, see Soofi (1994) for details.

If we think of a random variable X as the lifetime of a system then the entropy of X can be used for measuring the associated uncertainty. However, frequently, in reliability one has information about the current age of the system under consideration. In such cases, the age must be taken into account when measuring uncertainty. Obviously, the measure $H(F)$ in (1) is not suitable in such situations. A modified measure which takes the age consideration into account is described below; see also Ebrahimi (1996).

Given that the system has survived up to time t, a dynamic version of the entropy is defined by

$$\begin{aligned} H(F;t) &= -\int_t^\infty \frac{f(x)}{\bar{F}(t)} \log \frac{f(x)}{\bar{F}(t)} dx t \geq 0 \\ &= 1 - \frac{1}{\bar{F}(t)} \int_t^\infty (\log \lambda_F(x))f(x)dx, \end{aligned} \tag{2}$$

where $\bar{F}(x) = 1 - F(x)$, and $\lambda_F(x) = \frac{f(x)}{\bar{F}(x)}$ is the hazard function of X. (Throughout this paper, log will denote the natural logarithm and X will denote a nonnegative random variable having an absolutely continuous distribution.) After the system has survived for time $t, H(F;t)$ basically measures the expected uncertainty contained in the conditional density of $X - t$ given $X > t$ about the predictability of remaining life time of the system. That is, $H(F;t)$ measures concentration of conditional probabilities.

On the basis of the measure $H(F;t)$, Ebrahimi (1996) also defined and studied the following two nonparametric classes of life distributions.

Definition 2.1: A survival function $\bar{F}$ (with $\bar{F}(0) = 1$) is said to have decreasing (increasing) uncertainty of residual life (DURL(IURL)) if $H(F;t)$ is decreasing (increasing) in $t \geq 0$.

Ebrahimi and Kirmani (1996a) have explored further the properties and implications of the dynamic measure $H(F;t)$ and DURL(IURL) classes of life distributions.

Ebrahimi and Pellerey (1995) used the measure $H(F;t)$ to introduce a new partial ordering for comparing the uncertainties associated with two nonnegative random variables. The results obtained in this paper and the paper by Ebrahimi and Kirmani (1996a) help in establishing the uncertainty ordering between survival functions.

2.2 The dynamic version of Kullback-Leibler discrimination information

Let X and Y be two nonnegative random variables representing times to failure of two systems. Let $F(x) = P(X \leq x)$ and $G(y) = P(Y \leq y)$ be the failure distributions of X and Y with survival functions $\bar{F} = 1 - F$ and $\bar{G} = 1 - G$ respectively, with $\bar{F}(0) = \bar{G}(0) = 1$. We assume that F and G are differentiable and $f(t)$ and $g(t)$ denote the probability density functions of X and Y respectively.

The Kullback-Leibler discrimination information $K(X,Y)$ or $K(F,G)$ is defined as

$$K(F,G) = K(X,Y) = \int_0^\infty f(x) log \, \frac{f(x)}{g(x)} \, dx. \tag{3}$$

It is sometimes known as the relative entropy, or cross-entropy. $K(F,G) \geq 0$ with equality holding if and only if $f(x) = g(x)$ almost everywhere, and $K(F,G)$ is defined for $f(x) \neq 0$. Although $K(F,G)$ is not a distance function in the usual sense, it is often thought of as a pseudo-distance or a "closeness" measure between F and G. $K(F,G)$ is not symmetric in its two arguments; g is called the reference distribution.

$K(F,G)$ is a measure of the inefficiency of assuming that the distribution is G when the true distribution is F and in statistics it arises as an expected logarithm of the likelihood ratio.

In situations for which age must be taken into account when comparing or discriminating between two systems, we can replace F and G by distributions of the corresponding residual lifetimes. Given that both systems have survived up to time t, we therefore define Kullback-Leibler discrimination information at t by

$$
\begin{aligned}
K(X,Y;t) &= K(F,G;t) = \int_t^\infty \frac{f(x)}{\bar{F}(t)} log \left\{ \frac{\frac{f(x)}{\bar{F}(t)}}{\frac{g(x)}{\bar{G}(t)}} \right\} dx \\
&= log \, \bar{G}(t) - H(F;t) - \int_t^\infty \frac{f(x)}{\bar{F}(t)} log \, (g(x)) dx, \tag{4}
\end{aligned}
$$

where $H(F;t)$ is given by the equation (2).

The following example demonstrates computation and usefulness of $K(F,G;t)$.

Example 2.1.

Consider a parallel system of two independent components, each one with lifetime uniformly distributed over $[0,1]$. More specifically, suppose Y_1 and Y_2 are the lifetimes of components 1 and 2, respectively, with common probability density function g. If X is the lifetime of the system, then $X = \max(Y_1, Y_2)$ and the true probability density function of X is

$$f(x) = \begin{cases} 2x, & 0 < x < 1 \\ 0, & \text{elsewhere} \end{cases}.$$

One can easily verify that $K(F, G; t) = log\, 2 - \frac{1}{2} - log(1+t) - \frac{t^2}{1-t^2}\, log\, t$ which is non-increasing in t. Furthermore $lim_{t \to 1} K(F, G; t) = 0$. This simply means that as the system gets older and older, the distribution of remaining lifetime of the system gets closer and closer to the distribution of the remaining lifetime of each one of the components. Intuitively speaking, after a certain age the redundancy has negligible effect on the performance of the system.

The discrimination function in (4) is a measure of disparity between $F_t(x) = P(X - t \leq x | X > t) = \dfrac{F(t+x)}{\bar{F}(t)}$ and $G_t(x) = P(Y - t \leq x | Y > t) = \dfrac{G(t+x)}{\bar{G}(t)}$. If we have a system with true survival function $\bar{F}$, then $K(F, G; t)$ can also be interpreted as a measure of distance between G_t and the true distribution F_t.

As a dynamic measure of discriminatory information $K(F, G; t)$ turns out to be rather useful. Ebrahimi and Kirmani (1996b) proved the fact that $K(F, G; t)$ is constant in t if and only if F and G satisfy a proportional hazards model $\bar{G}(x) = (\bar{F}(x))^\beta$, for all $x \geq 0$, where $\beta > 0$.

Ebrahimi and Kirmani (1996b,c) have further studied properties of this measure. Also, they have applied the minimum discrimination principle to obtain the proportional hazards model.

3 Information diagnostics for reliability modeling

The maximum entropy (ME) principle of inference considers a class of distributions

$$\Omega_\theta = \{f(x|\theta) : E_f[T_j(X|\theta)] = \theta_j, j = 0, 1, \cdots, m\}, \tag{5}$$

where the T_j's are absolutely integrable functions with respect to f and $\theta = (\theta_1, \cdots, \theta_m)$. The inference is based on the model that maximizes the entropy

$$H(f(x|\theta)) = -\int (f(x|\theta))log\, f(x|\theta)dx, \tag{6}$$

subject to the information constraints that define Ω_θ. The ME model $f^*(x|\theta)$ in Ω_θ, if it exists, is of the form

$$f^*(x|\theta) = C(\theta)exp[\eta_1(\theta)T_1(x) + \cdots + \eta_m(\theta)T_m(x)], \tag{7}$$

where $C(\theta)$ is a normalizing constant. Many well known distributions are ME subject to various types of constraints. Table 1 presents some examples. The ME method is quite versatile for developing models that capture information as well as diagnostics that measure information in various statistical problems. We refer you to the paper by Soofi (1994) and references therein.

Entropy-based tests and diagnostics for reliability modeling are basically those that are proposed for measuring departures of data distribution from an ME distribution. In ME modeling, the information discrepancy between two distributions is measured in terms of their entropy difference. In discrimination information statistics the information discrepancy between two distributions is measured in terms of the Kullback-Leibler information. Soofi et al. (1995) have discussed the equivalence of these two functions and concluded that they are equivalent if the reference distribution is the ME distribution. Based on this equivalence Soofi et al. (1995) have introduced the concept of information discrimination (ID) distinguishability which is a unifying framework for the two methods of measuring information discrepancy between two distributions and obtained ID distinguishability of some location-scale families and the student-t distribution relative to normality and ID distinguishability of some lifetime distributions relative to exponentiality. The ID indices of distributions are useful for planning robustness and power studies; see Soofi et al. (1995).

An immediate consequence of the equivalence between entropy difference and relative entropy methods of measuring disparity between distributions is that one can estimate the discrimination function between members of Ω_θ and f^* by estimating the entropy of $F^*(x|\theta)$ and $f(x|\theta)$. When the value of θ is determined externally according to the nature of the problem, or when it is hypothesized to be a particular value θ_0, then the entropy of $f^*(x|\theta)$ is known. In many other situations, when θ is determined internally, based on the data, the estimation of the entropy of $f^*(x|\theta)$ is known. In many other situations, when θ is determined internally, based on the data, the estimation of the entropy of $f^*(x|\theta)$ becomes a parametric estimation problem. Given a sample of size n, the entropy of $f^*(x|\theta)$ may be estimated using a Bayesian or sampling theory procedure. The ME modeling procedure also requires estimating the entropy of the data generating distribution, the entropy of $f(x|\theta)$, according to a nonparametric procedure and comparing it with the entropy estimate of $f^*(x|\theta)$. Various nonparametric entropy estimation procedures are available, see Ebrahimi et al. (1992) and Soofi et al. (1995) for a frequentist approach, and Maz-

Table 1. Examples of Continuous Maximum Entropy Distributions

$f(x) > 0$	$T(X)$	ME Distribution
(a, b)	none	uniform
$(0,1)$	ln X, ln(1-X)	beta
$(0, \infty)$	X	exponential
$(0, \infty)$	X, ln X	gamma
$(0, \infty)$	X^β ln X, $(\beta \neq 1)$	Weibull
$(a, \infty), a > 0$	ln X	Pareto
$(-\infty, \infty)$	$|X|$	Laplace
$(-\infty, \infty)$	X^2	normal (mean = 0)
$(-\infty, \infty)$	X, X^2	normal
$(-\infty, \infty)$	ln $(1 + X^2)$	generalized Cauchy
$(-\infty, \infty)$	X,ln$(1 + e^{-\lambda X})$	generalized logistic
$(-\infty, \infty)$	$X, e^{-\lambda X}$	generalized extreme value
$(-\infty, \infty)$	X, X^2, X^3, X^4	quartic exponential

zuchi et al. (1996) for a Bayesian approach. Using a parametric estimate of the entropy of $f^*(x|\theta)$ and a non-parametric entropy estimate we get the ID statistic

$$ID_n(f : f^*|\theta_n) = 1 - exp\{H_n(f) - H(f^*(x|\theta_n))\},$$

where θ_n is the estimate of θ, $H_n(f)$ is the non-parametric estimator and $H(f^*(x|\theta_n))$ is the parametric estimator of the entropy. If an ID_n is not zero, then a search for additional or different types of constraints is in order. Soofi et al. (1995) obtained ID_n when the data are generated from increasing failure rate (IFR) distributions. They also developed information diagnostics that are useful for assessing departure of the data generating distribution from exponentiality in the IFR class.

Mazzuchi, Soofi, and Soyer (1996) have developed Bayesian analysis of the ID index. From a Bayesian viewpoint, the uncertainty about the ID index is captured via the uncertainty about the types of moments and about the moment parameters. Thus, given a sample of n observations $x = (x_1, \cdots, x_n)$, one can estimate the ID index and infer about the adequacy of the ME likelihood function f^*. If indeed the specified moments are adequate descriptions of the data generating distribution, then the distribution of the posterior ID index $ID(f : f^*; \theta|x)$ will be concentrated near zero for any likelihood function f chosen in Ω_θ. The posterior mean of $ID(f : f^*; \theta)$ is used as the Bayes estimate of ID and is referred to as the *Bayesian Information Discrimination (BID) index.*

The parametric BID index is computed as

$$BID_P(f : f^*; \theta|x) = E\{1 - exp[H(f^*; \theta) - H(f_P; \theta)]|x\}, \qquad (8)$$

where $H(f_P; \theta)$ is the entropy of a known density $f_P(x|\theta)$ that satisfies the moment constraints (5). The parametric BID_P index (8) can be used for inference about discriminating between the likelihood function $f_P(x|\theta)$ chosen from Ω_θ and the ME model $f^*(x|\theta)$.

The nonparametric BID index is computed as

$$BID_{NP}(f : f^*; \theta|x) = E\{1 - exp[H(f^*; \theta) - H(f_{NP}; \theta)]|x\}, \qquad (9)$$

where $H(f_{NP}; \theta)$ is the entropy of a nonparametric density estimate $f_{NP}(x|\theta)$ that satisfies the moment constraints (5). The nonparametric BID_{NP} index (9) is used for inference about the adequacy of the specified moments in (5) for describing the likelihood function.

Uncertainty about the types of moments in (5) induces uncertainty about the ME likelihood function f^*, the nonparametric distribution f_{NP}, and the ID indices all of which depend on the types of moments $T_j(X)$ and their values $\theta_j, j = 1, \cdots, m$. Thus, to make Bayesian inference about the adequacy of a likelihood function, we need both parametric and nonparametric Bayesian procedures for estimating entropy and the ID indices. Mazzuchi et al. (1996, 1997) have developed such Bayesian machinery.

4 Information Coefficients

When studying models of systems developing in time, an important notion in the predictability of the short term evolution of a system is the amount of information about the immediate future contained in the past.

Let $X_1, X_2, \cdots$, be a sequence of random variables on some probability space and let $X_1, X_2, \cdots, X_n$ have joint probability density function $f_n(x_1, \cdots, x_n)$. Ebrahimi and Kirmani (1997) defined the sequence

$$V_1 = \int_0^\infty f_1(x) \ log \ f_1(x)dx,$$
$$V_n = E[I_n(X_1, \cdots, X_{n-1})], n \geq 2, \tag{10}$$

where $I_n(x_1, \cdots, x_{n-1}) = \int_0^\infty h_n(x_n|x_1, \cdots, x_{n-1})$ $log(h_n(x_n|x_1, \cdots, x_{n-1}))$. They referred to V_n as the nth information coefficient for the sequence $\{X_n; n \geq 1\}$.

To clarify this definition we give the following example.

Example 4.1 (Ebrahimi and Kirmani (1997):

Consider the first order autoregressive time series model, which we express as

$$X_n = \rho X_{n-1} + e_n, \tag{11}$$

where $e_1, e_2, \cdots$ are independent normal with mean 0 and variance σ^2 random variables and $|\rho| < 1$. In general, if $X_1, \cdots, X_n$ is multivariate normal with mean vector μ_n and variance covariance matrix B_n, $n = 1, 2, \cdots$, then one can easily verify that the information coefficient V_n is

$$V_n = \frac{1}{2}(log2\pi) - \frac{1}{2} \ log \ \frac{||B_n||}{||B_{n-1}||}, \tag{12}$$

where $|| \cdot ||$ stands for the determinant. Under the model (11), the information coefficient becomes

$$V_n = \frac{1}{2}(log2\pi) - \frac{1}{2} \ log \ (1 - \rho^2). \tag{13}$$

That is, V_n is free from n. Intuitively speaking, observing the series longer does not improve predictability of the future. It should be mentioned that V_n is increasing in ρ. This simply means that the information we get from $X_1, \cdots, X_{n-1}$ about X_n increases as ρ increases.

Ebrahimi and Kirmani (1997) used $\{V_n; n \geq 1\}$ and introduced the notions of stable and more (less) predictable sequences. They also studied various properties of the sequence $\{V_n; n \geq 1\}$ and gave several applications. In particular they applied those notions to several replacement policies.

In reliability, engineers are usually interested in the properties of a counting process. A reason is that if repair times are negligible then successive failures of a repairable system can be modeled as the arrivals of a counting process. Intuitively speaking, if information in the interarrival times increases over time, then the future failures will be more predictable, predicted with more precision, than the past failures. If we think of X_i as being the time between the ith and $(i - 1)$st failure, $X_0 = 0$, then Ebrahimi (1993) used the V_n to measure information in the nth interarrival time X_n about future interarrivals. He also defined notions of a counting process being stable, less predictable and more predictable and applied these concepts to the reliability of repairable systems.

References

Ebrahimi, N. (1993), "More predictable, less predictable and stable counting processes", *J. Appl. Probab, 30*, 341-352.

Ebrahimi, N. (1996), "How to measure uncertainty in the residual life time distributions", *Sankhya A, 58*, 48-57.

Ebrahimi, N., Habibullah, M., and Soofi, E.S. (1992), "Testing Exponentiality Based on Kullback-Leibler Information", *Journal of the Royal Statistical Society,* Ser. B., 54, 739-748.

Ebrahimi, N., Habibullah, M., and Soofi, E.S. (1993), "The Role of information theory in reliability analysis", Advances in Reliability edited by A.P. Basu North Holland, 89-105.

Ebrahimi, N. and Kirmani, S.N.U.A. (1996a), "Some results on ordering of survival functions through uncertainty", *Stat. and probab. letters, 29*, 167-176.

Ebrahimi, N. and Kirmani, S.N.U.A. (1996b), "A characterization of the proportional hazards model through a measure of discrimination between two residual life distribution", *Biometrika, 83*, 233-235.

Ebrahimi, N. and Kirmani, S.N.U.A. (1996c), "A measure of discrimination between two residual lifetime distributions and its applications", *Ann. Inst. Statist. Math, 48*, 257-265.

Ebrahimi, N. and Kirmani, S.N.U.A. (1997), "Information coefficients for sequence of random variables", submitted for publication.

Ebrahimi, N. and Pellerey, F. (1995), "New partial ordering of survival functions based on notion of uncertainty", *J. Appl. Probab.*, 202-211.

Ebrahimi, N., Pflughoeft, K., and Soofi, E.S. (1994), "Two Measures of Sample Entropy," *Statistics and Probability Letters, 20*, 225-234.

Mazzuchi, T. A., Soofi, E.S. and Soyer, R. (1996) "Bayesian Analysis of Uncertainty in Information", submitted for publication.

Mazzuchi, T. A., Soofi, E.S. and Soyer, R. (1997) "Computations and Applications of Maximum Entropy Dirichlet Procedure", submitted for publication.

Singpurwalla, N. (1996). "Entropy and Information in Reliability", *Bayesian Analysis in Statistics and Econometrics*, Edited by Donald A. Berry, M. Chaloner, and John K. Geweke, John Wiley.

Soofi, E.S. (1994), "Capturing the intangible concept of information," *J. Amer. Statist. Ass., 89*, 1243-1254.

Soofi, E.S., Ebrahimi, N. and Habibullah, M. (1995), "Information distinguishability and its application in failure data", *J. Amer. Statist. Ass., 90*, 657-668.

INFERENCE ABOUT SHARP CHANGES IN HAZARD RATES

JAYANTA K. GHOSH

Indian Statistical Institute, 203 B.T.Road,
Calcutta 700035, India
and
Department of Statistics, Purdue University,
West Lafayette, IN 47907

SHRIKANT N. JOSHI and CHIRANJIT MUKHOPADHYAY

Indian Statistical Institute, 8-th Mile Mysore Road,
Bangalore 560059, India

It is common for mechanical or biological systems to experience a high hazard rate early in their lifetime and then a constant or steady hazard rate after a threshold time τ. In this article we review the available methods for estimating τ, which has been described as a change-point in a hazard rate in the literature. Considerable amount of work has been done on the model assuming the hazard rate to be different constants before and after τ. Estimates of τ are also available without any specific algebraic assumption about the form of the hazard rate before τ. Semiparametric, likelihood based, and Bayesian estimates of τ are recapitulated. A new result regarding the boundedness of the likelihood function is proved. Some simulation is also done comparing the available estimates of τ, supplementing earlier work done by the authors.

1 Introduction

In studies on hazard rates it is often the case that initially the hazard rate is high and then after a rapid or abrupt fall, it stabilizes at a lower value. This phenomenon can arise in both survival analysis and reliability studies. If a treatment is given to patients and their "survival" up to a relapse or some other identified episode is studied, then patients who "survive" the initial shock of a new treatment like chemotherapy will develop low hazard rates. A similar situation in reliability problems has led to the so called "burn-in" techniques to screen out defective electrical or electronic items and thus improve the performance of the remaining items. An alternative "burn-in" technique is to subject the items to an electrical or thermal shock and sell only those which survive.

Suppose T is a random variable representing survival. In a study on time to relapse after remission induction for leukemia patients (under a particular therapy) Mathews and Farewell (1982) introduced a hazard rate change-point model in which the rate moves from a high value α to a low value β at the change-point τ. Mathematically,

$$h(t) = \begin{cases} \alpha & \text{if } 0 \leq t \leq \tau \\ \beta & \text{if } t > \tau \end{cases} \tag{1}$$

This differs from the usual change-point models where a parameter in the model may change over time *e.g.* if $T_1, \ldots, T_i$ are i.i.d. with parameters (α, β) and $T_{i+1}, \ldots, T_n$ are i.i.d. with parameters (α', β') then $(i+1)$ is a change-point.

Around the same time in unpublished work in 1982 modeling lifetime of switches bought by the Indian Space Agency, Basu and Ghosh introduced what they called a truncated bath-tub model in which there is initially, *i.e.* up to time τ, a non-increasing failure rate α which stabilizes at a lower value β from τ onwards. For a precise definition of this semiparametric interval see Basu, Ghosh and Joshi (1988), henceforth abbreviated as BGJ. A variant of this model is considered in Kulasekara and Saxena (1991). They propose a kernel estimate of the hazard rate and estimate a parameter like τ essentially by a point where the estimated hazard rate is minimum. An even more flexible non-parametric formulation due to Müller and Wang (1990) merely assumes that the study period is bounded and at some point τ in this bounded interval $|h'(x)|$ attains a strict maximum, $h'(x)$ being the derivative of the hazard rate. In this formulation τ is the point where the change is most rapid. For the set of data on leukemia patients reported in Müller and Wang (1990) this formulation seems preferable to either the parametric or the semiparametric formulation. For this data the hazard rate is high but not non-increasing before τ and after the main change-point τ there seems to be another second small change-point. However, we believe the estimates for τ appropriate for the parametric or semiparametric model will do well for this data also even though the model assumptions are not fully met. It would be interesting to have some numerical comparison of all the estimates for τ in these different models at least for the data of Müller and Wang (1990).

For the parametric change-point model (1) it is well-known that the maximum likelihood estimate for τ has problems. This is discussed in more detail in section 2. An interesting different estimate was proposed by Nguyen, Rogers and Walker (1984). Its asymptotic distribution is shown to be a split normal in BGJ. The method of proof is generalized in Ghosh and Joshi (1992). Applications include M-estimates under non-standard conditions. Modified maximum likelihood estimates are proposed in Yao (1986) as well as BGJ.

For the semiparametric model, BGJ provide two estimates of τ, which compare well with parametric estimates of τ even under (1). However BGJ assume that a conservative upper bound, p_0, to $F(\tau)$ will be available to the statistician. The value for p_0 is taken to be 0.5 in their simulations. This may be interpreted as requiring that not more than 50% of the items are bad in the sense of having an unacceptably large hazard rate. In practice the values of $F(\tau)$ would have to be quite small, at least in reliability problems, if producing these items is to be profitable. So $F(\tau) \leq 0.5$ is a weak assumption in such cases.

Our focus in this review will be on the likelihood function associated with (1) and inference based on likelihood alone or Bayesian analysis. In section 2 we discuss properties of the likelihood and prove what appears to be a new result, namely, that the likelihood is bounded if the usual constraint $\alpha \geq \beta$ is imposed. Section 2 also contains a review of maximum likelihood estimates. Bayesian methods for this problem are reviewed in section 3. Bayesian analysis for (1) was introduced in Ghosh, Joshi and Mukhopadhyay (1993), referred to below as GJM. Data based priors were introduced earlier by Achcar and Bolfarine (1989). In most of this exposition inference will usually mean estimation. Estimation of τ will be given more importance than that of α and β. It may be mentioned that estimating τ is a special case of estimation of the point of discontinuity in a density. A general asymptotic theory of this is given in Chernoff and Rubin (1956) and Ibragimov and

Has'minskii (1981).

2   The Likelihood Function and Likelihood Based Methods

Let $T_1, \ldots, T_n$ be i.i.d. with the common hazard given by (1). The corresponding density will be referred to as $f(t|\alpha, \beta, \tau)$. Clearly $\alpha > 0$, $\beta > 0$, $\tau > 0$. Unless otherwise stated $\alpha \geq \beta$. Let $0 < T_{(1)} < \cdots < T_{(n)} < \infty$ be the order statistics. The log-likelihood function may be written in the form

$$\log L(\alpha, \beta, \tau) = R(\tau) \log \alpha - \alpha Q(\tau) + (n - R(\tau)) \log \beta - \beta(S - Q(\tau)) \quad (2)$$

where, vide GJM,

$$\begin{array}{ll} R(\tau) = \sum_{i=1}^{n} I_{[T_i \leq \tau]} & M(\tau) = \sum_{i=1}^{n} T_i I_{[T_i \leq \tau]} \\ Q(\tau) = M(\tau) + (n - R(\tau))\tau & S = \sum_{i=1}^{n} T_i \end{array} \quad (3)$$

It is well-known that without the constraint $\alpha \geq \beta$, the likelihood is unbounded. This may be shown as indicated in Nguyen, Rogers and Walker (1984). Set any arbitrary positive value to α and set $\beta = (T_{(n)} - \tau)^{-1}$, where $T_{(n-1)} < \tau < T_{(n)}$. It is now easy to verify from (2) that the likelihood tends to infinity as $\tau \uparrow T_{(n)}$. The fact that $\beta \uparrow \infty$ in this argument suggests the likelihood may be bounded under the constraint $\alpha \geq \beta$. The following simple proposition shows this is indeed the case.

Proposition 1 *Under the constraint $\alpha \geq \beta$ the likelihood is bounded.*

Proof: First assume that $T_{(1)} \leq \tau \leq T_{(n-1)}$. Then using (2),

$$\log L(\alpha, \beta, \tau) \leq \{(n \log \alpha)^+ - \alpha T_{(1)}\} + \{(n \log \beta)^+ - \beta(T_{(n)} - T_{(n-1)})\}$$

which is bounded. Now consider the case $0 < \tau < T_{(1)}$. Then

$$\log L(\alpha, \beta, \tau) = n \log \beta - \beta \sum_{i=1}^{n} (T_i - \tau)^+ = n \log \beta - \beta(S - nT_{(1)})$$

which is also bounded. If $\tau \geq T_{(n)}$, then $\log L(\alpha, \beta, \tau) = n \log \alpha - S\alpha$ is bounded. Finally consider the case $T_{(n-1)} < \tau < T_{(n)}$. Then using $\alpha \geq \beta$,

$$\log L(\alpha, \beta, \tau) = (n-1) \log \alpha + \log \beta - \alpha(S - T_{(n)} + \tau) - \beta(T_{(n)} - \tau) \leq n \log \alpha - \alpha(S - T_{(n)} + \tau)$$

which is bounded since $S > T_{(n)}$. This completes the proof.

Note that the constraint $\alpha \geq \beta$ is used only for $T_{(n-1)} < \tau < T_{(n)}$.

Unfortunately even though the likelihood is bounded, problems remain. It is shown in GJM (Proposition 3.1) that the likelihood can have a local maximum far to the right of τ. For $\alpha = 1.0$, $\beta = 0.5$, $\tau = 2.0$ and $n = 25$, the probability of a local maximum at $T_{(n)}$ or $T_{(n-1)}$ is greater than 0.3. Simulations from $F(\cdot|3.0, 2.0, 0.15)$ show the same phenomenon when α, β are known.

The following additional argument indicates when the likelihood at $\tau = T_{(n)}$ may be equal or comparable to the maximized likelihood for other τ. For fixed

τ with $T_{(m)} \leq T_{(m+1)}$ the maximum likelihood estimates for α, β (without any constraint) are, vide GJM,

$$\hat{\alpha}(\tau) = \frac{m}{n} \frac{1}{\overline{U}_n(\tau)}, \qquad \hat{\beta}(\tau) = \frac{n-m}{n} \frac{1}{\overline{V}_n(\tau)}$$

where $n\overline{U}_n(\tau) = \sum_{i=1}^n (T_i I_{[T_i \leq \tau]} + \tau I_{[T_i > \tau]})$ and $n\overline{V}_n(\tau) = \sum_{i=1}^n (T_i - \tau)I_{[T_i > \tau]}$. Under the constraint $\alpha \geq \beta$, the constrained MLE for α, β (for fixed τ) will be as above if $\hat{\alpha}(\tau) \geq \hat{\beta}(\tau)$. If $\hat{\alpha}(\tau) < \hat{\beta}(\tau)$ then the concavity of log-likelihood implies the constrained maximum will be attained on $\alpha = \beta$. The resulting maximum likelihood estimate for $\alpha = \beta$ is $(S/n)^{-1}$. More important, the constrained maximum for fixed τ is the same as the maximum attained for $\tau = T_{(n)}$.

All these arguments indicate that the constraint $\alpha \geq \beta$ bounds the likelihood but does not prevent it from having a local maximum at or near $T_{(n)}$ of magnitude comparable with the global maximum. Finally, note that if the likelihood is maximized at $\tau = T_{(n)}$, then it is maximized at all $\tau \geq T_{(n)}$.

Thus to get sensible modifications of maximum likelihood estimates one must bound it above. The modified maximum likelihood estimates of τ of Yao (1986) and BGJ satisfy this condition. Yao (1986) restricts to $\tau \leq T_{(n)}$, when maximizing the likelihood. BGJ force τ to lie in a closed bounded interval and impose identifiability conditions $\tau \geq \delta_1 > 0$, $|\alpha - \beta| \geq \delta_2 > 0$. Yao (1986) and BGJ show their modified maximum likelihood estimates have the limiting distribution described in Chernoff and Rubin (1956) and Ibragimov and Has'minski (1981). This limiting distribution of $n(\hat{\tau} - \tau_0)$ is explicitly given in Yao (1986).

Yao (1986) also shows the maximum likelihood estimates of α, β are asymptotically normal and asymptotically independent of $n(\hat{\tau} - \tau_0)$. Analogous results within Bayesian context are discussed in the next section.

The semiparametric and modified maximum lieklihood estimate of BGJ are compared with the estimate of Nguyen, Rogers and Walker (1984) in BGJ. The semiparametric estimates do quite well - probably because they use the weak but extra information $F(\tau) \leq 0.5$. The asymptotic split normal distribution captures quite well the variance of the estimate of Nguyen, Rogers and Walker (1984). The modified maximum likelihood estimate, in spite of having the best norming constant n, does not do very well. It tends to have a high positive bias indicating the difficulties, which remain with the likelihood, even though the likelihood is maximized over a restricted set.

It is clear from the simulations that as expected the performance of an estimate is better the larger the amount of discontinuity $\alpha - \beta$ when α(or β) is held fixed.

A likelihood ratio test for the null hypothesis of no change point (i.e. $\tau = 0$ or $\alpha = \beta$) against the model (1) is proposed and studied via simulations in Matthews and Farewell (1982). A very elegant asymptotic theory of the test statistic is developed in Matthews, Farewell and Pyke (1985). An alternative test based on one of the semiparametric estimates of BGJ is proposed in Ghosh and Joshi (1993). They study its properties through simulation and asymptotics. The sample size n being fairly large, the asymptotic results are close to the simulated values.

An accurate approximation to the distribution of the test statistic of Matthews and Farewell (1982) is provided by Loader (1991). Loader (1991) also provides a method for constructing an approximate confidence interval for τ.

It may be mentioned that the likelihood being a function of the quantiles defined in (??) is right continuous as a function of τ and has jumps at $\tau = T_{(i)}$, $i = 1, \ldots, n$. As observed by Chernoff and Rubin (1956), this is typical whenever the parameter in question locates a discontinuity in the density. It is also worth mentioning that (??) may be expressed as a mixture of two densities, vide BGJ, but so far this fact has not found any use in inference. Another interesting fact, vide GJM (Property 5), is that if m is such that $T_{(m)} \leq T_{(m+1)}$, then given m, $(T_{(m+1)} - \tau), \ldots, (T_{(n)} - \tau)$ can be regarded as an ordered sample of size $(n - m)$ from an exponential density with failure rate β. This is often convenient in calculation.

3 Bayesian Analysis

Achcar and Bolfarine (1989) consider a data dependent discrete uniform prior for τ supported on the order statistics $T_{(1)}, \ldots, T_{(n)}$. No compelling justification for this choice is made. Given τ, non-informative priors are chosen for α, β. The data analyzed are censored, but censoring is easily handled in Bayesian analysis.

We discuss in more details the priors chosen by GJM. GJM propose an automatic Bayesian analysis based on non-informative priors, also called default or non-subjective priors. Common choices are the uniform distribution over a bounded or unbounded region, which goes back to Laplace, the Jeffreys' prior, or the probability matching priors studied by Tibshirani (1989), Ghosh and Mukherjee (1992) and others. An expository review is available in chapter 9 of Ghosh (1994).

Since appropriate regularity conditions do not hold for τ and the posterior for τ does not converge, vide Ghosh, Ghosal and Samanta (1994) and Ghosal, Ghosh and Samanta (1995), the only choice left from the above catalogue is the uniform distribution. A uniform distribution over a bounded τ-interval was chosen, because an unbounded interval would lead to an improper posterior. Note in this connection that the likelihood function is free of τ for $\tau > T_{(n)}$. Given τ GJM chose $\pi(\alpha, \beta | \tau) = \frac{1}{\alpha\beta}$, because α, β are close to being scale parameters. The joint improper prior becomes

$$\pi(\alpha, \beta, \tau) = \frac{1}{\alpha\beta} \quad 0 < a < \tau < b < \infty, \ 0 < \beta \leq \alpha < \infty$$

Here a is chosen > 0 to avoid non-identifiability. GJM (Proposition 4.3) point out that the (marginal) posterior mode of τ would always be 0 if $a = 0$. This is another reason for requiring $a > 0$. GJM (page 148) also show that one needs $0 < \beta_0 < \beta$ for the posterior to be proper. It is shown in GJM that the posterior is proper for

$$\pi(\alpha, \beta, \tau) = \frac{1}{\alpha\beta} \quad 0 < a < \tau < b < \infty, \ 0 < \beta_0 \leq \beta \leq \alpha < \infty \tag{4}$$

GJM also consider priors of the above form with an upper bound α_0 on α. They hoped this and the constraint $\alpha \geq \beta$ might prevent likelihood related problems to influence the posterior too much. (A typographical error in the posterior of τ when $\alpha \leq \alpha_0$ in GJM should be pointed out viz. a factor of $\alpha_0^j e^{-P(\tau)\alpha_0}$ is missing in the second term within the third bracket of the expression for $\pi'(\tau | \mathbf{D})$ for

$T_i \leq \tau < T_{i+1}$.) GJM made a detailed simulation study comparing the performance of Bayes estimates for different choices of α_0, β_0 and the semiparametric estimates of BGJ. It appears that α_0 has very little effect. The effect of β_0 is important, casting some doubt on the role of the prior chosen as a default prior. Of course α, β are not really scale parameters. Possibly a uniform prior should have also been tried. Another possibility that GJM tried but gave up because of computational difficulty was to take $\pi(\alpha, \beta | \tau)$ as the Jeffreys' or reference priors.

Simulation showed the posterior mode performs better than the posterior mean. The reason for this is that, as explained in section 2, the likelihood often has a local maximum far to the right of the true value of τ. This leads to a heavy tail for the marginal for τ. See for example the plots of posteriors for three priors for the same data in GJM (Figures 4, 5, 6). Note that the true value of τ was 0.15. Compare these plots with the plots of the log-likelihood function, also for the same data, in Figure 3 of GJM. In such situations the posterior mean would often be much bigger than the true value of τ. The posterior mode had smaller bias and mean square but the posterior median might have been better still. GJM did not do any calculations with the posterior median due to computational difficulties.

A somewhat surprising fact was that the Bayes estimates were not doing as well as the semiparametric estimates of BGJ, which did not use the knowledge about the model but had the information $F(\tau) \leq p_0$. Unfortunately in retrospect it appears the choice of p_0 in the simulations was not satisfactory. Some of the α, β, τ chosen for simulations were such that $F(\tau)$ was greater than $p_0 = 0.5$. So somewhat routinely p_0 was taken too close to the actual value of $P(T \leq \tau | \alpha, \beta, \tau)$, so that p_0 was not conservative. A better alternative would have been to screen out such α, β, τ or choose a conservative p_0 suitable to such α, β, τ. To cope with the heavy tail, the upper bound b of τ was chosen in a data-dependent manner as $(T_{(n-1)} + T_{(n)})/2$. But truncation at $(T_{(n-1)} + T_{(n)})/2$ did not help much.

So a fresh set of simulation is done comparing the two estimates proposed by BGJ, a "smoothed" version of the second estimate of BGJ (this is the estimate considered by BGJ in their simulation study, while GJM used the definition of the estimate), and the posterior mean and the (marginal) posterior mode of τ based on the prior given in (4) together with using the extra knowledge $F(\tau) \leq p_0 = 0.5$. Thus let,

$$\hat{\tau}_{BGJ1} = \text{Inf}\left\{ t > 0 : y_n(t + h_n) - y_n(t) \leq h_n \hat{\lambda}_0 + \epsilon_n \right\}$$

$$\hat{\tau}_{BGJ2} = \text{Inf}\left\{ t > 0 : -y_n(t) - \log(1 - p_0) \leq \hat{\lambda}_0(\hat{\xi}_{p_0} - t) + \epsilon_n \right\}$$

$$\hat{\tau}_{BGJ2}^{(s)} = \begin{cases} \text{Inf}\left\{ t > 0 : -y_n(t) + \frac{1}{k}\sum y_n(T_{(i)}) \leq \hat{\lambda}_0\left(\frac{1}{k}\sum T_{(i)} - t\right) + \epsilon_n \right\} & \text{if} \leq \hat{\xi}_{p_0} \\ \hat{\xi}_{p_0} & \text{otherwise} \end{cases}$$

where $y_n(t) = -\log \overline{F}_n(t)$, $\overline{F}_n(t)$ being the empirical survival function, $\hat{\lambda}_0$ is a least-square estimate of the steady-state hazard rate λ_0 based on the k order statistics $T_{([np_0]+1)}, \ldots, T_{([np_1])}$ $(p_0 < p_1 < 1)$ and the corresponding $y_n(\cdot)$ values, $\hat{\xi}_{p_0}$ is the p_0-th sample quantile, $h_n = n^{-1/4}$, $\epsilon_n = cn^{-1/2}\log n$, and the range of summation in the formula of $\hat{\tau}_{BGJ2}^{(s)}$ is from $[np_0] + 1$ to $[np_1]$. Also let $\hat{\tau}_1$ and $\hat{\tau}_2$ denote the posterior mode and mean of τ based on the prior in (4). The upper-bound b of τ is

Table 1: Mean (MSE) Comparing Different Estimates of τ

(α, β, τ)	$\hat{\tau}_{BGJ1}$	$\hat{\tau}_{BGJ2}$	$\hat{\tau}^{(s)}_{BGJ2}$	$\hat{\tau}_1$	$\hat{\tau}_2$
$(3, 2.0, 0.10)$	0.078140 (0.012753)	0.057305 (0.006746)	0.093246 (0.011817)	0.102316 (0.001961)	0.137340 (0.002129)
$(3, 2.0, 0.15)$	0.104914 (0.015100)	0.077579 (0.009921)	0.111559 (0.011182)	0.124727 (0.002981)	0.146290 (0.000555)
$(3, 2.0, 0.20)$	0.111652 (0.018660)	0.095320 (0.015337)	0.116434 (0.014677)	0.151055 (0.004419)	0.153734 (0.002565)
$(3, 1.0, 0.10)$	0.088946 (0.009513)	0.083324 (0.008579)	0.130174 (0.026171)	0.100355 (0.001143)	0.122947 (0.001527)
$(3, 1.0, 0.15)$	0.123784 (0.005938)	0.108089 (0.005647)	0.141982 (0.014854)	0.145499 (0.001257)	0.157840 (0.000774)
$(3, 1.0, 0.20)$	0.172275 (0.006420)	0.148868 (0.004280)	0.170287 (0.006243)	0.184991 (0.001023)	0.188625 (0.000637)
$(2, 1.0, 0.10)$	0.066162 (0.012013)	0.075726 (0.014784)	0.137726 (0.039181)	0.118568 (0.005721)	0.188821 (0.011579)
$(2, 1.0, 0.15)$	0.105740 (0.013299)	0.094265 (0.013921)	0.159268 (0.031041)	0.143342 (0.003090)	0.191047 (0.003816)
$(2, 1.0, 0.20)$	0.153471 (0.014977)	0.128155 (0.014681)	0.170236 (0.022642)	0.201996 (0.004524)	0.231774 (0.003046)
$(2, 0.5, 0.10)$	0.078410 (0.006312)	0.142973 (0.051890)	0.287965 (0.194533)	0.106296 (0.003670)	0.156138 (0.009588)
$(2, 0.5, 0.15)$	0.113983 (0.004772)	0.145358 (0.034068)	0.256493 (0.118936)	0.149228 (0.000833)	0.173840 (0.002145)
$(2, 0.5, 0.20)$	0.169569 (0.009478)	0.165469 (0.018590)	0.256829 (0.067213)	0.191345 (0.000978)	0.209948 (0.001100)

taken to be minimum of $\left\{\hat{\xi}_{p_0}, c_0/\beta_0\right\}$, where $c_0 = \log(1/(1-p_0))$. c_0/β_0 arises from incorporating the information that $F(\tau) \leq p_0$, while b is sharpened further with $\hat{\xi}_{p_0}$ in the same spirit as $\hat{\tau}^{(s)}_{BGJ2}$. This truncation is sharper than $(T_{(n-1)} + T_{(n)})/2$ considered by GJM. Note that the knowledge of $F(\tau) \leq p_0$ is further used in the computation of $\hat{\tau}_1$ and $\hat{\tau}_2$ in the range of integration for α, which is from β to c_0/τ.

For each combination of parameter values (α, β, τ) 1000 samples of size $n = 100$ are drawn and for each sample $\hat{\tau}_{BGJ1}$, $\hat{\tau}_{BGJ2}$, $\hat{\tau}^{(s)}_{BGJ2}$, $\hat{\tau}_1$, and $\hat{\tau}_2$ are computed with $h_n = n^{-1/4}$, $\epsilon_n = 0.05$, $p_0 = 0.5$, $p_1 = 0.9$, $a = 0.05$, $\beta_0 = 0.05$. The upper-bound b of τ has already been discussed in the preceding paragraph. Table 1 gives the mean and the MSE (in parentheses) of these estimates for different values of α, β, τ, computed across 1000 simulations.

There has been some theoretical study of the posterior distribution of (α, β, τ). Ghosh, Joshi and Mukhopadhyay (1996) have shown that the posterior of α, β approach a normal distribution and asymptotically $\sqrt{n}(\alpha - \hat{\alpha}_0)$ and $\sqrt{n}(\beta - \hat{\beta}_0)$ are independent of $n(\tau - \tau_0)$, where τ_0 is the true unknown value of τ. They also provide approximations for the posterior mean and posterior variance of α, β. Ghosh, Ghosal, and Samanta (1994) and Ghosal, Ghosh, and Samanta (1995) have obtained a general necessary and sufficient condition for a suitably normed posterior to converge to a limiting ditribution. They show this general result implies the posterior distribution of $n(\tau - \tau_0)$ cannot converge to a general limiting distribution under τ_0. For a precise statement see the cited references.

In a recent thesis Zhang (1996) considers a sequential version of estimating τ. Fix a threshold ϵ and continue sampling until the posterior variance is less than or equal to ϵ for the first time. The behavior of the expected stopping time as $\epsilon \downarrow 0$ is interesting. Using results from classical sequential asymptotics, Zhang finds asymptotic upper bounds for the expected stopping time but simulations show this is quite conservative. Under certain restrictions on the stopping times, a rate of convergence to infinity for the expected stopping time is obtained but the rate is much faster than what appears in the simulations. The corresponding study for α, β is much simpler. Using the convergence of posteriors for α, β Zhang (1996) finds the exact rate of convergence.

Zhang (1996) also considers a model introduced in unpublished work of Basu and Ghosh where the hazard rate decreases from α to β in a linear way rather than by a jump. He also considers non-parametric Bayesian methods.

4 Concluding Remarks

The two most pressing Bayesian problems are study of Bayes estimates for other priors and a better way of dealing with the heavy-tailed likelihood. A plausible alternative prior is the uniform distribution for α, β (over $0 < \beta \le \alpha < \infty$ or $0 < \beta \le \alpha < \alpha_0$) and uniform or some other prior for τ. Using a posterior median (trimmed mean) may make the tail of the likelihood exert less influence on the inference for moderate sample sizes. Of course for sufficiently large samples the posterior for τ puts most of the mass close to the true value of τ. In fact given $T_1, \ldots, T_n$, $n(\tau - \tau_0)$ is $O_p(1)$, except for $(T_1, \ldots, T_n)$ in a set of small probability vide Ghosh, Joshi and Mukhopadhyay (1996) for a precise statement.

In analysis via maximum likelihood estimates, the various modified maximum likelihood estimates need to be compared with each other and other frequentist estimates. At present in the light of the new simulations, we recommend the Bayesian estimates with the prior given in (??) above, with the upper bound of τ at the minimum of the sample median (since it is assumed that $F(\tau) \le p_0 = 0.5$) and $0.69314/\beta_0$ (β_0 being the lower bound of β, this is the natural upper bound for τ from $F(\tau) \le p_0 = 0.5$), and an upper bound for α (from $F(\tau) \le p_0 = 0.5$) at a given value of τ to be $0.69314/\tau$.

Finally, the likelihood function and the model need further study. For example it would be nice to identify easily calculated diagnostic tools which can show if a given sample is likely to have a maximum far from the true value of τ. Similarly one would want to know which combinations of α, β, τ lead to high risk for estimates.

References

Achcar J.A. and Bolfarine H. (1989). "Constant Hazard Against a Change-Point Alternative: A Bayesian Approach with Censored Data", *Communications in Statistics - Theory and Methods*, **18**, pp. 3801-3819 .

Basu A.P., Ghosh J.K. and Joshi S.N. (1988). "On Estimating Change-Point in a Hazard Rate", *Statistical Decision Theory and Related Topics,* (S.S.Gupta and J.O.Berger eds.) Springer-Verlag, New York, pp. 239-252.

Berger J.O. and Bernardo J.M. (1992). "On the Development of Reference Prior Method", Bayesian Statistics Vol. 4. (J.M.Bernardo, J.O.Berger, A.P.Dawid and A.F.M. Smith eds.) Oxford University Press, (with discussions) pp. 35-60.

Chernoff H. and Rubin H. (1956). "The Estimation of the Location of a Discontinuity in Density", *Proceedings of the Third Berkeley Symposium on Mathematical Statistics and Probability,* University of California Press, pp. 19-37.

Ghosal S., Ghosh J.K. and Samanta T. (1995). "On Convergence of Posterior Distributions", *Annals of Statistics,* **23**, pp. 2145-2152.

Ghosh J.K. (1994). *Higher Order Asymptotics,* IMS and ASA, Hayward, California.

Ghosh J.K., Ghosal S. and Samanta T. (1994). "Stability and Convergence of Posterior in Non-Regular Problems", *Statistical Decision Theory and Related Topics.* (S.S.Gupta and J.O.Berger eds.) Springer-Verlag, pp. 183-201.

Ghosh J.K. and Joshi S.N. (1992). "Asymptotic Distribution of an Estimate of the Change-Point in a Hazard Rate", *Communications in Statistics - Theory and Methods,* **21** pp. 3571-3588.

Ghosh J.K. and Joshi S.N. (1993). "On Asymptotic Distribution of Generalized M-Estimates", *Sankhya Series A,* **55**, pp. 312-320.

Ghosh J.K. and Mukherjee R. (1992). "Non-informative Priors", Bayesian Statistics vol. 4. (J.M.Bernardo, J.O.Berger, A.P.Dawid and A.F.M. Smith eds.) Oxford University Press, (with discussions) pp. 195-210.

Ghosh J.K., Joshi S.N. and Mukhopadhyay C. (1993). "A Bayesian Approach to the Estimation of the Change-Point in a Hazard Rate", *Advances in Reliability,* (A.P.Basu ed.) Elsevier Science Publishers, North-Holland, pp. 141-170.

Ghosh J.K., Joshi S.N. and Mukhopadhyay C. (1996). "Asymptotics of a Bayesian Approach to the Estimation of the Change-Point in a Hazard Rate", *Communications in Statistics - Theory and Methods,* **25**, pp. 3147-3166.

Ibragimov I.A. and Has'minskii R.Z. (1981). *Statistical Estimation Asymptotic Theory,* Springer-Verlag, New York.

Kulasekara K.B. and Lal Saxena K.M. (1991). "Estimation of Change-Point in Failure Rate Models", *Journal of Statistical Planning and Inference,* **29**, pp. 111-124.

Loader C.R. (1991). "Inference for a Hazard Rate Change Point", *Biometrika,* **78**, pp. 749-757.

Matthews D.E. and Farewell V.T. (1982). "On Testing for a Constant Hazard against a Change-Point Alternative", *Biometrics,* **38**, pp. 463-468.

Matthews D.E., Farewell V.T. and Pyke R. (1985). "Asymptotic Score-Statistics Process and Tests for Constant Hazard against a Change-Point Alternative,", *An-*

nals of Statistics, **13**, pp. 583-591.

Müller H.G. and Wang J.L. (1990). "Nonparametric Analysis of Chanages in Hazard Rates for Censored Survival Data: An Alternative to Change-Point Models", *Biometrika,* **77**, pp. 305-314.

Nguyen H.T., Rogers G.S. and Walker E.A. (1984). "Estimation in Change-Point Hazard Rate Models", *Biometrika,* **71**, pp. 299-304.

Tibshirani R.J. (1989). "Non-informative Priors for One Parameter of Many", *Biometrika,* **76** pp. 604-608.

Yao Y.C. (1986). "Maximum Likelihood Estimation in Hazard Rate Models with a Change-Point", *Communications in Statistics - Theory and Methods,* **15** pp. 2455-2466.

Zhang P.H. (1996). "Contributions to Bayesian Survival Analysis", *Ph.D. Thesis,* Department of Statistics, Purdue University.

RECENT DEVELOPMENTS OF BAYESIAN INFERENCE FOR STRESS-STRENGTH MODELS

MALAY GHOSH

Department of Statistics
University of Florida-Gainsville
223 Griffin-Floyd Hall
Gainsville, FL 32611-8545, USA
E-mail: ghoshm@stat.ufl.edu

DONGCHU SUN

Department of Statistics
University of Missouri-Columbia
323 Math Science Building
Columbia, MO 65211, USA
E-mail: dsun@stat.missouri.edu

The problem of estimating the reliability of a stress-strength system, where a system fails at the moment the applied stress is greater than the strength, dates from Mann and Whitney's pioneering work in the 1940's. The Bayesian approach, especially using noninformative priors, did not receive much attention until recently. This paper gives a review on the development of Bayesian parametric and nonparametric methods for stress-strength models. In particular, Bayesian credible intervals using matching and reference priors are compared for normal, exponential and Weibull stress-strength models. Various examples are used to illustrate the results. Finally, a nonparametric Bayesian approach and empirical Bayes analysis are also considered.

1 Introduction

Consider the following stress-strength system (SSS), where Y, the strength of the system, is subject to stress X. The system fails at any moment the applied stress (or load) is greater than the strength (or resistance). Reliability of the system is then given by

$$\theta = P(X \leq Y). \tag{1}$$

This model has found applications in many areas, especially in structural and aircraft industries. For example, a solid propellant rocket engine is successfully fired provided the chamber pressure X generated by ignition stays below the burst pressure Y of the rocket chamber. When $X > Y$, the engine blows up, and the operation is a failure. As a second example, Kececioglu (1972) discusses a case where a torsion stress is the most critical type of stress for a rotating steal shaft on a computer. The quantity θ could be of interest in other areas of application as well, for example in comparing two drugs, fertilizers, chemical compounds etc..

The stress-strength model was first considered by Birnbaum (1956). The nonparametric UMVUE of θ is the Mann-Whitney U−statistic (cf. Mann and Whitney (1947), Lehmann (1951)). Distribution free upper confidence bounds for θ are given in Birnbaum (1956) and Birnbaum and McCarty (1958). Similar results are given in Owen, Craswell and Hansen (1964). Two sided asymptotic confidence bounds for θ are given in Sen (1967) and Govindarajulu (1968). Normal theory confidence bounds for θ are given in Owen *et al.* (1964) and Govindarajulu (1967). The UMVUE of θ in the normal case is given in Downton (1973). Comprehensive reviews of frequentist inference for stress-strength models are given by Basu (1985) and Johnson (1988), although Johnson's review contains a small Bayesian component.

The objective of the review is to provide instead the Bayesian inference for stress-strength models. The topic began with Enis and Geisser (1971). Their analysis is provided for the exponential and normal distributions. More recently, Basu and Tarmast (1987), and Thompson and Basu (1993) have considered the Bayesian analysis of exponential distributions. Ghosh and Yang (1996), Lee (1996), and Lee, Sun and Basu (1997) have provided credible intervals for θ for the normal distributions. Credible intervals for θ for the Weibull model are given in Sun, Ghosh and Basu (1996).

Nonparametric Bayesian analysis of θ began with Ferguson (1973). Empirical Bayes analysis in the same framework was considered by Hollander and Korwar (1976), and Ghosh and Lahiri (1992).

The present article reviews some of the above work. The outline of the remaining sections is as follows. Section 2 discusses the normal model. Section 3 considers the exponential model and the Weibull model. Some numerical examples are also given for illustration. Finally, nonparametric Bayesian inference and empirical Bayes analysis for θ are investigated in Section 4.

2 The Normal Case

2.1 Equal Variance Case

We begin with the two sample normal problem when $X_1, \cdots, X_m$, and $Y_1, \cdots, Y_n$ are independent random samples from $N(\mu_1, \sigma^2)$ and $N(\mu_2, \sigma^2)$ distributions, the parameters μ_1, μ_2 and σ being all unknown. Then

$$\theta = \Phi[-(\mu_1 - \mu_2)/(\sqrt{2}\sigma)],$$

which is one-to-one with $(\mu_1 - \mu_2)/\sigma = \eta$ (say). In the above, and in what follows, Φ is the standard normal distribution function.

Enis and Geisser (1971) considered the prior

$$\pi_J(\mu_1, \mu_2, \sigma) \propto \sigma^{-1}$$

in order to find the posterior of η. Writing $\bar{x} = m^{-1}\sum_{i=1}^{m} x_i, \bar{y} = n^{-1}\sum_{j=1}^{n} y_j$ and $s^2 = \sum_{i=1}^{m}(x_i - \bar{x})^2 + \sum_{j=1}^{n}(y_j - \bar{y})^2$, the posterior of η is given by

$$\pi(\eta|\bar{x}, \bar{y}, s) \propto \int_0^\infty v^{m+n-3} \exp\left[-\frac{1}{2}\left\{v^2 + \frac{mn}{m+n}(vz - \eta)^2\right\}\right] dv, \qquad (2)$$

where $z = (\bar{x} - \bar{y})/s$. Since this posterior is analytically intractable, so is the posterior of θ.

The above prior is based on Jeffreys' recommendation, although it is different from Jeffreys' general rule prior which is the positive square root of the determinant of the Fisher information matrix, and in this case is given by $\pi_*(\mu_1, \mu_2, \sigma) \propto \sigma^{-3}$. Despite the general appeal of the prior $\pi_J(\mu_1, \mu_2, \sigma) \propto \sigma^{-1}$, we shall see now that this prior fails to avoid the marginalization paradox (cf. Stone and Dawid (1972), Dawid, Stone and Zidek (1973)) in this example.

The marginalization paradox occurs when the posterior distribution of a parameter under a certain prior depends on the data only through a certain statistic, and yet this posterior cannot be obtained under *any* prior when one begins with the sampling distribution of the said statistic. A fundamental explanation of this phenomenon as given for example in Robert (1994, p. 131) is that an improper prior, say $\pi(\xi_1, \xi_2) \propto g(\xi_2)$ does not induce the marginal $\pi_2(\xi_2) \propto g(\xi_2)$. A much deeper discussion appears in Dawid *et al.* (1973).

To see the marginalization paradox in this example, first notice that the posterior given in (2) depends on the minimal sufficient statistic $(\overline{X}, \overline{Y}, S)$ only through $Z = (\overline{X} - \overline{Y})/S$. The marginal pdf of Z in this example is given by

$$f(z|\mu_1, \mu_2, \sigma) \propto \int_0^\infty v^{m+n-2} \exp\left[-\frac{1}{2}\left\{v^2 + \frac{mn}{m+n}(vz - \eta)^2\right\}\right] dv. \tag{3}$$

Thus there does not exist any prior for (μ_1, μ_2, σ) based on the likelihood given in (3) which can produce the posterior given in (2).

It is easy to fix the marginalization paradox in this example. For example, any prior $\pi(\mu_1, \mu_2, \sigma) \propto \sigma^{-2} h(\eta)$ avoids the paradox. Dawid *et al.* (1973) recommended the choice $h(\eta) = 1$.

We now develop another class of priors for this problem as given for example in Ghosh and Yang (1996), Lee (1996), and Lee, Sun and Basu (1997). These priors, usually referred to as "probability matching priors" were first developed by Welch and Peers (1963), and Peers (1965). The recent population of such priors owes much to Stein (1985), Tibshirani (1989), Ghosh and Mukerjee (1992), Datta and Ghosh (1995a), and Datta (1996) among others. According to this criterion, for a scalar parameter of interest, one matches asymptotically the coverage probabilities of one-sided Bayesian credible intervals with the corresponding frequentist probabilities. When this matching is achieved via posterior quantiles, and one requires a remainder term which is $o_p(n^{-\frac{1}{2}})$, where n is the sample size, these matching priors are usually referred to as 'first order' probability matching priors, and are found as solutions of certain differential equations given for example in Welch and Peers (1963), Stein (1985), and Datta and Ghosh (1995a). Ghosh and Yang (1996) have shown (see their (2.11)) that for the given problem, writing $m = \rho n$, a class of first order matching priors is given by

$$\pi_{(c)}(\mu_1, \mu_2, \sigma) \propto \sigma^{2c}[2(1 + \rho^{-1})(1 + \rho) + (\mu_1 - \mu_2)^2 \sigma^{-2}]^{c+\frac{1}{2}}, \tag{4}$$

where c is a constant. Lee, Sun and Basu (1997) showed that the general class of first order probability matching priors for θ is given by

$$\pi(\mu_1, \mu_2, \sigma) \propto \sigma^{-1} g[2(1 + \rho^{-1})(1 + \rho)\sigma^2 + (\mu_1 - \mu_2)^2], \tag{5}$$

where g is an arbitrary function. Clearly (4) is a special case of (5) when $g(t) = t^{c+\frac{1}{2}}$.

Ghosh and Yang (1996) also showed that the only choice of c in the class (4) which avoids the marginalization paradox is $c = -1$. Thus, Jeffreys' prior $\pi_J(\mu_1, \mu_2, \sigma) \propto \sigma^{-1}$ obtained by putting $c = -1/2$ in (4) is a first order matching prior, but fails to avoid the marginalization paradox. Jeffreys' general rule prior $\pi_*(\mu_1, \mu_2, \sigma) \propto \sigma^{-3}$ is neither a first order matching prior, nor avoids the marginalization paradox. Ghosh and Yang (1996) have also shown that the entire class of priors given in (4) leads to proper posteriors. For the particular choice $c = -1/2$, Enis and Geisser (1971) have given expressions for the posterior mean of θ, and credible intervals for θ which require numerical integration.

A general class of noninformative priors called "reference priors" was proposed by Bernardo (1979), and were subsequently extended in a series of articles by Berger and Bernardo (1989, 1992a, 1992b). Ghosh and Yang (1996) provided a complete catalog of reference priors when the parameter of interest is θ or η $(= (\mu_1 - \mu_2)/\sigma)$. The following table lists the reference priors. For a general algorithm of finding reference priors, we refer to Berger and Bernardo (1992a).

Table 1.   Catalog of reference priors for Normal models

Grouped parameters in their order of importance	Priors distributions
$\{\eta, \mu_2, \sigma\}$	$\pi_*(\eta, \mu_2, \sigma) \propto \sigma^{-2}$ (Jeffreys' general rule prior)
$\{\mu_2, \sigma\}, \{\eta\}$	$\pi_*(\eta, \mu_2, \sigma) \propto \sigma^{-2}$
$\{\eta\}, \{\mu_2\}, \{\sigma\}$	$\pi_1(\eta, \mu_2, \sigma) \propto \sigma^{-1}[2(1 + \rho^{-1})(1 + \rho) + \eta^2]^{-\frac{1}{2}}$ (one at a time Reference Prior)
$\{\eta\}, \{\sigma\}, \{\mu_2\}$	$\pi_1(\eta, \mu_2, \sigma) \propto \sigma^{-1}[2(1 + \rho^{-1})(1 + \rho) + \eta^2]^{-\frac{1}{2}}$ (one at a time Reference Prior)
$\{\eta\}, \{\mu_2, \sigma\}$	$\pi_2(\eta, \mu_2, \sigma) \propto \sigma^{-2}[2(1 + \rho^{-1})(1 + \rho) + \eta^2]^{-\frac{1}{2}}$

Since $\pi_{(c)}(\eta, \mu_2, \sigma) \propto \sigma^{2c+1}[2(1 + \rho^{-1})(1 + \rho) + \eta^2]^{c+\frac{1}{2}}$ from (4), it follows from the above table that neither Jeffreys' general rule prior nor the two-group reference prior of Bernardo meets either the probability matching criterion or avoids the marginalization paradox. The one-at-a-time reference prior of Berger and Bernardo (irrespective of the ordering of μ_2 and σ) meets the probability matching criterion and avoids the marginalization paradox. Thus, the one-at-a-time reference prior is the recommended choice in this example.

Lee (1996) did some numerical studies to see the performance of the four priors π_*, π_1, π_2 and π_J, when sample sizes are small. Based on these four priors, he computed the frequentist coverage probabilities of α-posterior quantiles of θ. Table 2 gives numerical values when $\alpha = 0.05$ and 0.95. The computations are based on random samples from two normal distributions when $(\mu_1, \mu_2, \sigma) = (0, 0, 1)$. It is seen that the two matching priors π_1 and π_J perform better than Jeffreys' priors π_* and a two group reference prior π_2, with π_J being the best among them.

Table 2. Frequentist coverage probabilities of 0.05 (0.95) posterior quantiles of θ for normal models with equal variance.

m	n	π_*		π_2		π_1		π_J	
		.05	.95	.05	.95	.05	.95	.05	.95
2	2	.128	.871	.113	.885	.072	.924	.034	.963
3	3	.089	.902	.080	.913	.061	.934	.046	.948
5	5	.072	.929	.067	.934	.055	.945	.049	.951
10	10	.061	.943	.059	.945	.053	.949	.051	.951

2.2 Unequal Variances

We now examine the two sample normal problem when $X_1, \cdots, X_m$, and $Y_1, \cdots, Y_n$ are independent random samples from $N(\mu_1, \sigma_1^2)$ and $N(\mu_2, \sigma_2^2)$ distributions, the parameters $\mu_1, \mu_2, \sigma_1, \sigma_2$ being all unknown. Then

$$\theta = \Phi\left[-(\mu_1 - \mu_2)\Big/\sqrt{\sigma_1^2 + \sigma_2^2}\right],$$

which is one-to-one with $(\mu_1 - \mu_2)/\sqrt{\sigma_1^2 + \sigma_2^2} = \eta$ (say). Jeffreys' general rule prior is of the form $\pi(\mu_1, \mu_2, \sigma_1, \sigma_2) \propto 1/(\sigma_1^2 \sigma_2^2)$.

Lee (1996) found various reference priors and a first order probability matching prior when $m = n$. Using the reparameterization

$$\left(\eta, \ \xi = \frac{\mu_1 + \mu_2}{\sqrt{\sigma_1^2 + \sigma_2^2}}, \ \omega_1 = \sqrt{\sigma_1^2 + \sigma_2^2}, \ \omega_2 = \frac{\sigma_2}{\sqrt{\sigma_1^2 + \sigma_2^2}}\right),$$

the reference prior and the reverse reference prior, with grouped parameters $\{\theta\}, \{\xi, \omega_1, \omega_2\}$ in their order of importance, are identical and given by

$$\pi(\eta, \xi, \omega_1, \omega_2) \propto \frac{[2 + \eta^2 - 2\eta^2\omega_2^2(1 - \omega_2^2)]^{\frac{1}{2}}}{(\eta^2 + 2)\omega_1\omega_2^2(1 - \omega_2^2)^{3/2}}. \tag{6}$$

It is interesting to note that (η, ω_2), ξ, and ω_1 are independent, while the marginal prior for ξ is a constant and the marginal prior for ω_1 is proportional to $1/\omega_1$. Transforming back to original parameters $(\mu_1, \mu_2, \sigma_1, \sigma_2)$, the resulting reference prior is then

$$\pi(\mu_1, \mu_2, \sigma_1, \sigma_2) \propto \frac{\sqrt{2(\sigma_1^2 + \sigma_2^2)^3 + (\mu_1 - \mu_2)^2(\sigma_1^4 + \sigma_2^4)}}{\sigma_1^2 \sigma_2^2 \sqrt{\sigma_1^2 + \sigma_2^2} \, [2(\sigma_1^2 + \sigma_2^2) + (\mu_1 - \mu_2)^2]}.$$

Lee (1996) showed that neither Jeffreys' prior, nor the reference prior (6) is a first order probability matching prior for η. Lee also found the following first order probability matching prior

$$\pi(\mu_1, \mu_2, \sigma_1, \sigma_2) \propto (\sigma_1^3 \sigma_2^3)^{-1} \sqrt{2(\sigma_1^2 + \sigma_2^2)^3 + (\mu_1 - \mu_2)^2(\sigma_1^4 + \sigma_2^4)}$$

when θ is the parameter of interest.

3   Weibull and Exponential Models

A commonly used distribution for analyzing lifetime data is the Weibull distribution with pdf

$$f(x|\eta, \beta) = \frac{\beta x^{\beta-1}}{\eta^\beta} \exp(-(x/\eta)^\beta) 1_{[x>0]}, \tag{7}$$

where 1 is the usual indication function, $\eta > 0$ and $\beta > 0$. Such a distribution will be denoted by $W(\mu, \beta)$. When $\beta = 1$, the distribution is exponential. The Weibull distribution generalizes the exponential distribution by allowing increasing or decreasing failure rates. Assume that the stress $X \sim W(\eta_1, \beta)$ and $Y \sim W(\eta_2, \beta)$. If β is known, X^β and Y^β are independent exponential with scale parameters η_1 and η_2, and $\theta = P(X \leq Y) = P(X^\beta \leq Y^\beta) = \eta_2/(\eta_1 + \eta_2)$, and the problem reduces to inference for θ based on two independent exponentials with scale parameters η_1 and η_2.

3.1   Noninformative Priors for Exponential Models

We first consider the case where β is known, and without loss of generality is taken as 1. A conjugate prior for this case as given in Enis and Geisser (1971) is

$$\pi(\eta_1, \eta_2) \propto \prod_{i=1}^{2} [\eta_i^{-(b_i+1)} \exp(-a_i/\eta_i)], \tag{8}$$

which are independent inverse gammas, and are written as IG(a_1,b_1) and IG(a_2,b_2) respectively. Enis and Geisser (1971) have found an exact expression for the posterior mean of θ in this case. (See (3.14) of Enis and Geisser). The marginal posterior of θ is given by

$$\pi(\theta|t_1, t_2) \propto \theta^{m+b_1-1}(1-\theta)^{n+b_2-1}(1-c\theta)^{-(m+n+b_1+b_2)}, \tag{9}$$

where $t_1 = \sum_{i=1}^{m} x_i, t_2 = \sum_{j=1}^{n} y_j$ and $c = 1-(a_1+t_1)/(a_2+t_2) < 1$. This is used to find $E(\theta|t_1, t_2)$ as the ratio of two integrals. To find a credible interval for θ use the one-to-one transformation $r = (1-\theta)/(1-c\theta)$. Then r has the Beta $(n+b_2, m+b_1)$ posterior. Thus, credible intervals for r (or θ) can be easily constructed using incomplete beta integrals.

As a limiting case of the above prior, Enis and Geisser (1971) considered the prior $p(\eta_1, \eta_2) \propto (\eta_1 \eta_2)^{-1}$. This is also Jeffreys' general rule prior in this case. The information matrix is $I(\eta_1, \eta_2) = \text{Diag}(m\eta_1^{-2}, n\eta_2^{-2})$ so that $|I(\eta_1, \eta_2)|^{1/2} \propto (\eta_1 \eta_2)^{-1}$.

We shall now point out an optimality of this prior according to the probability matching criterion. To this end, first notice that θ is a one-to-one function of $\phi = \eta_2/\eta_1$ so that it suffices to find the probability matching prior for ϕ. We make the one-to-one transformation from (η_1, η_2) to (ϕ, ψ), where $\psi = \eta_1^{(m+n)/(2n)} \eta_2^{(m+n)/(2m)}$. The resulting Fisher information matrix is $I(\phi, \psi) = \text{Diag} \left(\frac{mn}{(m+n)\phi^2}, \frac{4m^2n^2}{(m+n)^3\psi^2} \right)$. With the notation $m = \rho n$, as in the previous section,

$$I(\phi, \psi) = n\text{Diag}\left(\frac{\rho}{(1+\rho)\phi^2}, \frac{4\rho^2}{(1+\rho)^3\psi^2} \right) = n\text{Diag}(a_1, a_2) \text{ (say)}. \tag{10}$$

Hence, following Tibshirani (1989), the class of first order probability matching priors in characterized by

$$\pi(\phi, \psi) \propto \phi^{-1} g(\psi), \tag{11}$$

where g is an arbitrary function.

Clearly, there are infinitely many first order matching priors, since g is arbitrary. It is possible to narrow down the selection of priors within this class by requiring second order matching as given in Mukerjee and Dey (1993) or Sun and Ye (1996). For second order matching one requires the remainder term to be $o_p(n^{-1})$ rather than $o_p(n^{-\frac{1}{2}})$. Following Mukerjee and Dey (1993), in this example, a first order matching prior is also second order matching if and only if $g(\psi)$ given in (11) satisfies the differential equation

$$\frac{1}{6}g(\psi)\frac{\partial}{\partial\phi}(a_1^{-\frac{3}{2}}L_{1,1,1}) + \frac{\partial}{\partial\psi}(a_1^{-\frac{1}{2}}L_{112}a_2^{-1}g(\psi)) = 0. \tag{12}$$

In the above

$$L_{1,1,1} = E[\frac{\partial}{\partial\phi} \log L(\phi, \psi)]^3 \text{ and } L_{112} = E[\frac{\partial^3}{\partial\phi^2\partial\psi} \log L(\phi, \psi)],$$

$L(\phi, \psi)$ being the likelihood function of (ϕ, ψ) after reparametrization from the original joint pdf. Here,

$$L(\phi, \psi) = \psi^{-\frac{2mn}{m+n}} \exp\left[-\psi^{-\frac{2mn}{m+n}}\left(\phi^{\frac{n}{m+n}}\sum_{i=1}^{m}x_i + \phi^{-\frac{m}{m+n}}\sum_{j=1}^{n}y_j\right)\right]. \tag{13}$$

Straight forward calculations yield $L_{1,1,1}$ is a constant multiple of ϕ^{-3}. Also $a_1^{-\frac{3}{2}}$ is a constant multiple of ϕ^3. Hence, $\frac{\partial}{\partial\phi}(a_1^{-\frac{3}{2}}L_{1,1,1}) = 0$. Also, on simplification $\frac{\partial}{\partial\psi}\{a_1^{-\frac{1}{2}}L_{112}a_2^{-1}g(w)\}$ is a constant multiple of $\frac{\partial}{\partial\psi}(\phi^{-1}\psi g(\psi)) = \phi^{-1}\frac{d}{d\psi}\{\psi g(\psi)\}$.

Thus, the solution of (12) is given by $g(\psi) \propto \psi^{-1}$, and the unique second matching prior is given by $\pi(\phi, \psi) \propto \phi^{-1}\psi^{-1}$. This is also the reference prior of Bernardo (1979). Back to the (η_1, η_2) formulation this prior transforms to $\pi(\eta_1, \eta_2) \propto (\eta_1\eta_2)^{-1}$ which is Jeffreys' prior. The invariance of first order matching priors is proved in Datta and Ghosh (1996), and that of first and second order matching priors is proved in Mukerjee and Ghosh (1997).

3.2   Noninformative Priors for Weibull Models

We conclude this section with the discussion of Bayesian analysis of Weibull stress-strength models. To this end, suppose $X_1, \cdots, X_m$, and $Y_1, \cdots, Y_n$, are mutually independent with $X_1, \cdots, X_m$ iid $W(\eta_1, \beta)$ and $Y_1, \cdots, Y_n$ iid $W(\eta_2, \beta)$. However, now η_1, η_2 and β are all unknown, and the parameter of interest is $\theta = P(X \le Y) = \eta_2^\beta/(\eta_1^\beta + \eta_2^\beta)$. Sun, Ghosh and Basu (1996) have derived the class of first order probability matching priors as well as the different reference priors of Berger and Bernardo for this problem. Writing

$$\gamma = 1 + \int_0^\infty \log(z)e^{-z}dz, \quad \gamma^* = \int_0^\infty (\log z)^2 e^{-z}dz - \left\{\int_0^\infty \log(z)e^{-z}dz\right\}^2.$$

The general class of first order probability matching priors is given by

$$\pi(\eta_1, \eta_2, \beta) \propto \frac{\sqrt{\gamma^* + \frac{mn}{(m+n)^2}\beta^2 \log^2\left(\frac{\eta_2}{\eta_1}\right)}}{\eta_1\eta_2\beta^2} \times$$
$$h\left(\eta_1 \exp\left\{\frac{\gamma}{\beta} + \sqrt{\frac{n\gamma}{m}}\frac{i}{\beta}\right\}, \; \eta_2 \exp\left\{\frac{\gamma}{\beta} - \sqrt{\frac{m\gamma}{n}}\frac{i}{\beta}\right\}\right), \quad (14)$$

where h is an arbitrary function and $i = \sqrt{-1}$. An interesting class of matching priors can be obtained when $h(s, t) = g(s^m t^n)$, for some function $g > 0$, for which (14) becomes

$$\pi(\eta_1, \eta_2, \beta) \propto \frac{\sqrt{\gamma^* + \frac{mn}{(m+n)^2}\beta^2 \log^2\left(\frac{\eta_2}{\eta_1}\right)}}{\eta_1\eta_2\beta^2} g\left(\eta_1^m \eta_2^n \exp\left\{\frac{(m+n)\gamma}{\beta}\right\}\right). \quad (15)$$

There are infinitely many first order matching priors for θ. It is not clear which one gives a higher order probability mathing. The simple choice is to use the one with $g \equiv 1$. Use the reparameterization $\xi = (\eta_1\eta_2)^\beta/(\eta_1^\beta + \eta_2^\beta)$. Writing $m = \rho n$, under the reparameterization (θ, ξ, β), the matching prior (15) when $g \equiv 1$ becomes

$$\pi_M(\theta, \xi, \beta) \propto \frac{\{\gamma^*(1 + \rho)^2 + \rho \log^2[\theta/(1 - \theta)]\}^{\frac{1}{2}}}{(1 - \theta)\theta \xi \beta^4}. \quad (16)$$

The complete catalog of reference priors, with the reparameterization (θ, ξ, β), for this problem is given in Table 3.

Table 3.   Catalog of reference priors for Weibull models

Grouped parameters in their order of importance	Priors distributions
$\{\theta, \xi, \beta\}$	$\pi_0(\theta, \xi, \beta) \propto \frac{1}{(1-\theta)\theta\xi\beta}$ (Jeffrey's general rule prior)
$\{\beta\}, \{\theta\}, \{\xi\}$	$\pi_0(\theta, \xi, \beta) \propto \frac{1}{(1-\theta)\theta\xi\beta}$
$\{\beta\}, \{\theta, \xi\}$	$\pi_0(\theta, \xi, \beta) \propto \frac{1}{(1-\theta)\theta\xi\beta}$
$\{\theta\}, \{\xi, \beta\}$	$\pi_1(\theta, \xi, \beta) \propto \frac{1}{(1-\theta)\theta\xi\beta\{\gamma^*(1+\rho)^2 + \rho\log^2[\theta/(1-\theta)]\}^{\frac{1}{2}}}$
$\{\theta\}, \{\beta\}, \{\xi\}$	$\pi_1(\theta, \xi, \beta) \propto \frac{1}{(1-\theta)\theta\xi\beta\{\gamma^*(1+\rho)^2 + \rho\log^2[\theta/(1-\theta)]\}^{\frac{1}{2}}}$
$\{\xi, \beta\}, \{\theta\}$	$\pi_2(\theta, \xi, \beta) \propto \frac{1}{(1-\theta)\theta\xi\beta\{\rho(1-\theta)^2 + \theta^2\}^{\frac{1}{2}}}$

To see the difference among the three priors in Table 3 and a matching prior π_M, given by (16), note that the marginal priors are all independent, the marginal densities for ξ are all the same, and the priors for β are quite simple. We then plot the four marginal densities of θ in Figure 1, with two choices of ρ being 1 and 6, respectively. Even when $\rho = 6$, the marginal priors for θ are still quite symmetric.

The details are available in Ghosh, Sun and Basu (1996). These authors have also proved the propriety of the posteriors for all the matching priors for $m + n \geq 4$ and of all the reference priors π_0, π_1, π_2 for $m + n \geq 3$. It turns out though that none of the reference priors of Berger and Bernardo is even first order matching.

Sun, Ghosh and Basu (1996) performed simulation studies to compare the marginal posterior densities of θ. A random sample of sizes $(m, n) = (2, 4)$ when $\eta_1 = 3, \eta_2 = 2, \beta = 4.0$ gives $x = (1.83, 3.58)$ and $y = (2.72, 2.55, 1.75, 3.50)$. The posterior densities $\pi(\theta|x, y)$ under the four priors are plotted in Figure 2 (a). A second random sample of sizes $(m, n) = (5, 5)$ when $\eta_1 = 3, \eta_2 = 2$ and $\beta = 0.5$ gives $x = (2.00, 1.50, 0.53, 2.06, 3.06)$ and $y = (6.40, 15.66, 4.64, 1.67, 0.10)$. The corresponding $\pi(\theta|x, y)$ are plotted in Figure 2 (b). We can see that in both cases $\pi(\theta|x, y)$ under the matching prior π_M tends to be symmetric about 0.5.

The estimated frequentist coverage probabilities of $\alpha-$posterior $\alpha-$quantiles of θ when these four priors are used can be found in Table 4. Here $\alpha = 0.05$ and 0.95. The computation of these numerical values was based on the 10,000 samples for each of fixed true values given in the table. For the cases presented in Table 4, we see that the matching prior π_M is much better than the reference priors π_0, π_1 and π_2. In conclusion, we recommend the use of π_M for the Weibull stress-strength model.

(a) $\rho = m/n = 1$

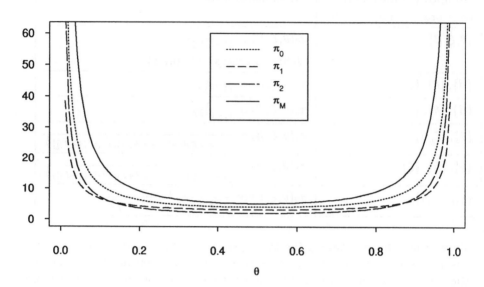

(b) $\rho = m/n = 6$

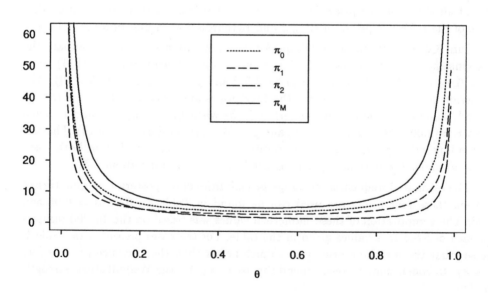

Figure 1: Marginal reference and matching priors of θ for Weibull models

(a) (m, n) = (2, 4)

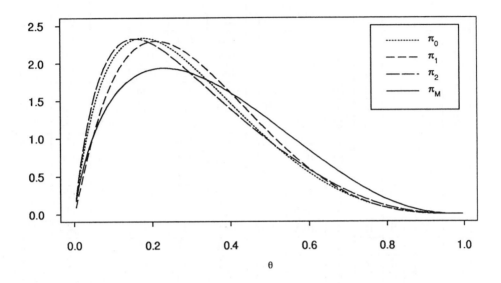

(b) m = n = 6

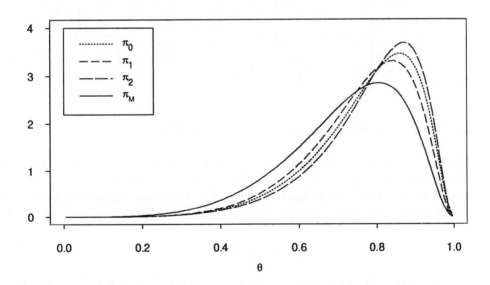

Figure 2: Marginal posterior densities of θ for Weibull models

Table 4. Frequentist coverage probabilities of 0.05 (0.95) posterior quantiles of θ for Weibull models.

η_1	η_2	β	m	n	π_0		π_1		π_2		π_M	
					.05	.95	.05	.95	.05	.95	.05	.95
3	2	0.5	2	2	.132	.840	.116	.871	.140	.823	.012	.997
3	2	0.5	2	3	.108	.875	.096	.896	.107	.852	.029	.973
3	2	0.5	5	5	.069	.918	.066	.928	.073	.909	.042	.959
3	2	0.5	10	10	.061	.935	.060	.939	.063	.930	.050	.953
5	5	0.5	2	2	.146	.853	.121	.877	.156	.840	.008	.992
5	5	0.5	2	3	.116	.884	.098	.901	.118	.865	.029	.973
5	5	0.5	5	5	.075	.923	.070	.929	.083	.916	.042	.958
5	5	0.5	10	10	.065	.941	.063	.944	.068	.936	.050	.954
3	2	1.0	2	2	.118	.829	.107	.867	.124	.808	.014	.999
3	2	1.0	2	3	.100	.864	.092	.892	.095	.840	.029	.974
3	2	1.0	5	5	.065	.913	.064	.926	.065	.901	.042	.959
3	2	1.0	10	10	.058	.932	.058	.938	.057	.925	.049	.951
5	5	1.0	2	2	.146	.853	.121	.877	.156	.840	.008	.992
5	5	1.0	2	3	.116	.884	.098	.901	.118	.865	.029	.973
5	5	1.0	5	5	.075	.923	.070	.929	.083	.916	.042	.958
5	5	1.0	10	10	.065	.941	.063	.944	.068	.936	.050	.954
3	2	1.5	2	2	.103	.817	.099	.866	.107	.792	.015	.999
3	2	1.5	2	3	.091	.854	.088	.887	.085	.827	.029	.975
3	2	1.5	5	5	.059	.907	.060	.925	.058	.894	.043	.960
3	2	1.5	10	10	.053	.930	.055	.936	.051	.922	.048	.953
5	5	1.5	2	2	.146	.853	.121	.877	.156	.840	.008	.992
5	5	1.5	2	3	.116	.884	.098	.901	.118	.865	.029	.973
5	5	1.5	5	5	.075	.923	.070	.929	.083	.916	.042	.958
5	5	1.5	10	10	.065	.941	.063	.944	.068	.936	.050	.954

4 Nonparametric Bayesian Estimation of $P(X \leq Y)$

As before, we consider the two sample problem with $X_1, \cdots, X_m$, constituting the first sample, and $Y_1, \cdots, Y_n$ constituting the second. However, now the X_i are sampled from an unknown distribution function F, and the Y_j are sampled from an unknown distribution function G. Interest continues in inference for the parameter

$$\theta = \int F \, dG.$$

As mentioned in the introduction, the Mann-Whitney U-statistic is the nonparametric UMVUE of θ, and has played the leading role virtually in any classical inference procedure regarding θ. (See, e.g. Lehmann (1951), Sen (1967) or Govindarajulu (1968)).

Nonparametric Bayes estimation of θ is of more recent origin. Ferguson (1973) initialized such a study based on Dirichlet process priors. A random probability measure P defined on a measurable $(\mathcal{X}, \mathcal{B})$ is said to have a Dirichlet process prior with parameter α if for any measurable partition $(A_1, \cdots, A_k)$ of $\mathcal{X}$, $(P(A_1), \cdots, P(A_k))$ has a Dirichlet distribution with parameters $(\alpha(A_1), \cdots, \alpha(A_k))$.

For the given two sample problem, $X_1, \cdots, X_m$ constitutes a random sample of size m from the random probability measure P_1, while $Y_1, \cdots, Y_n$ constitutes an independent random sample of size n from a random probability measure P_2. Assume that P_1 and P_2 are independent Dirichlet process priors with parameters α_1 and α_2. Then the posterior mean of θ is given by

$$
\begin{aligned}
e_B \;=\;& D_1 D_2 \theta_0 + D_1(1 - D_2)F_0^* + (1 - D_1)D_2(1 - G_0^*(-)) \\
& + (1 - D_1)(1 - D_2)U,
\end{aligned} \tag{17}
$$

where

$$
\begin{aligned}
D_1 &= \alpha_1(\mathbb{R})/(\alpha_1(\mathbb{R}) + m); \\
D_2 &= \alpha_2(\mathbb{R})/(\alpha_2(\mathbb{R}) + n); \\
F_0(x) &= \alpha_1(-\infty, x]/\alpha_1(\mathbb{R}); \\
G_0(y) &= \alpha_2(-\infty, y]/\alpha_2(\mathbb{R}); \\
\theta_0 &= \int F_0 dG_0; \\
F_0^* &= n^{-1} \sum_{j=1}^{m} F_0(Y_j); \\
1 - G_0^*(-) &= \frac{1}{m} \sum_{i=1}^{m} [1 - G_0(X_i-)],
\end{aligned}
$$

and U is the Mann-Whitney U-statistic, defined by

$$
U \;=\; \frac{1}{mn} \sum_{i=1}^{m} \sum_{j=1}^{n} 1_{[X_i \le Y_j]}.
$$

In an empirical Bayes (EB) analysis, at the current, say kth stage of the experiment, information is available not only for the current stage, but also for the previous $k - 1$ stages. We shall denote the random sample by $X_{i1}, \cdots, X_{im_i}, Y_{i1}, \cdots, Y_{in_i}$ at the ith stage $(i = 1, \cdots, k)$. Then Hollander and Korwar (1976) obtained the EB estimator e_{Hk}^k of θ where

$$
\begin{aligned}
e_{Hk}^k \;=\;& D_{1k} D_{2k} \left\{ \frac{1}{(k-1)^2} \sum_{i=1}^{k-1} \sum_{j=1}^{k-1} U_{ij} \right\} + D_{1k}(1 - D_{2k}) \left\{ \frac{1}{k-1} \sum_{j=1}^{k-1} U_{ij} \right\} \\
& + (1 - D_{1k})D_{2k} \left\{ \frac{1}{k-1} \sum_{j=1}^{k-1} U_{kj} \right\} + (1 - D_{1k})(1 - D_{2k})U_{kk}, \tag{18}
\end{aligned}
$$

where

$$U_{ij} = \sum_{l=1}^{m_i} \sum_{l'=1}^{n_j} 1_{[X_{il} \leq Y_{jl'}]} \quad (i = 1, \cdots, k; \ j = 1, \cdots, k),$$

and

$$D_{1k} = \frac{\alpha_1(\mathbb{R})}{\alpha_1(\mathbb{R}) + m_k}, \quad D_{2k} = \frac{\alpha_2(\mathbb{R})}{\alpha_2(\mathbb{R}) + n_k}.$$

Thus, for estimating F_0^* and G_0^*, Hollander and Korwar (1976) used only the past data and not the current data. Ghosh and Lahiri (1992) obtained instead estimates of F_0^* and G_0^* based on both the past and the current data. This led to the estimator,

$$\begin{aligned}
e_{EB}^k &= D_{1k} D_{2k} \widetilde{\Delta}_{0k} + D_{1k}(1 - D_{2k}) \widetilde{F}_{0k}^* + (1 - D_{1k}) D_{2k} \widetilde{G}_{0k}^* \\
&+ (1 - D_{1k})(1 - D_{2k}) U_{kk},
\end{aligned} \tag{19}$$

where

$$\begin{aligned}
\widetilde{\Delta}_{0k} &= \sum_{i=1}^{k} \sum_{j=1}^{k} (1 - D_{1i})(1 - D_{2j}) U_{ij} \Big/ \sum_{i=1}^{k} \sum_{j=1}^{k} (1 - D_{1i})(1 - D_{2j}); \\
\widetilde{F}_{0k}^* &= \sum_{i=1}^{k} (1 - D_{1i}) U_{ik} \Big/ \sum_{i=1}^{k} (1 - D_{1i}); \\
1 - \widetilde{G}_{0k}^*(-) &= \sum_{j=1}^{k} (1 - D_{2j}) U_{jk} \Big/ \sum_{j=1}^{k} (1 - D_{2j}); .
\end{aligned}$$

Ghosh and Lahiri (1992) have established the superiority of the ED estimator e_{EB}^k over the previous EB estimator e_{Hk}^k. These authors have also shown that the difference between the Bayes risks of the EB estimator e_{EB}^k and the optimal Bayes estimator e_B^k under a subjective Dirichlet prior converges to zero as $k \longrightarrow \infty$. This is the so-called "asymptotic optimality" of the EB produce in the sense of Robbins (1956).

Acknowledgments

Ghosh's research was partially supported by NSF Grant Number SES - 9201210 and SBR - 9423996. Sun's research was partially supported by the National Security Agency grant MDA904-96-1-0074. The manuscript was prepared using computer facilities supported in part by the NSF Grants DMS 95-08296 awarded to the Department of Statistics at the University of Missouri–Columbia.

References

1. A. P. Basu in *The Frontiers of Modern Statistical Inference Procedures*, ed. E.J. Dudewicz (American Science Press, Columbus, 1985).
2. A. P. Basu and G. Tarmast in *Probability and Bayesian Statistics*, ed. R. Viertl (Plenum Publishing Corporation, 1987).
3. J. O. Berger and J. M. Bernardo, *J. Amer. Statist. Assoc.* **84**, 200 (1989).
4. J. O. Berger and J. M. Bernardo, *Biometrika* **79**, 25 (1992a).
5. J. O. Berger and J. M. Bernardo, in *Bayesian Statistics* 4, Eds. J.M. Bernardo, J.O. Berger, A.P Dawid and A.F.M. Smith. (Oxford Univ. Press, 1992).
6. J. M. Bernardo, *J. Roy. Statist. Soc. Ser. B* **41**, 113 (1979).
7. Z. W. Birnbaum, in *Proceedings of the Third Berkeley Symposium on Mathematical Statistics and Probability, Volume I: Contributions to the Theory of Statistics and Probability*, (University of California Press Berkeley, California, 1956).
8. Z. W. Birnbaum and R. C. McCarty, *Ann. Math. Statist.*295581958.
9. G. S. Datta, *Biometrika* **83**, 287 (1996).
10. G. S. Datta and J.K. Ghosh, *Biometrika* **82**, 37 (1995a).
11. G. S. Datta and M. Ghosh, *J. Amer. Statist. Assoc.* **90**, 1357 (1995b).
12. G. S. Datta and M. Ghosh, *Ann. Statist.* **24**, 141 (1996).
13. A. P. Dawid, M. Stone, and J. V. Zidek, *J. Roy. Statist. Soc. Ser. B* **35**, 189 (1973).
14. F. Downton, *Technometrics* **15**, 551 (1973).
15. P. Enis and S. Geisser, *J. Amer. Statist. Assoc.* **66**, 162 (1971).
16. T. S. Ferguson, *Ann. Statist.* **1**, 209 (1973).
17. M. Ghosh and P. Lahiri in *Order Statist. Nonparam.: Theory Appl.* eds. P. K. Sen and I. A. Salama (Elsevier/North-Holland, New York; Amsterdam, 1992).
18. J.K. Ghosh and R. Mukerjee in *Bayesian Statistics* 4, Eds. J.M. Bernardo, J.O. Berger, A.P Dawid and A.F.M. Smith. (Oxford Univ. Press, 1992).
19. M. Ghosh and M. C. Yang, *Test* **5**, 145 (1996).
20. Z. Govindarajulu, *Sankhyā Ser. B* **29**, 35 (1967).
21. Z. Govindarajulu, *Ann. Inst. Statist. Math.* **20**, 229 (1968).
22. Z. Govindarajulu, *Ann. Inst. Statist. Math.* **28**, 307 (1976).
23. M. Hollander and R. M. Korwar, *Comm. Statist. A—Theory Methods* **5**, 1369 (1976).
24. R. A. Johnson in *Handbook of Statistics, Volume 7: Quality Control and Reliability*, eds. P.R. Krishnaiah and C.R. Rao (Elsevier Science Pub. Co., Inc., New York, 1988).
25. Kececioglu, D. *Nuclear Eng. Des.* **9**, 257 (1972).
26. G. Lee, *Noninformative priors for some models useful in reliability and survival analysis*. Ph. D. dissertation, The University of Missouri-Columbia, (1996).
27. G. Lee, D. Sun, and A. P. Basu, *Noninformative priors for estimating $P(X <$

Y) *in normal distributions,* to be submitted. (1997).

28. E. L. Lehmann, *Ann. Math. Statist.* **22**, 165 (1951).

29. H. B. Mann and D. R. Whitney, *Ann. Math. Statist.* **18**, 50 (1947).

30. R. Mukerjee and K. D. Dey, *Biometrika* **80**, 499 (1993).

31. D. B. Owen, K. J. Craswell, and D. L. Hanson, *J. Amer. Statist. Assoc.* **59**, 906 (1964).

32. H. W. Peers, *J. Royal Statist. Soc., Ser. B* **27**, 9 (1965).

33. C. P. Robert, *The Bayesian Choice, A Decision-Theoretic Motivation* (Springer-Verlag New York, 1994).

34. H. Robbins in *Proceedings of the Third Berkeley Symposium on Mathematical Statistics and Probability, Volume I: Contributions to the Theory of Statistics and Probability* (University of California Press, Berkeley, California, 1956).

35. P. K. Sen, *Sankhyā Ser. A* **29**, 157 (1967).

36. C. Stein, Sequential Math. Statist.: Banach Center Publication **16**, 485 (1985).

37. D. Sun, M. Ghosh, and A.P. Basu, *Bayesian analysis for a stress-strength system via noninformative priors, Canadian Journal of Statistics,* in press, (1997).

38. D. Sun and K. Ye, *Biometrika* **83**, 55 (1996).

39. M. Stone and A. P. Dawid, *Biometrika* **59**, 369 (1972).

40. R. D. Thompson and A. P. Basu in *Advances in Reliability,* ed. A. Basu (North-Holland, Amsterdam, 1993).

41. R. Tibshirani, *Biometrika* **76**, 604 (1989).

42. Welch, B. N. and Peers, B. (1963). *J. Roy. Statist. Soc. Ser. B* **35**, 318 (1963).

DETERMINING AN OPTIMAL PERFORMANCE CONDITION IN A MIXTURE EXPERIMENT

Subir Ghosh and Thomas Liu

University of California, Riverside

In this paper, we consider a situation where a product is formed by mixing together two or more ingredients in different proportions. The problem is to find a combination of mixture proportions optimizing a quality characteristic of the product. We determine analytically such optimal performance conditions in a real experiment performed at Spillers Milling Limited in the United Kingdom [see Draper, et. al. (1993,1994)].

1 Introduction

Consider an experiment performed at Spillers Milling Limited in the United Kingdom to study the effect of mixtures of four different varieties of wheat in producing a good-quality bread [see Draper, et. al (1993, 1994)]. Table 1 presents the design and data collected from the experiment. Four flours obtained from four selected best varieties of wheat are to be mixed in proportions x_1, x_2, x_3, and x_4 to prepare dough for making bread. The quality characteristic of bread thus made is evaluated by observing the specific volume (ml/100g). Higher specific volume means better quality. The problem is to find an optimum combination (x_1, x_2, x_3, x_4) of mixture proportions giving the highest response y, the specific volume of bread. The optimum combination is determined from a fitted canonical second order mixture model to the data given in Table 1.

In the study of dependence of y on the mixture proportions x_1, x_2, x_3, and x_4, the unknown response surface is approximated by a first or a second order polynomial in the region of interest. If the first order mixture model gives a significant lack of fit indicating the presence of a surface curvature, a second order mixture model is then fitted. Suppose that x_{u1}, x_{u2}, x_{u3}, and x_{u4} are the proportions of mixture components in the uth run, u = 1,..., 36. We define the indicator variable Z_{uj}, j = 1, 2, 3, 4, so that

$$Z_{uj} = \begin{cases} 0 \text{ if the uth run is not in Block j} \\ 1 \text{ if the uth run is in Block j} \end{cases}$$

Let $y(x_{u1}, x_{u2}, x_{u3}, x_{u4}) = y_u$, be the observation for the uth run, u = 1,..., 36. In this paper, the y_u's are assumed to be uncorrelated normal random variables with constant variance σ^2 and mean

$$
\begin{aligned}
E(y_u) = {} & \beta_1 x_{u1} + \beta_2 x_{u2} + \beta_3 x_{u3} + \beta_4 x_{u4} \\
& + \beta_{12} x_{u1} x_{u2} + \beta_{13} x_{u1} x_{u3} + \beta_{14} x_{u1} x_{u4} \\
& + \beta_{23} x_{u2} x_{u3} + \beta_{24} x_{u2} x_{u4} + \beta_{34} x_{u3} x_{u4} \\
& + \gamma_2 (z_{u2} - 0.25) + \gamma_3 (z_{u3} - 0.25) + \gamma_4 (z_{u4} - 0.25),
\end{aligned}
\tag{1}
$$

where the β's, γ's and σ^2 are unknown constants. In matrix notation, we write

$$
E(\underline{y}) = X\beta, \quad V(\underline{y}) = \sigma^2 I,
\tag{2}
$$

where $\underline{\beta} = (\beta_1, \beta_2, \beta_3, \beta_4; \beta_{12}, \beta_{13}, \beta_{14}, \beta_{23}, \beta_{24}, \beta_{34}; \gamma_2, \gamma_3, \gamma_4)'$, Rank X=13. The expression of $E(y_u)$ in (1) is known as the second order canonical representation. The parameters γ's are the block parameters. It can be checked that the columns of X for γ's are in fact orthogonal to all other columns of X. Therefore, the mixture design in Table 1 has orthogonal blocks.

Table 1: Mixture Experiment Data [Draper et. al. (1993, 1994)]

Blocks	Mixture Proportions				Responses from Block	
	x_1	x_2	x_3	x_4	1	3
	0	.25	0	.75	403	381
	.25	0	.75	0	425	422
	0	.75	0	.25	442	412
1 and 3	.75	0	.25	0	433	413
	0	.75	.25	0	445	398
	.25	0	0	.75	435	412
	0	0	.75	.25	385	371
	.75	.25	0	0	425	428
	.25	.25	.25	.25	433	393
Blocks	x_1	x_2	x_3	x_4	2	4
	0	.75	0	.25	423	404
	.25	0	.75	0	417	425
	0	.25	0	.75	398	391
	.75	0	.25	0	407	426
2 and 4	0	0	.25	.75	388	362
	.25	.75	0	0	435	427
	0	.25	.75	0	379	390
	.75	0	0	.25	406	411
	.25	.25	.25	.25	439	409

The fitted model under (1) is

$$\hat{y} = \hat{y}(x_1, x_2, x_3, x_4)$$
$$= 386.208x_1 + 435.125x_2 + 384.708x_3 + 389.292x_4$$
$$+ 96.444x_1x_2 + 189.556x_1x_3 + 150.667x_1x_4 \qquad (3)$$
$$- 36.889x_2x_3 - 29.111x_2x_4 - 56x_3x_4$$
$$- 14.889(z_2 - 0.25) - 21.778(z_3 - 0.25) - 20.111(z_4 - 0.25)$$

Ignoring the terms with nonsignificant t values, the fitted model becomes

$$\hat{y} = 383.396x_1 + 430.283x_2 + 375.202x_3 + 381.583x_4$$
$$+ 107.827x_1x_2 + 217.889x_1x_3 + 169.693x_1x_4 \qquad (4)$$
$$- 14.889(z_2 - 0.25) - 21.778(z_3 - 0.25) - 20.111(z_4 - 0.25)$$

In this paper, we keep six decimal points in our calculations but present only up to three decimal points. We use both fitted models (3) and (4) in finding the optimum proportions of ingredients. Considering the parts of (3) and (4) involving x_1, x_2, x_3, and x_4, we define

$$\phi_1 = \phi_1(x_1, x_2, x_3, x_4)$$
$$= 386.208x_1 + 435.125x_2 + 384.708x_3 + 389.292x_4$$
$$+ 96.444x_1x_2 + 189.556x_1x_3 + 150.667x_1x_4 \qquad (5)$$
$$- 36.889x_2x_3 - 29.111x_2x_4 - 56x_3x_4,$$

$$\phi_2 = \phi_2(x_1, x_2, x_3, x_4)$$
$$= 383.396x_1 + 430.283x_2 + 375.202x_3 + 381.583x_4 \qquad (6)$$
$$+ 107.827x_1x_2 + 217.889x_1x_3 + 169.693x_1x_4.$$

The problem is now to maximize ϕ_1 and ϕ_2 with respect to $x_1, x_2, x_3,$ and x_4 subject to the conditions that

$$x_1 + x_2 + x_3 + x_4 = 1, \quad 0 \le x_1, x_2, x_3, x_4 \le 1. \qquad (7)$$

We show that the function ϕ_1 attains its maximum subject to the conditions in (7) when $x_1 = 0.246$, $x_2 = 0.754$, $x_3 = x_4 = 0$ and the maximum value of ϕ_1 is 440.980. Again, the function ϕ_2 attains its maximum subject to the conditions in (7) when $x_1 = 0.283$, $x_2 = 0.717$, $x_3 = x_4 = 0$ and the maximum value of ϕ_2 is 438.893. The recommended optimum combination is therefore $(0.246, 0.754, 0, 0) \simeq (0.25, 0.75, 0, 0)$ based on the fitted model (3) and $(0.283, 0.717, 0, 0) \simeq (0.28, 0.72, 0, 0)$ based on the fitted model (4). It is to be noted that we prefer to make the recommendation for the optimum combination $(0.25, 0.75, 0, 0)$ based on the fitted model (3). The validation experiment must be performed to check our recommendation for the optimum combination. Table 2 presents the average response over blocks for thirteen distinct runs. We observe that the run $(0.25, 0.75, 0, 0)$ is giving the highest value 431.00 of the average response over blocks. This is a crude partial check on the validity of our recommended optimum combination because of different block replications for runs.

Table 2: Average Response Over Blocks

Mixture Proportions				Responses from Block				Average Response
x_1	x_2	x_3	x_4	1	2	3	4	
0	.25	0	.75	403	398	381	391	393.25
.25	0	.75	0	425	417	422	425	422.25
0	.75	0	.25	442	423	412	404	420.25
.75	0	.25	0	433	407	413	426	419.75
0	.75	.25	0	445		398		421.50
0	.25	.75	0		379		390	384.50
.25	0	0	.75	435		412		423.50
.75	0	0	.25		406		411	408.50
0	0	.75	.25	385		371		378.00
0	0	.25	.75		388		362	375.00
.75	.25	0	0	425		428		426.50
.25	.75	0	0		435		427	431.00
.25	.25	.25	.25	433	439	393	409	418.50

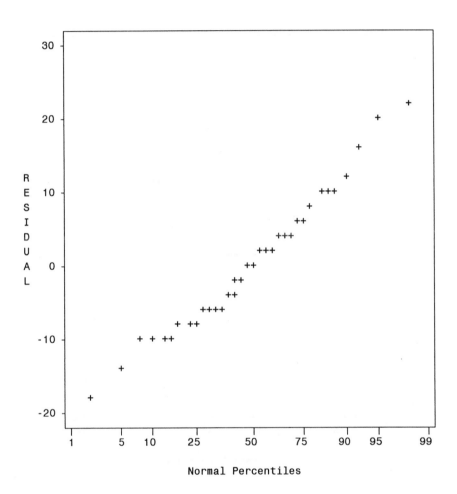

Figure 1. Normal Probability Plot for the residuals
under the Fitted Model (3)

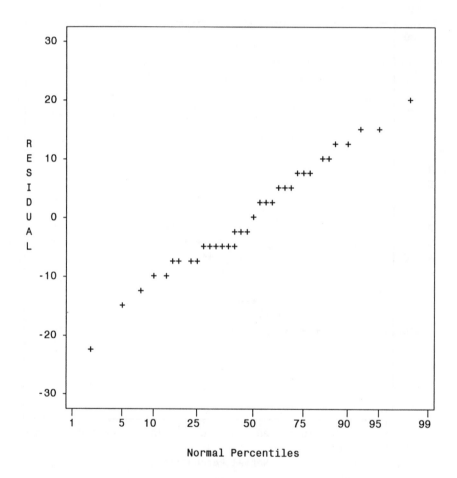

Figure 2. Normal Probability Plot for the residuals
under the Fitted Model (4)

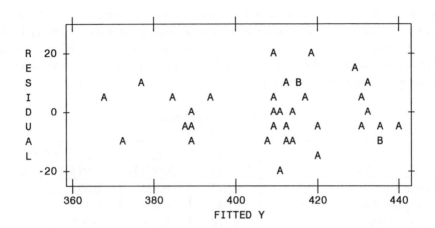

Figure 3. Residual Plot under the Fitted Model (3)

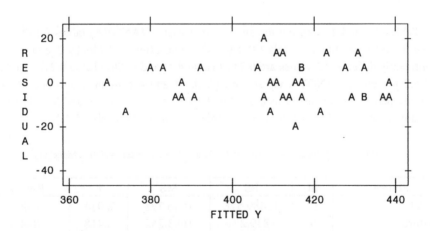

Figure 4. Residual Plot under the Fitted Model (4)
Note: B stands for two A's.

2 Model Assumptions

We now present the checking of the model assumptions. Table 3 gives the Shapiro-Wilk, Kolmogorov-Smirnov, Anderson-Darling, Cramer-von Mises, and Chi-square test statistics for testing the normality assumption for the residuals under the fitted models (3) and (4). The large P-values of test statistics indicate no significant departure from normality. Figures 1 and 2 showing the normal probability plots for the fitted models (3) and (4), respectively, are also supportive of normality. Figures 3 and 4 showing the residual plots under the fitted model (3) and (4), respectively, indicate no violation in the homogeneity of variance assumption.

Table 3: Tests for Normality

	Test Statistic		P-value	
Name	Model (3)	Model (4)	Model (3)	Model (4)
Shapiro-Wilk	0.973	0.985	0.603	0.929
Kolmogorov-Smirnov	0.082	0.099	> 0.15	> 0.15
Anderson-Darling	0.310	0.256	> 0.5	> 0.5
Cramer-von Mises	0.046	0.046	> 0.5	> 0.5
Chi-square	5.536	1.060	0.327	0.787
	(df = 4)	(df = 3)		

Tables 4.1 and 4.2 present the analysis of variance (ANOVA) tables for the data in Table 1 under the fitted models (3) and (4). The effects of blocks 2 and 4 are not significantly different [P-values are 0.341 (Table 4.1) and 0.326 (Table 4.2)]. Tables 5.1 and 5.2 present the ANOVA tables with blocks 2 and 4 combined into one block and therefore permitting the test of lack of fit under the fitted models (3) and (4). The lack of fit is not significant [P-values are 0.645 (Table 5.1) and 0.705 (Table 5.2)].

Table 4.1: Analysis of Variance for the Data in Table 1 Under the Fitted Model (3)

Source	df	SS	MS	F	P-value
Model	12	12489.576	1040.798	8.014	0.000
Mixture	9	9839.271	1093.252	8.418	0.000
Block	3	2650.306	883.435	6.802	0.002
1 vs. 3	1	2134.222	2134.222	16.433	0.000
2 vs. 4	1	122.722	122.722	0.945	0.341
(1+3) vs. (2+4)	1	393.361	393.361	3.029	0.095
Error	23	2987.174	129.877		
Total	35	15476.750			

Table 4.2: Analysis of Variance for the Data in Table 1 Under the Fitted Model (4)

Source	df	SS	MS	F	P-value
Model	9	12298.007	1366.445	11.177	0.000
Mixture	6	9647.701	1607.950	13.152	0.000
Block	3	2650.306	883.435	7.226	0.001
1 vs. 3	1	2134.222	2134.222	17.457	0.000
2 vs. 4	1	122.722	122.722	1.004	0.326
(1+3) vs. (2+4)	1	393.361	393.361	3.217	0.085
Error	26	3178.743	122.259		
Total	35	15476.750			

Table 5.1: Analysis of Variance with Blocks 2 and 4 Combined Under the Fitted Model (3)

Source	df	SS	MS	F	P-value
Model	11	12366.854	1124.259	8.676	0.000
Mixture	9	9839.271	1093.252	8.437	0.000
Block	2	2527.583	1263.792	9.753	0.001
1 vs. 3	1	2134.222	2134.222	16.470	0.001
(1+3) vs. (2+4)	1	393.361	393.361	3.036	0.094
Error	24	3109.895	129.579		
Lack-of-Fit	15	1799.396	119.960	0.824	0.645
Pure Error	9	1310.500	145.611		
Total	35	15476.750			

Table 5.2: Analysis of Variance with Blocks 2 and 4 Combined Under the Fitted Model (4)

Source	df	SS	MS	F	P-value
Model	8	12175.285	1521.911	12.446	0.000
Mixture	6	9647.701	1607.950	13.150	0.000
Block	2	2527.583	1263.792	10.336	0.001
1 vs. 3	1	2134.222	2134.222	17.454	0.000
(1+3) vs. (2+4)	1	393.361	393.361	3.217	0.084
Error	27	3301.465	122.276		
Lack-of-Fit	18	1990.965	110.609	0.760	0.705
Pure Error	9	1310.500	145.611		
Total	35	15476.750			

3 Optimum Condition

We now determine the optimum condition (x_1, x_2, x_3, x_4) that maximizes ϕ_1 in (5) subject to the conditions in (7). Substituting $x_1 = 1 - x_2 - x_3 - x_4$ in ϕ_1, we express it as a quadratic function in x_4.

$$
\begin{aligned}
\phi_1 = &(-150.667)x_4^2 \\
&+ (153.750 - 276.222x_2 - 396.222x_3)x_4 \\
&+ (386.208 + 145.361x_2 - 96.444x_2^2 \\
&+ 188.056x_3 - 189.556x_3^2 - 322.889x_2x_3)
\end{aligned}
\tag{8}
$$

Note the coefficient of x_4 in (8) $(153.750 - 276.222x_2 - 396.222x_3)$ is nonnegative if and only if (iff)

$$
x_2 \leq \frac{153.750 - 396.222x_3}{276.222} = A(x_3), \text{ say.}
\tag{9}
$$

Again the right hand side of (9) is nonnegative iff

$$
x_3 \leq 153.750/396.222 = 0.388.
\tag{10}
$$

We now consider the following admissible cases

$$
\begin{aligned}
&\text{Case I:}   && x_2 \leq A(x_3), x_3 \leq 0.388, \\
&\text{Case II:} && x_2 > A(x_3), x_3 \leq 0.388, \\
&\text{Case III:} && x_2 > A(x_3), x_3 > 0.388.
\end{aligned}
\tag{11}
$$

Note that the case $x_2 \leq A(x_3)$ and $x_3 > 0.388$ is inadmissible because it violates the condition in (7) that x_2 is nonnegative. It can be checked that the function ϕ_1 in (8) attains its maximum with respect to x_4 subject to the conditions in (7) when

$$
\text{Case I:} \quad x_4{}^{max} = \frac{153.750 - 276.222x_2 - 396.222x_3}{2 \times 150.667},
$$

$$
\text{Cases II \& III:} \quad x_4{}^{max} = 0.
\tag{12}
$$

For Case I, the function ϕ_1 in (8), at $x_4 = x_4{}^{max}$, becomes

$$\phi_1^I = \left(x_4^{max}\right)$$

$$= 30.157x_2^2 + \left(4.424 + 40.315x_3\right)x_2 \tag{13}$$

$$+ \left(425.432 - 14.110x_3 + 70.940x_3^2\right)$$

The function $\phi_1^I\left(x_4^{max}\right)$ has a minimum, and therefore the function $\phi_1^I\left(x_4^{max}\right)$ attains its maximum with respect to x_2 subject to the conditions in (7) at either $x_2 = 0$ or $x_2 = A(x_3)$ in (9). However, the first two terms in $\phi_1^I\left(x_4^{max}\right)$ are monotonically increasing in x_2 and $A(x_3)$ is nonnegative for Case I. Thus $\phi_1^I\left(x_4^{max}\right)$ attains its maximum with respect to x_2 when $x_2^{max} = A(x_3)$ and the maximum value is

$$\phi_1^I\left(x_4^{max}, x_2^{max}\right) = 437.238 - 46.172x_3 + 75.163x_3^2. \tag{14}$$

Again, the above function has a minimum and therefore $\phi_1^I\left(x_4^{max}, x_2^{max}\right)$ attains its maximum with respect to x_3 subject to the condition in (7) at either $x_3 = 0$ or $x_3 = 0.388$. The numerical values of $\phi_1^I\left(x_4^{max}, x_2^{max}\right)$ are equal to 437.238 at $x_3 = 0$ and 430.639 at $x_3 = 0.388$. Thus $\phi_1^I\left(x_4^{max}, x_2^{max}\right)$ is maximum at $x_3^{max} = 0$ with the value $\phi_1^I\left(x_4^{max}, x_2^{max}, x_3^{max}\right) = 437.238$. In other words, the function ϕ_1 for Case I has its maximum value 437.238 at $x_3 = 0$, $x_2 = 0.557$, $x_4 = 0$ and $x_1 = 0.443$.

For Cases II and III, the function ϕ_1 in (8), at $x_4 = x_4^{max} = 0$ becomes

$$\phi_1^u\left(x_4^{max}\right)$$

$$= -96.444x_2^2 + \left[145.361 - 322.889x_3\right]x_2 \tag{15}$$

$$+ \left[386.208 + 188.056x_3 - 189.556x_3^2\right]u = II, III.$$

It can be seen that the function $\phi_1^u\left(x_4^{max}\right)$ in (15) attains its maximum with respect to x_2 subject to the conditions in (7) when

1. $x_2^{max} = \dfrac{145.361 - 322.889x_3}{2 \times 96.444}$, for $x_3 \leq 0.450$,

2. $x_2^{max} = 0$, for $x_3 > 0.450$. $\tag{16}$

In 1 for $x_3 \leq 0.450$, the function $\phi_1^u\left(x_4^{max}\right)$ at $x_2 = x_2^{max}$ in (16) becomes

$$\phi_1^{u1}\left(x_4^{max}, x_2^{max}\right)$$

$$= 440.980 + 80.696x_3\left(x_3 - 0.685\right).$$

(17)

Clearly the function of $\phi_1^{u1}\left(x_4^{max}, x_2^{max}\right)$ attains its maximum value 440.980 when $x_3 = 0$. In other words, the function ϕ_1 for Cases II and III and $x_3 \leq 0.450$, has its maximum value 440.980 at $x_3 = 0$, $x_2 = 0.754$, $x_4 = 0$, and $x_1 = 0.246$. In 2 for $x_3 > 0.450$, the function $\phi_1^u\left(x_4^{max}\right)$ at $x_2 = x_2^{max} = 0$ becomes

$$\phi_1^{u2}\left(x_4^{max}, x_2^{max}\right)$$

$$= 386.208 + 188.056x_3 - 189.556x_3^2.$$

(18)

The function $\phi_1^{u2}\left(x_4^{max}, x_2^{max}\right)$ in (18) attains its maximum with respect to x_3 subject to the conditions in (7) when $x_3 = x_3^{max} = 0.496$ and the maximum value is 432.850. In other words, the function ϕ_1 for Cases II and III and $x_3 > 0.450$, has its maximum value 432.850 at $x_2 = 0$, $x_3 = 0.496$, $x_4 = 0$, and $x_1 = 0.504$.

The maximum values of ϕ_1 are 437.238 for Case I, 440.980 and 432.850 for two subcases of Cases II and III. Therefore the function ϕ_1 in (5) attains its maximum value 440.980 subject to the conditions in (7) when $x_1 = 0.246$, $x_2 = 0.754$, and $x_3 = x_4 = 0$.

We next determine the optimum condition $\left(x_1, x_2, x_3, x_4\right)$ that maximizes ϕ_2 in (6) subject to the conditions in (7). Substituting $x_1 = 1 - x_2 - x_3 - x_4$ in ϕ_2, we get

$$\phi_2 = (-169.693)x_4^2$$
$$+\left(167.880 - 277.520x_2 - 387.583x_3\right)x_4$$
$$+\left(-107.827x_2^2 + \left(154.713 - 325.716x_3\right)x_2\right.$$
$$+\left.\left(383.396 + 209.695x_3 - 217.889x_3^2\right)\right).$$

(19)

Let

$$B(x_3) = \frac{167.880 - 387.583x_3}{277.520}.$$

(20)

As before, we have the three admissible cases

Case I: $x_2 \leq B(x_3), \ x_3 \leq 0.433,$

Case II: $x_2 > B(x_3), \ x_3 \leq 0.433,$

Case II: $x_2 > B(x_3), \ x_3 \geq 0.433.$

The function ϕ_2 in (19) attains its maximum with respect to x_4 when

Case I: $x_4{}^{max} = \dfrac{167.880 - 277.520x_2 - 387.583x_3}{2 \times 169.693}$,

Case II & III: $x_4{}^{max} = 0.$ (21)

For Case I, the function ϕ_2 in (19), at $x_4 = x_4{}^{max}$, becomes

$$\phi_2{}^I\left(x_4{}^{max}\right)$$

$$= 5.639x_2^2 + (17.436 - 8.786x_3)x_2 \tag{22}$$

$$+ \left(424.918 + 17.975x_3 + 3.422x_3^2\right)$$

Again, $\phi_2{}^I\left(x_4{}^{max}\right)$ attains its maximum with respect to x_2 when $x_2{}^{max} = B(x_3)$ and the maximum value is

$$\phi_2{}^I\left(x_4{}^{max}, x_2{}^{max}\right) = 437.529 - 21.218x_3 + 26.691x_3^2. \tag{23}$$

The above function attains its maximum with respect to x_3 at either $x_3 = 0$ or $x_3 = 0.433$. The values of $\phi_2{}^I\left(x_4{}^{max}, x_2{}^{max}\right)$ at $x_3 = 0$ and 0.433 are 437.529 and 433.346, respectively. In other words, the function ϕ_2 for Case I has its maximum value 437.529 at $x_3 = 0$, $x_2 = 0.605$, $x_4 = 0$, and $x_1 = 0.395$. For Cases II and III, the function ϕ_2 at $x_4 = x_4{}^{max} = 0$ becomes

$$\phi_2{}^u\left(x_4{}^{max}\right)$$

$$= -107.827x_2^2 + \left[154.713 - 325.716x_3\right]x_2 \tag{24}$$

$$+ \left[383.396 + 209.695x_3 - 217.889x_3^2\right] \quad u = II, \ III.$$

Again, $\phi_2{}^u\left(x_4{}^{max}\right)$ in (24) attains its maximum with respect to x_2 when

1. $x_2{}^{max} = \dfrac{154.713 - 325.716x_3}{2 \times 107.827}$, for $x_3 \leq 0.475,$

2. $x_2{}^{max} = 0$, for $x_3 > 0.475$.   (25)

In 1 for $x_3 \leq 0.475$, the function $\phi_2{}^{u}\left(x_4{}^{max}\right)$ at $x_2 = x_2{}^{max}$ in (25) becomes

$$\phi_2{}^{u1}\left(x_4{}^{max}, x_2{}^{max}\right)$$
$$= 438.893 + 28.086x_3\left(x_3 - 0.854\right)$$
(26)

The above function attains its maximum value 438.893 when $x_3 = 0$. In other words, the function ϕ_2 for Cases II and III and $x_3 \leq 0.475$ has its maximum value 438.893 at $x_3 = 0$, $x_2 = 0.717$, $x_4 = 0$, and $x_1 = 0.283$. In 2 for $x_3 > 0.475$, the function $\phi_2{}^{u}\left(x_4{}^{max}\right)$ at $x_2 = x_2{}^{max} = 0$ becomes

$$\phi_2{}^{u2}\left(x_4{}^{max}, x_2{}^{max}\right)$$
$$= 383.396 + 209.695x_3 - 217.889x_3^2.$$
(27)

The above function attains its maximum when $x_3 = x_3{}^{max} = 0.481$ and the maximum value is 433.849. Therefore the function ϕ_2 in (6) has its maximum value 438.893 at $x_1 = 0.283$, $x_2 = 0.717$, and $x_3 = x_4 = 0$.

4 Conclusions

A mixture experiment was performed in the investigation of bread-making flours at Spillers Milling Limited in the United Kingdom. In this paper, we find out analytically optimum mixture proportions of four bread-making flours giving the highest specific volume of bread. Two optimum combinations are (0.25, 0.75, 0, 0) under the first fitted model and (0.28, 0.72, 0, 0) under the second fitted model. The recommended optimum combination is (0.25, 0.75, 0, 0) because it is based on the fitted model without ignoring the nonsignificant terms.

Acknowledgment

This research is supported by the Air Force Office of Scientific Research under grant F49620-96-1-0094.

References

Cornell, J.A. *Experiments with mixtures* (2nd Edition). John Wiley, New York (1990).

Draper, N.R., Prescott, P., Lewis, S.M., Dean, A.M., John, P.W.M. and Tuck, M.G. Mixture designs for four components in orthogonal blocks. *Technometrics* 35(3):268-276 (1993); with correction: *Technometrics* 36(2):234 (1994).

Ghosh, S. *Statistical design and analysis of industrial experiments*. Marcel Dekker, New York (1990).

Ghosh, S., Schuchany, W.R. and Smith, W.B. *Statistics of Quality*. Marcel Dekker, New York (1996).

RESIDUAL LIFE FUNCTION IN RELAIBILITY STUDIES

RAMESH C. GUPTA
Department of Mathematics, University of Maine
Orono, Maine 04469-5752, USA

S.N.U.A. KIRMANI
Department of Mathematics, University of Northern Iowa
Cedar Falls, Iowa 50614, USA

In life testing situations, the additional lifetime given that a component has survived until time t is called the residual life function (RLF) of the component. More specifically, if X is the life of a component, then the random variable $X_t = X - t \mid X > t$ is called the RLF. In this paper the ageing properties of X_t are studied in relation to the ageing properties of X. The classes of distributions having decreasing mean residual life (DMRL), decreasing variance residual life (DVRL) and decreasing percentile residual life are studied and some characterizations of these classes are obtained. It is shown that the DVRL distributions are intimately connected to the behavior of the mean residual life function of the equilibrium distribution. Some characterization results based on the general moments of the RLF are obtained. Finally, some stochastic order relations dealing with the residual life distribution are presented.

1 Introduction

In life testing situations, the additional lifetime given that a component has survived until time t is called the residual life function (RLF) of the component. More specifically, if X is the life of a component then the random variable (r.v.) $X_t = X - t \mid X > t$ is called the RLF. In the insurance business this random variable represents the amount of claim if the deductible for a particular policy is t and X is a random variable representing the loss.

As mentioned in Berger et.al. (1988), the variable of interest need not be a lifetime and t = 0 need not correspond to "birth or a new component". For example, consider a cancer patient undergoing chemotherapy. The variable of interest would be in knowing how long he would live given that his treatment began six months ago.

Another example of interest is given by Watson and Wells (1961) in reference to the possibility of improving the useful life of items by eliminating those with short lives. Suppose that the life of a product is denoted by an r.v. X when operated in a certain well defined manner. Now suppose that every item of the product is put in operation and run until either the item fails or a time t elapses, whichever comes first. A fraction F(t) of the product, in the long run, will fail and the remaining lifetimes are the fraction 1 - F(t) that do not fail and will be denoted by X_t.

The quantity $\mu_F(t) = E(X - t \mid X > t)$ is called the mean residual life function (MRLF) or the life expectancy at age t and has been employed in life length studies by various authors eg. Hollander and Proschan (1975), Bryson and Siddiqui (1969) and Muth

(1977). Chen (1983) et. al. have developed tests for monotone MRLF using randomly censored data; Bhattacharjee (1982) and Hall and Wellner (1981) have characterized the class of mean resi- dual life functions. Limiting properties of the MRLF have been studied by Meilijson (1972) and Balkema and DeHann (1974). It has also been shown by Gupta (1975, 1981) that the MRLF determines the distribution function uniquely. In particular, it is well known that a constant MRLF characterizes the exponential distribution. For a more comprehensive review of MRLF, see Guess and Proschan (1988). Yang (1977, 1978) studied the estimate $\hat{\mu}_F(t)$ on a fixed finite interval [0,T] and showed that this estimator is asymptotically unbiased, uniformly strongly consistent and converges in distribution to a Gaussian process on [0,b] with b = F(T) < 1. She assumed F to be absolutely continuous with probability density function f positive on R^+. Hall and Wellner (1979) reformulated Yang's results, identifying her limiting process in terms of Brownian motion and dropping her absolute continuity assumption, and extended her results to the half line. They also studied the variance of the limiting process and provided simultaneous confidence bands for the MRLF. The discritized life table version of the MRLF has been considered by Chiang (1960, 1968). He also explains the construction of the complete and abridged life table.

Since the exponential distribution is characterized by the constancy of the MRLF, Hollander and Proschan (1975) derived tests of (1) H_0:F is exponential versus H_1:F is DMRL (decreasing mean residual life). The null distribution of the Hollander-Proschan statistic for DMRL has been discussed by Langenberg and Srinivasan (1979).

Another quantity which has generated interest in recent years is the variance of the RLF viz $\sigma_F^2(t) = Var(X - t \mid X > t)$. $\sigma_F^2(t)$ appears in the formula for $Var(\hat{\mu}_n(t))$ where $\hat{\mu}_n(t)$ is an estimator of the MRLF, see Hall and Wellner (1981). It also appears in the expression of weights assigned for censored observations, Schmee and Hahn (1979). Karlin (1982) has studied the monotonic behavior of $\sigma_F^2(t)$ when the density is log-convex (log concave). Defining the residual coefficient of variation as $\gamma_F(t) = \sigma_F(t)/\mu_F(t)$, Gupta (1987) has characterized the monotonic behavior of $\sigma_F^2(t)$ in terms of $\gamma_F(t)$.

Launer (1984) introduced the class of life distributions having decreasing (increasing) variance residual life (DVRL, IVRL). Gupta, Kirmani and Launer (1987) have shown that the DVRL (IVRL) distributions are intimately connected to the behavior of the MRLF of the equilibrium distribution and have given some more characterizations of these classes of distributions. They have also presented some counterexamples to demonstrate the lack of relationship between DVRL (IVRL) and NBUE (NWUE) distributions.

Gupta and Gupta (1983) have studied the higher moments $\phi_r(t) = E((X - t)^r \mid X > t)$

of RLF and showed that the constancy of $\phi_r(t)$ for one real positive number r is enough to guarantee that the distribution is exponential. They have given a method to determine the distribution if the ratio of two consecutive moments is known. Defining the concept of partial moments as $\psi_r(t) = E((X - t)^+)^r = \int_t^\infty r(x - t)^{r-1}\overline{F}(x)dx$, where $\overline{F}(x) = 1 - F(x)$, it was shown that one partial moment is enough for the determination of the distribution.

Schmittlein and Morrison (1981) pointed out that the MRLF has a number of practical shortcomings especially in situations where the data are censored. In such cases the empirical mean residual life cannot be calculated. Moreover, even in the case of complete data, the estimated mean residual life will tend to be unstable owing to its strong dependence on the very few long durations. As an alternate, they recommend the median residual life function representing the median additional time to failure given no failure by time t. Calculation of this quantity poses no difficulty as long as one is able to receive half of the observations. More generally, Haines and Singpurwalla (1974) defined the αth percentile of the RLF ($0 < \alpha < 1$) and studied the decreasing (increasing) α-percentile Residual Life class abbreviated as DPRL-α (IPRL-α). Joe and Proschan (1983(a), 1983(b)) developed some tests for these classes. Gupta and Langford (1984) have studied how closely one can determine a distribution when its α-percentile residual life function is known. They actually studied a more general functional equation f($\phi(t)$) = sf(t) called Schröder's equation and showed that the α-percentile residual life function for one specific value of α, $0 < \alpha < 1$, is not enough to determine the distribution function uniquely. However, under certain conditions, the knowledge of two percentiles α_1, α_2 of the residual life function uniquely determines the distribution, see Arnold and Brockett (1983) and Joe (1985).

The organization of the paper is as follows: Section 2 contains a review of some ageing criteria and describes how these ageing properties of the original distribution are transformed onto the ageing properties of the residual life. The MRLF and its various aspects are discussed in section 3. Section 4 deals with the variance residual life function while the general moments of the RLF are studied in section 5. The percentile of the RLF is contained in section 6. Finally, in section 7, we present some stochastic order relations dealing with the residual life distribution.

2 Some criteria of Ageing for the RLF

In this section we review some of the ageing criteria and their relationships. We also describe how these ageing properties of the original distribution are transformed into the ageing properties of the distribution of the residual life.

Let X be a continuous positive random variable representing the life of a component unit. Let F be the cumulative distribution function of X and $\overline{F}(x) = 1 - F(x)$ be the reliability function or survival function of X. Then $\overline{F}_t(x) = P(X > x + t \mid X > t)$ is the survival function of unit of age t. Evidently, any study of the phenomenon of ageing should be based on $\overline{F}_t(x)$ and functions related to it. Thus

(1) F is said to be PF_2 if lnf(x) is concave, where $f(\cdot)$ is the density corresponding to $F(\cdot)$.

(2) F is said to have increasing (decreasing) failure rate IFR (DFR) if

$\overline{F}_t(x) = \overline{F}(x + t)/\overline{F}(t)$ is decreasing (increasing) in t. If F is absolutely continuous with density f, then F is in IFR (DFR) class if the failure rate $r_F(t) = f(t)/\overline{F}(t)$ is increasing (decreasing).

(3) F is said to have increasing (decreasing) failure rate average IFRA (DFRA) if

$\int_0^t r_F(x)dx/t$ is increasing (decreasing).

(4) F is said to have new better (worse) than used, NBU (NWU) if $\overline{F}_t(x) \leq (\geq)$

$\overline{F}(x)$ for $x \geq 0$, $t \geq 0$.

(5) F is said to have decreasing (increasing) mean residual life DMRL (IMRL) if

the mean residual life $\mu_F(t) = \int_t^\infty \overline{F}(x)dx/\overline{F}(t)$ is decreasing (increasing) assuming that the mean $\mu_F(0)$ exists.

(6) F is said to have new better (worse) than used in expectation, NBUE (NWUE) if $\mu_F(t) \leq (\geq) \mu_F(0)$ for all $t \geq 0$. The chain of implications between these classes of distributions is

$$PF_2 \Rightarrow IFR \Rightarrow IFRA \Rightarrow NBU$$
$$\Downarrow \qquad \Downarrow$$
$$\Rightarrow DMRL \qquad \Rightarrow NBUE.$$

See also Barlow and Proschan (1975).

The reverse implications are not true, for counterexamples, see Bryson and Siddiqui (1969). Some extensions of these classes of distributions are contained in Klefsjö (1982, 1983), Shaked (1981), Singh and Deshpande (1985), Deshpande et.al. (1986), Basu and Ebrahimi (1984, 1985), Abouammoh (1988), Abouammoh and Ahmed (1988), and Loh (1984). It can be readily verified that

(a) $r_{F_t}(y) = r_F(t + y)$ for every $y > 0, t > 0$

(b) $\mu_{F_t}(y) = \mu_F(t + y)$ for every $y > 0, t > 0$

The following theorem describes the relationship between the ageing properties of $F(\cdot)$ and $F_t(\cdot)$.

Theorem 2.1. (Gupta, 1987).

(a) F is IFR iff F_t is NBU for every $t > 0$.

(b) F is DMRL iff F_t is NBUE for every $t > 0$.

(c) F is NBU does not imply F_t is NBUE.

(d) F is NBUE does not imply F_t is NBUE.

3 Mean Residual Life Function

The MRLF of a nonnegative r.v. X is defined as $\mu_F(t) = E(X - t^3 X > t)$

$$= \int_t^\infty \overline{F}(x)dx / \overline{F}(t).$$

It can be easily verified that

(i) $r_F(t) = [1 + \mu'_F(t)]/\mu_F(t)$

(ii) $\overline{F}(t) = \dfrac{\mu_F(0)}{\mu_F(t)} \exp\left\{ -\int_0^t \dfrac{dx}{\mu_F(x)} \right\}.$

Thus $\overline{F}(t)$, $r_F(t)$ and $\mu_F(t)$ are all equivalent in the sense that given one of them, the other two can be determined. Hence in the analysis of survival data, one sometimes estimates $r_F(t)$ or $\mu_F(t)$ instead of $\overline{F}(t)$ according to the convenience of the procedure available. It is easy to see that the constancy of $r_F(t)$ or $\mu_F(t)$ characterizes the exponential distribution. Note that both $r_F(t)$ and $\mu_F(t)$ are conditional concepts. The failure rate function at t provides information about the immediate future after time t while the MRLF provides information about the whole interval after t. Guess and Proschan (1988) remark that it is possible for the MRLF to exist but the failure rate function not to exist. Likewise it is possible for the failure rate function to exist while the MRLF does not exist.

The following theorem gives a characterization for a function to represent the MRLF of a r.v. nondegenerate at 0.

Theorem 3.1. Let g be a real valued function satisfying

(i) $g:[0,\infty) \to [0,\infty)$

(ii) $g(0) > 0$

(iii) g is right continuous

(iv) $g(t) + t$ is increasing on $[0,\infty)$

(v) If there exists a t_0 such that $g(t_0^-)$

$$= \lim_{t \to t_0^-} g(t) = 0,$$

then g(t)=0 for all $t \geq t_0$. If such a t_0 does not exist, then $\int_0^\infty \dfrac{1}{g(u)} du = \infty$ holds.

A function g satisfies the above condition iff it is the MRLF of a r.v. nondegenerate at 0; see Hall and Wellner (1981), Bhattacharjee (1982) and Guess and Proschan (1985). Bhattacharjee (1982) also gives some interesting examples of MRLFs which are of independent interest and demonstrate the richness of the MRLFs.

Since the MRLF determines the distribution uniquely, the moments of the distribution can be determined in terms of the MRLF, see Gupta (1981b) for details. Also one would expect that, in a mixture situation, given the original distribution and the MRLF of the mixture, one should be able to obtain the mixing distribution. Such a problem was investigated by Gupta (1982) for the exponential type distribution studied by Elandt-Johnson (1976). The case of the exponential distribution was considered by Morrison (1978) and reproved by Bhattacharjee (1980).

3.1 Interpretation of MRLF in Renewal Theory

Suppose a component operating in a system is replaced upon failure by another component possessing the same life distribution, so that the sequence of component life lengths forms a renewal process. At any time t, the component in operation is identified for study. Let u_t be the age of the component in use at t and v_t be the remaining life of the component. The quantities u_t and v_t are known as the backward and forward recurrence times, respectively, in renewal theory. For large values of t, assuming the life distribution is nonlattice, the limiting pdf of u_t or v_t is given by

$$g(y) = \bar{F}(y)/\mu,$$

where $\mu = E(X) < \infty$ and $F(0) = 0$. Let Y be a r.v. having this pdf. We shall denote its distribution function by $G(\cdot)$ and call it the induced distribution of F. It can be easily verified that

$$\mu_F(t)r_G(t) = 1$$

This implies that F is DMRL iff G is IFR. For the exponential distribution, F and G coincide. The following theorem characterizes the exponential distribution by the equality of the mean values of F and G.

Theorem 3.2. Suppose the renewal distribution is NBUE (or NWUE). Then $E(Y) = E(X)$ iff X has an exponential distribution.

The following theorem shows that if we drop the condition of NBUE, then the equality of the mean values of X and Y characterizes the exponential distribution in the one parameter exponential family.

Theorem 3.3.   Suppose the renewal distribution belongs to the one parameter exponential family with pdf

$$f(x; \theta) = c(\theta)h(x)e^{\theta x}.$$

If $E(Y) = E(X)$ for all θ in some interval I, then X has an exponential distribution for all θ in I, see Gupta (1981a, 1984) for details.

4 Variance Residual Life Function

The variance residual life function of a non-negative r.v. X is defined as

$$\sigma_F^2(t) = \text{Var}(X - t \mid X > t).$$

Launer (1984) introduced the class of life distributions having decreasing (increasing) variance residual life as follows.

A distribution function F is said to have decreasing (increasing) variance residual life (DVRL, IMRL) if $\sigma_F^2(t)$ is a nonincreasing (nondecreasing) function of t on $(0,\infty)$.

It can be verified that

$$\sigma_F^2(t) = \frac{2}{\overline{F}(x)} \int_t^\infty \mu_F(x)dx - \mu_F^2(t) \text{ and}$$

$$\frac{d}{dt}\sigma_F^2(t) = r_F(t)(\sigma_F^2(t) - \mu_F^2(t)).$$

Denoting the residual coefficient of variation $s_F(t)/m_F(t)$ by $\gamma_F(t)$, the following characterization is obtained, see Gupta (1987) and Gupta et. al. (1987).

<u>Theorem 4.1.</u> (i) $\sigma_F^2(t) = c^2 \Longleftrightarrow \mu_F(t) = c \Longleftrightarrow$ F is exponential.

(ii) $\sigma_F^2(t)$ is increasing iff $\gamma_F(t) \geq 1$

(iii) $\sigma_F^2(t)$ is decreasing iff $\gamma_F(t) \leq 1$.

The above theorem and Theorem 2.1 (b) yield the following important result.

<u>Theorem 4.2.</u> F is DMRL (IMRL) $\Rightarrow \sigma_F^2(t)$ is decreasing (increasing). This theorem shows that if a distribution is DMRL (IMRL), it is also DVRL (IVRL).

Although both NBUE and DVRL classes are extensions of DMRL, there is no connection between them in the sense that one does not imply the other. For example, see Launer (1984) and Gupta (1987). We now present the following useful characterizations of DVRL (IVRL) distributions.

<u>Theorem 4.3.</u> Suppose F is a strictly increasing life distribution. Then F is DVRL (IVRL) iff $E((X - t)^2 | X > t)/\mu_F(t)$ is a decreasing (increasing) function of t.

<u>Theorem 4.4.</u> Suppose F is a strictly increasing life distribution. Then F is DVRL (IVRL) iff the induced distribution corresponding to F is DMRL (IMRL).

The above theorem helps us to study the relationship between u_t (the age of the component) and v_t (the residual life of the component), see section 3, as follows.

<u>Theorem 4.5.</u> Suppose F is a strictly increasing life distribution. If F is DVRL (IVRL), then

$$\text{Cov}(u_t, v_t) \leq (\geq) 0.$$

<u>Remark.</u> Theorem 4.5 extends a result of Gupta (1984) for IFR distributions.

Finally, the following result gives a bound on the moments for DVRL (IVRL) distributions.

<u>Theorem 4.6.</u> Suppose F is a strictly increasing life distribution such that $E(X^n) < \infty$. If F is DVRL (IFRL), then

$$E(X^n) \leq (\geq) \frac{n! \, \mu_2^{n-1}}{2^{n-1} \, \mu^{n-2}}, \, n = 2, 3, \dots .$$

Note that for DMRL (IMRL) distributions, this bound is sharper than the bound given by Marshall and Proschan (1972).

5. General Moments of Residual Life Functions

The general moments of RLF are defined as

$$\phi_r(t) = E((X - r)^r \mid X > t)$$

$$= \int_t^\infty r(x - t)^{r-1} \overline{F}(x) dx / \overline{F}(t),$$

see Gupta and Gupta (1983).

Note that $\phi_1(t) = \mu_F(t)$ is the MRLF and it is well known that $\mu_F(t)$ determines the distribution function uniquely. The question, therefore, arises whether $\phi_r(t)$ for one specific value of r characterizes the distribution. The following theorem gives a differential difference equation satisfied by these moments.

Theorem 5.1. For a continuous random variable X, the residual moments satisfy

$$\phi_{r-1}'(t) = - (r - 1) \phi_{r-2}(t)$$

$$+ \frac{\phi_r'(t)}{\phi_r(t)} \phi_{r-1}(t) + \frac{r \, \phi_{r-1}^2(t)}{\phi_r(t)}.$$

Using the above result Gupta and Gupta (1983) have shown that the ratio of two consecutive residual moments is enough to determine
the distribution uniquely. The result is contained in the following theorem.

Theorem 5.2. If $\dfrac{\phi_r(t)}{\phi_{r-1}(t)}$ = g(t), then the distribution of X is known. In particular if

g(t) = constant = c for one specific integer r, then the distribution of X is exponential.

The following theorem shows that the residual coefficient of variation $\gamma_F^2(t)$ determines the distribution.

Theorem 5.3. $\gamma_F^2(t)$ determines the distribution. In particular if

(i) $\gamma_F^2(t)$ = constant $\neq$ 1, then X has a power distribution and if

(ii) $\gamma_F^2(t)$ = 1, then X has an exponential distribution.

The following theorem shows that the constancy of any residual moment implies that the distribution is exponential.

Theorem 5.4. If the rth residual moment of X is constant for any positive real number r, then, under mild conditions, X has an exponential distribution and conversely.

Define the partial moments as

$$\psi_r(t) = E[((X - t)^+)^r] = \int_t^\infty r(x - t)^{r-1}\, \overline{F}(x)dx.$$

Note that $\overline{F}(t)\phi_r(t) = \psi_r(t)$.

The following theorem shows that $\psi_r(t)$ determines the distribution for any positive real number r.

<u>Theorem 5.5.</u> Suppose X is a continuous random variable with pdf f(x). Then

$$\psi^*_{\mu-1}(s) = \frac{\Gamma(\mu)\Gamma(s)}{\Gamma(\mu+s)}\, f^*(s + \mu)$$

where $f^*(s) = \int_0^\infty x^{s-1}f(x)dx$ is the Mellin transform of f at s and μ is a positive real number.

In case μ is a positive integer, say r, the following theorem gives a simpler method of determining the distribution.

<u>Theorem 5.6.</u> For any positive integer r, $\psi_r(t)$ determines all of the partial moments that exist.

<u>Remark.</u> In case r is a positive integer, a recurrence relation exists between two consecutive partial moments. This recurrence relation determines all of the partial moments and the distribution.

6 Percentile of the Residual Life Function

The αth ($0 < \alpha < 1$) percentile of the RLF is defined as

$$q_{\alpha,F}(t) = F_t^{-1}(\alpha)$$
$$= \inf\{y : F_t(y) \geq \alpha\}$$
$$= \inf\{y : \overline{F}(t + y) \leq (1 - \alpha)\overline{F}(t)\}$$
$$= \inf\{y : \overline{F}(y) \leq (1 - \alpha)\overline{F}(t)\} - t$$

i.e. $\overline{F}(t + q_\alpha(t)) = (1 - \alpha)\overline{F}(t)$.

Define $q_{\alpha,F}(t)$ to be 0 for $t \geq \overline{F}^{-1}(1)$.

Like the MRLF, the question now arises whether $q_\alpha(t)$ determines the distribution function uniquely. For that, one has to solve the functional equation

$$\overline{F}(t + q_{\alpha,F}(t)) = (1 - \alpha)\overline{F}(t).$$

Gupta and Langford (1984) consider a more general function equation called Schröder's equation and prove the following result.

<u>Theorem 6.1.</u> Consider Schröder's equation

$$f(\phi(t)) = sf(t),\ 0 \leq t < \infty$$

where $0 < s < 1$, and ϕ is a continuous, strictly increasing function on $[0,\infty)$ and satisfies

$\phi(t) > t$ for every $t \geq 0$. The general solution of this equation is of the form

$$f(t) = f_0(t)k(\ln f_0(t))$$

where k is a periodic function with period $|\ln s|$ and f_0 is a particular solution which is positive, strictly decreasing, continuous and which satisfies $f_0(0) = 1$.

Using the above theorem, it can be seen that $q_\alpha(t)$ for one specific a does not determine the distribution function uniquely, see also Arnold and Brockett (1983) and Joe (1985). However, the following theorem shows that the knowledge of the α and ß ($\alpha \neq$ ß) percentile residual life functions determine the distribution function uniquely.

Theorem 6.2. Let F and G be continuous life distributions. If $q_{\alpha,F} \equiv q_{\alpha,,G}$ and $q_{\beta,F} \equiv \alpha_{\beta,G}$ for α and ß in (0,1) such that $\ln(1 - \alpha)/\ln(1 - \text{ß})$ is irrational then $F \equiv G$.

Some illustrative examples in terms of percentile residual life functions are contained in Gupta and Langford (1984) and in Joe (1985).

Analogous to the MRLF, Haines and Singpurwalla (1974) define some classes of distributions for the percentile RLF and study the relationship between these classes and other ageing criteria.

7 Stochastic Order Relations

Let X and Y be nonnegative absolutely continuous random variables with probability density functions f(x) and g(x) and survival functions $\overline{F}(x)$ and $\overline{G}(x)$. Then

1. X is said to be larger than Y in likelihood ratio ordering, written as

    $X \underset{\geq}{^{LR}} Y$ if f(x)/g(x) is nondecreasing in x.

2. X is said to be larger than Y in failure rate ordering, written as

    $X \underset{\geq}{^{FR}} Y$ if $r_F(x) \geq r_G(x)$ for all x.

3. X is said to be larger than Y in stochastic ordering, written as

    $X \underset{\geq}{^{st}} Y$ if $\overline{F}(x) \geq \overline{G}(x)$ for all x.

4. X is said to be larger than Y in mean residual life ordering.

    Written as $X \underset{\geq}{^{MRL}} Y$ if $\mu_F(x) \geq \mu_G(x)$ for all x. Deshpande et.al. (1990)

    show that $X \underset{\geq}{^{MRL}} Y$ iff $\int_x^\infty \overline{F}(u)du / \int_x^\infty \overline{G}(u)du$ is increasing in x.

5. X is said to be larger than Y in variability ordering written as $X \underset{\geq}{^{v}} Y$ if

    $\int_a^\infty \overline{F}(x)dx \geq \int_a^\infty \overline{G}(x)dx$ for all a ≥ 0. Note that in the literature variability ordering has also been called stop loss ordering (Hesselager, 1995) and ST2 ordering, see Belzunce et.al. and Shaked and Shanthikumar (1994). For some generalized variability ordering, see Zarek (1995), Li and Zhu (1994), Bhattacharjee (1991) and Bhattacharjee and Sethuraman (1990).

6. X is said to be larger than Y in variance residual life ordering, written as

$$X \overset{VR}{\underset{\geq}{}} Y \text{ if } \sigma_F^2 (x) \leq \sigma_G^2 (x) \text{ for all x.}$$

It is well known that

$$X \overset{LR}{\underset{\geq}{}} Y \Rightarrow X \overset{FR}{\underset{\geq}{}} Y \Rightarrow X \overset{st}{\underset{\geq}{}} Y \Rightarrow X \overset{v}{\underset{\geq}{}} Y$$

and

$$X \overset{FR}{\underset{\geq}{}} Y \Rightarrow X \overset{MRL}{\underset{\geq}{}} Y \Rightarrow X \overset{v}{\underset{\geq}{}} Y.$$

For reverse implications and counterexamples, see Gupta and Kirmani (1987) and Singh (1989). The following theorem gives some characterization results for various orderings to hold between the original variable X and the residual life variable X_t.

Theorem 7.1.

(a) $X \overset{LR}{\underset{\geq}{}} X_t$ iff X is PF_2

(b) $X \overset{FR}{\underset{\geq}{}} X_t$ iff X is IFR

(c) $X \overset{st}{\underset{\geq}{}} X_t$ iff X is NBU

(d) $X \overset{MRL}{\underset{\geq}{}} X_t$ iff X is DMRL

(e) $X \overset{VR}{\underset{\geq}{}} X_t$ iff X is DVRL

For details and proofs, see Fagiuoli and Pellery (1993), Deshpande et.al. (1990) and Singh (1989).

We now present some characterizations of the DMRL and DVRL classes in terms of residual life distributions.

Theorem 7.2.

(a) X is DMRL iff $X_t \overset{v}{\underset{\geq}{}} X_{t'}$ for all t < t'.

(b) X is DVRL iff $E(X_t^2)/E(X_t)$ is a nonincreasing function of t.

For details and proofs, see Cao and Wang (1991, 1992) and Gupta and Kirmani (1987).

Let us now define equilibrium distribution corresponding to a distribution function F (•) as

$$H_F(t) = \frac{1}{\mu_F} \int_0^t \overline{F} (x)dx.$$

We will call the random variable corresponding to $H_F(t)$ as X_E. Let F and G be two probability distributions with the same mean. Then the following result is contained in Deshpande et.al. (1990).

Theorem 7.3.

(a) $X \overset{FR}{\geq} Y$ iff $X_E \overset{LR}{\leq} Y_E$

(b) $X \overset{MRL}{\leq} Y$ iff $X_E \overset{FR}{<} Y_E$

For some other characterizations of IFR, DMRL and DVRL in terms of the order relations between F and its equilibrium distribution, see Deshpande et.al. (1990).

Let us now define the cumulative hazard corresponding to the failure rate $r_F(t)$ by

$$R_F(x) = \int_0^x r_F(t)dt.$$

Then the following are equivalent.

(a) $X_t \overset{st}{\geq} Y_t$ for all t

(b) $R_F(x) - R_G(x)$ is increasing in $x \geq 0$
(c) $R_F R_G^{-1}(x) - x$ is increasing in $x \geq 0$

(d) $X \overset{FR}{\geq} Y$

For proof and discussion, see Bartoszewicz (1987). We now define dispersive ordering as follows.

The random variable Y is more dispersed than the random variable X, written as $X \overset{disp}{<} Y$ if $F_X^{-1}(q) - F_X^{-1}(p) \leq F_Y^{-1}(q) - F_Y^{-1}(p)$ for all $0 < p < q < 1$. For this definition and some equivalent formulations, see Lewis and Thompson (1981), Shaked (1982) and Shaked and Shanthikumar (1994). The following result gives a characterization of the IFR class in terms of the residual distribution.

Theorem 7.4. The following are equivalent.

(a) X is IFR

(b) $X \overset{FR}{\geq} X_t$ for all t

(c) $X_s \overset{FR}{\geq} X_t$ for all $s \leq t$

For proof and discussion, see Belzunce et.al. (1996) and Pellerey and Shaked (1995b). Hickey (1986) gives a more general concept of dispersive ordering as follows.

A random variable Y is more dispersed in dilation than the random variable X, written as $X \underset{dil}{<} Y$ if

E[ϕ(X - E(X))] $\leq$ E[ϕ(Y - E(Y))] for every convex function ϕ. This partial ordering can be characterized in terms of variability ordering as

$$X \underset{\text{dil}}{<} Y \text{ iff } X - E(X) \overset{v}{\geq} Y - E(Y).$$

Other properties and characterizations have been given by Munoz-Perez and Sanches-Gomez (1990) and Pellerey and Shaked (1995a). The following theorem gives a characterization of the DMRL class in terms of the dispersion in dilation between the residual life variables.

<u>Theorem 7.5.</u> X is DMRL iff $X_s \underset{\text{dil}}{\overset{>}{}} X_t$ for all s < t. For proof see Belzunce et.al. (1996).

References

Abouammoh, A. M. and Ahmed, A. N. (1988). The new better than used failure rate class of life distributions. *Adv. Appl. Prob.* **20**, 237-40.

Abouammoh, A. M. (1988). On the criteria of the mean remaining life. *Statistics and Probability Letters* **6**, 205-211.

Arnold, B. C. and Brockett, P. L. (1983). When does the ßth percentile residual life function determine the distribution? *Operations Research* **31**, 391-396.

Balkema, A. A. and DeHann, L. (1974). Residual life at great age. *Annals of Probability* **2**, 792-804.

Barlow, R. E. and Proschan, F. (1975). Statistical Theory of Reliability and Life Testing: Probability Models. Holt, Rinehart and Winston, Inc. New York.

Bartoszewicz, J. (1987). A note on dispersive ordering defined by hazard functions. *Statistics and Probability Letters* **6**, 13-16.

Basu, A. P. and Ebrahimi, N. (1984). On k-order harmonic new better than used in expectation distribution. *Ann. Inst. Statist. Math.* **A 36**, 87-100.

Basu, A. P. and Ebrahimi, N. (1985). Corrections to "On k-order harmonic new better than used in expectation distributions". *Ann. Inst. Statistic Math.* **A 37**, 365-366.

Belzunce, F., Candel, J. and Ruiz, J. M. (1996). Dispersive orderings and characterizations of ageing classes. *Statistics and Probability Letters* **28**, 321-327.

Berger, R. L., Boos, D. D. and Guess, F. M. (1988). Tests and confidence sets for comparing two mean residual life functions. *Biometrics* **44**, 103-115.

Bhattacharjee, M. C. (1980). On a characterization of gamma distributions via exponential mixtures. *J. Appl. Prob.* **17**, 574-576.

Bhattacharjee, M. C. (1982). The class of mean residual lives and some consequences. *Siam J. Alg. Disc. Math.* **3 (1)**, 56-65.

Bhattacharjee, M. C. and Sethuraman, J. (1990). Families of life distributions characterized by two moments. *J. of Applied Probability* **27**, 720-725.

Bryson, C. and Siddiqui, M. M. (1969). Some criteria for ageing. *J. of American Statistical Association* **64**, 1472-83.

Cao, J. and Wang, Y. (1991). The NBUE and NWUC classes of life distributions. *J. of Applied Probability* **28**, 473-479.

Cao, J. and Wang, Y. (1992). Correction *J. of Applied Probability* **29**, 753.

Chen, Y. Y., Hollander, M. and Langberg, N. A. (1983). Tests for monotone mean residual life, using randomly censored data. *Biometrics* **39**, 119-127.

Chiang, C. L. (1960). A stochastic study of life table and its applications: I Probability distributions of the Biometric functions. *Biometrics* **16**, 618-635.

Chiang, C. L. (1968). Introduction to stochastic processes in Biostatistics. John Wiley and Sons, New York.

Deshpande, J. V., Kochar, S.C. and Singh, H. (1986). Aspects in positive ageing. *J. Appl. Prob.* **23**, 74-758.

Deshpande, J. V., Singh, H., Bagai, I. and Jain, K. (1990). Some partial orders describing positive ageing. *Communications in Statistics - Stochastic Models* **6(3)**, 471-481.

Elandt-Johnson, R. C. (1976). A class of distributions generated from distributions of exponential type. *Naval Research Logistic Quarterly* **23(1)**, 131-138.

Fagiuoli, E. and Pellerey, F. (1993). New partial orderings and applications. *Naval Research Logistics Quarterly* **40**, 829-842.

Guess, F. and Proschan, F. (1988). Mean residual life: Theory and Applications. Handbook of Statistics, ed. P. R. Krishnaiah, North Holland, Amsterdam, 215-224.

Gupta, P. L. and Gupta, R. C. (1983). On the moments of residual life in reliability and some characterization results. *Communications in Statistics* **12(4)**, 449-461.

Gupta, R. C. (1975). On characterizations of distributions by conditional expectations. *Communications in Statistics* **4(1)**, 99-103.

Gupta, R. C. (1981a). On the mean residual life function in survival studies. Statistical Distributions in Scientific Work 5, *D-Reidel Publishing Co.*, Boston, 327-334.

Gupta, R. C. (1981b). Moments in terms of mean residual life function. *IEEE Transactions on Reliability* **R30(5)**, 450-451.

Gupta, R. C. (1982). On the determination of mixing density in reliability studies. *J. of Statistical Planning and Inference* **7**, 101-105.

Gupta, R. C. (1984). Some characterizations of renewal densities with emphasis in reliability. *Math Operations Forschung and Statistik* **15**, 571-579.

Gupta, R. C. and Langford, E. S. (1984). On the determination of a distribution by its median residual life function: A functional equation. *J. Appl. Prob.* **21**, 120-128.

Gupta, R. C. (1987). On the monotonic properties of the residual variance and the applications in reliability. *Jour. Statist. Plann. Inference* **16**, 329-335.

Gupta, R. C., Kirmani, S.N.U.A. and Launer, R. L. (1987). On life distributions having monotone residual variance. *Probability in the Engineering and Informational Sciences* **1**, 299-307.

Gupta, R.C. and Kirmani, S.N.U.A. (1987). On order relations between reliability measurements. *Communications in Statistics, Stochastic Models* **3(1)**, 149-156.

Haines, A. L. and Singpurwalla, N. D. (1974). Some contributions to the stochastic characterization of wear. Reliability and Biometry, *SIAM*, Philadelphia, 47-80.

Hall, W. J. and Wellner, J. A. (1979). Estimation of mean residual life. Technical Report, University of Rochester.

Hall, W. J. and Wellner, J.A. (1981). Mean residual life. In Statistics and Related Topics, ed. M. Csorgo, D. A. Dawson, J.N.K. Rao and A.K.M.D.E. Saleh, North Holland Publishing Co., 169-184.

Hesselager, D. (1995). Order relations for some distributions. Insurance: *Mathematics and Economics* **16**, 129-134.

Hickey, R. J. (1986). Concepts of dispersions in distributions: A comparative note. *J. of Applied Probability* **23**, 914-921.

Hollander, W. and Proschan, F. (1975). Tests for mean residual life. *Biometrika* **62**,

85-593.

Joe, H. and Proschan, F. (1983a). Tests for properties of the percentile residual life function. *Communications in Statistics* **12(10)**, 1087-1119.

Joe, H. and Proschan, F. (1983b). Tests for properties of residual life. *Communications in Statistics* **12(10)**, 1121-1134.

Joe, H. (1985). Characterizations of life distributions from percentile residual life times. *Ann. Inst. Statist. Math.* **37**, 165-172.

Karlin, S. (1982). Some results on optimal partitioning of variance and monotonically with truncation level. In Statistics and Probability:Essays in honor of C. R. Rao, ed. G. Kallianpur, P. R. Krishnaiah and J. K. Ghosh, North Holland Publishing Co., 375-382.

Klefsjö, B. (1982). The HNBUE and HNWUE classes of life distributions. *Naval Research Logistics Quarterly* **29**, 331-344.

Klefsjö, B. (1983). A useful ageing property based on Laplace transform. *J. Appl. Prob.* **20**, 615-626.

Langenberg, P. and Srinivasan, R. (1979). Null distribution of the Hollander-Proschan statistic for decreasing mean residual life. *Biometrika* **66**, 679-680.

Launer, R. L. (1984). Inequalities for NBUE and NWUE life distributions. *Operations Research* **32**, 660-667.

Lewis, T. and Thompson, J. W. (1981). Dispersive distributions and the connection between dispersivity and strong unimodality. *J. of Applied Probability* **18**, 76-90.

Li, H. and Zhu, H. (1994). Stochastic equivalence of ordered random variables with applications in reliability theory. *Statistics and Probability Letters* **20**, 383-393.

Loh, W. Y. (1984). A new generalization of the class of NBU distributions. *IEEE Trans. Reliability* **R-33**, 419-422.

Marshall, A. W. and Proschan, F. (1972). Classes of distributions applicable in replacement with renewal theory implications. Proceedings of the 6th Berkeley Symposium on Mathematical Statistics and Probability, Vol. 1, ed. L. LeCam, J. Neyman and E. L. Scott, 395-415, University of California Press, Berkeley.

Meilijson, I. (1972). Limiting properties of the mean residual lifetime function. *Ann. Math. Statist.* **43**, 354-357.

Morrison, D.G. (1978). On linearly increasing mean residual lifetimes. *J. Appl. Prob.* **15**, 617-620.

Munoz-Perez, J. and Sanchez-Gomez, A. (1990). Dispersive ordering by dilation. *J. of Applied Probability* **27**, 440-444.

Muth, E. J. (1977). Reliability models with positive memory derived from the mean residual life function. In *Theory and Applications of Reliability*, ed. C. P. Tsokos and I. N. Shimi, Academic Press, 401-434.

Pellerey, F. and Shaked, M. (1995a). The dilation order and its relationship to other stochastic orders. Technical Report, University of Arizona.

Pellerey, F. and Shaked, M. (1995b). Characterizations of the IFR and DFR ageing notions by means of dispersive order. Technical Report, University of Arizona.

Schmee, J. and Hahn, G. J. (1979). A simple method for Regression Analysis with censored data. *Technometrics* **21(4)**, 417-434.

Schmittlein, D. and Morrison, D. (1981). The median residual lifetime: A characterization theorem and an application. *Operations Research* **29**, 392-399.

Shaked, M. (1982). Dispersive ordering of distributions. *J. of Applied Probability* **19**, 310-320.

Shaked, M. and Shanthikumar, J. G. (1994). Stochastic orders and their applications.

Academic Press, New York.

Shaked, M. (1981). Extensions of IHR and IHRA ageing notions. *Siam J. of Appl. Math.* **40(3)**, 542-550.

Singh, H. and Deshpande, J. V. (1985). On some new ageing properties. *Scand. J. Statist.*, **12**, 213-220.

Singh, H. (1989). On partial orderings of life distributions. *Naval Research Logistics Quarterly* **36**, 103-110.

Watson, G. S. and Wells, W. T. (1961). On the possibility of improving the mean useful life of items by eliminating those with short lives. *Technometrics* **3**, 281-298.

Yang, G. L. (1977). Life expectancy under random censorship. *Stochastic Processes and their Applications.* **6**, 33-39.

Yang, G. L. (1978). Estimation of a biometric function. *Annals of Statistics* **6(1)**, 112-116.

Zarek, M. (1995). Some generalized variability orderings among life distributions with application to Weibull's and gamma distribution. *Fasciculi Mathematics* **25**, 197-209.

PARAMETRIC RANDOM EFFECTS MODELS BASED ON THE EXPONENTIAL AND WEIBULL DISTRIBUTIONS

ALAN C. KIMBER, CHANGQING ZHU

Department of Mathematical and Computing Sciences, University of Surrey,
Guildford GU2 5XH, England
E-Mail: a.kimber@surrey.ac.uk

The exponential distribution, or more generally the Weibull distribution, is a natural starting point in the modelling of lifetime data arising in reliability and materials strength data. There has been a growing interest in recent years in allowing for heterogeneity, or frailty, within this context. To achieve parsimonious parametric models it is natural to consider continuous mixtures of exponential or Weibull distributions. Moreover, for certain mixing distributions tractable results may be obtained. In this paper some results for such exponential-based and Weibull-based random effects models are discussed.

1 Introduction

The Weibull distribution has long been used as a natural starting point for the parametric analysis of lifetime data in reliability and for materials strength data, and its properties are well known [1][2]. For example, the Weibull distribution is an extreme value distribution, which gives it theoretical plausibility in situations where a 'weakest link' failure mechanisms is thought to act. The Weibull distribution is flexible in terms of shape (positive or negative skewness) and hazard function (decreasing, flat, increasing) and has a closed form survivor function which is particularly covenient for dealing with censored observations, estimating quantiles, and constructing probability plots. On the log-scale the Weibull distribution can be parametrised simply in terms of a location parameter and a scale parameter. Moreover, covariate information may be incorporated in a natural way, and the Weibull model is both an accelerated life model and a proportional hazards model [1].

In recent years there has been considerable interest in random effects or frailty models in the context of lifetime data [3][4][5][6][7][8][9][10][11]. A natural approach is to consider a mixture, either discrete or continuous, of Weibull distributions in the following way. Suppose that a positive random variable Y has, conditional on a positive quantity η, the survivor function

$$S(y|\eta) = \exp(-\eta s), \tag{1}$$

where

$$s = \xi y^\phi. \tag{2}$$

That is Y, conditionally on η, is a Weibull random variable with positive parameters ξ and ϕ. Note that when $\phi = 1$ the exponential distribution is obtained. If η is allowed to vary with distribution function G on $(0, \infty)$, then the unconditional survivor function of Y is

$$S(y) = \int_0^\infty \exp(-\eta s) dG(\eta), \tag{3}$$

where s is as above.

Various choices of G lead to a tractable unconditional survivor function for Y. For example, if G is an appropriate discrete distribution, then a finite mixture distribution for Y is obtained. However, note that even a two-component mixture, the simplest possible such finite mixture, has four parameters. In the interests of parsimony we shall concentrate on simple continuous mixtures that give rise to three-parameter models in the case of a Weibull-based random effects model and to two-parameter models in the case of an exponential-based random effects model. In particular we shall discuss the situations in which G corresponds to a gamma distribution, an inverse Gaussian distribution and a positive stable distribution.

¿From a practical point of view such models may make good scientific or engineering sense. For example, in manufacturing an item whose lifetime distribution is of interest, raw materials may be of variable quality, there may be variations in the fabrication process, or a potentially important covariate may have been missed. Any of these might cause variability over and above that anticipated for a Weibull model.

In section 2 we discuss the properties of these models and basic statistical methods. In section 3 we focus on methods for detecting heterogeneity and present some new results for the special case of exponential-based random effects models. Multivariate generalisations are discussed in section 4, followed by some concluding remarks in section 5.

2   Some Tractable Random Effects Models

2.1   The Weibull-Positive Stable Mixture

Suppose that mixing distribution G corresponds to a positive stable distribution with characteristic exponent σ (with $0 < \sigma \le 1$) [12]. Then, using standard results for Laplace transforms, it follows immediately that

$$S(y) = \exp(-s^{\sigma}). \tag{4}$$

That is, the unconditional distribution of Y is another Weibull distribution. Hence the Weibull distribution may be regarded as a proper mixture of Weibull distributions. Thus, without additional information, it is impossible to detect heterogeneity of this type in general, purely on the basis of observed data. Note that if we start by assuming an exponential model (so that $\phi = 1$), the resulting random effects model is Weibull with shape parameter σ. Hence a Weibull model with decreasing hazard may be thought of as proper mixture of exponential variables. Note also that the case $\sigma = 1$ leads to a degenerate mixture which leaves the basic Weibull or exponential distribution unchanged.

2.2   The Weibull-Gamma Mixture

Suppose that the mixing distribution G corresponds to a gamma distribution with positive shape parameter α and unit scale parameter. Then direct evaluation of (3) shows [4] that

$$S(y) = (1 + s)^{-\alpha}. \tag{5}$$

Note that this is not a Weibull distribution: it is sometimes known as a Burr distribution. Its hazard function,

$$h(y) = \frac{\alpha \xi \phi y^{\phi-1}}{1 + \xi y^{\phi}},$$ (6)

is either decreasing ($\phi \leq 1$) or upturned bathtub shaped ($\phi > 1$). Note that α acts as a scaling parameter on the hazard function.

To find moments it is natural to work on the log scale since the underlying Weibull distribution is a location-scale model on this scale. Crowder[4] shows that

$$E(\log Y) = -\phi^{-1}(\log \xi + \gamma + \psi(\alpha)) \text{ and } Var(\log Y) = \phi^{-2}(\psi'(\alpha) + \frac{\pi^2}{6}),$$ (7)

where γ is Euler's constant and ψ and ψ' are, respectively, the digamma function and its derivative. The key property here is the the variance does not depend on ξ. Thus, in line with traditional, Normal-theory random effects models, it is a natural starting point to include any covariate information in the ξ parameter. In particular, a log-linear structure is usually assumed:

$$\log \xi = \mathbf{x}^T \beta,$$ (8)

where $\mathbf{x}$ is a vector of known covariates and β is a vector of regression coefficients. This model is a natural generalisation of the standard Weibull regression model[4].

Let $Y_1, \ldots, Y_n$ be n independent and identically distributed positive random variables and define V by

$$V = min(Y_1, \ldots, Y_n).$$ (9)

Then if each of the Y_i has survivor function (5), it follows[1] that V also has survivor function of the same form as (5) but with α replaced by $n\alpha$. Thus, the Weibull-gamma mixture is closed under the operation of finding a univariate minimum.

2.3 The Weibull-Inverse Gaussian Mixture

Suppose that G corresponds to an inverse Gaussian distribution. Then we have

$$S(y) = \exp[\theta - (\theta^2 + s)^{\frac{1}{2}}]$$ (10)

for positive parameter θ. Note that (10) is a special case of the distribution of Crowder[12] with, in his notation,

$$S(y) = \exp[\kappa^{\nu} - (\kappa + s)^{\nu}].$$ (11)

Clearly $\nu = \frac{1}{2}$ and $\kappa = \theta^2$. It is interesting to note that survivor function (11) arises as a generalisation of the Weibull-positive stable mixture. Returning now to the Weibull-inverse Gaussian mixture with survivor function (10), it follows that the hazard function is given by

$$h(y) = \frac{\xi \phi y^{\phi-1}}{2(\theta^2 + s)^{\frac{1}{2}}}.$$ (12)

The effect of the additional parameter θ in this case on the hazard function is more complex than in the Weibull-gamma hazard function. Note also that a wider variety of hazard shapes is possible here than in the Weibull-gamma case: $h(y)$ is decreasing ($\phi \leq 1$), upturned bathtub shaped ($1 < \phi < 2$), and increasing ($\phi \geq 2$).

Moments of $\log Y$ are not particularly illuminating but covariate information may be introduced as in (8) for the Weibull-gamma model.

With V as given in (9) and with the Y_i with survivor function (10), it is easy to show that V has survivor function of the same form as (10) but with θ and ξ replaced by $n\theta$ and $n^2\xi$ respectively, thereby confirming a similar closure property to that of Weibull-gamma minima.

3 Detecting Heterogeneity

3.1 Preliminary Remarks

Many statistical methods are available for detecting departures from an underlying Weibull (or exponential) model. These range from informal methods, such as probability plotting [1], to more formal goodness-of-fit tests [13]. These methods are potentially useful but are not specifically aimed at detecting heterogeneity. In this section we shall review an approach, based on an appropriate score test, that is designed to detect heterogeneity in the present context.

3.2 A Score Test for Heterogeneity

Suppose that a random sample $y_1, \ldots, y_n$ is available from a distribution with survivor function (3). We would like to be able to test whether there is any heterogeneity or not. This is complicated by the results of subsection 2.1. However, if we restrict attention to mixing distributions with finite variance, then this effectively rules out the very special form of heterogeneity induced by the positive stable mixing distribution, which has infinite variance. To fix ideas, suppose that the underlying Weibull parameters, ξ and ϕ, are known. Then Crowder and Kimber [11] show that a score test statistic for this type of heterogeneity is

$$T_n = \sum_{i=1}^{n} (s_i - \frac{s_i^2}{2}), \tag{13}$$

where

$$s_i = \xi y_i^{\phi}. \tag{14}$$

Under the null hypothesis that the observations are independent Weibull observations, the s_i are independent unit exponential quantities. Thus, it follows that, under the null hypothesis, as $n \to \infty$

$$\frac{T_n}{\tau_n} \to N(0,1), \tag{15}$$

where $\tau_n^2 = 2n$. Large negative values of T_n indicate the presence of heterogeneity. Non-null properties of T_n are derived by Crowder and Kimber [11] in the Weibull-gamma mixture case.

3.3 The Weibull Case with Nuisance Parameters

In practice it is more usual that the underlying Weibull parameters, ξ and ϕ, are unknown. Let $\hat{\xi}$ and $\hat{\phi}$ denote, respectively, the maximum likelihood estimates of ξ and ϕ under the null hypothesis and let

$$\hat{s}_i = \hat{\xi} y_i^{\hat{\phi}}. \tag{16}$$

Then a natural generalisation of T_n is

$$\hat{T}_n = \sum_{i=1}^{n} (\hat{s}_i - \frac{\hat{s}_i^2}{2}). \tag{17}$$

The asymptotic null distribution of $\hat{T}_n$ was found by Kimber [10], who applied the results of Pierce [14], to be such that, as $n \to \infty$

$$\frac{\hat{T}_n}{\hat{\tau}_n} \to N(0, 1), \tag{18}$$

with

$$\hat{\tau}_n^2 = \tau_n^2 - n(1 + \frac{6}{\pi^2}) = n(1 - \frac{6}{\pi^2}). \tag{19}$$

Thus, naive use of $\hat{T}_n$ in place of T_n without adjusting for parameter estimation would lead to a seriously conservative procedure because

$$\tau_n^2 > 5\hat{\tau}_n^2.$$

Suppose now that there is the possibility of some observations being right censored. To fix ideas, suppose that all lifetimes y_i are subject to right censoring at a lifetime $c > 0$. Define

$$d = \hat{\xi} c^{\hat{\phi}} \tag{20}$$

and let

$$\Gamma(d, a, m) = \int_0^d (\log u)^m u^{a-1} \exp(-u) du. \tag{21}$$

Then define the matrix $\mathbf{j}$ and the vector $\mathbf{b}$ by

$$\mathbf{j} = \begin{pmatrix} \Gamma(d,1,0) & \Gamma(d,1,1)+\Gamma(d,1,0) \\ \Gamma(d,1,1)+\Gamma(d,1,0) & \Gamma(d,1,2)+2\Gamma(d,1,1)+\Gamma(d,1,0) \end{pmatrix} \tag{22}$$

and

$$\mathbf{b}^T = \begin{pmatrix} \Gamma(d,2,0) & \Gamma(d,2,1)+\Gamma(d,2,0) \end{pmatrix} \tag{23}$$

respectively.

Then under the null hypothesis

$$\frac{\hat{T}_n}{\hat{\tau}_{nc}}$$

is asymptotically standard Normal with

$$\hat{\tau}_{nc}^2 = n[\Gamma(d,3,0) - \mathbf{b}^T \mathbf{j}^{-1} \mathbf{b}]. \tag{24}$$

Note that as $c \to \infty$, which is equivalent to the case of no censoring, $\hat{\tau}_{nc} \to \hat{\tau}_n$ as expected.

3.4 The Exponential Case with Nuisance Parameters

Suppose now that $\phi = 1$ is known so that the underlying distribution is exponential and that there is censoring at c. In the null case the maximum likelihood estimate for ξ is

$$\tilde{\xi} = \frac{r}{\sum_{i=1}^{n} y_i}, \tag{25}$$

where r is the number of uncensored items. Let $\tilde{s}_i = \tilde{\xi} y_i$ and

$$\tilde{T}_n = \sum_{i=1}^{n} (\tilde{s}_i - \frac{\tilde{s}_i^2}{2}). \tag{26}$$

Then to find the asymptotic null distribution of $\tilde{T}_n$, it follows that only the entry $(1, 1)$ of $\mathbf{j}$ and the first entry of $\mathbf{b}$ need be considered, yielding

$$\tilde{\tau}_{nc}^2 = n \left[\Gamma(\tilde{d}, 3, 0) - \frac{\Gamma(\tilde{d}, 2, 0)^2}{\Gamma(\tilde{d}, 1, 0)} \right], \tag{27}$$

where $\tilde{d} = \tilde{\xi} c$.

After some simplification, it follows that

$$\tilde{\tau}_{nc}^2 = n \left[1 - \exp(-\tilde{d}) - \frac{\tilde{d}^2 \exp(-\tilde{d})}{1 - \exp(-\tilde{d})} \right]. \tag{28}$$

Then, in the null case, asymptotically

$$\frac{\tilde{T}_n}{\tilde{\tau}_{nc}} \rightarrow N(0, 1). \tag{29}$$

If we let $c \rightarrow \infty$ in the above, we obtain the variance of $\tilde{T}_n$ to be n, which corresponds to there being no censoring.

The modification to allow for more general known values of ϕ is entirely straightforward.

These new results cover a gap in the literature and enable testing for heterogeneity in important cases with known ϕ, such as the exponential ($\phi = 1$) and the Rayleigh ($\phi = 2$) underlying distributions.

4 Multivariate Random Effects Models

4.1 Motivation

Whilst in many practical situations in reliability one deals with univariate lifetimes, multivariate lifetime distributions have been used and have the potential for wider application. For example, one might be interested in component lifetimes with p components sampled from each of n batches. Given the batch it may be that component lifetimes are thought to be independent Weibull quantities. However, if there is variability between batches, then a p-variate version of a Weibull-based random effects model may be useful. Various applications are given in the literature [1 3 4 5 8 9 10 11].

4.2   Some General Results

It is entirely straightforward to generalise results (1) and (3) to the p-variate situation. Let $\mathbf{Y} = (Y_1, \ldots, Y_p)^T$ be a p-variate observation such that, conditional on a quantity η, the p components are independent Weibull random variables. That is $\mathbf{Y}$ has joint conditional survivor function

$$S(\mathbf{y}|\eta) = \exp(-\eta s_p), \tag{30}$$

where

$$s_p = \sum_{j=1}^{p} \xi_j y^{\phi_j}. \tag{31}$$

Of course, s_1 is identical to s given in (2) earlier.

With G as before, it follows immediately that the joint unconditional survivor function of $\mathbf{Y}$ is

$$S(\mathbf{y}) = \int_0^\infty \exp(-\eta s_p) dG(\eta). \tag{32}$$

By choosing positive stable, gamma, and inverse Gaussian mixing distributions as in the univariate case, the joint survivor function of $\mathbf{Y}$ will be identical in form to (4), (5), and (10) respectively except that in each case s must be changed to s_p.

The marginal properties are, of course, the same as for the univariate situation, and all three Weibull mixtures have an equi-correlated structure with positive association between components. Note, however, that the Crowder generalisation[12], see (11), of the Weibull-positive stable mixture can have negative association between components.

Consider now the variable $U = min(Y_1, \ldots, Y_p)$. This variable might be of interest within a competing risks framework where only the time to failure of the 'weakest' of p components is observed. Suppose also that all of the ϕ_j are equal. Then for the Weibull-positive stable mixture, U has survivor function of the same form as (4) but with ξ replaced by $\sum \xi_j$. Similar results hold for the Weibull-gamma mixture and for the Weibull-inverse Gaussian mixture.

4.3   Detection of Heterogeneity in the Multivariate Context

The score test for heterogeneity discussed in the univariate framework may be generalised to cope with the p-variate case as follows. Suppose that an uncensored random sample $\mathbf{y}_1, \ldots, \mathbf{y}_n$ of p-variate observations is available. Under the null hypothesis, the p components are independent Weibull observations, with component j having Weibull parameters ξ_j and ϕ_j. Then a score test statistic for heterogeneity in the case where all the Weibull parameters are known is

$$T_{np} = \sum_{i=1}^{n} (ps_{pi} - \frac{s_{pi}^2}{2}) - \frac{np(p-1)}{2}, \tag{33}$$

where s_{pi} is s_p for observation i. Crowder and Kimber[11] show that this statistic, under the null hypothesis, has mean zero, variance $np(p+3)/2$, and is asymptotically Normal. As before, large negative values indicate the presence of heterogeneity.

If $\hat{T}_{pn}$ is the corresponding score statistic where the Weibull parameters have been estimated by maximum likelihood under the null model, then Crowder and Kimber [11] show that similar null properties are obtained except that the variance is now

$$\frac{np(p+3)}{2} - np\left(1 - \frac{6}{\pi^2}\right).$$

Likewise, if $\tilde{T}_{pn}$ is the corresponding score statistic where the ϕ_j are known (for example, all the ϕ_j equal to 1) but where the ξ_j are estimated by maximum likelihood under the null model, it may be shown, using an approach similar to that adopted in subsection 3.4, that the corresponding variance term is

$$\frac{np(p+1)}{2}.$$

If right censoring is allowed, then with nuisance parameters the variance calculations become algebraically complicated, though they may be obtained numerically relatively straightforwardly. Crowder and Kimber [11] give details and an example in the bivariate case ($p = 2$).

These score statistics essentially try to detect both marginal departures from the Weibull assumption and correlation between components. Since the Weibull-positive stable mixture has Weibull marginal behavious, it follows that this type of score statistic will be inappropriate in this situation. As yet unpublished work by the authors [15] makes a start at addressing this particular problem.

5   Concluding Remarks

The models presented here give simple and parsimonious ways for modelling heterogeneity in Weibull, exponential, or related data. Because of the closed form survivor functions for these models, it is easy to fit these distributions by maximum likelihood with or without right censored observations.

The score tests for heterogeneity discussed here require only maximum likelihood fitting under the null (Weibull) model. This aspect may be particularly valuable for working engineers who have good, robust software for fitting Weibull models but perhaps little in the way of support for fitting more complex models. Here the score tests would warn of heterogeneity. Once alerted to this, further investigation on the basis of scientific or engineering knowledge might enable the data modelling problem to be brought back into the Weibull framework by inclusion of appropriate covariates.

Acknowledgements

The authors thank Martin Crowder for helpful discussions about random effects models. Changqing Zhu is supported by a University of Surrey Research Scholarship and a CVCP Overseas Research Studentship award.

References

1. M. J. Crowder, A. C. Kimber, R. L. Smith and T. J. Sweeting *Statistical Analysis of Reliability Data*, (Chapman and Hall, London, 1991).
2. N. L. Johnson and S. Kotz *Continuous Univariate Distributions-1*, (Wiley, New York, 1970).
3. P. Hougaard, Lifetable methods for heterogeneous populations: distributions describing the heterogeneity *Biometrika* **71** 75 (1984).
4. M. J. Crowder, A distributional model for repeated failure time methods *Journal of the Royal Statistical Society* **B47** 447 (1985).
5. A. C. Kimber and M. J. Crowder, A repeated measures model with applications in psychology *British Journal of Mathematical and Statistical Psychology* **43** 283 (1990).
6. P. Hougaard, Modelling heterogeneity in survival data *Journal of Applied Probability* **28** 695 (1991).
7. N. P. Jewell, A. C. Kimber, M-L. T. Lee and G. A. Whitmore (eds) *Lifetime Data: Models in Reliability and Survival Analysis*, (Kluwer, Dordrecht, 1995).
8. J. H. Shih and T. A. Louis, Assessing gamma frailty models for clustered failure time data *Lifetime Data Analysis* **1** 205 (1995).
9. J. T. Wassell, G. W. Kulczycki and E. S. Moyer, Frailty models of manufacturing effects *Lifetime Data Analysis* **1** 161 (1995).
10. A. C. Kimber, A Weibull-based score test for heterogeneity *Lifetime Data Analysis* **2** 63 (1996).
11. M. J. Crowder and A. C. Kimber, A score test for the multivariate Burr and other Weibull mixture distributions. To appear, Scandinavian Journal of Statistics (1997).
12. M. J. Crowder, A multivariate distribution with Weibull connections *Journal of the Royal Statistical Society* **B51** 93 (1989).
13. R. B. D'Agostino and M. A. Stephens *Goodness-of-Fit Techniques*, (Marcel Dekker, New York, 1986).
14. D. A. Pierce, The asymptotic effect of substituting estimators for parameters in certain types of statistics *Annals of Statistics* **10** 475 (1982).
15. C. Zhu and A. C. Kimber, A score test for a Weibull-based random effects model. University of Surrey Technical Report in Statistics, 1997.

STOCHASTIC COMPARISONS OF SPACINGS AND ORDER STATISTICS

SUBHASH C. KOCHAR

Indian Statistical Institute
7, SJS Sansanwal Marg
New Delhi-110016, India
E-mail: kochar@isid.ernet.in

It is well known that the normalized spacings based on a random sample from a life distribution are independent and identically distributed if and only if the underlying life distribution is exponential. It is of interest to investigate the stochastic properties of the spacings when the parent observations do not necessarily constitute a random sample from an exponential distribution. In this paper we review some of the results obtained recently in this area. We discuss the cases when the parent observations are identically as well as non-identically distributed. But most of the time we shall be assuming that the observations are independent. The case of independent exponentials with unequal scale parameters is discussed in detail. In particular, the following result and its consequences are discussed. Let $D_1, \ldots, D_n$ be the normalized spacings associated with independent exponential random variables $X_1, \ldots, X_n$, where X_i has hazard rate λ_i, $i = 1, \ldots, n$. Let $D_1^\star, \ldots, D_n^\star$ be the normalized spacings of a random sample $Y_1, \ldots, Y_n$ of size n from the exponential distribution with hazard rate $\overline{\lambda} = \sum_{i=1}^n \lambda_1/n$. Then for any $n \geq 2$, the random vector $(D_1, \ldots, D_n)$ is greater than the random vector $(D_1^\star, \ldots, D_n^\star)$ in the sense of multivariate likelihood ratio ordering. It is also true that for any j between 2 and n, the survival function of $X_{j:n} - X_{1:n}$ is Schur convex in $(\lambda_1, \ldots, \lambda_n)$. We also discuss some of the recently obtained results on order statistics from non i.i.d. exponentials. It is proved that the hazard rate of $X_{n:n}$ is smaller than that of $Y_{n:n}$ and that $Y_{n:n}$ is less dispersed than $X_{n:n}$. While the survival function of $X_{n:n}$ is Schur convex in λ, Boland, El-Neweihi and Proschan [*J. Appl. Prob.*, **31** (1994), 180-192] have shown that for $n > 2$, the hazard rate of $X_{n:n}$ is not Schur concave. However, the reversed hazard rate of $X_{n:n}$ is Schur convex in λ.

1 Introduction

Order statistics and spacings play an important role in reliability theory in particular and in statistics, in general. An important motivation for considering order statistics arises in reliability theory while studying k-out-of-n systems. A k-out-of-n system functions if and only if at least k components function. Note that the time to failure of a k-out-of-n system corresponds to the $(n - k + 1)$th order statistic from the set of underlying random variables. Thus the time to failure of a series system corresponds to the first order statistic while that of a parallel system corresponds to the largest order statistic. Like order statistics, sample spacings are also of great importance in statistics. In the reliability context they correspond to times elapsed between successive failures of components in a system. Many of the tests of goodness-of-fit are based on functions of sample spacings. The sample range, which is commonly used in practice, is also a function of the sample spacings.

Let $X_1, \ldots, X_n$ be n random variables with marginal probability distribution functions $F_1, \ldots, F_n$, respectively. We shall denote by $X_{i:n}$ the ith order statistic of $X_1, \ldots, X_n$. The random variables $X_{i:n} - X_{i-1:n}$ are called sample spacings and $D_{i:n} = (n - i + 1)(X_{i:n} - X_{i-1:n})$ is called the ith normalized spacing, $i = 1, \ldots, n$, with $X_{0:n} \equiv 0$. To simplify notation, we shall drop the second suffix n in $D_{i:n}$ when there is no ambiguity.

The distribution theory of order statistics and spacings from a single underlying population has been extensively studied in the literature. There are excellent books by David [10] and Arnold, Balakrishnan and Nagaraja [1] on this subject. It is not uncommon in practice to encounter systems where the components' life times are not independent and identically distributed. However, not much attention has been given to this case partly because of the reason that the distribution theory becomes quite complicated in this case. The above mentioned books on order statistics only briefly touch upon this topic. The theory of permanents provides an effective tool to handle the distribution theory of order statistics and spacings of non-identical but independent observations (cf. Bapat and Beg [3]).

Partial orderings of probability distributions is an interesting branch of applied probability which helps in getting useful bounds on survival functions and other distributional properties of various statistics. Some interesting partial ordering results on order statistics from independent but non-identically distributed random variables have been obtained by Sen [24], Pledger and Proschan [21], Proschan and Sethuraman [22], Bapat and Kochar [4], Boland, El-Neweihi and Proschan [7], Boland, Hollander, Joag-Dev and Kochar [8], Shaked and Shanthikumar [26], and Kochar [15], among others. The topic of stochastic orders for order statistics has been surveyed in an excellent review article by Boland, Shaked and Shanthikuamr [9]. Many of these results can also be found scattered in the various chapters of the treatise of Shaked and Shanthikumar [25] on stochastic orders.

An excellent review of the work on spacings up to 1965 is given in an expository article by Pyke [23]. In the present article we shall consolidate the results on stochastic comparisons of spacings from life distributions obtained during the last few years. Important contributions in this area have been made by Barlow and Proschan [5], Pledger and Proschan [21], David and Groenveld [11], Kochar and Kirmani [16], Kochar and Korwar [17], and Kochar and Rojo [19], among others. In the next section we review the various partial orderings of distributions that we shall be using in this paper. The third section is devoted to general partial ordering results on sample spacings from certain nonparametric families of life distributions. The cases of i.i.d. as well as non-i.i.d. random variables are discussed. Section 4 is devoted to the study of stochastic comparisons of spacings when the original observations are independent exponentials with unequal scale parameters. As the results on stochastic orders for order statistics prior to 1996 have been discussed in detail in Boland, Shaked and Shanthikuamr [9], in the last section we briefly discuss some of the recently obtained results in this area.

2   Some useful partial orderings of probability distributions

Let X and Y be two random variables with density functions f and g, distribution functions F and G, and survival functions $\overline{F}$ and $\overline{G}$, respectively. Being lifetimes, we shall assume throughout this paper that all the random variables under consideration are non-negative. The failure (or hazard) rate of a random variable X with distribution function F is defined as $r_F(x) = f(x)/\overline{F}(x)$, and it is said to have a *decreasing failure rate* (DFR) distribution if its failure rate $r_F(t)$ is nonincreasing in t for $t \geq 0$. The dual of the DFR property is the IFR (*increasing failure rate*) property if the failure rate is nondecreasing on $[0, \infty)$. The *reversed hazard rate* of a life distribution F is defined as $\tilde{r}_F(x) = f(x)/F(x)$. The *mean residual life* function of a random variable X with distribution function F is defined as $\mu_F(t) \equiv E(X - t | X > t) = \int_0^\infty \overline{F}(x)\,dx / \overline{F}(t)$. We say that F is an *increasing mean residual life* (IMRL) distribution if its mean residual life function $\mu_F(t)$ is nondecreasing in t for all $t \geq 0$. The dual of the IMRL property is the DMRL (*decreasing mean residual life*) property if $\mu_F(t)$ is nonincreasing in t. See Barlow and Proschan [6] for more details regarding these nonparametric classes of life distributions.

Now we give some definitions of stochastic dominance between random variables.

Definition 2.1 *X is said to be smaller than Y according to likelihood ratio ordering (denoted by $X \leq_{lr} Y$) if*

$$f(x)/g(x) \quad \text{is nonincreasing in } x . \tag{1}$$

Definition 2.2 *X is said to be smaller than Y according to hazard (or failure) rate ordering (denoted by $X \leq_{hr} Y$) if*

$$r_G(x) \leq r_F(x) \quad \text{for } x \geq 0. \tag{2}$$

It can been seen that $X \leq_{hr} Y$ if and only if $\overline{F}(x)/\overline{G}(x)$ is nonincreasing in x .

Definition 2.3 *X is said to be smaller than Y according to reversed hazard (or failure) rate ordering (denoted by $X \leq_{rh} Y$) if*

$$\tilde{r}_F(x) \leq \tilde{r}_G(x) \quad \text{for all } x, \tag{3}$$

or equivalently, if

$$F(x)/G(x) \text{ is nonincreasing in } x.$$

Perhaps the most commonly used and easily understood notion of stochastic dominance is the notion of *usual stochastic order* as defined below.

Definition 2.4 *X is said to be smaller than Y according to usual stochastic order (denoted by $X \leq_{st} Y$) if*

$$\overline{F}(x) \leq \overline{G}(x) \quad \text{for all } x. \tag{4}$$

It can be seen that $X \leq_{st} Y$ if and only if

$$E[\phi(X)] \leq E[\phi(Y)] \quad \text{for all increasing functions } \phi : \mathcal{R} \to \mathcal{R}, \tag{5}$$

for which the expectations exist.

Definition 2.5 *X is said to be smaller than Y according to mean residual life ordering (denoted by $X \leq_{mrl} Y$) if*

$$\mu_F(t) \leq \mu_G(t) \quad \text{for all } t \geq 0 \text{ such that } \overline{F}(t) > 0. \tag{6}$$

Figure 1 depicts the chain of implications among the above stochatic orders.

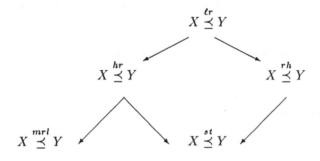

Figure 1 : Implications among notions of stochastic dominance

The above notions of stochastic dominance among univariate random variables can be extended to the multivariate case.

Definition 2.6 *The random vector $\mathbf{X} = (X_1, \ldots, X_n)$ is smaller than the random vector $\mathbf{Y} = (Y_1, \ldots, Y_n)$ in the multivariate stochastic order (denoted by $\mathbf{X} \overset{st}{\preceq} \mathbf{Y}$) if*

$$E[\phi(\mathbf{X})] \leq E[\phi(\mathbf{Y})], \tag{7}$$

for all increasing functions ϕ whenever the expectations exist.

Karlin and Rinott [14] introduced and studied the concept of multivariate likelihood ratio ordering. Let f and g denote the density functions of $\mathbf{X}$ and $\mathbf{Y}$, respectively.

Definition 2.7 *The random vector* **X** *is smaller than the random vector* **Y** *in the multivariate likelihood ratio order (denoted by* $\mathbf{X} \overset{lr}{\preceq} \mathbf{Y}$*) if*

$$f(\mathbf{x})g(\mathbf{y}) \leq f(\mathbf{x} \wedge \mathbf{y})g(\mathbf{x} \vee \mathbf{y}) \quad \text{for every } \mathbf{x} \text{ and } \mathbf{y} \text{ in } \mathcal{R}^n, \tag{8}$$

where $\mathbf{x} \wedge \mathbf{y} = (min(x_1, y_1), \ldots, min(x_n, y_n))$ *and* $\mathbf{x} \vee \mathbf{y} = (max(x_1, y_1), \ldots, max(x_n, y_n))$.

It is known that multivariate likelihood ratio ordering implies multivariate stochastic ordering, but the converse is not true. Also if two random vectors are ordered according to multivariate likelihood ratio ordering, then their corresponding subsets are also ordered accordingly. See Chapters 1 and 4 of Shaked and Shanthikuamr [25] for more details on various kinds of stochatic orders, their interrelationships and their properties.

We shall also need the concepts of majorization of vectors and Schur convexity of functions. Let $\{x_{(1)} \leq x_{(2)} \leq \cdots \leq x_{(n)}\}$ denote the increasing arrangement of the components of the vector $\mathbf{x} = (x_1, x_2, \cdots, x_n)$. The vector $\mathbf{y}$ is said to *majorize* the vector $\mathbf{x}$ (written as $\mathbf{x} \overset{m}{\preceq} \mathbf{y}$) if $\sum_{i=1}^{j} y_{(i)} \leq \sum_{i=1}^{j} x_{(i)}$, $j = 1, \ldots, n-1$ and $\sum_{i=1}^{n} y_{(i)} = \sum_{i=1}^{n} x_{(i)}$.

Definition 2.8 *A real-valued function* ϕ *defined on a set* $A \subset \mathcal{R}^n$ *is said to be Schur convex (Schur concave) on* A *if* $\mathbf{x} \overset{m}{\preceq} \mathbf{y} \Rightarrow \phi(\mathbf{x}) \leq (\geq)\phi(\mathbf{y})$.

When one is faced with the problem of stochastically comparing dependent random variables, it may not be enough to consider only their marginal distributions as the dependence information would be ignored in that case. Realizing this, Shanthikumar and Yao [27] introduced some new stochastic orders for comparing the components of a random vector. We focus our discussion on the extension of the idea of likelihood ratio ordering. For two *independent* random variables X_1 and X_2, it is known that $X_1 \leq_{lr} X_2$ if and only if

$$E[\phi(X_2, X_1)] \leq E[\phi(X_1, X_2)], \quad \forall \, \phi \in \mathcal{G}_{\ell r}, \tag{9}$$

where,

$$\mathcal{G}_{\ell r} : \{\phi : \phi(x_2, x_1) \leq \phi(x_1, x_2), \, \forall \, x_1 \leq x_2\}. \tag{10}$$

Motivated by the above characterization of likelihood ratio ordering, Shanthikumar and Yao [27] extended this concept to the bivariate case as follows:

Definition 2.9 *For a bivariate random variable* (X_1, X_2)*,* X_1 *is said to be smaller than* X_2 *according to joint likelihood ordering* $(X_1 \leq_{lr:j} X_2)$ *if and only if (9) holds.*

It can be seen that

$$X_1 \leq_{lr:j} X_2 \quad \Leftrightarrow \quad f \in \mathcal{G}_{lr},$$

where $f(\cdot, \cdot)$ denotes the joint density of (X_1, X_2).

A bivariate function $\phi \in \mathcal{G}_{\ell r}$ is called *arrangement increasing* (AI). In their seminal work on order relations between the components of a bivariate random vector, Yanagimoto and Sibuya [28] also consider this ordering, though they did not relate it to the notion of likelihood ratio ordering.

As pointed out by Shanthikumar and Yao [27], joint likelihood ratio ordering between two dependent random variables may not imply likelihood ratio ordering between their marginal distributions, but it does imply stochastic ordering between them (that is, $X_1 \leq_{lr:j} X_2 \Rightarrow X_1 \leq_{st} X_2$). However, in the case of independent random variables, $X_1 \leq_{lr:j} X_2 \Leftrightarrow X_1 \leq_{lr} X_2$.

The concept of joint likelihood ratio ordering can be extended to higher dimensional random vectors. Let $\mathbf{x} = (x_1, ..., x_n)$ and $\mathbf{y} = (y_1, ..., y_n)$ be two vectors. We say that $\mathbf{x}$ *is better arranged than* $\mathbf{y}$ $(\mathbf{x} \overset{a}{\succeq} \mathbf{y})$ if $\mathbf{x}$ can be obtained from $\mathbf{y}$ through successive pairwise interchanges of its components, with each interchange resulting in an increasing order of the two interchanged components, e.g. $(4, 1, 5, 3) \overset{a}{\succeq} (4, 3, 5, 1) \overset{a}{\succeq} (4, 5, 3, 1)$. A function $g : \mathcal{R}^n \to \mathcal{R}$ that preserves the ordering $\overset{a}{\succeq}$ is called an *arrangement increasing* function denoted by $g \in \mathcal{AI}$ if $\mathbf{x} \overset{a}{\succeq} \mathbf{y} \Rightarrow g(x) \geq g(y)$, (cf. Marshall and Olkin [20] p. 160 for the definition of an arrangement increasing function on $\mathcal{R}^n$).

Definition 2.10 *Let* $f(x_1, ..., x_n)$ *denote the joint density of* $\mathbf{X}$. *Then*

$$X_1 \leq_{lr:j} X_2 \leq_{lr:j} X_2, \ldots, \leq_{lr:j} X_n \Leftrightarrow f(x_1, ..., x_n) \in \mathcal{AI}. \qquad (11)$$

Hollander, Proschan and Sethuraman [13] call such a function as *decreasing in transportation* (DT) function. They also discuss many properties of such functions and give an extensive list of multivariate densities which are DT (or arrangement increasing).

3 Stochastic orders for normalized spacings

In this section we present some stochastic relations that exist among normalized spacings of independent random variables. First we consider the case of i.i.d. observations.

3.1 The i.i.d. case

It is well known that the normalized spacings $D_{1:n}, \ldots, D_{n:n}$ based on a random sample $X_1, \ldots, X_n$ from a continuous distribution are independent and identically distributed if and only if the random sample is from an exponential distribution. Barlow and Proschan [5] proved that the successive normalized spacings are stochastically increasing (decreasing) if the random sample is from a DFR (IFR) distribution. Their result is stated below.

Theorem 3.1 *Let* $X_1, \ldots, X_n$ *be a random sample of size* n *from a DFR distribution. Then*

(a)

$$D_{i:n} \leq_{st} D_{i+1:n}, \quad i = 1, \ldots, n-1, \qquad (12)$$

(b)
$$D_{i:n+1} \leq_{st} D_{i:n}, \quad n \geq i \text{ for fixed } i. \tag{13}$$

Similar results hold for the IFR case with the inequalities reversed in (a) and (b) above.

Kochar and Kirmani [16] strengthened the above result of Barlow and Proschan [5] from stochastic ordering to hazard rate ordering and it is stated below.

Theorem 3.2 Let $X_1, \ldots, X_n$ be a random sample of size n from a DFR distribution. Then

(a)
$$D_{i:n} \leq_{hr} D_{i+1:n} \quad for\ i = 1, \ldots, n-1; \tag{14}$$

(b)
$$D_{i:n+1} \leq_{hr} D_{i:n}, \quad n \geq i \text{ for fixed } i. \tag{15}$$

Barlow and Proschan [5] have shown that if $X_1, \ldots, X_n$ is a random sample from a DFR distribution, then their spacings $D_1, \ldots, D_n$ also have DFR distributions. Bagai and Kochar [2] proved that if $F \leq_{hr} G$ and if either F or G is DFR then, F is less dispersed than G ($F \overset{disp}{\leq} G$) in the sense that

$$F^{-1}(v) - F^{-1}(u) \leq G^{-1}(v) - G^{-1}(u) \text{ for } 0 \leq u \leq v \leq 1. \tag{16}$$

One of the consequences of $F \overset{disp}{\leq} G$ is that $var(X) \leq var(Y)$, where F and G are the distribution functions of the random variables X and Y, respectively. Their result is formally stated below.

Lemma 3.1 Let X and Y be two nonnegative random variables. If $X \leq_{hr} Y$ and X or Y is DFR, then $X \overset{disp}{\leq} Y$.

The proof of the next theorem follows from the above results and Theorem 3.2.

Theorem 3.3 Let $X_1, \ldots, X_n$ be a random sample from a DFR distribution. Then for $i = 1, \ldots, n-1$,

(a) $D_{i:n} \overset{disp}{\leq} D_{i+1:n}$,

(b) $var(D_{i:n}) \leq var(D_{i+1:n})$,

(c) $D_{i:n+1} \overset{disp}{\leq} D_{i:n}$,

(d) $var(D_{i:n+1}) \leq var(D_{i:n})$.

However, if the random sample is from a distribution with log-convex density, then the following stronger result holds (cf. Kochar and Kirmani [16]).

Theorem 3.4 *Let $X_1, \ldots, X_n$ be a random sample of size n from a distribution with log-convex density. Then*

$$D_{i:n} \leq_{lr} D_{i+1:n} \quad for \ i = 1, \ldots, n-1. \tag{17}$$

3.2 The non i.i.d. case

The next result proved in Kochar and Kirmani [16] establishes joint likelihood ratio ordering between any two consecutive normalized spacings when the joint density function of the observations $\{X_1, \ldots, X_n\}$ is convex. The random variables $X_1, \ldots, X_n$ need not be independent and identically distributed in this case.

Theorem 3.5 *Let the joint density $f_{\mathbf{X}}(x_1, \ldots, x_n)$ of $\mathbf{X} = (X_1, \ldots, X_n)$ be convex. Then*

$$D_{1:n} \leq_{lr:j} D_{2:n} \leq_{lr:j} \cdots \leq_{lr:j} D_{n:n}. \tag{18}$$

In particular if $X_i's$ are independent with log-convex densities, then the their joint density function is convex and as a consequence the conclusions of Theorem 3.6 will hold. This is stated formally in the next corollary.

Corollary 3.1 *Let $X_1, X_2, \ldots, X_n$ be independent random variables with log-convex densities. Then*

$$D_{1:n} \leq_{lr:j} D_{2:n} \leq_{lr:j} \cdots \leq_{lr:j} D_{n:n}. \tag{19}$$

As pointed out earlier, in general, joint likelihood ratio ordering between dependent random variables does not imply likelihood ratio ordering between their marginal distributions, but it does imply stochastic ordering between them.

4   On normalized spacings of non-identical exponential random variables

The exponential distribution plays an important role in statistics. Because of its non-aging property, it has many nice properties and it often gives very convenient bounds on survival probabilities and other characteristics of interest for systems with non-exponential components. In this section we study the stochastic properties of spacings from independent exponential distributions with possibly unequal scale parameters. The next theorem gives the joint as well as the marginal distributions of the normalized spacings from such distributions. Its proof can be found in Kochar and Korwar [17].

Theorem 4.1 *Let $X_1, \ldots, X_n$ be independent random variables with X_i having the exponential distribution with survival function $\overline{F}_i(t) = exp(-\lambda_i t)$, $t \geq 0$, for $i = 1, \ldots, n$. Then*

(a) *D_1 has exponential distribution with scale parameter $\sum_{i=1}^{n} \lambda_i / n$ and D_1 is independent of $(D_2, \ldots, D_n)$;*

(b) the joint p.d.f. of $(D_{i_1}, \ldots, D_{i_k})$, $2 \leq i_j \leq n$, $j = 1, \ldots, k$, $1 \leq k \leq n-1$ is

$$\left(\prod_{i=1}^n \lambda_i\right) \sum_{\mathbf{r}} \frac{1}{\prod_{i=1}^n \left(\sum_{j=i}^n \lambda(r_j)\right)} \prod_{\ell=1}^k \frac{\sum_{j=i_\ell}^n \lambda(r_j)}{n - i_\ell + 1} \; e^{-\frac{d_{i_\ell}}{n-i_\ell+1} \sum_{j=i_\ell}^n \lambda(r_j)}, \tag{20}$$

for $d_{i_\ell} \geq 0$, $\ell = 1, \ldots, k$; where $\mathbf{r}$ is a permutation of $(1, 2, \ldots, n)$ and $\lambda(i) = \lambda_i$;

(c) for $i \in \{2, \ldots, n\}$, the distribution of D_i is a mixture of independent exponential random variables with p.d.f.

$$\sum_S P(S) \frac{(s - \Lambda(S))}{n - i + 1} \; exp\{-(s - \Lambda(S))d_i / (n - i + 1)\}, \quad d_i \geq 0, \tag{21}$$

where the $\sum_S$ is over all subsets $S \subset \{1, \ldots, n\}$ of size $(i-1)$, $s = \sum_{i=1}^n \lambda_i$, $\Lambda(S) = \sum_{j \in S} \lambda_j$ and

$$P(S) = \sum_{\mathbf{r}} \left(\prod_{i \in S} \lambda_i\right) \left[\prod_{\ell=1}^{i-1} \left\{\sum_{j=\ell}^{i-1} \lambda(r(k_j)) + s - \Lambda(S)\right\}\right]^{-1}, \tag{22}$$

the $\sum_{\mathbf{r}}$ is being taken over all permutations $\mathbf{r} = (r_{k_1}, \ldots, r_{k_{i-1}})$ of the elements $k_j \in S$; $j = 1, \ldots, i-1$.

Since mixtures of log-convex densities are log-convex (and hence DFR), the proof of the next theorem immediately follows from Theorem 4.1(c) above.

Theorem 4.2 Let $D_1, \ldots, D_n$ be the normalized spacings based on n independent exponential distributions with possibly different scale parameters. Then each D_i has log-convex density for $i = 1. \ldots, n$.

Pledger and Proschan [21] proved that like in the case of a random sample from a DFR distribution, in this case also the normalized spacings are stochastically increasing. Their result is formally stated below.

Theorem 4.3 Let $X_1, \ldots, X_n$ be independent exponential random variables with possibly unequal scale parameters. Then

$$D_{i:n} \leq_{st} D_{i+1:n}, \quad i = 1, \ldots, n-1. \tag{23}$$

Kochar and Korwar [17] strengthened this result from stochastic ordering to joint likelihood ratio ordering. This follows from Corollary 3.1 because the density function of an exponential random variable is log-convex.

Theorem 4.4 *Let $X_1, \ldots, X_n$ be independent exponential random variables with possibly unequal scale parameters. Then*

$$D_{i:n} \leq_{lr:j} D_{i+1:n}, \quad i = 1, \ldots, n - 1. \tag{24}$$

This result is much stronger than the one given by Theorem 4.3.

Can the conclusion of Theorem 4.4 be further strengthened? Kochar and Korwar [17] conjecture that under the assumptions of Theorem 4.4,

$$D_{i:n} \leq_{hr} D_{i+1:n}, \quad i = 1, \ldots, n - 1.$$

However, they could prove it only for the case $n = 3$.

4.1 Comparisons with spacings from i.i.d. exponentials

Let $D_1, \ldots, D_n$ be the normalized spacings associated with independent exponential random variables $X_1, \ldots, X_n$, with X_i having hazard rate λ_i, $i = 1, \ldots, n$. Let $Y_1, \ldots, Y_n$ be a random sample of size n from an exponential distribution with common hazard rate $\bar{\lambda} = \sum_{i=1}^{n} \lambda_i / n$. Let $D_1^\star, \ldots, D_n^\star$ be their associated normalized spacings. Pledger and Proschan [21] proved that in this case

$$D_1^\star \stackrel{st}{=} D_1 \quad \text{and} \quad D_i^\star \leq_{st} D_i, \quad \text{for } i = 2, \ldots, n. \tag{25}$$

This means that the spacings are comparatively stochastically larger when the scale parameters are unequal than when they are all equal. Kochar and Rojo [19] strengthened the above result to establish multivariate likelihood ratio ordering between the vectors of spacings $(D_1, \ldots, D_n)$ and $(D_1^\star, \ldots, D_n^\star)$ and it is stated below.

Theorem 4.5 *Let $D_1, \ldots, D_n$ be the normalized spacings associated with independent exponential random variables $X_1, \ldots, X_n$, where X_i has hazard rate λ_i, $i = 1, \ldots, n$. Let $D_1^\star, \ldots, D_n^\star$ be the normalized spacings of a random sample $Y_1, \ldots, Y_n$ of size n from the exponential distribution with hazard rate $\bar{\lambda} = \sum_{i=1}^{n} \lambda_i / n$. Then for any $n \geq 2$,*

$$(D_1^\star, \ldots, D_n^\star) \stackrel{lr}{\preceq} (D_1, \ldots, D_n). \tag{26}$$

In particular,

$$D_1^\star \stackrel{st}{=} D_1 \quad \text{and} \quad D_i^\star \leq_{lr} D_i, \quad \text{for } i = 2, \ldots, n. \tag{27}$$

Since likelihood ratio ordering implies hazard rate ordering, the proof of the next theorem follows from (29), the DFR property of the spacings as established in Theorem 4.2 and the result of Bagai and Kochar [2] on connection between hazard rate ordering and dispersive ordering.

Theorem 4.6 *Let $D_1, \ldots, D_n$ be the normalized spacings associated with independent exponential random variables $X_1, \ldots, X_n$, where X_i has hazard rate λ_i, $i = 1, \ldots, n$. Let $D_1^\star, \ldots, D_n^\star$ be the normalized spacings of a random sample $Y_1, \ldots, Y_n$ of size n from the exponential distribution with hazard rate $\bar{\lambda} = \sum_{i=1}^{n} \lambda_i / n$. Then for any $n \geq 2$,*

(a)

$$D^\star_{i:n} \overset{disp}{\preceq} D_{i:n}, \quad for \; i = 2, \ldots, n, and$$

(b)

$$var(D_{i:n}) \leq var(D^\star_{i:n}) = \frac{1}{(\bar{\lambda})^2}.$$

Multivariate likelihood ratio ordering implies multivariate stochastic ordering, and the later is invariant under monotone transformations. Making use of these observations, we get the following corollary from Theorem 4.4.

Corollary 4.1 *Let $D_1, \ldots, D_n$ be the normalized spacings associated with independent exponential random variables $X_1, \ldots, X_n$, where X_i has hazard rate λ_i, $i = 1, \ldots, n$. Let $D^\star_1, \ldots, D^\star_n$ be the normalized spacings of a random sample $Y_1, \ldots, Y_n$ of size n from the exponential distribution with hazard rate $\bar{\lambda}$. Then for any $n \geq 2$,*

(a)

$$(D^\star_1, \ldots, D^\star_n) \overset{st}{\preceq} (D_1, \ldots, D_n); \tag{28}$$

(b) for $1 \leq i < j \leq n$,

$$(Y_{j:n} - Y_{i:n}) \overset{st}{\preceq} (X_{j:n} - X_{i:n}). \tag{29}$$

In particular, the above result gives a lower bound on the survival function of the sample range of a set of independent heterogeneous exponential random variables in terms of that of the sample range of a random sample of size of the same size from an exponential distribution with common hazard rate $\bar{\lambda}$. It is easy to see that in the later case the distribution of the sample range is the same as that of the largest order statistic in a sample of size $(n-1)$ from the exponential distribution with hazard rate $\bar{\lambda}$. From these results we get the following corollary.

Corollary 4.2 : *Let $X_1, \ldots, X_n$ be independent exponential random variables with X_i having hazard rate λ_i, for $i = 1, \ldots, n$. Then for $x > 0$,*

$$P[X_{n:n} - X_{1:n} \leq x] \leq [1 - exp(-\bar{\lambda}x)]^{n-1}. \tag{30}$$

4.2 Some Schur type results

The sample range and the generalized spacings of the type $X_{j:n} - X_{1:n}$ are of special interest in statistics. Kochar and Korwar [17] proved that for any $n > 1$, the survival function of $X_{2:n} - X_{1:n}$ is Schur convex in $\boldsymbol{\lambda}$. Kochar and Rojo [19] have strengthened this result and they have proved that the vector $(X_{2:n} - X_{1:n}, \ldots, X_{n:n} - X_{1:n})$ is stochastically larger when the λ_i 's are more dispersed in the sense of majorization. Their result is stated below.

Theorem 4.7 *Let $X_1, \ldots, X_n$ be independent exponential random variables with X_i having hazard rate λ_i, for $i = 1, \ldots, n$. Let $Y_1, \ldots, Y_n$ be another set of exponential random variables with $\lambda_i^\star$ as the hazard rate of Y_i, $i = 1, \ldots, n$. Let $\boldsymbol{\lambda} = (\lambda_1, \ldots, \lambda_n)$ and $\boldsymbol{\lambda}^\star = (\lambda_i^\star, \ldots, \lambda_n^\star)$, then $\boldsymbol{\lambda} \overset{m}{\succeq} \boldsymbol{\lambda}^\star$ implies*

$$(X_{2:n} - X_{1:n}, \ldots, X_{n:n} - X_{1:n}) \overset{st}{\succeq} (Y_{2:n} - Y_{1:n}, \ldots, Y_{n:n} - Y_{1:n}) \tag{31}$$

An important consequence of this result is that the sample range $X_{n:n} - X_{1:n}$ is stochastically larger when the λ_i's are more dispersed, a result more general than the one given by Corollary 4.2.

One may wonder whether one can extend Theorem 4.7 to other spacings. Pledger and Proschan [21] have shown with the help of an example that for $n = 3$, the survival function of the last spacing $D_{3:3}$ is not Schur convex. However, we have the following positive and even stronger result for $n = 2$. Its proof can be found in Kochar and Rojo [19].

Theorem 4.8 *Let X_1 and X_2 be two independent exponential random variables with hazard rates λ_1 and λ_2, respectively. Let $\{Y_1, Y_2\}$ be another set of independent exponential random variables with respective hazard rates $\lambda_1^\star$ and $\lambda_2^\star$. Then for $(\lambda_1, \lambda_2) \overset{m}{\succeq} (\lambda_1^\star, \lambda_2^\star)$,*

$$X_{2:2} - X_{1:2} \overset{lr}{\succeq} Y_{2:2} - Y_{1:2}. \tag{32}$$

5   Some new results on stochastic comparisons of order statistics

As mentioned earlier, Boland, Shaked and Shanthikuamr [9] gives an excellent review of the work done on stochastic orderings of order statistics prior to 1996. In this section we discuss some of the results obtained recently in this area.

Kochar [15] obtained the following result on dispersive ordering between successive order statistics in a random sample from a DFR distribution.

Theorem 5.1 *Let $X_1, \ldots, X_n$ be a random sample from a DFR distribution. Then for $i < j$,*

$$X_{i:n} \overset{disp}{\leq} X_{j:n}. \tag{33}$$

Consequently,

$$var(X_{i:n}) \leq var(X_{j:n}) \quad for \; i < j. \tag{34}$$

The DFR assumption is very crucial for the above result to hold. Boland, Shaked and Shanthikumar [9] have shown that if we have a random sample of size 2 from uniform distribution over $[0, 1]$ (which is not DFR), then $X_{1:2}$ is not less dispersed than $X_{2:2}$.

Now we consider the problem of stochastically comparing order statistics when the parent observations are independent but not necessarily identically distributed. Boland, El-Neweihi and Proschan [7] have shown that if $X_1, \ldots, X_n$ are independent random variables, then for $i < j$, $X_{j:n}$ has smaller hazard (failure) rate than $X_{i:n}$. Using this result together with Lemma 3.1 , we get the following theorem.

Theorem 5.2 *Let $X_1, \ldots, X_n$ be independent nonnegative random variables, then for $i < j$,*

$$X_{i:n} \overset{disp}{\leq} X_{j:n} \quad provided \ X_{i:n} \ is \ DFR.$$

Even if we sample from a DFR distribution, it may not be true that $X_{i:n}$ is DFR for every i. But the smallest order statistic $X_{1:n}$ will have DFR distribution in this case. This follows from the fact that the failure rate of a series system of independent components is the sum of the failure rates of the components. So if each component of the series system has decreasing failure (hazard) rate, the system will also have DFR distribution. This leads us to the following result.

Corollary 5.1 *Let $X_1, \ldots, X_n$ be independent DFR random variables, then for $j > 1$,*

$$X_{1:n} \overset{disp}{\preceq} X_{j:n}.$$

It follows from the above discussion that amongst all k-out-of n systems of n independent DFR components, the series system is least dispersed but has greatest hazard rate.

5.1 Order statistics from heterogeneous exponential distributions

Pledger and Proschan [21] proved the following result on order statistics from exponential distributions.

Theorem 5.3 *Let $X_1, \ldots, X_n$ be independent exponential random variables with X_i having hazard rate λ_i, $i = 1, \ldots, n$. Let $Y_1, \ldots, Y_n$ be another set of independent exponential random variables with $\lambda_i^\star$ as the hazard rate of Y_i, $i = 1, \ldots, n$. Let $\boldsymbol{\lambda} = (\lambda_1, \ldots, \lambda_n)$ and $\boldsymbol{\lambda}^\star = (\lambda_1^\star, \ldots, \lambda_n^\star)$. Then $\boldsymbol{\lambda} \overset{m}{\succeq} \boldsymbol{\lambda}^\star$ implies*

$$X_{1:n} \overset{st}{=} Y_{1:n} \quad and \ X_{i:n} \leq_{st} Y_{i:n}, \quad for \ i = 2, \ldots, n.$$

Proschan and Sethuraman [22] strengthened this result from component wise stochastic ordering to multivariate stochastic ordering. They proved that under the conditions of the above theorem

$$(X_{1:n}, \ldots, X_{n:n}) \overset{st}{\succeq} (Y_{1:n}, \ldots, Y_{n:n}). \tag{35}$$

It will be interesting to see to what extent these results can be strengthened. For the special case $n = 2$ and $i = 2$, Boland, El-Neweihi and Proschan [7] partially strengthened the above result from stochastic ordering to hazard rate ordering. Their result is stated below.

Theorem 5.4 *Let $r_{\lambda_1,\lambda_2}(t)$ be the hazard rate of a parallel system of two components whose lifetimes are independent exponential random variables with hazard rates λ_1 and λ_2, respectively. Then $r_{\lambda_1,\lambda_2}(t)$ is Schur-concave in (λ_1, λ_2).*

Dykstra, Kochar and Rojo [12] further strengthened the above result from hazard rate ordering to likelihood ratio ordering and their result is stated below.

Theorem 5.5 *Let X_1 and X_2 be two independent exponential random variables with hazard rates λ_1 and λ_2, respectively. Let Y_1 and Y_2 be another set of independent exponential random variables with respective hazard rates $\lambda_1^\star$ and $\lambda_2^\star$. Then for $(\lambda_1, \lambda_2) \overset{m}{\succeq} (\lambda_1^\star, \lambda_2^\star)$,*

$$X_{2:2} \leq_{lr} Y_{2:2} . \tag{36}$$

Boland, El-Neweihi and Proschan [7] conclude that the above result cannot be generalized to arbitrary n. They show with the help of an example that the hazard rate of a parallel system of three exponential components is *not* Schur concave in λ. Then the next natural problem is to compare the hazard rate of $X_{n:n}$ with that of $Y_{n:n}$, where $Y_1, \ldots, Y_n$ is a random sample of size n from the exponential distribution with hazard rate $\overline{\lambda} = \sum_{i=1}^n \lambda_i/n$. Dykstra, Kochar and Rojo [12] have proved the following result.

Theorem 5.6 *Let $X_1, \ldots, X_n$ be independent exponential random variables with X_i having hazard rate λ_i, $i = 1, \ldots, n$. Let $Y_1, \ldots, Y_n$ be a random sample of size n from the exponential distribution with common hazard rate $\overline{\lambda} = \sum_{i=1}^n \lambda_i/n$. Then*

(a)

$$Y_{n:n} \overset{disp}{\leq} X_{n:n}; \tag{37}$$

(b)

$$Y_{n:n} \leq_{hr} X_{n:n}. \tag{38}$$

This result gives a convenient upper bound on the hazard rate of a parallel system of heterogeneous exponential components in terms of that of a parallel system of identically distributed exponential random variables. It also gives a lower bound on the variance of $X_{n:n}$ in terms of that of $Y_{n:n}$. These results are stated in the form of the following corollary.

Corollary 5.2 *Under the conditions of Theorem 2.1,*

(a) the hazard rate $r_{X_{n:n}}$ of $X_{n:n}$ satisfies

$$r_{X_{n:n}}(x) \leq \frac{n\overline{\lambda}[1 - exp(-\overline{\lambda}x)]^{n-1}exp(-\overline{\lambda}x)}{1 - [1 - exp(-\overline{\lambda}x)]^n};$$

(b)

$$var(X_{n:n}) \geq \frac{1}{\overline{\lambda}^2}\sum_{i=1}^n \frac{1}{(n-i+1)^2}.$$

As discussed above, the hazard rate of $X_{n:n}$, the lifetime of a parallel system of n components is not Schur concave in λ for $n > 2$. However, as proved in Dykstra, Kochar and Rojo [12] the reversed hazard rate of $X_{n:n}$ is Schur convex in λ for any $n > 1$.

In this paper we have discussed only the topic of stochastic dominance and variability orderings among order statistics and spacings of random variables from life distributions. There are other types of partial orderings associated with them which are also of great interest. In a recent paper Boland, Hollander, Joag-Dev and Kochar [8] have studied the problem of dependence among order statistics of independent but non-identically distributed random variables. Kochar and Korwar [18] have studied the stochastic properties of order statistics while sampling from finite populations. However, due to the space limitations, we shall not pursue these topics further here.

References

1. B. Arnold, N. Balakrishnan and H.N. Nagaraja, *A First Course in Order Statistics* (Wiley : New York, 1992).
2. I. Bagai and S. C. Kochar, On tail ordering and comparison of failure rates *Commun. Statist. Theor. Meth.* **15** 1377-1388 (1986).
3. R. B. Bapat and M. I. Beg, Order statistics for nonidentically distributed random variables and permanents *Sankhyā Series A* **51** 79-93 (1989).
4. R. B. Bapat and S. C. Kochar, On likelihood ratio ordering of order statistics *Linear Algebra and Its Applications* **199** 281-291 (1994).
5. R. E. Barlow and F. Proschan, Inequalities for linear combinations of order statistics from restricted families *Ann. Math. Statist.* **37** 1574-1592 (1986).
6. R. E. Barlow and F. Proschan, *Statistical Theory of Reliability and Life Testing* (To Begin With : Silver Spring, Maryland (1981).
7. P. J. Boland, E. El-Neweihi and F. Proschan, Applications of the hazard rate ordering in reliability and order statistics *J. Appl. Prob.* **31** 180-192 (1994).
8. P. J. Boland, M. Hollander, K. Joag-Dev and S. Kochar, Bivariate dependence properties of order statistics *J. Mult. Anal.* **56** 75-89 (1996).
9. P. J. Boland, M. Shaked and J. G. Shanthikumar, *Technical Report* (Department of Statistics, University College, Dublin).
10. H. A. David, *Order Statistics* (Wiley : New York 1980).
11. H. A. David. and R. A. Groenveld, Measures of local variation in a distribution: expected length of spacings and variances of order statistics *Biometrika* **69** 227-232 (1982).
12. R. Dykstra, S. Kochar and J. Rojo, to appear in Stochastic comparisons of parallel systems of heterogeneous exponential components *J. Statist. Plan. Inf.* (1997).
13. M. Hollander, F. Proschan and J. Sethuraman, Functions decreasing in transposition and their applications in ranking problems *Ann. Statist.* **4** 722-733 (1977).
14. S. Karlin and Y. Rinott, Classes of orderings of measures and related correlation inequalities I. Multivariate totally positive distributions *J. Mult. Anal.*

10 467-498 (1980).

15. S. C. Kochar, Dispersive ordering of order statistics *Statistics & Probability Letters* **27** 271-274 (1995).

16. S. C. Kochar and S. N. U. A. Kirmani, Some results on normalized spacings from restricted families of distributions *J. Statist. Plan. Inf.* **46** 47-57 (1995).

17. S. C. Kochar and R. Korwar, Stochastic orders for spacings of heterogeneous exponential random variables *J. Mult. Anal.* **57** 69-73 (1996).

18. S. C. Kochar and R. Korwar, to appear in Stochastic properties of order statistics from finite populations *Sankhyā Series A* **59** (1997).

19. S. Kochar and J. Rojo, Some new results on stochastic comparisons of spacings from heterogeneous exponential distributions *J. Mult. Anal.* **59** 272-281 (1996).

20. A. W. Marshall and I. Olkin, *Inequalities : Theory of Majorization and Its Applications* (Academic Press : New York 1979).

21. P. Pledger and F. Proschan, Comparisons of order statistics and spacings from heterogeneous distributions, *Optimizing Methods in Statistics*, 89-113 (1971) ed. J.S. Rustagi (Academic Press : New York).

22. F. Proschan and J. Sethuraman, Stochastic comparisons of order statistics from heterogeneous populations, with applications in reliability *J.Mult. Analysis* **6** 608-616 (1976).

23. R. Pyke, Spacings *J. Roy. Statist. Soc. Ser. B.* **7** 395-499 (1965).

24. P. K. Sen, A note on order statistics for heterogeneous distributions *Ann. Math. Statist.* **41** 2137-2139 (1970).

25. M. Shaked and J. G. Shanthikumar, *Stochastic Orders and their Applications* (Academic Press, San Diego, CA 1994).

26. M. Shaked and J. G. Shanthikumar, Hazard rate ordering of k-out-of-n systems *Statistics & Probability Letters* **23** 1-8 (1995).

27. J. G. Shanthikumar and D. D.Yao, Bivariate characterization of some stochastic order relations *Adv. Appl. Probab.* **23** 642-659 (1991).

28. T. Yanagimoto and M. Sibuya, Stochastically larger component of a random vector *Ann. Inst. Statist. Math.* **24** 259-269 (1972).

BAYESIAN RELIABILITY MODELING AND INFERENCE UNDER ORDERING RESTRICTIONS

THOMAS A. MAZZUCHI

Department of Operations Research, The George Washington University,
Washington, D.C. 20052, USA
E-mail: mazzu@gwis2.circ.gwu.edu

REFIK SOYER

Department of Management Science, The George Washington University,
Washington, D.C. 20052, USA
E-mail: soyer@gwis2.circ.gwu.edu

In this paper we present a general framework for modeling reliability problems where ordering restrictions exist on failure characteristics. We consider reliability problems such as accelerated life testing, reliability growth testing, and minimal repair and introduce a common model that describes the ordering of the failure characteristics in these problems. The model enables us to deal with both the attribute and variable data sampling models. We also present a general Bayesian approach to make inferences from the model using Markov Chain Monte Carlo methods and discuss some extensions of the general model.

1 Introduction and Overview

Inherent in reliability analysis is the underlying fact that failure behaviour is affected by external environments in addition to time. Taking this fact into consideration is important in modeling for inference and decision making. One of the more basic effects of different environments is a simple ordering of the failure characteristics. While there are many examples of this in the literature, some of the more well known problems in reliability which exhibit this behaviour are accelerated life testing, reliability growth testing, and minimal repair.

In what follows we will first present an overview of these three problems and discuss how the ordering restrictions on failure characteristics arise in each case. In Section 2 we present a general model to describe the ordering restrictions in each of the three problems. The model uses an ordered Dirichlet distribution to describe the behavior of failure characteristics under different environments/testing stages and enables us to deal with both attribute and variable data sampling models. In Section 3, we develop a general Bayesian framework to make inferences for both attribute and variable data cases in each of the three problems. The use of the Markov Chain Monte Carlo methods facilitate the computations of the posterior and predictive densities of interest. Finally in Section 4 we consider some extensions of the general model by considering specific examples from reliability growth testing and step-stress accelerated life testing. In so doing, we show how the general model can be adopted to these specic cases and how the proposed Markov Chain Monte Carlo approach can be used for making inferences.

1.1 The Accelerated Life Testing Problem

It is common practice in dealing with high reliability items/systems to conduct a life test in a more severe environment than the actual use environment. Such a test is called an accelerated life test (ALT) and is usually undertaken to save the time and cost of testing. The statistical issue is to make inference about the life length of the item operating under use conditions based on failure data obtained under the more severe environments. Recently, Bayesian approaches have been proposed for making inference from ALT's; see for example [1], [2], and [3]. In dealing with intractable predictive and posterior distributions, most of these approaches used some sort of approximation or technique such as the *linear Bayesian* inference methods.

Typically an ALT is performed under m accelerated environments $E_1, \ldots, E_m$, which can be characterized by several stresses such as pressure, voltage, etc. The test can be performed either *simultaneously* at all m environments or in a sequential manner starting from more (less) severe environments. If $E_i > E_j$, where $E_i > E_j$ implies that E_i is more severe environment than E_j, then it is reasonable to expect that $R_i(t)$, reliability at mission time t under E_i, will be smaller than $R_j(t)$, reliability at mission time t under E_j. Thus, if $E_1 > E_2 > \ldots > E_m$, then

$$R_1(t) < R_2(t) < \ldots < R_m(t) < R_{m+1}(t), \qquad (1)$$

where $R_{m+1}(t)$ is the reliability at the actual use environment.

Under environment E_j, n_j items are put on test for τ_j units of time. If each stress environment produces r_j failures and $n_j - r_j$ succesful test completions, then let the times of failure for the ith item tested in environment E_j be given by t_{ij}, $i = 1, \ldots, r_j$; $j = 1, \ldots, m + 1$. The sampling model for the jth environment is proportional to

$$\left\{ \prod_{i=1}^{r_j} - \frac{d}{dt_{ij}} \left\{ R_j(t_{ij}) \right\} \right\} R_j(\tau_j)^{n_j - r_j}. \qquad (2)$$

In writing the above we denote the failure density as $-\frac{d}{dt_{ij}} \left\{ R_j(t_{ij}) \right\}$ and assume a variable data scenario, however, (2) can be easily modified for the attribute and single shot (time independent) cases as proportional to

$$\left(1 - R_j \right)^{r_j} \left(R_j \right)^{n_j - r_j}, \qquad (3)$$

by defining $R_j \equiv R_j(\tau_j)$.

1.2 The Reliability Growth Testing Problem

During the development phase of a new system, testing is performed in stages, and at the end of each test stage, design changes/modifications are made to the system with the intention of improving its performance. This process is termed "reliability growth" and has been the focus of much attention in the reliability literature.

During the reliability growth process, upon the discovery of failures, modifications are made to the system to remove the cause of failures and therefore increase system reliability. The test − modification scenario is repeated for some specified number of times. Let m denote the total number of test/modify stages, $R_i(t)$, $i = 1, \ldots, m$, denote the system reliability at mission time t for the ith stage of testing (that is, prior to the ith system modification), and $R_{m+1}(t)$ denote the field reliability at mission time t. Since modifications are made to the system at each stage, it is reasonable to assume that the mission reliabilities are defined as in (1).

The system undergoes an m-stage testing process where at stage j, n_j copies are tested for τ_j units of time producing r_j failures and $n_j - r_j$ succesful test completions. At the end of each stage, the failures are examined so that (when possible) modifications can be made to the system to remove failure modes and therefore increase system reliability. The test − modification scenario is repeated for some specified number of times, m. In the jth stage of testing, the sampling model is given by (2) where t_{ij} is the failure time of the ith unit in stage j. Accounting for attribute data and single shot items follows along the same lines as in Section 1.1. A slightly different version of this testing − modification scenario was considered in [4] where identical replications of the system were assumed to be tested until a failure was observed. This producees a sampling model in the stage j proportional to

$$-\frac{d}{dt_j}\left\{R_j(t_j)\right\} R_j(\tau_j)^{n_j-1}, \tag{4}$$

where t_j is the single observed failure time at stage j.

1.3 Minimal Repair Problem

It is common practice to use planned replacement strategies for systems (or components) that experience aging. One of the well known protocols for replacement, introduced in [5], is block replacement whereby items are minimally repaired upon failure but replaced at times $t_B, 2t_B, \ldots$, etc. Minimal repair means that the item can be repaired so that its failure characteristics (reliability, failure rate, etc.) are as they were just prior to failure. Consider a modification of this scenario, whereby replacement occurs after a specified number of repairs, m (i.e. the item is replaced at the time of them $m + 1$st failure). If $R_j(t)$ is the reliability at mission time t after $m + 1 - j$ repairs, then again (1) is a reasonable assumption.

Define stage j as the period after the $m + 1 - j$ th repair, then the sampling model in the jth stage is proportional to

$$-\frac{d}{dt_j}\Big\{R_j(t_j)\Big\}. \tag{5}$$

In this case neither the attribute or single shot model apply.

2   Modeling Ordering Restrictions

For simplicity we consider the exponential life model for the problems above, and thus (1) is equivalent to

$$\infty > \lambda_1 \geq \lambda_2 \cdots \geq \lambda_{m+1} > 0 \tag{6}$$

as $R_i(t) = e^{-\lambda_i t}$. Defining $R_i \equiv e^{-\lambda_i}$, we write equivalent to (6),

$$0 \leq R_1 \cdots \leq R_{m+1} \leq 1, \tag{7}$$

which can be used as the unifying relation for variable, attribute, and single shot testing scenarios in all the three problems.

   Following the development in [6], a natural and mathematically tractable model for $\underset{\sim}{R} = (R_1, \ldots, R_{m+1})$ is the ordered Dirichlet distribution given as

$$\Pi(\underset{\sim}{R} \mid D^{(0)}) = \frac{\Gamma(\beta)}{\prod\limits_{j=1}^{m+2} \Gamma(\beta\alpha_j)} \prod_{j=1}^{m+2} (R_j - R_{j-1})^{\beta\alpha_j - 1}, \tag{8}$$

where $R_0 \equiv 0$, $R_{m+2} \equiv 1$, and $D^{(0)}$ represents the information prior to any testing as captured by the prior parameters $\beta, \alpha_i > 0$, $\sum\limits_{i=1}^{m+2} \alpha_i = 1$. Note that the distribution is defined over the simplex $\Big\{\underset{\sim}{R} \mid 0 \leq R_1 \cdots \leq R_{m+1} \leq 1\Big\}$ and thus embodies the restrictions in (7) and imposes no additional restrictions in the analysis.

   An attractive feature of the ordered Dirichlet distribution is that incorporation of expert judgement is facilitated, as all relevant marginal and conditional distributions are beta distributions. For example, if we define $\alpha_i^* = \sum\limits_{j=1}^{i} \alpha_j$, then it can be shown that the marginal distributions are given by

$$[R_i \mid D^{(0)}] \equiv [e^{-\lambda_i} \mid D^{(0)}] \sim Beta\Big(\beta\alpha_i^*, \beta(1 - \alpha_i^*)\Big), \tag{9}$$

where $E[R_i \mid D^{(0)}] = \alpha_i^*$, $V[R_i \mid D^{(0)}] = \frac{\alpha_i^*(1 - \alpha_i^*)}{\beta + 1}$ and thus, β is a *degree of belief* parameter with lower values of β reflecting more spread in the distribution. Similarly, it can be shown that for $i < j$,

$$[R_j - R_i \mid D^{(0)}] \sim Beta\Big(\beta(\alpha_j^* - \alpha_i^*), \beta(1 - \alpha_j^* + \alpha_i^*)\Big) \tag{10}$$

and

$$[\frac{R_i}{R_j} \mid D^{(0)}] \equiv [e^{-(\lambda_i - \lambda_j)} \mid D^{(0)}] \sim Beta\left(\beta\alpha_i^*, \beta(\alpha_j^* - \alpha_i^*)\right), \text{ for } i < j. \quad (11)$$

It has been common practice to elicit measures such as the mean, mode, and/or variance in order to specify the prior distribution parameters. That is, the elicited quantities would be compared with the analytical expressions in order to specify the prior parameters. The expressions for mean, median, mode, variance, etc. for the beta distribution are well known and can be obtained in closed form for all of the above. A more detailed discussion of the elicitation procedures is given in [4] and [7]. In the case where prior information is not available, the ordered uniform distribution, obtained as (8) with $\beta\alpha_j = 1$ for all j, can be used. In the context of reliability growth testing, it was shown in [7] that the Bayesian reliability growth model of [8] is a special case of the above model with $\alpha_j = 1/(m+2)$, $j = 1, \ldots, m+2$, and $\beta = m+2$.

After the specification of the prior parameters, predictive inference can be made using (9), (10) and (11). Many of the predictive forms are in closed form; see for example, [4], [9] and [10].

3   A General Framework for Bayesian Inference with Ordering Restrictions

In discussing the inference framework for the three problems, we consider a sequential approach to processing the data. For the cases of reliability growth testing and minimal repair, data accumulate in a sequential manner as a result of the modifications and minimal repairs made to the system at different stages. Without loss of generality, we assume that during the ALT, testing is performed sequentially from high to low stress environments. Note that for the case where testing is performed simultaneously at all m environments, we can still process the data sequentially.

3.1 Inference for Attribute Data

We introduce the inference method using (3) and generalize to the other testing and problem scenarios. Defining $D^{(i)} \equiv \{D^{(0)}, n_1, r_1, \ldots, n_i, r_i\}$, and assuming that the r_j's are conditionally independent given $\underset{\sim}{R}$, the likelihood function of $\underset{\sim}{R}$ after i stages of testing is given by $\mathcal{L}(\underset{\sim}{R}; D^{(i)}) = \prod_{j=1}^{i} (1 - R_j)^{r_j} R_j^{n_j - r_j}$, and the posterior distribution of $\underset{\sim}{R}$ is obtained as

$$\Pi(\underset{\sim}{R} \mid D^{(i)}) \propto \left\{ \prod_{j=1}^{m+2} (R_j - R_{j-1})^{\beta\alpha_j - 1} \right\} \left\{ \prod_{j=1}^{i} (1 - R_j)^{r_j} R_j^{n_j - r_j} \right\}. \quad (12)$$

Note that in the case of the geometric sampling of [4], we only need to replace r_j by 1. As discussed in [6], it is possible to expand the $R_j^{n_j - r_j}$ terms in a binomial series and to integrate out $(R_1, \ldots, R_{i-1})$ over the region $0 < R_1 \leq \cdots \leq R_{i-1} \leq R_i$. Then the posterior distribution $\Pi(R_i, \ldots, R_{m+1} \mid D^{(i)})$ can be obtained as a mixture of ordered Dirichlet distributions. Thus, given $D^{(i)}$, marginal distributions for the current and future stage reliabilities R_j, $j = i, \ldots, m+1$, and predictive distributions can be obtained by taking weighted averages of the prior expressions evaluated for revised parameter values. However, the mixture involves nested sums of ratios of gamma functions, and when m is large (say, $m \geq 10$), the evaluation of the weights becomes cumbersome and the results become inaccurate due to the problems in evaluation of gamma functions. In addition, posterior distribution results for R_j, $j = 1, \ldots, i-1$ do not exist in closed form. In what follows, we will illustrate that the computations for all posterior and predictive forms can be done more efficiently and accurately by using Markov Chain Monte Carlo (MCMC) methods.

As illustrated in [4], for large values of m Gibbs sampling can easily be adopted to make inference in this type of problems with ordering restrictions. After i stages of testing, it is desired to obtain the posterior distribution $\Pi(\underset{\sim}{R} \mid D^{(i)})$ and its marginal distributions $\Pi(R_j \mid D^{(i)})$, $j = 1, \ldots, m+1$. The Gibbs sampler enables the drawing of samples from $\Pi(\underset{\sim}{R} \mid D^{(i)})$ without actually computing the exact distributional form. This is achieved by successive drawings from the full conditional distributions $\Pi(R_j \mid \underset{\sim}{R}^{(-j)}, D^{(i)})$ where $\underset{\sim}{R}^{(-j)} = \{ R_\ell \mid \ell \neq j, \ell = 1, 2, \ldots, m+1 \}$.

For $j = 1, \ldots, i$, the exact form of $\Pi(R_j \mid \underset{\sim}{R}^{(-j)}, D^{(i)})$ is not known, and thus to facilitate the implementation of the Gibbs sampler, some random variable generation method must be employed. In [4], the authors adopted the standard rejection sampling method of the type described in [11]. Following the development in [4], in order to sample from the conditional posterior $\Pi(R_j \mid \underset{\sim}{R}^{(-j)}, D^{(i)})$, $j = 1, \ldots, i$, we can proceed as follows:

Step 1: For $j = 1, \ldots, i$, generate R_j from the prior conditional distribution $\Pi(R_j \mid \underset{\sim}{R}^{(-j)}, D^{(0)})$. It can be shown from (8) that the prior conditional distribution

$$\Pi(R_j \mid \underset{\sim}{R}^{(-j)}, D^{(0)}) = \Pi(R_j \mid R_{j-1}, R_{j+1}, D^{(0)}) \tag{13}$$

is given by

$$[R_j \mid R_{j-1}, R_{j+1}, D^{(0)}] \sim Beta\left(\beta \alpha_j, \beta \alpha_{j+1}; (R_{j-1}, R_{j+1}) \right), \tag{14}$$

a truncated beta density over (R_{j-1}, R_{j+1}).

Step 2: Generate independently a 0-1 uniform random variate u.

Step 3: Compute the ratio

$$\frac{\mathcal{L}(R_j; \underset{\sim}{R}^{(-j)}, D^{(i)})}{\mathcal{L}(\widehat{R}_j; \underset{\sim}{R}^{(-j)}, D^{(i)})}, \tag{15}$$

where for $j = 1, \ldots, i$ the conditional likelihood

$$\mathcal{L}(R_j; \underset{\sim}{R}^{(-j)}, D^{(i)}) \propto \begin{cases} (1 - R_j)^{r_j} R_j^{n_j - r_j} & R_{j-1} \le R_j \le R_{j+1} \\ 0 & \text{otherwise} \end{cases} \tag{16}$$

and $\widehat{R}_j$, the maximum of (16), is available analytically as

$$\widehat{R}_j = \begin{cases} R_{j-1} & \text{if } \frac{n_j - r_j}{n_j} \le R_{j-1} \\ \frac{n_j - r_j}{n_j} & \text{if } R_{j-1} < \frac{n_j - r_j}{n_j} < R_{j+1} \\ R_{j+1} & \text{if } R_{j+1} \le \frac{n_j - r_j}{n_j} \end{cases} \tag{17}$$

For the case $n_j = r_j$, $\widehat{R}_j$ is the value of R_{j-1} generated at the previous iteration of the Gibbs sampler.

Step 4: Accept R_j if

$$u \le \frac{\mathcal{L}(R_j; \underset{\sim}{R}^{(-j)}, D^{(i)})}{\mathcal{L}(\widehat{R}_j; \underset{\sim}{R}^{(-j)}, D^{(i)})},$$

otherwise reject R_j and go to step 1 and repeat the process.

For $j = i + 1, \ldots, m + 1$, the exact form of the full conditionals is known as they are identical to the prior conditionals

$$\Pi(R_j| \underset{\sim}{R}^{(-j)}, D^{(i)}) = \Pi(R_j| R_{j-1}, R_{j+1}, D^{(0)}), \text{ for } j = i + 1, \ldots, m + 1, \tag{18}$$

whose distribution is given by (14). This simplifies drawing from these distributions for $j = i + 1, \ldots, m + 1$, as the standard rejection sampling method need not be employed. Once the sample $\underset{\sim}{R}^1, \underset{\sim}{R}^2, \ldots\ldots, \underset{\sim}{R}^r$ is obtained from the posterior distribution $\Pi(\underset{\sim}{R}|D^{(i)})$, all the posterior predictive and marginal distributions and expressions can be obtained.

3.2 Inference for Variable Data

Inference for the exponential variable-data case with sampling model (2) follows along the same lines as in the previous section by using the transformation $R_i \equiv e^{-\lambda_i}$. In this case, the conditional likelihood of R_j used in the standard rejection sampling for $j = 1, \ldots, i$ is given by

$$\mathcal{L}(R_j; \underset{\sim}{R}^{(-j)}, D^{(i)}) \propto \begin{cases} \left(ln(1/R_j)\right)^{r_j} R_j^{T_j} & R_{j-1} \le R_j \le R_{j+1} \\ 0 & \text{otherwise,} \end{cases} \tag{19}$$

where $T_j = \sum_{i=1}^{r_j} t_{ij} + (n_j - r_j)\tau_j$ and the maximum of the likelihood is given by

$$\widehat{R}_j = \begin{cases} R_{j-1} & \text{if } e^{-(r_j/T_j)} \le R_{j-1} \\ e^{-(r_j/T_j)} & \text{if } R_{j-1} < e^{-(r_j/T_j)} < R_{j+1}. \\ R_{j+1} & \text{if } R_{j+1} \le e^{-(r_j/T_j)} \end{cases} \tag{20}$$

As before, for $j = i+1,\ldots,m+1$, the exact form of the full conditionals is given by (14).

In addressing the variable data models, as discussed in [4], a Weibull failure model can be easily accomodated by considering the following form of the failure density

$$f(t_j \mid \lambda_j, \phi) = \lambda_j \phi t_j^{\phi-1} exp(-\lambda_j t_j^\phi), \tag{21}$$

where $\lambda_j > 0$, $\phi > 0$. Note that this model assumes the aging characteristics, as modeled through ϕ, are constant from stage to stage, but that the scale parameter, λ_j, may change from one stage to another. Ordering is still defined using (6) on the λ_j's. Again using the parameterization $R_j \equiv e^{-\lambda_j}$, the ordered Dirichlet prior (8) can be employed. Note that, since ϕ is fixed over the stages, this implies that $\{\lambda_j\}$, and therefore the failure rates over the stages, $\lambda_j \phi t_j^{\phi-1}$, are ordered (nonincreasing) in j. However within each stage, the failure characteristics can be either decreasing ($\phi < 1$), constant ($\phi = 1$), or increasing ($\phi > 1$). This provides more flexibility in the modeling process.

Analogous to (19), the likelihood of R_j and ϕ is obtained as

$$\mathcal{L}(R_j, \phi; T_j(\phi), r_j) = \left(\phi \, ln(1/R_j)\right)^{r_j} \left(\prod_{i=1}^{r_j} t_{ij}^{\phi-1}\right) R_j^{T_j(\phi)}, \tag{22}$$

where $T_j(\phi) = \sum_{i=1}^{r_j} t_{ij}^\phi + (n_j - r_j)\tau_j^\phi$. In implementing the Gibbs sampler for the Weibull model, it is necessary to draw from the full conditional distributions $\Pi(R_j | \underset{\sim}{R}^{(-j)}, \phi, D^{(i)})$, for $j = 1,\ldots,m+1$, and $\Pi(\phi \mid \underset{\sim}{R}, D^{(i)})$. As before, for stages $j = 1,\ldots,i$, the conditional distribution for R_j is not of a familiar form, and therefore the standard rejection sampling method will be employed. It can be shown that the conditional likelihoods for R_j, $j = 1,\ldots,i$, are given by (22) with maximums (20) where T_j is replaced by $T_j(\phi)$. For stages, $j = i+1,\ldots,m+1$, the full conditionals are given by (14), as ϕ and $\underset{\sim}{R}$ are assumed independent, a priori.

An additional step for this model is needed to complete the Gibbs sequence. This step involves generating ϕ from its full conditional posterior distribution, $\Pi(\phi \mid \underset{\sim}{R}, D^{(i)}) = \Pi(\phi \mid R_1, R_2,\ldots,R_i, D^{(i)})$. In order to draw a sample from $\Pi(\phi \mid \underset{\sim}{R}, D^{(i)})$, one can again use the standard rejection sampling method with conditional likelihood

$$\mathcal{L}(\phi; \underset{\sim}{R}, D^{(i)}) \propto \prod_{j=1}^{i} \phi^{r_j} \left(\prod_{i=1}^{r_j} t_{ij}^{\phi-1}\right) R_j^{T_j(\phi)}. \tag{23}$$

The maximum of the above conditional likelihood must be obtained numerically. This is a straightforward task, since $\mathcal{L}(\phi; \underset{\sim}{R}, D^{(i)})$ is concave in ϕ. The more efficient adaptive rejection procedure of [12] can also be used.

4 Extension to Other Models

The general inference framework of Section 3 can be easily adopted to other inference problems in reliability with ordering restrictions. This can be accomplished through either a further modification of the prior form given in (8) or the development of a new likelihood expression. We provide an example of both cases. In what follows we consider the well known attribute data reliability growth model of [13] and the step-stress ALT model of [2] and discuss how inference can be made in both of these models using our framework.

4.1 Inference for Barlow-Scheur Reliability Growth Model

In [9] the authors presented a Bayesian approach for the attribute data reliability growth model of Barlow and Scheuer [13]. The Barlow-Scheuer model is an extension of the model presented Section 3.1 which assumes that failures at all stages are *assignable cause failures* implying that they can be identified and corrected. The Barlow-Scheuer model categorizes the failures at any stage as assignable cause and *inherent* failures where the inherent failures can only be eliminated with an advancement of technology. Thus, the authors assume that the probability of assignable cause failures are nonincreasing over the test stages. The test results at each stage is described by a trinomial model with the inherent failure probability q, and assignable cause failure probability, p_i.

In [9] the authors presented a Bayesian analysis of this model by using an ordered Dirichlet type prior on q and assignable cause failure probabilities for all the stages, p_j, $j = 1, \ldots, m+1$, which preserves the desired ordering $p_1 \geq \cdots \geq p_{m+1}$. The authors showed that the model proposed in [14] was a special case of this model. By using $R_j = 1 - q - p_j$, we can write

$$\Pi(\underset{\sim}{R}, q \mid D^{(0)}) \propto \left[\prod_{j=1}^{m+1} (R_j - R_{j-1})^{\beta \alpha_j - 1} \right] (1 - q - R_{m+1})^{\beta \alpha_{m+2} - 1} q^{\beta \alpha_0 - 1}, \quad (24)$$

defined over $0 \leq R_1 \cdots \leq R_{m+1} \leq 1 - q$, implying that

$$[q \mid D^{(0)}] \sim Beta\Big(\beta \alpha_0, \beta(1 - \alpha_0)\Big) \quad (25)$$

and

$$[q \mid \underset{\sim}{R} , D^{(0)}] = [q \mid R_{m+1}, D^{(0)}] \sim Beta\Big(\beta\alpha_0, \beta\alpha_{m+2}; (0, 1 - R_{m+1})\Big). \qquad (26)$$

It follows from the above that $\Pi(R_j \mid \underset{\sim}{R}^{(-j)}, q, D^{(0)}) = \Pi(R_j \mid R_{j-1}, R_{j+1}, D^{(0)})$ for $j = 1, \ldots, m$, as given by (14) and that $\Pi(R_{m+1} \mid \underset{\sim}{R}^{(-(m+1))}, q, D^{(0)}) = \Pi(R_{m+1} \mid R_m, q, D^{(0)})$ with

$$[R_{m+1} \mid R_m, q, D^{(0)}] \sim Beta\Big(\beta\alpha_{m+1}, \beta\alpha_{m+2}; (R_m, 1 - q)\Big). \qquad (27)$$

Full distributional results for p_i's and stage reliabilities, R_i's, were obtained in [9] as mixtures of ordered Dirichlet distributions of the form as in (8). As before, the weights in these mixtures are dependent on gamma functions and computational difficulties arise when m is large.

Let a_j denote the number of inherent failures, and r_j the number of assignable cause failures, at test stage j and after i stages of testing define $D^{(i)} \equiv \Big\{D^{(0)}, a_1, n_1, a_2, n_2, \cdots, a_i, n_i\Big\}$. To implement the Gibbs sampler we need to draw from the full conditional distributions $\Pi(R_j \mid \underset{\sim}{R}^{(-j)}, q, D^{(i)})$, for $j = 1, \ldots, m + 1$, and $\Pi(q \mid \underset{\sim}{R}, D^{(i)})$. As in the attribute reliability growth of Section 3.1, for stages $j = 1, \ldots, i$, $\Pi(R_j \mid \underset{\sim}{R}^{(-j)}, q, D^{(i)})$ is not of a familiar form, and therefore we need to use the rejection method of Section 3.1. It can be shown that the conditional likelihoods for $j = 1, \ldots, i$, are given by

$$\mathcal{L}(R_j; \underset{\sim}{R}^{(-j)}, q, D^{(i)}) = \mathcal{L}(R_j; R_{j-1}, R_{j+1}, q, a_j, n_j)$$
$$\propto (1 - q - R_j)^{r_j} q^{a_j} R_j^{n_j - a_j - r_j}, \qquad (28)$$

defined over $R_{j-1} < R_j < R_{j+1}$. Given (28), we can use the rejection algorithm with the acceptance probability in Step 3 as

$$\frac{\mathcal{L}(R_j; R_{j-1}, R_{j+1}, q, a_j, n_j)}{\mathcal{L}(\widehat{R}_j; R_{j-1}, R_{j+1}, q, a_j, n_j)}, \qquad (29)$$

where, for $(n_j - a_j) > r_j$, the maximum of the conditional likelihood is given by

$$\widehat{R}_j = \begin{cases} R_{j-1}, & \text{if } \frac{(n_j - a_j - r_j)(1-q)}{n_j - a_j} \leq R_{j-1} \\ \frac{(n_j - a_j - r_j)(1-q)}{n_j - a_j}, & \text{if } R_{j-1} < \frac{(n_j - a_j - r_j)(1-q)}{n_j - a_j} < R_{j+1} \\ R_{j+1}, & \text{if } R_{j+1} \leq \frac{(n_j - a_j - r_j)(1-q)}{n_j - a_j} \end{cases} \qquad (30)$$

We note that (30) implies $\widehat{R}_j < (1 - q)$. For the case $(n_j - a_j) = r_j$, again $\widehat{R}_j$ can be replaced by the value of R_j generated at the previous iteration of the Gibbs sampler, and the acceptance probability is computed accordingly.

For testing stages $j = i + 1, \ldots, m$, the conditional distributions are obtained as $\Pi(R_j \mid \underset{\sim}{R}^{(-j)}, q, D^{(i)}) = \Pi(R_j \mid R_{j-1}, R_{j+1}, D^{(0)})$ which are given as truncated beta densities. For stage $m+1$, $\Pi(R_{m+1} | \underset{\sim}{R}^{(-(m+1))}, q, D^{(i)}) = \Pi(R_{m+1}|R_m, q, D^{(0)})$, which is again a truncated beta density of the form (14).

To obtain the full conditional posterior of q, we write

$$\Pi(q \mid \underset{\sim}{R}, D^{(i)}) \propto \left[\prod_{j=1}^{i} (1 - q - R_j)\, q^{a_j} \right] (1 - q - R_{m+1})^{\beta \alpha_{m+2} - 1} q^{\beta \alpha_0 - 1}, \qquad (31)$$

defined over $0 \le R_1 \cdots \le R_{m+1} \le 1 - q$, implying that

$$\Pi(q \mid \underset{\sim}{R}, D^{(i)}) = \Pi(q \mid R_1, R_2, \ldots, R_i, R_{m+1}, D^{(i)}). \qquad (32)$$

To draw from $\Pi(q \mid \underset{\sim}{R}, D^{(i)})$, we use again the rejection algorithm of Section 3.1. The maximum of the conditional likelihood

$$\mathcal{L}(q; \underset{\sim}{R}, D^{(i)}) \propto \left[\prod_{j=1}^{i} (1 - q - R_j)\, q^{a_j} \right]$$

is obtained as $\widehat{q}$ such that

$$\sum_{j=1}^{i} a_j = \sum_{j=1}^{i} \frac{\widehat{q}}{(1 - \widehat{q} - R_j)} \qquad (33)$$

and $0 \le R_1 \cdots \le R_i \le R_{m+1} \le 1 - \widehat{q}$. We note that the value of q solving (33) must be smaller than $(1 - R_{m+1})$.

4.2 Step Stress ALT Model

Consider a modification of the accelerated life testing scenario of Section 1.1 where a test is conducted in such manner that each test unit is subjected to each of the m accelerated environments as well as the use environment in a sequential manner from least severe (use) environment to the most severe environment. This is called a progressive step stress test. Such a test is analyzed in [2] under the assumption that the failure behavior under each stress environment is governed by the exponential life distribution and the parameters of these distributions conform to the ordering given by (6) or using the transformation $R_j = e^{-\lambda_j}$, (7). We present a special case of the testing scenario considered in [2] and illustrate that this more elaborate testing scenario can be modeled using (8) with only a slight modification of the likelihood function.

The failure rate of a test item under such a test is the step function. It is also possible, especially when the stress environment involves temperature, that the stress conditions cannot be changed instantaneously, and thus one must account for a gradual change from one set of stress conditions to the next. This is known as a ramping phenomenon. Here we develop the likelihood for the step function case. The ramping case is considered in [2].

For the development of the likelihood, consider the test for each of n_0 items starting at time τ_0 at use stress with failure rate λ_{m+1}. At time τ_i, the item is subjected to environment E_{m+1-i} for $i = 0, ..., m$. If the operational status of the test items is observable only at times τ_i, $i = 0, ..., m$, and if n_i denotes the number of items at risk of failure at time τ_i and s_i denotes the number of items failing in $[\tau_i, \tau_{i+1}]$, the likelihood function after completing the $m + 1$ time intervals can be written as

$$\mathcal{L}(\underset{\sim}{R} ; D^{(m+1)}) \propto \prod_{i=0}^{m} \left((R_{m+1-i})^{(\tau_{i+1}-\tau_i)} \right)^{n_i-s_i} \left(1 - (R_{m+1-i})^{(\tau_{i+1}-\tau_i)} \right)^{s_i}, \quad (34)$$

where $D^{(m+1)} = \{n_0, s_0, ..., s_m\}$, $s_m = n_m$, and $\tau_{m+1} \equiv \infty$.

After the test is completed, posterior inference for $\underset{\sim}{R}$ may proceed as in Section 3.1 using the Gibbs sampler. Again the full conditionals are not available and thus we use rejection sampling method of Section 3.1 with the conditional likelihood, $\mathcal{L}(R_j ; \underset{\sim}{R}^{(-j)}, D^{(m+1)})$, given as

$$\propto \begin{cases} \left((R_{m+1-i})^{(\tau_{i+1}-\tau_i)} \right)^{n_i-s_i} \left(1 - (R_{m+1-i})^{(\tau_{i+1}-\tau_i)} \right)^{s_i} & R_{j-1} \le R_j \le R_{j+1} \\ 0 & \text{otherwise} \end{cases}. \quad (35)$$

The maximum of the conditional likelihood in (35) is given by

$$\widehat{R}_j = \begin{cases} R_{j-1} & \text{if } \left[\frac{n_i-s_i}{n_i} \right]^{1/(\tau_{i+1}-\tau_i)} \le R_{j-1} \\ \left[\frac{n_i-s_i}{n_i} \right]^{1/(\tau_{i+1}-\tau_i)} & \text{if } R_{j-1} < \left[\frac{n_i-s_i}{n_i} \right]^{1/(\tau_{i+1}-\tau_i)} < R_{j+1} \\ R_{j+1} & \text{if } R_{j+1} \le \left[\frac{n_i-s_i}{n_i} \right]^{1/(\tau_{i+1}-\tau_i)} \end{cases}. \quad (36)$$

We note that closed form results have been developed in [2] for the ramping case with attribute data. But as in the previously discussed cases, these closed form results are subject to error as m and n_0 get even moderately large. In addition, for the variable case, no closed form expression exists, however, the MCMC method of Section 3.2 can again be adopted in the following manner. Letting s_i remain as previously defined and t_{ij} denote the time of the jth failure in interval $[\tau_i, \tau_{i+1}]$, $i = 0, ..., m, j = 1, ..., s_i$, the likelihood of $\underset{\sim}{\lambda} = (\lambda_1, ..., \lambda_{m+1})$ given all the observed test data, $D^{(m+1)}$, is given by

$$\mathcal{L}(\underset{\sim}{\lambda} ; D^{(m+1)}) \propto \prod_{i=0}^{m} \lambda_{m+1-i}^{s_i} \, exp\left(- \sum_{k=0}^{i-1} \lambda_{m+1-k}(\tau_{k+1} - \tau_k)s_i - (\lambda_{m+1-i}T_i - \tau_i s_i) \right),$$

$$(37)$$

where $D^{(m+1)} \equiv \{t_{ij}, i = 0, ..., m, j = 1, ..., s_i\}$ and $T_i = \sum_{j=1}^{s_i} t_{ij}$.

Posterior inference may proceed as in Section 3.1 using the Gibbs sampler. Again the full conditionals are not available and thus we use rejection sampling method of Section 3.1 with the conditional likelihood, $\mathcal{L}(\lambda_{m+1-j}; \underset{\sim}{\lambda}^{(-(m+1-j))}, D^{(m+1)})$, given as

$$\lambda_{m+1-j}^{s_j} exp\left(-\lambda_{m+1-j}\left[(\tau_{j+1} - \tau_j)\sum_{k=j+1}^{m} s_k + T_j - \tau_j s_j\right]\right). \tag{38}$$

As before, the maximum of (38) exists in closed form as

$$\widehat{\lambda}_{m+1-j} = \begin{cases} \lambda_{m+2-j} & \text{if } (s_j/K_j) \leq \lambda_{m+2-j} \\ (s_j/K_j) & \text{if } \lambda_{m+2-j} < (s_j/K_j) < \lambda_{m-j}, \\ \lambda_{m-j} & \text{if } \lambda_{m-j} \leq (s_j/K_j) \end{cases} \tag{39}$$

where $K_j = (\tau_{j+1} - \tau_j)\sum_{k=j+1}^{m} s_k + T_j - \tau_j s_j$.

Acknowledgements

The work of Refik Soyer was partially supported by 1996-1997 Dupont Educational Aid Grant.

References

1. T. A. Mazzuchi and R. Soyer, "A Dynamic General Linear Model for Inference from Accelerated Life Tests", *Naval Research Logistics Quarterly*, **39**, 757-773, (1992).

2. R. van Dorp, T. A. Mazzuchi, G. Fornell and L. R. Pollock, "A Bayes Approach to Step-Stress Accelerated Life Testing", *IEEE Transactions on Reliability*, **R- 45**, 491-498, (1996).

3. T. A. Mazzuchi, R. Soyer and A. L. Vopatek, "Linear Bayesian Inference for Accelerated Weibull Model", *LifeTime Data Analysis*, **13**, 1-13, (1997).

4. A. Erkanli, T. A. Mazzuchi and R. Soyer, "Bayesian Computations for a Class of Reliability Growth Models", *Technometrics*, tentatively accepted, (1997).

5. R. E. Barlow and L. Hunter, "Optimum Preventive Maintenance Policies", *Operations Research*, **8**, 90-100, (1960).

6. T. A. Mazzuchi and R. Soyer, "Reliability Assessment and Prediction during Product Development", *Proceedings of the Annual Reliability and Maintainability Symposium*, 468-474, (1992).

7. T. A. Mazzuchi and R. Soyer, "Determination of Stopping Criteria During Product Development", *Advances in Reliability*, 269-280, ed. A. P. Basu (North-Holland, 1993).

8. A. F .M. Smith, "A Bayesian Note on Reliability Growth during a Development Testing Program", *IEEE Transactions on Reliability*, **R-26**, 346-347, (1977).

9. T. A. Mazzuchi and R. Soyer, "A Bayes Method for Assessing Product-Reliability During Development Testing", *IEEE Transactions on Reliability*, **R-42**, 503-510, (1993).

10. R. van Dorp, T. A. Mazzuchi and R. Soyer, "Sequential Inference and Decision Making for Single Mission Systems Development", *Journal of Statistical Planning and Inference* , forthcoming, (1997).

11. A. F. M. Smith and A. E. Gelfand, "Bayesian Statistics without Tears: A Sampling-Resampling Perspective", *The American Statistician*, **46**, 84-88, (1992).

12. W. R. Gilks P. Wild, "Adaptive Rejection Sampling for Gibbs Sampling", *Applied Statistics*, **41**, 337-348, (1992).

13. R. E. Barlow and R. M. Scheuer, "Reliability Growth during a Development Test Program", *Technometrics*, **8**, 53-60, (1966).

14. C. M. Weinrich and A. J. Gross, "The Barlow-Scheuer Reliability Growth Model from a Bayesian Viewpoint", *Technometrics,* **20**, 249-254, (1978).

STRESS-STRENGTH RELIABILITY IN THE WEIBULL CASE

S. P. MUKHERJEE, SUDHANSU S. MAITI

Department of Statistics, Calcutta University

35, Ballygunge Circular Road

Calcutta-700 019

India

Inference on R = P(X < Y), when X and Y are independent Weibull variables with unknown shape parameters, is of special interest. Assuming a common shape parameter for X and Y, approximate lower bounds for R are obtained from complete as well as censored (type-II) samples. The delta method as well as Jack-knife and bootstrap procedures for complete samples and the delta method and Jack-knife procedure for censored samples have been used here to derive estimates of the scale parameters and the common shape parameter. Findings of a simulation study have also been reported.

1 Introduction

Estimation of reliability in stress-strength models has engaged the attention of many reliability analysts. Let Y be a random variable denoting strength built into a product and let the stress to which it is subjected be denoted by the random variable X. Then the stress-strength reliability is defined as R = P(X < Y), i.e., the probability of strength exceeding stress.

Problems of estimating reliability R, of determining lower bounds for R, and of testing hypotheses relating to R have been studied by a host of authors. Because of the wide applicability of the Weibull distribution in the context of reliability analysis, inferences on R when X and Y have independent Weibull distributions with unknown parameters is of special interest. Assuming a common shape parameter, approximate confidence limits for R have been obtained in terms of its maximum likelihood estimate by McCool (1991). A greater emphasis has been given on lower confidence limits as this is the one of practical importance – engineers want to assert that the system is at least as reliable as R_L, for some lower limit R_L.

If the probability density functions of X and Y are

$$f_X(x) = \frac{p}{\theta_1^p} x^{p-1} \exp\left\{-\left(\frac{x}{\theta_1}\right)^p\right\}; x > 0, \theta_1 > 0, p > 0 \quad \text{and}$$

$$g_Y(y) = \frac{p}{\theta_2^p} y^{p-1} \exp\left\{-\left(\frac{y}{\theta_2}\right)^p\right\}; y > 0, \theta_2 > 0, p > 0,$$

then reliability is given by

$$R = \frac{\theta_2^p}{\theta_1^p + \theta_2^p}. \tag{1}$$

Various estimates of θ_1, θ_2, and p yield different estimates of R. Approximate lower bounds for R are obtained from complete as well as censored (type-II) samples. The delta method as well as Jack-knife and bootstrap procedures for complete samples and the delta method and Jack-knife procedures for censored samples have been used to derive estiamtes of the scale parameters and the common shape parameter. Some numerical comparisons with respect to reliability confidence bounds have been made among these procedures using simulated data.

2 Inferences from Complete Samples

2.1 Delta Method

Suppose the samples $(x_1, x_2, \ldots, x_{n_1})$ on x and $(y_1, y_2, \ldots, y_{n_2})$ on y are of sizes n_1 and n_2, respectively. Then the log-likelihood function is given by

$$\ln L = \left(n_1 + n_2\right)\ln p - p\left\{n_1 \ln \theta_1 + n_2 \ln \theta_2\right\} + \left(p - 1\right)\left\{\sum_{i=1}^{n_1}\ln x_i + \sum_{j=1}^{n_2}\ln y_j\right\}$$

$$-\frac{\sum\limits_{i=1}^{n_1}x_i^p}{\theta_1^p} - \frac{\sum\limits_{j=1}^{n_2}y_j^p}{\theta_2^p}.$$

We have the likelihood equations as

$$\theta_1 = \left\{\frac{1}{n_1}\sum_i x_i^p\right\}^{\frac{1}{p}}, \tag{2}$$

$$\theta_2 = \left\{\frac{1}{n_{21}}\sum_j y_j^p\right\}^{\frac{1}{p}}, \text{ and} \tag{3}$$

$$\frac{n_1 + n_2}{p} + \left(\sum_i \ln x_i + \sum_j \ln y_j\right) - \left\{\frac{n_1 \sum\limits_i x_i^p \ln x_i}{\sum\limits_i x_i^p} + \frac{n_2 \sum\limits_j y_j^p \ln y_j}{\sum\limits_j y_j^p}\right\} = 0. \tag{4}$$

Let the m.l.e.'s (solutions of these equations) be $\hat{\theta}_1$, $\hat{\theta}_2$, and $\hat{p}$. An estimate $\hat{R}$ is obtained by replacing θ_1, θ_2, and p by $\hat{\theta}_1$, $\hat{\theta}_2$, and $\hat{p}$ in (1). Now, the asymptotic variance-covariance matrix of $(\hat{\theta}_1, \hat{\theta}_2, \hat{p})$ is given by

$$V = \begin{pmatrix} a_{11} & a_{12} & a_{13} \\ a_{12} & a_{22} & a_{23} \\ a_{13} & a_{23} & a_{33} \end{pmatrix}^{-1},$$

where

$$-a_{11} = \frac{\partial^2 \ln L}{\partial \theta_1^2} = \frac{n_1 p}{\theta_1^2} - p(p+1)\frac{\sum x_i^p}{\theta_1^{p+2}},$$

$$-a_{22} = \frac{\partial^2 \ln L}{\partial \theta_2^2} = \frac{n_2 p}{\theta_2^2} - p(p+1)\frac{\Sigma y_j^p}{\theta_2^{p+2}},$$

$$-a_{33} = -\frac{(n_1 + n_2)}{p^2} - \Sigma_i \left(\frac{x_i}{\theta_1}\right)^p \left\{\ln\left(\frac{x_i}{\theta_1}\right)\right\}^2 - \Sigma_j \left(\frac{y_j}{\theta_2}\right)^p \left\{\ln\left(\frac{y_j}{\theta_2}\right)\right\}^2,$$

$$-a_{12} = 0,$$

$$-a_{13} = \frac{\partial^2 \ln L}{\partial \theta_1 \partial p} = -\frac{n_1}{\theta_1} + \frac{1}{\theta_1}\left\{\Sigma_i \left(\frac{x_i}{\theta_1}\right)^p + p\Sigma_i \left(\frac{x_i}{\theta_1}\right)^p \ln\left(\frac{x_i}{\theta_1}\right)\right\}, \text{ and}$$

$$-a_{23} = \frac{\partial^2 \ln L}{\partial \theta_2 \partial p} = -\frac{n_2}{\theta_2} + \frac{1}{\theta_2}\left\{\Sigma_j \left(\frac{y_j}{\theta_2}\right)^p + p\Sigma_j \left(\frac{y_j}{\theta_2}\right)^p \ln\left(\frac{y_j}{\theta_2}\right)\right\}.$$

Let $G = \left(\frac{\partial R}{\partial \theta_1}, \frac{\partial R}{\partial \theta_2}, \frac{\partial R}{\partial p}\right)^T$ yield the asymptotic variance of $\hat{R}$ as $S_\Delta^2(\hat{R}) = G^T V G$. We have

$$\frac{\partial R}{\partial \theta_1} = -\frac{p\theta_1^{p-1}\theta_2^p}{\left(\theta_1^p + \theta_2^p\right)^2},$$

$$\frac{\partial R}{\partial \theta_2} = \frac{p\theta_1^p\theta_2^{p-1}}{\left(\theta_1^p + \theta_2^p\right)^2}, \text{ and}$$

$$\frac{\partial R}{\partial p} = -\frac{(\theta_1\theta_2)^p(\ln\theta_1 - \ln\theta_2)}{\left(\theta_1^p + \theta_2^p\right)^2}.$$

Here $\hat{p}$ fails to satisfy regularity condition for converging to asymptotic normality. In spite of this, by approximating $\dfrac{R - \hat{R}}{S_\Delta^2(\hat{R})}$ as a standard normal variate, a lower confidence bound to R can be constructed.

2.2 Further Transformations

Instead of approximating $\hat{R}$ as asymptotically normal, one may consider some normalizing transformations g(R) of R and assume $\left\{g(\hat{R}) - g(R)\right\}/\sqrt{\text{Var } g(\hat{R})}$ as N(0,1). It is known that $\text{Var } g(\hat{R}) = \left\{g'(R)\right\}^2 \text{Var}(\hat{R})$. The following two transformations are generally used.

(a) Logit transformation: $g(R) = \ln\left(\dfrac{R}{1-R}\right)$, with $g'(R) = \dfrac{1}{R(1-R)}$.

(b) Arcsine transformation: $g(R) = \mathrm{Sin}^{-1}\left(\sqrt{R}\right)$, with $g'(R) = \dfrac{1}{2\sqrt{R(1-R)}}$.

2.3 Initial Value of p

An initial value of the shape parameter p is necessary to estimate this parameter by any iteration method. Since the square of the population coefficient of variation,

$$\frac{\Gamma\left(\dfrac{p+2}{2}\right) - \Gamma^2\left(\dfrac{p+1}{p}\right)}{\Gamma^2\left(\dfrac{p+1}{p}\right)},$$

is a function of p only, equating it to the sample coefficient of variation one can have an estimate of p. The value of p can also be approximately obtained from the figure given by Cohen (1965). This way two starting values of p can be obtained from the two samples. One can pool these estimates in some way to get an initial value of p. Another way is to pool the means and variances of the two samples and hence get the sample coefficient of variation and equate it to the population coefficient of variation.

If one assumes that the combined sample of X and Y observations arise from a mixture of the two pdfs, i.e. from $h_x(x) = \alpha_1 f_x(x; \theta_1, p) + \alpha_2 g_x(x; \theta_2, p)$, then the square of the population coefficient of variation is

$$\frac{\mu_2}{\mu_1} = \frac{2\Gamma\left(\dfrac{p+2}{p}\right) - \Gamma^2\left(\dfrac{p+1}{p}\right)}{\Gamma^2\left(\dfrac{p+1}{p}\right)} - \frac{4\theta_1\theta_2}{\left(\theta_1 + \theta_2\right)^2} \cdot \frac{\Gamma\left(\dfrac{p+2}{p}\right)}{\Gamma^2\left(\dfrac{p+1}{p}\right)}. \tag{5}$$

We assume here the samples are of equal size and for that reason we take $\alpha_1 = \alpha_2 = \tfrac{1}{2}$. The above mentioned equation is not a function of p only. It involves the scale parameters also. Therefore one can't have an estimate of p equating the population coefficient of variation with the combined sample coefficient of variation. Since $\dfrac{4\theta_1\theta_2}{\left(\theta_1 + \theta_2\right)^2} < 1$, if θ_1 and θ_2 are close to each other then $\dfrac{4\theta_1\theta_2}{\left(\theta_1 + \theta_2\right)^2} \approx 1$, and hence

$$\frac{\mu_2}{\mu_1} \approx \frac{\Gamma\left(\dfrac{p+2}{p}\right) - \Gamma^2\left(\dfrac{p+1}{p}\right)}{\Gamma^2\left(\dfrac{p+1}{p}\right)}. \tag{6}$$

So using this approximation, an estimate of p can be found.

2.4 Use of Resampling Procedures

Resampling methods are used to gain an idea about the distribution of the statistics under study. Here we use the jack-knife and boot-strap methods to derive estimates of the variance-covariance matrix V, and then compute lower confidence bounds on R based on these estimates.

2.4.1 Jack-knife Method

In this method we take the ith sample by excluding x_i from the sample of size n, i.e. n samples each of size n-1. We have $\hat{\theta}_1^{(i)}$, $\hat{\theta}_2^{(i)}$, and $\hat{p}^{(i)}$ from (2), (3), and (4), respectively, excluding x_i and y_i and replacing n by (n−1). Now the Jack-knife estimator of any parameter θ is obtained as

$$\tilde{\theta} = n\hat{\theta} - (n-1)\overline{\theta}, \tag{7}$$

where $\overline{\theta} = \dfrac{1}{n}\sum_{i=1}^{n}\hat{\theta}^{(i)}$ and $\hat{\theta}$ is an estimate of the original sample.

The variance and covariance expressions are obtained as

$$\hat{\sigma}_j^2(\theta) = \left(\frac{n-1}{n}\right)\sum_{i=1}^{n}\left(\hat{\theta}^{(i)} - \overline{\theta}\right)^2 \quad \text{and} \tag{8}$$

$$\hat{\sigma}_j(\theta_1,\theta_2) = \left(\frac{n-1}{n}\right)\sum_{i=1}^{n}\left(\hat{\theta}_1^{(i)} - \overline{\theta}_1\right)\left(\hat{\theta}_2^{(i)} - \overline{\theta}_2\right). \tag{9}$$

Using these expressions we have an estimate of the variance-covariance matrix V, and then we proceed similarly as indicated in Sections 2.1 and 2.2 to obtain the lower confidence bound for R.

2.4.2 Bootstrap Standard Error Method

In this procedure we take samples of size n from the n observations with a replacement scheme. Let there be k such samples. By using these we can find k estimates of each parameter from (2), (3), and (4), and from these k values we have pooled estimates for the variance-covariance matrix. Finally, the lower confidence bound on R can be found. Here,

$$\overline{\overline{\theta}} = \frac{1}{k}\sum_{i=1}^{k}\hat{\theta}^{(i)}, \tag{10}$$

$$\hat{\sigma}_B^2(\theta) = \frac{1}{k-1}\sum_{i=1}^{k}\left(\hat{\theta}^{(i)} - \overline{\overline{\theta}}\right)^2, \quad \text{and} \tag{11}$$

$$\hat{\sigma}_B(\theta_1,\theta_2) = \frac{1}{k-1}\sum_{i=1}^{k}\left(\hat{\theta}_1^{(i)} - \overline{\overline{\theta}}_1\right)\left(\hat{\theta}_2^{(i)} - \overline{\overline{\theta}}_2\right). \tag{12}$$

3 Inferences From Censored (Type-II) Samples

Suppose we are given random samples of size n_1 and n_2 drawn from the distributions of X and Y censored at the r_1^{th} and r_2^{th} order statistics, respectively. The observations are so indexed that x_i and y_j denote uncensored observations for $i \leq r_1$, $j \leq r_2$, and $x_i = x_{r_1}$ for $i > r_1$, $y_j = y_{r_2}$ for $j > r_2$.

3.1 Delta Method

From the log-likelihood function

$$\ln L = (r_1 + r_2) \ln p - p\{r_1 \ln \theta_1 + r_2 \ln \theta_2\} + (p-1)\left\{\sum_{i=1}^{r_1} \ln x_i + \sum_{j=1}^{r_2} \ln y_j\right\} -$$

$$\left\{\frac{\sum_{i=1}^{r_1} x_i^p + (n_1 - r_1)x_{r_1}^p}{\theta_1^p}\right\} - \left\{\frac{\sum_{j=1}^{r_2} y_j^p + (n_2 - r_2)y_{r_2}^p}{\theta_2^p}\right\},$$

we have the likelihood equations given by

$$\theta_1 = \left\{\frac{\sum_{i=1}^{r_1} x_i^p + (n_1 - r_1)x_{r_1}^p}{r_1}\right\}^{1/p}, \tag{13}$$

$$\theta_2 = \left\{\frac{\sum_{j=1}^{r_2} y_j^p + (n_2 - r_2)y_{r_2}^p}{r_2}\right\}^{1/p}, \tag{14}$$

and

$$\frac{r_1 + r_2}{p} + \left(\sum_{i=1}^{r_1} \ln x_i + \sum_{j=1}^{r_2} \ln y_j\right) - \left\{\frac{r_1 \sum_{i=1}^{r_1} x_i^p \ln x_i + (n_1 - r_1)x_{r_1}^p \ln x_{r_1}}{\sum_{i=1}^{r_1} x_i^p + (n_1 - r_1)x_{r_1}^p}\right\} -$$

$$\left\{\frac{r_2 \sum_{j=1}^{r_2} y_j^p \ln y_j + (n_2 - r_2)y_{r_2}^p \ln y_{r_2}}{\sum_{j=1}^{r_2} y_j^p + (n_2 - r_2)y_{r_2}^p}\right\} = 0. \tag{15}$$

Approximate expressions for V, $\dfrac{\partial^2 \ln L}{\partial^2 \theta_1^2}$, $\dfrac{\partial^2 \ln L}{\partial^2 \theta_2^2}$, $\dfrac{\partial^2 \ln L}{\partial^2 p^2}$, $\dfrac{\partial^2 \ln L}{\partial \theta_1 \, \partial \theta_2}$, $\dfrac{\partial^2 \ln L}{\partial \theta_1 \, \partial p}$, and

$\dfrac{\partial^2 \ln L}{\partial \theta_2 \, \partial p}$ are obtained by replacing n_i by r_i, $i = 1, 2$, in the corresponding expressions obtained in Section 2.

3.2 Jack-knife Method

For simplicity we assume that $n_1 = n_2 \, (= n)$ and $r_1 = r_2 \, (= r)$. Now for $i = 1(1)r$, the estimates of $\hat\theta_1^{(i)}$, $\hat\theta_2^{(i)}$, and $\hat{p}^{(i)}$ derived from the sample of size $n - 1$ by excluding x_i (of the original sample of size n), are given by

$$\hat\theta_1^{(i)} = \left\{ \frac{\displaystyle\sum_{j(\neq i)=1}^{r} x_j^{\hat p} + (n-r) x_r^{\hat p}}{r-1} \right\}^{\!\!1/\hat p}, \tag{16}$$

$$\hat\theta_2^{(i)} = \left\{ \frac{\displaystyle\sum_{j(\neq i)=1}^{r} y_j^{\hat p} + (n-r) y_r^{\hat p}}{r-1} \right\}^{\!\!1/\hat p}, \tag{17}$$

and

$$\frac{2}{\hat p} + \frac{\displaystyle\sum_{j(\neq i)=1}^{r} \ln x_j + \sum_{j(\neq i)=1}^{r} \ln y_j}{r-1} - \left\{ \frac{\displaystyle\sum_{j(\neq i)=1}^{r} x_j^{\hat p} \ln x_j + (n-r) x_r^{\hat p} \ln x_r}{\displaystyle\sum_{j(\neq i)=1}^{r} x_j^{\hat p} + (n-r) x_r^{\hat p}} \right\} - $$

$$\left\{ \frac{\displaystyle\sum_{j(\neq 1)=1}^{r} y_j^{\hat p} \ln y_j + (n-r) y_r^{\hat p} \ln y_r}{\displaystyle\sum_{j(\neq i)=1}^{r} y_j^{\hat p} + (n-r) y_r^{\hat p}} \right\} = 0. \tag{18}$$

For $i = (r + 1)(1)n$, $\hat\theta_1^{(i)}$, $\hat\theta_2^{(i)}$, and $\hat{p}^{(i)}$ will be the same and are given by

$$\hat\theta_1^{(i)} = \left\{ \frac{\displaystyle\sum_{i=1}^{r} x_i^{\hat p} + (n-r-1) x_r^{\hat p}}{r} \right\}^{\!\!1/\hat p}, \tag{19}$$

$$\hat{\theta}_2^{(i)} = \left\{ \frac{\sum\limits_{i=1}^{r} y_i^{\hat{p}} + (n-r-1) y_r^{\hat{p}}}{r} \right\}^{1/\hat{p}}, \tag{20}$$

and

$$\frac{2}{\hat{p}} + \frac{\sum\limits_{i=1}^{r} \ln x_i + \sum\limits_{i=1}^{r} \ln y_i}{r-1} - \left\{ \frac{\sum\limits_{i=1}^{r} x_i^{\hat{p}} \ln x_i + (n-r-1) x_r^{\hat{p}} \ln x_r}{\sum\limits_{i=1}^{r} x_i^{\hat{p}} + (n-r-1) x_r^{\hat{p}}} \right\} -$$

$$\left\{ \frac{\sum\limits_{i=1}^{r} y_i^{\hat{p}} \ln y_i + (n-r-1) y_r^{\hat{p}} \ln y_r}{\sum\limits_{i=1}^{r} y_i^{\hat{p}} + (n-r-1) y_r^{\hat{p}}} \right\} = 0. \tag{21}$$

4 Some Simulation Results

We consider equal sample sizes n for X and Y. A simulation study has been conducted for different n. Observations are generated from a Weibull distribution for different sets of parameters using the Monte Carlo simulation technique. Results are presented in Tables 1-6 at the end of the paper. From this study the following important findings emerge. A few simulation results have been shown in the tables and a graphical presentation has also been included.

1) For combining estimates of p (from samples on X and Y) to offer a pooled estimate, among Arithmetic, Geometric, and Harmonic means, the Harmonic mean gives slightly better results than the Arithmetic or Geometric mean in the sense that deviation from the true value is smaller. If the pooled estimates of the mean and standard deviation (from the two samples) are used to calculate the coefficient of variation, and p is estimated from (6), then the estimate is sometimes more efficient than the previous ones.

2) The computational complexity and effort involved in the Jack-knife method is more than that in the delta method. Moreover, the Jack-knife procedure does not yield a lower confidence bound nearer the true reliability when compared to the delta method.

3) If $\hat{\theta}_1 > \hat{\theta}_2$, the Arcsine transformed lower bound (for reliability) is greater, and if $\hat{\theta}_1 < \hat{\theta}_2$, then the Logit transformed lower bound is greater. A graphical presentation also reveals this fact.

4) Lower confidence bounds obtained in untransformed cases are the least favorable.

5 Graphical Presentation of Simulation Results

Rather than computing confidence intervals for a single confidence level, a p-value was computed as $p = \Phi\left\{\left(\hat{R} - R\right)/S_\Delta\left(\hat{R}\right)\right\}$. If this p-value is less than 0.05, then a one-sided 95% lower confidence limit would fail to include the true R. If the confidence intervals have the correct coverage, then $P(p < \alpha) = \alpha$, for every α, $0 < \alpha < 1$. Therefore, an exact confidence interval procedure would result in a curve of the cdf of p which differs from the diagonal line.

Random samples of the same size n have been drawn from the X and Y populations (both Weibull) for different θ_1, θ_2, and p by the Monte Carlo Technique, and using these samples $\hat{R}$ has been computed. 1000 such repetitions have been considered. We plot the empirical cumulative distribution function of p-values in two different situations:

Figure 1: $n = 16$, $\theta_1 = 4$, $\theta_2 = 2$, $p = 1.6$ and

Figure 1: $n = 15$, $\theta_1 = 2$, $\theta_2 = 4$, $p = 2$.

Both the figures justify the claims in Section 4. Moreover, in Figure 1 we see that the Logit transformation gives the correct coverage of the true R but the delta method and the Arcsine transformation fail to do so. In Figure 2, though the delta method and the Arcsine transformation give correct coverages, since the deviation of the delta method is more from the true R, it is least favorable.

6 Example

To illustrate the computation of the lower confidence bound on R, let us consider the example of Basu (1981). Fifteen items of random strength y_1, y_2, ..., y_{15} are subject to random stresses x_1, x_2, ..., x_{15}. They are as follows:

Stress	Strength
0.0352	1.7700
0.0397	0.9457
0.0677	1.8985
0.0233	2.6121
0.0873	1.0929
0.1156	0.0362
0.0286	1.0615
0.0200	2.3895
0.0793	0.0982
0.0072	0.7971
0.0245	0.8316
0.0251	3.2304
0.0469	0.4373
0.0838	2.5648
0.0796	0.6377

Here $\hat{\theta}_1 = 0.05505049$, $\hat{\theta}_2 = 1.510287$, $\hat{p} = 1.44956$, Estimate of R = 0.9918428, and S.E. of estimated R = 0.006615272.

Lower Bound for R	97.5%	95%
Delta method untransformed	0.9788769	0.9809342
Logit transformation	0.9918419	0.9918421
Arcsine transformation	0.9917376	0.9917543

References

1. A. P. Basu (1981). The estimation of P(X < Y) for distributions useful in life testing. *Naval Research Logistic Quarterly*, **28**, 383-392.

2. A. C. Cohen (1965). Maximum likelihood estimation in the Weibull distribution based on complete and censored samples. *Technometrics*, **7**(4), 579-588.

3. J. I. McCool (1991). Inference on P(Y < X) in the Weibull case. *Communications in Statistics - Simulation and Computation*, **20**(1), 129-148.

Table 1 : Simulation results in Delta Method

(θ_1, θ_2, p)	R	n	$(\hat{\theta}_1, \hat{\theta}_2, \hat{p})$	$\hat{R}$	95% Lower Confidence bound		
					Untrans-formed	Logit	Arcsin
10 5 1.5	0.2612039	35	9.177041 5.145133 1.396308	0.3083202	0.2213718	0.2683436	0.2899371
		25	8.030105 3.869039 1.451786	0.2572940	0.1618383	0.2082306	0.2392696
		40	7.627628 4.053379 1.567847	0.2706684	0.1910749	0.2304622	0.2551028
		35	7.830444 5.313658 1.434352	0.3644379	0.2704630	0.3273839	0.3428173
		30	10.390020 5.068871 1.569316	0.2448369	0.1571793	0.1983390	0.2288148
5 10 1.5	0.7387961	30	4.205185 10.073640 1.452784	0.7805985	0.7015819	0.7767713	0.7669183
		35	4.372204 12.238490 1.632932	0.8430114	0.7779064	0.8414001	0.8342998
		40	5.990225 12.014880 1.579539	0.7501449	0.6736133	0.7453368	0.7356662
		35	4.638559 10.408340 1.590776	0.7834136	0.7061680	0.7797681	0.7701656
		25	5.415222 11.924600 1.777692	0.8027057	0.7135940	0.7992140	0.7884058
5 15 1.2	0.7889046	20	4.968807 12.902060 1.209598	0.7602785	0.6753606	0.7553646	0.7446341
		25	4.348282 12.532690 1.243442	0.7881056	0.6963320	0.7839557	0.7725810
		40	4.327200 13.678350 1.387938	0.8316503	0.7713770	0.8299351	0.8231280
		30	4.060207 13.748510 1.183634	0.8090230	0.7360597	0.8063477	0.7976243

Table 2 : Simulation results in Jack-Knife Method

(θ_1, θ_2, p)	R	n	$(\hat{\theta}_1, \hat{\theta}_2, \hat{p})$	$\hat{R}$	95% Lower Confidence bound		
					Untrans-formed	Logit	Arcsin
10 5 1.5	0.2612039	35	9.307953 5.249695 1.448525	0.3037362	0.2293507	0.2689247	0.2881229
		25	7.958603 3.820404 1.356754	0.2697841	0.1892996	0.2290712	0.2540788
		40	7.900849 4.152664 1.699814	0.2509872	0.1648408	0.2057988	0.2349697
		35	7.762177 5.269608 1.355885	0.3716503	0.2910925	0.3404380	0.3529399
		30	10.282870 5.030884 1.478535	0.2578870	0.1825855	0.2186292	0.2436097
5 10 1.5	0.7379610	30	4.167091 9.959137 1.343691	0.7632761	0.6592086	0.7573949	0.7442211
		35	4.411179 12.393740 1.667877	0.8485148	0.7860185	0.8470750	0.84033949
		40	5.902176 11.734100 1.425739	0.7270560	0.6407660	0.7205805	0.7096860
		35	4.67247 10.52347 1.609646	0.7869945	0.7054980	0.7832735	0.7731757
		25	5.342285 11.735290 1.561241	0.7735760	0.6933650	0.7694358	0.7593700
5 15 1.2	0.7889046	20	4.982498 12.959900 1.194058	0.7579428	0.6693163	0.7527121	0.7415010
		25	4.386917 12.788180 1.265799	0.7948270	0.7024400	0.7909105	0.7795592
		35	4.546646 16.907530 1.203354	0.8292680	0.7595616	0.8272265	0.8192866
		40	4.224198 13.304510 1.237049	0.8052183	0.7217228	0.8020308	0.7919582
		30	3.901039 10.350920 1.037502	0.7818506	0.6977390	0.7778213	0.7673374

Table 3 : Simulation results in Delta Method for Censored (Type-II) Case

(θ_1, θ_2, p)	R	n	r	$(\hat{\theta}_1, \hat{\theta}_2, \hat{p})$	$\hat{R}$	95% Lower Confidence bound		
						Untrans-formed	Logit	Arcsin
10 5 1.5	0.2612039	35	15	10.474930 4.305822 1.416792	0.2210511	0.1008689	0.1566914	0.2007126
		25	10	7.448239 2.722684 1.616005	0.1643404	0.0372445	0.0934215	0.1472055
		40	15	7.343067 4.633752 1.586224	0.3251329	0.1846350	0.2646559	0.2497055
		35	35	7.108619 5.115411 1.557955	0.3745736	0.2481676	0.3265774	0.3452137
		30	15	8.071806 4.530628 1.932871	0.2467048	0.1200592	0.1819833	0.2235579
5 10 1.5	0.7387961	30	15	3.473822 11.199260 1.514129	0.8547592	0.7578355	0.8527028	0.8425221
		35	15	5.564235 10.764920 1.492352	0.7280701	0.5967890	0.7182531	0.7017003
		40	25	6.315957 10.666250 1.697327	0.7087679	0.6047049	0.6998635	0.6870623
		35	15	4.799750 8.131873 1.852600	0.7264613	0.5931038	0.7163704	0.6995767
		25	12	5.725024 9.752001 2.066830	0.7504186	0.6075283	0.7414124	0.7231951
5 15 1.2	0.7889046	30	12	3.794328 16.543290 1.164104	0.8473688	0.7314282	0.8446482	0.8320761
		25	10	4.953728 9.923921 1.303723	0.7121475	0.5450729	0.6981080	0.6773243
		35	15	3.587253 16.977960 1.334049	0.8883323	0.8032931	0.8872675	0.8797584
		40	15	3.719695 11.077690 1.711992	0.8662539	0.7718983	0.8645625	0.8551401
		20	15	2.998886 14.170270 1.384608	0.8956842	0.8167548	0.8948222	0.8881951

Table 4 : Simulation results in Jack-Knife Method for censored (Type-II) Case

(θ_1, θ_2, p)	R	n	r	$(\hat{\theta}_1, \hat{\theta}_2, \hat{p})$	$\hat{R}$	95% Lower Confidence bound		
						Untrans-formed	Logit	Arcsin
10 5 1.5	0.2612039	35	15	10.087400 4.133789 1.307279	0.2375371	0.1286155	0.1800729	0.2180989
		25	10	7.119797 2.600350 1.472931	0.1848865	0.0677035	0.1191806	0.1675587
		40	15	7.065796 4.464005 1.470051	0.3373612	0.2590211	0.3038671	0.2518233 0.3199639
		35	20	7.035034 5.066010 1.421219	0.3854081	0.3255459	0.3630582	0.3712793
		30	15	7.931275 4.450775 1.811112	0.2599280	0.1702658	0.2138907	0.2428701
5 10 1.5	0.7387961	30	15	3.385689 10.900880 1.373692	0.8328885	0.7090078	0.8294003	0.8152968
		35	15	5.381119 10.446260 1.386887	0.7150396	0.6244478	0.7076266	0.6964063
		40	25	6.270523 10.594060 1.588425	0.6969936	0.6383710	0.6915845	0.6845430
		35	15	4.684708 7.939759 1.740334	0.7146680	0.6259498	0.7073905	0.6964094
		25	12	5.593903 9.559631 1.946056	0.7393971	0.6525928	0.7334588	0.7225011
5 15 1.2	0.7889046	30	12	3.476692 15.281100 1.038586	0.8231247	0.6656350	0.8181436	0.7996261
		25	10	4.556816 9.134598 1.270989	0.7076303	0.6365097	0.7015140	0.6928103
		35	15	3.416466 16.195370 1.192890	0.8648628	0.7207382	0.8622091	0.8475822
		40	15	3.607208 10.760560 1.585007	0.8497112	0.7391374	0.8471966	0.8353215
		30	15	2.903275 13.698000 1.289101	0.8807891	0.7838169	0.8794041	0.8704208

Table 5 : Simulation results in Bootetrap Approach

(θ_1, θ_2, p)	R	n	k	$(\hat{\theta}_1, \hat{\theta}_2, \hat{p})$	$\hat{R}$	95% Lower Confidence bound		
						Untrans-formed	Logit	Arcsin
10 5 1.5	0.2612039	35	18	9.320240 5.117357 1.432036	0.2976366	0.2224917	0.261941	0.2820502
		25	15	7.737086 3.830669 1.454218	0.2645808	0.1870443	0.2248171	0.2496345
		40	25	7.595061 4.016991 1.559267	0.2702809	0.1756401	0.2229228	0.2518233
		35	25	7.771570 5.189700 1.455074	0.3571964	0.2884798	0.3293305	0.3414987
		30	20	10.715050 5.068871 1.569316	0.2296655	0.1409118	0.1812510	0.2141553
5 10 1.5	0.7387961	30	15	4.180240 9.647929 1.471952	0.7740116	0.6589263	0.7680801	0.7535716
		35	30	4.444758 12.238490 1.699038	0.8492204	0.7796387	0.8476317	0.8402034
		40	30	5.940320 12.005890 1.669489	0.7639954	0.6892008	0.7598040	0.7503786
		35	25	4.622545 10.393000 1.669489	0.7878916	0.7058804	0.7841785	0.7740268
		25	20	5.515654 11.174940 1.986688	0.8026227	0.7104935	0.7990089	0.7878275
5 15 1.2	0.7889046	30	18	5.163789 14.094520 1.261981	0.7802588	0.6941946	0.7760748	0.7653279
		25	15	4.163798 12.690540 1.301283	0.8100273	0.6898939	0.8056539	0.7912036
		35	25	4.829187 16.763240 1.292991	0.8332920	0.7529626	0.8310475	0.8219853
		40	30	4.452333 13.664900 1.432892	0.8329748	0.7587386	0.8310475	0.8225201
		20	30	3.923079 13.621540 1.210896	13.6215400	0.7360597	0.8063477	0.7976243

Table 6 : Estimation of p in different approaches using simulation data

(θ_1, θ_2, p)	n	Estimate of $\overset{*}{p}$			
		A.M.	G.M.	H.M.	* *
$5, 10, 1.5$	35	1.725	1.702939	1.68116	1.15
	40	1.575	1.565248	1.555556	1.275
	35	1.6	1.576190	1.552734	1.225
	25	1.9	1.802776	1.710526	1.25
$10, 5, 1.5$	30	1.4375	1.436141	1.434783	1.225
	35	1.55	1.548184	1.546371	1.325
	40	1.4125	1.409787	1.407080	1.145
	45	1.45	1.449138	1.448276	1.250
$4, 6, 1.0$	25	1.1375	1.137431	1.137363	1.100
	30	0.8500	0.849632	0.849265	0.925
	35	1.1125	1.106797	1.101124	1.075
	40	1.0500	1.049702	1.049405	1.050
$6, 4, 1.0$	35	1.0125	0.999375	0.986419	0.975
	20	1.1625	1.157044	1.151613	1.125
	40	1.0625	1.060660	1.058824	1.025
	30	0.9000	0.894427	0.888889	0.975

* * is the estimate of p using (2.3.2)

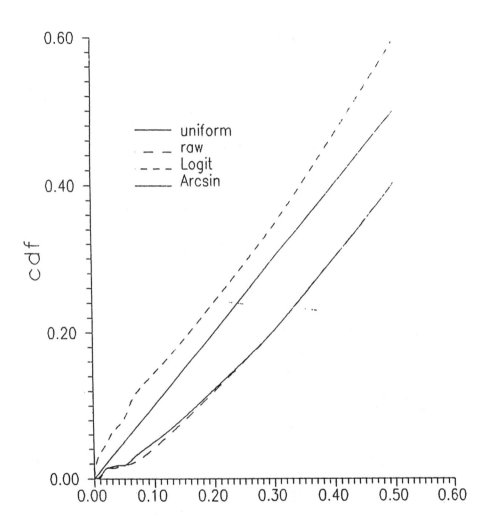

Figure 1. Empirical CDF of p-values

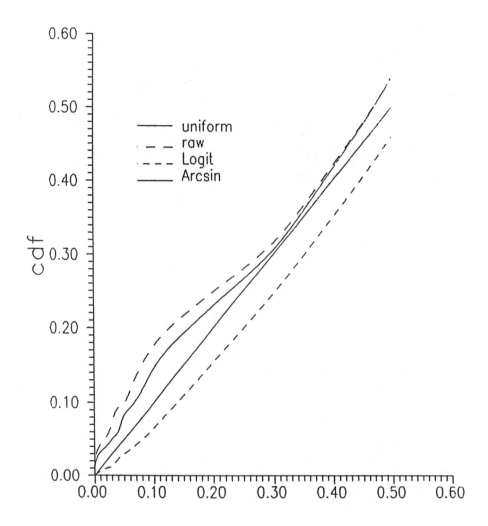

Figure 2. Empirical CDF of p-values in Delta Method

MIXTURE OF WEIBULL DISTRIBUTIONS – PARAMETRIC CHARACTERIZATION OF DENSITY FUNCTION

D.N.P. MURTHY

Department of Mechanical Engineering, The University of Queensland, St. Lucia,
Q 4072, Australia
Email: murthy@mech.uq.edu.au

R. JIANG

Department of Automobile Engineering, Changsha Communication Institute, changsha, Hunan,
P.R. China

A mixture distribution involving two two-parameter Weibull distributions is characterized by five parameters. The shape of the density function for the mixture depends on the parameter values and can be one of four distinct shapes. A complete parametric characterization of the shape of the density function for the mixture is presented.

1 Introduction

Mixtures of Weibull distributions have been used to model a variety of failure data. Jiang and Murthy[1] discuss the use of the Weibull probability paper to determine if a given data set can be adequately modeled by a mixture of two two-parameter Weibull distributions and review some of the relevant earlier literature on the topic. The estimation of parameters for such mixtures has also received a lot of attention and there are several papers which deal with this topic – see, for example, Kao[2], Rider[3], Falls[4], Chen et al[5], Jiang and Kececioglu[6], and Kececioglu and Sun[7].

The mixture of two 2-parameter Weibull distributions is characterized by five parameters – the scale parameters (η_i, i = 1,2) and shape parameters (β_i, i = 1,2) for the two distributions and the mixture parameter (p). The shape of the density funciton can be one of four distinct shapes (shown in Figure 1 and labeled Types A–D) depending on the parameter values. Several papers have discussed some of these shapes – for example, Sichart and Vollertsen[8], Yuan and Shih[9], Jiang and Kececioglu[10], Drapella[11], and Gera[12]. There is no paper which discusses all the different possible shapes for the density function or gives a complete parametric characterization. In this paper, we carry out a complete parametric characterization of the density function. Such a study is of interest in the modeling of failure data, where based on a histogram plot, one must decide if the given failure data set can be modeled adequately by a mixture of Weibull distributions.

The outline of this paper is as follows. We commence with the parametric characterization of the density function for a single Weibull distribution in Section 2. In Section 3 we carry out some preliminary analysis and outline the approach used to carry out the parametric characterization of the density function for a mixture of two Weibull distributions. This is done in Sections 4 – 8 where each section looks at the characterization for a different region in the parameter space. Finally, we conclude with some comments in Section 9.

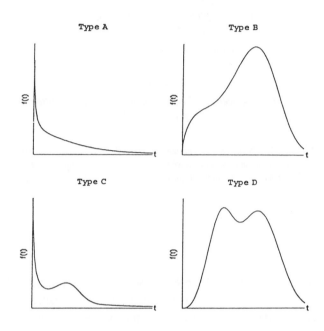

Figure 1: Four Basic Shapes (Type A − D) for Density Function of a Mixture of Two Weibull Distributions.

Notation

$F(t)$, $f(t)$:	Cumulative distribution function and probability density function
$R(t)$:	Reliability function ($= 1 - F(t)$)
$F_i(t)$, $f_i(t)$:	Cumulative distribution function and probability density function for sub-population i, i = 1,2
$R_i(t)$:	Reliability function for sub-population i, i = 1,2
η_i, β_i:	Scale and shape parameters of $F_i(t)$, i = 1,2
p [q]:	Mixture parameter ($0 < p < 1$ and $p + q = 1$)
η:	Scale parameter for single Weibull or $= \eta_2/\eta_1$ (for mixture of Weibull)
β:	Shape parameter for single Weibull or $= \beta_2/\beta_1$ (for mixture of Weibull)
z:	$= \left(t/\eta_2 \right)^{\beta_2}$
z_{pi}:	$= 1 - 1/\beta_i$, i = 1,2
k:	$= \left(\eta \right)^{\beta_1}$

2 Single Weibull Distribution

The reliability and density functions for a 2-parameter Weibull distribution are given by

$$R(t) = \exp\left\{-\left(\frac{t}{\eta}\right)^{\beta}\right\} \tag{1}$$

and

$$f(t) = -\, dR(t) / dt = -\, \beta\, R(t)\, ln(R(t)) / t \tag{2}$$

with $0 < \beta < \infty$ and $0 < \eta < \infty$.

When $\beta \leq 1$, the shape is Type A and when $\beta > 1$, it is Type B. In the latter case, the peak of $f(t)$ occurs at t_p given by

$$t_p = \eta(1 - 1/\beta)^{1/\beta}. \tag{3}$$

Note that the shape is influenced only by the shape parameter β and the scale parameter η has no influence. The result for the 3-parameter Weibull distribution is the same as that for the 2-parameter case except that when $\beta > 1$, the peak of $f(t)$ occurs at

$$t_p = \gamma + \eta(1 - 1/\beta)^{1/\beta}. \tag{4}$$

3 Mixture of Two Weibull Distributions

The reliability function for a mixture of 2-Weibull distributions is given by

$$R(t) = p\, R_1(t) + q\, R_2(t) \tag{5}$$

where $q = (1 - p)$ and

$$R_i(t) = \exp\left\{-\left(\frac{t}{\eta_i}\right)^{\beta_i}\right\} \tag{6}$$

for $i = 1,2$. η_i and β_i $(i = 1,2)$ are the scale and shape parameters for the two sub-populations, and p $(0 < p < 1)$ is the mixture parameter. As a result, the reliability function is characterized by five parameters.

When $\eta_1 = \eta_2$ and $\beta_1 = \beta_2$, the mixture reduces to a single Weibull distribution. Hence, when $\beta_1 = \beta_2$, we assume, without loss of generality, that $\eta_1 > \eta_2$. In this case, the reliability function is characterized by 4 parameters. When $\beta_1 \neq \beta_2$, we assume, without loss of generality, that $\beta_1 < \beta_2$.

From (5), we have on differentiation

$$f(t) = p\, f_1(t) + q\, f_2(t) \tag{7}$$

and

$$f'(t) = df(t)/dt = p\, df_1(t)/dt + q\, df_2(t)/dt. \tag{8}$$

The shape of the density function can be characterized by studying the zeros of its derivative. Let $f'(t) = df(t)/dt$. Define

$$\begin{cases} \eta = \dfrac{\eta_2}{\eta_1}, \ \beta = \dfrac{\beta_2}{\beta_1}, \ k = \eta^{\beta_1} \\ z = \left(\dfrac{t}{\eta_2}\right)^{\beta_2}; \ z_{pi} = 1 - \dfrac{1}{\beta_i}, \ i = 1, 2 \end{cases} \tag{9}$$

Using (9) in (8), and after some simplification, we have

$$f'(t) = p\,(\beta_1/t)^2\,h(z), \tag{10}$$

where

$$h(z) = kz^{1/\beta}(z_{p1} - kz^{1/\beta})/\exp(kz^{1/\beta}) + \beta^2 qz(z_{p2} - z)/[p\,\exp(z)], \tag{11}$$

which can be rewritten as

$$h(z) = kz^{1/\beta}\exp(-kz^{1/\beta})\,h_0(z) \tag{12}$$

where

$$h_0(z) = (z_{p1} - kz^{1/\beta}) + \beta^2 qz^{1\,-\,1/\beta}(z_{p2} - z)\exp(kz^{1/\beta} - z)/(pk). \tag{13}$$

Note that z is a monotonic transformation of t. Since $[p\,(\beta_1/t)^2] > 0$, the number of zeros of $f'(t)$ over the interval $(0, \infty)$ is the same as the number of zeros of $h(z)$ over the interval $(0, \infty)$. Again, since $k\,z^{1/\beta}\exp(-kz^{1/\beta}) > 0$ in the open region $(0, \infty)$, the number of zeros of $h(z)$ over the interval $(0, \infty)$ is the same as that for $h_0(z)$. Hence, in the remainder of the paper, we will use either $h(z)$ or $h_0(z)$ to determine the number of zeros of $f'(t)$.

3.1 Parametric characterization of pdf

For the general case there are five parameters. The shape of the density function is dependent on the values of two shape parameters β_1 and β_2, the value of the ratio of the scale parameters η (and not the individual values of the scale parameters) and finally, on the value of the mixing parameter p. The ranges for these parameters are as follows:

$$0 < \beta_1 < \infty, 0 < \beta_2 < \infty, 0 < \eta < \infty, \text{ and } 0 < p < 1.$$

As a result, the parametric characterization of pdf needs to be done in a four dimensional parameter space. We shall display the results in the two dimensional plane of β_1 and β_2 for different combinations of p and η.

For the two special cases – namely, (i) $\beta_1 = \beta_2\ (= \beta_0)$ and (ii) $\beta_1 = 1$ and $\beta_2 > 1$ – we have only three parameters. In this case, we carry out the characterization in a two dimensional space of p and β_0 or β_2 for different values of η.

The complete parametric characterization of the density function is carried out by considering the following five cases:

Case (a) $\beta_1 = \beta_2 = \beta_0$ [Section 4]
Case (b) $\beta_1 = 1$ and $\beta_2 > 1$ [Section 5]
Case (c) $0 < \beta_1 < \beta_2 \le 1$ [Section 6]
Case (d) $0 < \beta_1 < 1 < \beta_2$ [Section 7]
Case (e) $1 < \beta_1 < \beta_2$ [Section 8]

The regions in the β_1–β_2 plane corresponding to these five cases are shown in Figure 2.

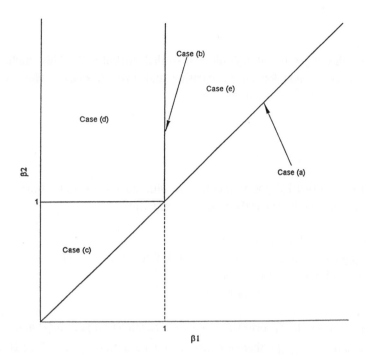

Figure 2: Different Sub-Regions for Parametric Characterization of the Density Function in the β_1–β_2 Plane.

4 Case (a): [$\beta_1 = \beta_2 = \beta_0$]

Define

$$k_0 = \left(\eta_2 \Big/ \eta_1 \right)^{\beta_0} \quad \text{and} \quad z_{p0} = 1 - 1/\beta_0. \tag{14}$$

Since η $(= \eta_2/\eta_1)$ < 1 and β_0 > 0, we have k_0 < 1. Using (14) in (13) and after some simplification, we can write $h_0(z)$ as

$$h_0(z) = h_1(z) - h_2(z), \tag{15}$$

where

$$h_1(z) = (q/p/k_0)(z_{p0} - z)\exp\{-(1 - k_0)z\} \tag{16}$$

and

$$h_2(z) = k_0 z - z_{p0}. \tag{17}$$

The two cases, $\beta_0 \le 1$ and $\beta_0 > 1$, need to be considered separately.

4.1 $\beta_0 \leq 1$

In this case, $df_1(t)/dt < 0$ and $df_2(t)/dt < 0$, so that $df(t)/dt < 0$. This implies that the density function is monotonically decreasing and Type A shape. Note that $f(0)$ is infinite when $\beta_0 < 1$ and finite when $\beta_0 = 1$.

4.2 $\beta_0 > 1$

From (17) we see that $h_2(z)$ is a straight line with slope $k_0 > 0$ and $h_2(0) = -z_{p0} < 0$. From (16) we see that $h_1(z)$ has the following properties:

(a) $h_1(0) > 0$ and $h_1(z) \to 0$ as $z \to \infty$,
(b) it passes through the points $(0, qz_{p0}/(pk_0))$ and $(z_{p0}, 0)$, and
(c) the minimum value occurs at

$$z_p = z_{p0} + 1/(1 - k_0). \tag{18}$$

Note that the zeros of $h_0(z)$ correspond to the intersections between $h_1(z)$ and $h_2(z)$. For a given η and β_0 ($> \hat{\beta}_0(\eta)$), there is either one intersection (implying Type B shape) or three intersections (implying Type D shape).

Figure 3 shows the parametric characterization of the pdf in the β_0–p plane for $\eta = 0.6$. The demarcation between the regions is given by two curves Γ_1 and Γ_2. They meet at $\beta_0 = \beta_0(\eta)$ and as $\beta_0 \to \infty$, Γ_1 and Γ_2 have asymptotes corresponding to $p = 1$ and 0, respectively. Inside the region bounded by Γ_1 and Γ_2 the shape is Type D and elsewhere it is Type B. The curves Γ_1 and Γ_2 were obtained from the relationship that at the transition from Type B to Type D (or vice versa) the two curves ($h_1(z)$ and $h_2(z)$) meet and have the same tangent. As a result, they are obtained by solving the following two equations:

$$h_0(z) = 0 \text{ and } dh_0(z)/dz = 0.$$

We omit giving the details and these can be found in Jiang (1996). [Note: This approach will also be used in Sections 5 – 7 to obtain the demarcation between regions with different shapes for the pdf.]

Figure 4 shows the demarcation curves Γ (comprising of Γ_1 and Γ_2 of Figure 3) for a range of η values. As can be seen, the curve moves to the right as η increases. For a given η, the shape is Type B to the left of the curve and is Type D to the right of the curve.

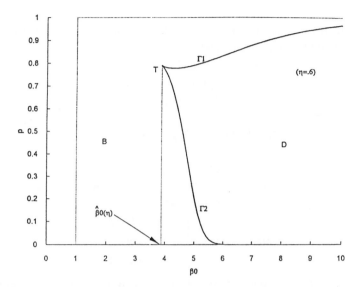

Figure 3: [Case (A)] Parametric Characterization of Density Function in the β_0–p Plane for a Fixed η.

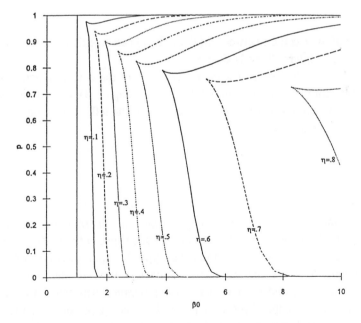

Figure 4: [Case (A)] Parametric Characterization of Density Function in the β_0–p Plane for Different η.

5 Case (b): [$\beta_1 = 1$ and $\beta_2 > 1$]

Since $\beta_1 = 1$, this case corresponds to a mixture of an exponential distribution and a Weibull distribution. From (9), we have

$$\beta = \beta_2, \, k = \eta, \text{ and } z_{p1} = 0. \tag{19}$$

Using (19) in (11) and after some simplification, we have

$$h(z) = z^{2/\beta}(h_1(z) - h_2(z)) \tag{20}$$

with

$$h_1(z) = \beta^2(q/p)z^{(1-2/\beta)}(z_{p2} - z)\exp(-z) \tag{21}$$

and

$$h_2(z) = k^2 \exp\{-kz^{1/\beta}\}. \tag{22}$$

Note that $h_2(z)$ is a decreasing curve with $h_2(0) = k^2$ and $h_2(z) \to 0$ as $z \to \infty$. $h_1(z)$ passes through the point $(z_{p2}, 0)$, with $h_1(z) \to 0$ as $z \to \infty$ and

$$\lim_{z \to 0} h_1(z) = \begin{cases} \infty & when \, \beta_2 < 2 \\ 2q / p & when \, \beta_2 = 2 \, . \\ 0 & when \, \beta_2 > 2 \end{cases}$$

This implies that h_1 has three different shapes depending on β_2, and these need to be examined separately.

5.1 $\beta_2 < 2$

The number of intersections between $h_1(z)$ and $h_2(z)$ can be either one (implying Type B shape) or three (implying Type D shape).

Figure 5 shows the parametric characterization of the density function in the η–p plane for different values of β_2. For a given β_2, the shape is Type D in the region to the right of the demarcation curve Γ and is of Type B to the left. The demarcation curve Γ moves to the left with β_2 increasing.

5.2 $\beta_2 = 2$

This corresponds to a mixture of an Exponential distribution and a Rayleigh distribution. In this case, the number of intersections between the $h_1(z)$ and $h_2(z)$ can be zero (implying Type A shape), one (implying Type B shape), or two (implying Type C shape). Figure 6 shows the parametric characterization in the η–p plane. The demarcation curves Γ_1 and Γ_2 characterize the separation between shapes Type B and Type D and between shapes Type C and Type D. The shapes of these two curves are different – Γ_1 is nearly convex and Γ_2 is monotonically decreasing with η. As $\eta \to \infty$, Γ_1 asymptotes to the horizontal line through p = 1 and Γ_2 asymptotes to the horizontal line through p = 0.

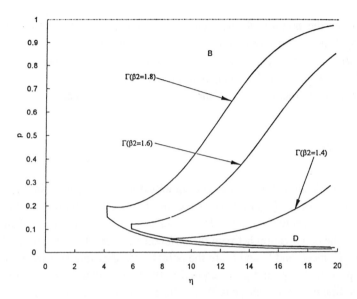

Figure 5: [Case (B)] Parametric Characterization of Density Function in the η–p Plane for Different β_2 (< 2).

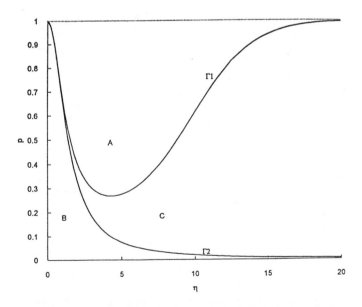

Figure 6: [Case (B)] Parametric Characterization of Density Function in the η–p Plane for $\beta_2 = 2$.

5.3 $\beta_2 > 2$

In this case, the number of intersections between $h_1(z)$ and $h_2(z)$ can be either zero (implying Type A shape) or two (implying Type C shape).

Figure 7 shows the parametric characterization in the η–p plane for different values of β_2. For a given β_2, in the region above the demarcation curve Γ, the shape is Type A and Type C in the region below. Γ moves upwards as β_2 increases.

5.4 Summary

The results of Sections 5.1 – 5.3 can be combined to show the plots in the β_2–p plane for different values of η. Figure 8 shows the plot for $\eta = 8$. The regions for the different shapes are as shown in the figure. The curves Γ_1 and Γ_2 and the vertical line through β_2 = 2 define the demarcation between the different regions. T is the point where curve Γ_1 and curve Γ_2 meet, and U and L are the points were they intersect with the vertical line through β_2 = 2. Let p_u and p_l correspond to the p-coordinates for U and L respectively. Along the line β_2 = 2, the shape is Type B for $p < p_l$, Type C for $p_l < p < p_u$, and Type A for $p > p_u$.

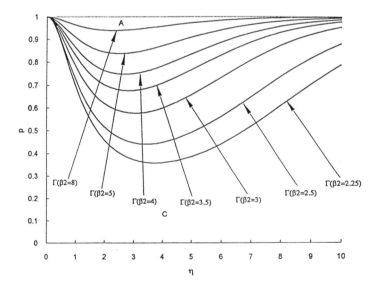

Figure 7: [Case (B)] Parametric Characterization of Density Function in the η–p Plane for Different β_2 (> 2).

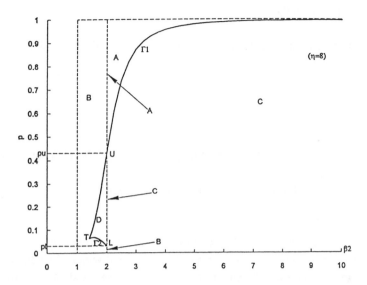

Figure 8: [Case (B)] Parametric Characterization of Density Function in the β_2–p Plane for η = 8.

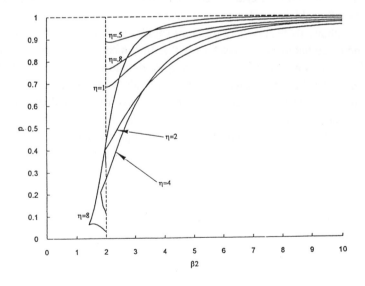

Figure 9: [Case (B)] Parametric Characterization of Density Function in the β_2–p Plane for Different η.

Figure 9 shows the curves Γ_1 and Γ_2 for different values of η. Note that as η gets smaller, T, L, and U come closer to each other, and they finally merge into a single point for $\eta \leq 1$.

6 Case (c): $[0 < \beta_1 < \beta_2 \leq 1]$

Since $\beta_1 < \beta_2 \leq 1$, we have $df_i(t)/dt < 0$, for $i = 1,2$. Hence $df(t)/dt < 0$ implying Type A shape.

7 Case (d): $[0 < \beta_1 < 1 < \beta_2]$

Since $\beta_1 < 1$ and $\beta_2 > 1$, we have from (9) $z_{p1} < 0$ and $z_{p2} > 0$. (12) can be rewritten in the form of (15) with

$$h_1(z) = [\beta^2 q \, z^{(1-1/\beta)} \, (z_{p2} - z) \exp(kz^{1/\beta} - z)]/pk \tag{23}$$

and

$$h_2(z) = -(z_{p1} - kz^{1/\beta}). \tag{24}$$

Note that $h_2(z) > 0$ with $h_2(0) = -z_{p1}$, and the slope at the origin being infinite. $h_1(z)$ is positive for $z < z_{p2}$ and negative for $z > z_{p2}$. It is unimodal with $h_2(0) = h_2(z_{p2}) = 0$. The number of intersections can be either zero (implying Type A shape) or two (implying Type C shape).

Figure 10 shows the regions for Type A and Type C in the β_1-β_2 plane for a fixed p and η. The demarcation curves Γ_1 and Γ_2 change with p and η. Γ_1 is obtained using the approach mentioned in Section 4. Unfortunately, it is not possible to obtain Γ_2 by this approach. It is obtained by a grid search over the β_1-β_2 plane. This approach is also used in Section 8 to obtain the demarcation curves.

Figure 11 shows these curves for different values of p with $\eta = 4.0$. Note that for p > 0.2, Γ_1 and Γ_2 do not meet whereas for p ≤ 0.2, they meet. As p increases, the region for Type A increases as the curve Γ_1 moves upwards and Γ_2 moves towards the right. Figure 12 shows the Γ_1 and Γ_2 curves for different values of η with p $= 0.7$. Note that the region where the shape is Type A increases initially with η and then decreases as can be seen from the curves for $\eta = 1, 4,$ and 8. Also when β_2 is large, the shape is always Type C. This is to be expected since the sub-population 2 has a very high peak.

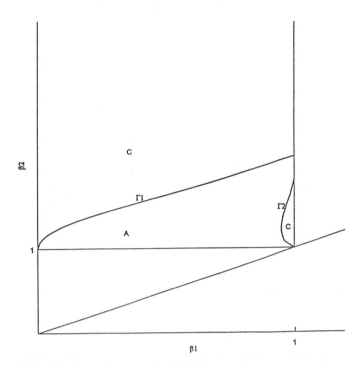

Figure 10: Characterization of Density Function in the $\beta_1-\beta_2$ Plane for $\beta_1 < 1 < \beta_2$ and Fixed η and p.

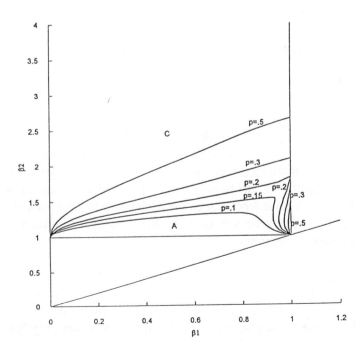

Figure 11: Characterization of Density Function in the $\beta_1-\beta_2$ Plane for $\beta_1 < 1 < \beta_2$, Fixed η, and Different p.

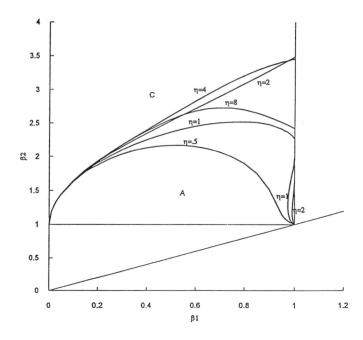

Figure 12: Characterization of Density Function in the β_1–β_2 Plane for $\beta_1 < 1 < \beta_2$, Fixed p, and Different η.

8 Case (e): $[1 < \beta_1 < \beta_2]$

Since $\beta_1 > 1$, we have from (9) $z_{p2} > z_{p1} > 0$. Also, (12) can be rewritten in the form of (15) with

$$h_1(z) = [\beta^2 q\, z\, (z_{p2} - z)]/[p \exp(z)] \tag{25}$$

and

$$h2(z) = -[kz^{1/\beta}\, (z_{p1} - kz^{1/\beta})] / \exp\{kz^{1/\beta}\}. \tag{26}$$

$h_1(z)$ is positive for $0 < z < z_{p2}$, negative for $z_{p2} < z < \infty$, $h_1(0) = h_1(z_{p2}) = 0$, and $h_1(z) \to 0$ as $z \to \infty$. In contrast, $h_2(z)$ is negative for $0 < z < (z_{p1}/k)^\beta$, positive for $(z_{p1}/k)^\beta < z < \infty$, $h_2(0) = h_2((z_{p1}/k)^\beta) = 0$, and $h_2(z) \to 0$ as $z \to \infty$. Hence the number of intersections between $h_1(z)$ and $h_2(z)$ is dependent on the relationship between z_{p2} and $(z_{p1}/k)^\beta$. Note that $z = z_{p2}$ $[(z_{p1}/k)^\beta]$ corresponds to $t = t_{p2}$ $[t_{p1}]$. As such, we need to examine the following three cases separately.

(i)　$t_{p1} > t_{p2}$ $[(z_{p1}/k)^\beta > z_{p2}]$

This occurs when η or k is small. In this case, the number of intersections between $h_1(z)$ and $h_2(z)$ can be one (implying Type B shape) or three (implying Type D shape).

(ii) $t_{p1} < t_{p2}$ $[(z_{p1}/k)^\beta < z_{p2}]$

This occurs when η or k is large. In this case, the shapes can be either Type B or Type D depending on the parameter values.

(iii) $t_{p1} = t_{p2}$ $[(z_{p1}/k)^\beta = z_{p2}]$

In this case, there is a single intersection between $h_1(z)$ and $h_2(z)$. This occurs at $z = (z_{p1}/k)^\beta = z_{p2}$. As a result, the shape is Type B.

From the above discussion, one can see that η has a significant effect on the shape of the density function as t_{p1} and t_{p2} are functions of η.

In the β_1–β_2 plane, the curve corresponding to $t_{p1} = t_{p2}$ is given by

$$\left(1 - \frac{1}{\beta_1}\right)^{\frac{1}{\beta_1}} = \eta\left(1 - \frac{1}{\beta_2}\right)^{\frac{1}{\beta_2}}. \tag{27}$$

Since $\beta_2 > \beta_1$, (27) is satisfied only for $\eta < 1$. Figure 13 shows plots of this curve (which we shall denote by S in later figures) for a range of η values. For a given η, it passes through the point $(1,1)$ and as $\beta_2 \to \infty$, it asymptotes to the vertical line through $\beta = \beta_{1\max}$ given by

$$\left(1 - \frac{1}{\beta_{1\max}}\right)^{\frac{1}{\beta_{1\max}}} = \eta. \tag{28}$$

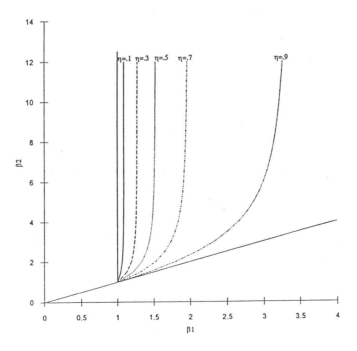

Figure 13: Plots of Curve S for Different η.

For a given η, for β_1 and β_2 values to the left of S, we have $t_{p1} < t_{p2}$ and to the right of S, we have $t_{p1} > t_{p2}$. Note that as $\eta \to 1$, $\beta_{1max} \to \infty$. As a result, when $\eta < 1$ one can have either $t_{p1} < t_{p2}$ or $t_{p1} > t_{p2}$, and when $\eta > 1$ we always have $t_{p1} < t_{p2}$.

We first consider the case $\eta < 1$. For a given η and p, the parametric characterization of the pdf is as shown in Figure 14. The demarcation curves are given by Γ_1 and Γ_2. To the left of Γ_1 and to the right of Γ_2 the shape is Type D, and in the region between Γ_1 and Γ_2, it is of Type B. The curves Γ_1 and Γ_2 approach the curve S as β_2 tends to infinity.

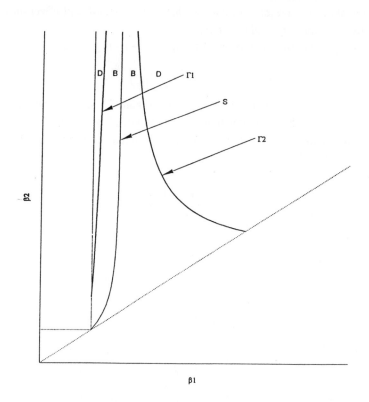

Figure 14: [$\eta < 1$] Parametric Characterization of Density Function in the β_1–β_2 Plane for $1 < \beta_1 < \beta_2$ and Fixed η and p.

Figure 15 shows the demarcation curve for a fixed η (= 0.5) and a range of different values for p. The shape of Γ_2 for small p is different from that for large p as can be seen from the figure.

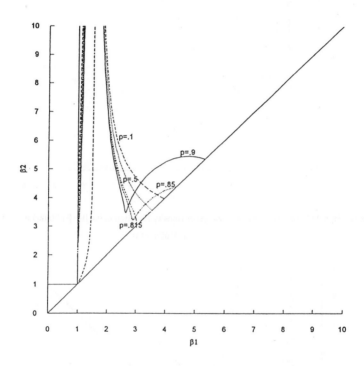

Figure 15: [$\eta < 1$] Parametric Characterization of Density Function in the $\beta_1-\beta_2$ Plane for $1 < \beta_1 < \beta_2$, Fixed η, and Different p.

Figure 16 shows the demarcation curves for different values of η with $p = 0.15$. The curve Γ_1 is very close to the vertical line through $\beta_1 = 1$. Figure 17 shows the results for the case $p = 0.85$. In this case, we see that the curve Γ_2 has kinks when η becomes bigger.

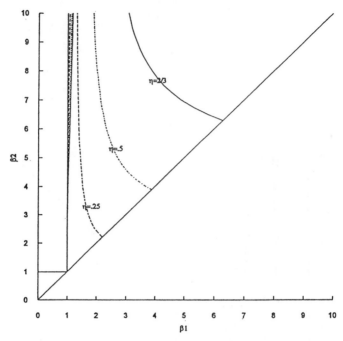

Figure 16: [η < 1] Parametric Characterization of Density Function in the β₁–β₂ Plane for $1 < \beta_1 < \beta_2$, p = 0.15, and Different η.

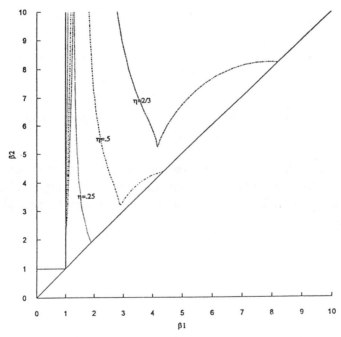

Figure 17: [η < 1] Parametric Characterization of Density Function in the β₁–β₂ Plane for $1 < \beta_1 < \beta_2$, p = 0.85, and Different η.

For the case $\eta > 1$, Figure 18 shows a typical plot of the parametric characterization for the density function for a given value of p and η. The demarcation is characterized by a single curve Γ. The shape of the pdf is of Type B below Γ and Type D above.

Figure 19 shows the demarcation curve Γ for $\eta = 2$ and a range of p values. For large or small p, Γ has a simple shape. It becomes more complex for intermediate values of p. Also, as p increases, the area below Γ (which corresponds to the region where the pdf shape is Type D) first decreases and then increases as can be seen from the figure.

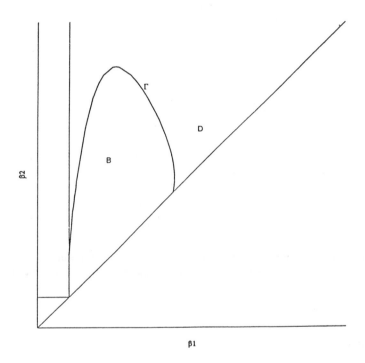

Figure 18: [$\eta > 1$] Parametric Characterization of Density Function in the β_1–β_2 Plane for $1 < \beta_1 < \beta_2$ and Fixed η and p.

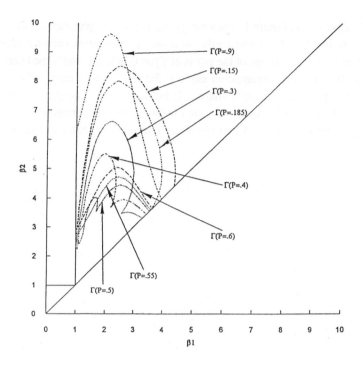

Figure 19: $[\eta > 1]$ Parametric Characterization of Density Function in the $\beta_1 - \beta_2$ Plane for $\beta_1 < 1 < \beta_2$, Fixed η, and Different p.

Results for p = 0.15 and different values of η are shown in Figure 20. Note that the area below Γ (which corresponds to the region where the shape is Type D) decreases as η increases.

9 Conclusions

In this paper we have shown that the shape of the density function for a mixture of two Weibull distributions can be one of four different types – decreasing (Type A), unimodal (Type B), decreasing followed by unimodal (Type C), and bimodal (Type D). The shape depends on the two shape parameters, the ratio of the two scale parameters and the parameter representing the mixing. We have given a complete parametric characterization of the pdf in the four-dimensional parameter space. A similar study for the failure rate of a mixing of two Weibull Distributions is given in Jiang and Murthy[13]. The parametric study of the pdf for a mixture of three or more Weibull distributions is still an open problem.

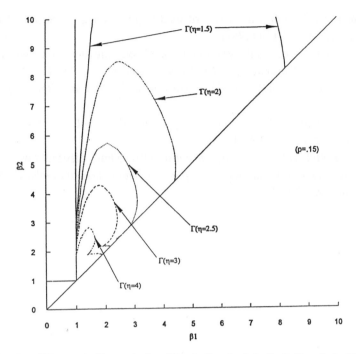

Figure 20: [η > 1] Parametric Characterization of Density Function in the β_1–β_2 Plane for $1 < \beta_1 < \beta_2$, Fixed p, and Different η.

References

1. R. Jiang and D.N.P. Murthy, Modeling failure data by mixture of two Weibull distributions: A graphical approach, *IEEE Tran. on Rel.* **44**, 477-488 (1995).

2. J.H.K. Kao, A graphical estimation of mixed Weibull parameters in life-testing of electron tubes, *Technometrics* **1**, 389-407 (1959).

3. P.R. Rider, Estimating the parameters of mixed Poisson, binomial, and Weibull distributions by the method of moments, *Bul. In. Stat. Inst.* **39**, 225-232 (1961).

4. L.W. Falls, Estimation of parameters in compound Weibull distribution, *Technometrics* **12**, 399-407 (1970).

5. K. Chen, A.S. Papdopoulos, and P. Tamer, On Bayes estimation for mixtures of two Weibull distributions, *Micro. & Rel.* **29**, 609-386 (1989).

6. S. Jiang and D. Kececioglu, Graphical representation of two mixed-Weibull distributions, *IEEE Tran. on Rel.* **41**, 241-247 (1992).

7. D. Kececioglu and F. Sun, Mixed-Weibull parameter estimation for burn-in data using the Bayesian approach, *Micro. & Rel.* **34**, 1657-1679 (1993).

8. K.V. Sichart and R.P. Vollertsen, Bimodal lifetime distributions of dielectrics for integrated circuits, *Qual. and Rel. Eng. Int.* **7**, 299-305 (1991).

9. J. Yuan and S. Shih, Mixed hazard models and their applications, *Rel. Eng. and Sys. Safety* **33**, 115-129 (1991).

10. S. Jiang and D. Kececioglu, Maximum likelihood estimates, from censored data, for mixed-Weibull distributions, *IEEE Trans. on Rel.* **41**, 248-255 (1992).

11. A. Drapella, Bimodal distribution. Are you sure?, *Qual. and Rel. Eng. Int.* **9**, 385-386 (1993).

12. A.E. Gera, Unimodality and bimodality of mixed Weibull distributions, *Qual. and Rel. Eng. Int.* **10**, 355-358 (1994).

13. R. Jiang and D.N.P. Murthy, Mixture of Weibull distributions: Parametric characterization of failure rate; *Accepted for publication in Stochastic Models and Data Analysis* (1996).

STATISTICAL APPROACHES TO MODELING AND ESTIMATING SOFTWARE RELIABILITY

Tapan K. Nayak
Department of Statistics
George Washington University
Washington, DC 20052
E-mail: tapan@gwis2.circ.gwu.edu

This paper gives an overview of statistical models and methods for software reliability estimation. Most commonly used models for analyzing software failure data are classified into two groups and comparative features of those two groups of models are discussed. Difficulties in estimating the unknown parameters from standard debugging data are explained and an alternative experiment, called recapture debugging, which generates additional statistical information is described. Some methods for analyzing recapture debugging data are reviewed. It is suggested that further statistical research should be conducted not only on analysis of routinely collected data but also on the design of the debugging experiment with the goal of generating increased statistical information.

1 Introduction

Software engineering is an important part of modern computer technology. In recent years the quality of computer hardware has improved significantly, and the cost has decreased. As a result, computerized systems are now very common. The success of such systems depends heavily on the quality and reliability of the software in the systems. Failure-free operation of software is crucial for computerized systems in navigation and control of spacecrafts, air traffic control, communication networks, medical care, banking, and other businesses. As failure of such systems result in serious economic and human consequences, large corporations such as IBM, AT&T, and many government agencies have made efforts to measure and improve software quality.

There are some basic differences between software and hardware reliability. While deterioration of material is a main cause of hardware failure, software failure is due to human errors in the design and programming of the software. In recent years, rapid technological advances in material sciences and micro electronics have greatly improved hardware quality in terms of speed, capacity, and reliability. However, the advances in physical sciences do not help much in improving software reliability.

Unlike hardware, software does not age with time; once an error is removed, it is removed forever. Also, whereas hardware can be replicated to introduce redundancy and increase reliability, software replication just replicates the errors. To introduce redundancy in software and make it fault tolerant, one needs to redesign the software and insert additional code. As software failures are due to errors in human logic, software quality can be enhanced by improvement in design, development, and testing techniques. We refer to [35, 38] for further discussion of software engineering techniques.

As the formal theory of proving correctness of software is still inadequate, in practice software quality is assessed by testing it under varying operating environments. The design of the testing experiment, analysis of data generated from testing, estimation of software reliability at every stage of testing, and the decision whether to continue or stop testing requires appropriate statistical methodology. This paper reviews some common approaches to software reliability estimation. The discussion focuses on the rationale and interpretations of those approaches. Specific inferences for various approaches can be obtained from the literature cited. Section 2 presents two general classes of models for analyzing the usual software failure data and discusses their similarities and differences. In Section 3, a more informative experiment, called recapture debugging, is described and inferences based on recapture debugging data are reviewed. Connections between the statistical problems in software reliability and in traditional capture-recapture sampling are also discussed.

2 Models for Analyzing Standard Debugging Data

2.1 Standard Debugging Data

Once a piece of software is developed and all obvious errors are removed, it is run with different inputs in order to detect the nontrivial errors (or bugs) and to assess the reliability of the software. When the software is executed with a specific input, a particular path in the software is followed and the software fails only if that path contains at least one error. Thus, the software fails for some but not all inputs. When a failure occurs, efforts are made to detect and remove the error that caused the failure. When the testing time, τ, is fixed, the standard debugging data consist of the observed values of R, the number of failures during testing, and $0 \le T_{(1)} \le \cdots \le T_{(R)} \le \tau$, the successive failure times, measured by execution or calendar time. See [26] for further discussion of the choice of time.

As selection of the inputs typically involve much randomness, $R, T_{(1)}, \cdots, T_{(R)}$ are regarded as random variables. For analyzing the data one uses a model, which specifies the joint density of the observable random variables that constitute the data (viz. $R, T_{(1)}, T_{(2)}, \cdots, T_{(R)}$), and the reliability measures of the software as functions of some unknown parameters. Roughly speaking, a model allows us to go from the observed data to the reliability measures via the parameters of the model. Of course, a postulated model may not describe the true situation correctly. Further, the choice of the reliability measures (called reliability metrics in software engineering) may depend on the proposed model. Two types of models commonly found in the literature are discussed below.

2.2 General Order Statistics Models

The general order statistics (GOS) models are based on the following assumptions.
Assumption 1. Whenever the software fails, the error causing the failure is detected and corrected completely without inserting any new errors, i.e., the debugging process is perfect.
Assumption 2. The software initially contains an unknown number of errors, ν, and the detection times of those errors are independently and identically distributed (iid) with a common density $f_\theta(x)$, where θ is an unknown parameter, possibly vector valued.

Clearly, different choices of the density function $f_\theta(x)$ lead to different models. Raftery [32] called a model based on Assumptions 1 and 2 a GOS model. Here, R is binomial with parameters ν and $p = F_\theta(\tau) = \int_0^\tau f_\theta(x)dx$ and given R, the successive failure times $T_{(1)} \le \cdots \le T_{(R)}$ are the order statistics of R observations from the truncated density $f_\theta^*(x) = f_\theta(x)/F_\theta(\tau), 0 \le x \le \tau$. The joint density of the data is

$$
\begin{aligned}
f_\theta(r, t_{(1)}, \cdots, t_{(r)}) &= f_\theta(r)f_\theta(t_{(1)}, \cdots, t_{(r)}|r) \\
&= [\binom{\nu}{r}\{F_\theta(\tau)\}^r \{1 - F_\theta(\tau)\}^{\nu-r}] \times [r! \prod_{i=1}^r \frac{f_\theta(t_{(i)})}{F_\theta(\tau)}] \\
&= \frac{\nu!}{(u-r)!}[\prod_{i=1}^r f_\theta(t_{(i)})][1 - F_\theta(\tau)]^{\nu-r}.
\end{aligned}
$$

In GOS models, the initial number of errors in the software (ν) and the remaining number of errors after debugging ($\nu - R$) are regarded as important measures of software reliability, and hence estimation of ν has received considerable attention in the literature. Estimation of the reliability (the probability of failure-free operation for a specified amount of time k) of the debugged program at time τ, the expected time to the next failure, and the reliability of the software if debugging is continued for an additional time t are also important. The corresponding reliability metrics are: the probability of failure-free operation of the debugged program between time τ and $\tau + k$, given by

$$
\gamma_1(k) = \{[1 - F_\theta(\tau + k)]/[1 - F_\theta(\tau)]\}^{(\nu-R)};
$$

the mean time to the next failure (MTTF) of the debugged program, given by

$$
\text{MTTF} = \int_0^\infty \gamma_1(k)dk = \int_0^\infty \{[1 - F_\theta(\tau + k)]/[1 - F_\theta(\tau)]\}^{(\nu-R)}dk;
$$

and the probability of failure-free operation of the debugged program between time units $\tau + t$ and $\tau + t + k$ if the debugging process is continued for an additional time t, which is useful for deciding when to stop debugging and is given by

$$
\gamma_2(t, k) = \{[1 - F_\theta(\tau + t + k) + F_\theta(\tau + t) - F_\theta(\tau)]/[1 - F_\theta(\tau)]\}^{(\nu-R)}.
$$

The most well known GOS model is the Jelinski and Moranda [18] model, which assumes that the common density is exponential given by

$$
f_\theta(x) = \theta e^{-\theta x}, \quad x > 0, \quad \theta > 0.
$$

For this model, Blumenthal and Marcus [3] derived the maximum likelihood estimates (MLE) and the Bayes estimates based on a gamma prior for θ and a uniform prior for ν. Forman and Singpurwalla [10] showed that the likelihood function can be very unstable and the MLE of ν can be infinite with positive probability. Littlewood and Verrall [24] gave necessary and sufficient conditions for the MLE of ν to be finite. Goudie and Goldie [15] showed that a nonnegative unbiased estimator of ν does not exist and concluded that ν should be estimated by an interval estimate.

Joe and Reid [19] found that the harmonic mean of the end points of the 50% likelihood interval is the most effective finite estimator of ν among several methods they tried. Raftery [32] presented a Bayes empirical Bayes analysis. Estimation of ν and θ is similar to estimation of both n and p of a binomial distribution, which has been discussed in [2, 16, 33], and it has been noted that the maximum likelihood and moment estimators of n can behave in unstable ways.

Use of GOS models in software reliability have often been criticized. The assumption of perfect debugging may be unrealistic. The definition of an error is also a bit problematic as often the scope of a software is not clearly defined, see Musa et al. [26]. The assumption that the detection times of the errors are iid have also been questioned. For example, to allow the failure rates of the errors to be different, Littlewood [23] modified the Jelinski-Moranda model by considering θ to be a random variable with gamma distribution. But, Joe and Reid [20] noted that the same likelihood can be obtained from assuming that the detection times are iid with Pareto type 2 density. Actually, as explained next, questioning the assumption that the detection times are identically distributed is not well justified (cf. Nayak [28]) when one assumes perfect debugging. Let T_i denote the detection time of the ith error. As the errors are not known, the labels are arbitrary and T_i are not observable; one can observe only their order statistics. Thus, only the joint density of the order statistics is needed for modeling the data. However, any specified density of the order statistics can be obtained from an exchangeable joint density of $T_1, \cdots, T_\nu$, which implies that the marginal distributions of T_i are identical and pair-wise correlations are all equal. Specifically, if a model specifies the joint density of $T_{(1)}, \cdots T_{(\nu)}$ by $f_\theta(t_{(1)}, \cdots, t_{(\nu)})$, without any loss of generality one may assume that the joint density of $T_1, \cdots T_\nu$ is $\frac{1}{\nu!} f_\theta(t_{(1)}, \cdots, t_{(\nu)})$, which is exchangeable. Nayak [28] discussed an approach for obtaining models with correlated $T_1, \cdots T_\nu$. However, the standard data are inadequate for reliable estimation of the unknown correlation (and the reliability measures).

2.3 Nonhomogeneous Poisson Process Models

A common alternative to GOS models is to model the failure times by a nonhomogeneous Poisson process (NHPP) with a rate function $\lambda(t)$ that depends on some unknown parameters. The mixture of the GOS model with respect to Poisson (μ) distribution for ν yields an NHPP model (cf. Langberg and Singpurwalla [22], Musa et al. [26], p. 269, or Kuo and Yang [21]) with rate function $\lambda(t) = \mu f_\theta(t)$ and the mean function $m(t) = \mu F_\theta(t)$. Conversely, if $\int_0^\infty \lambda(t)dt$ is finite (or equivalently $m(\infty)$ is finite), an NHPP model can be represented as a Poisson mixture of GOS models. Kuo and Yang [21] related the case where $m(\infty)$ is infinite to a record value statistics model. We shall restrict our discussion to NHPP models with finite $m(\infty)$. In NHPP models, μ is the expected number of failures in $[0, \infty)$. The total number of failures, say M, in $[0, \infty)$ is a random variable with distribution Poisson (μ), the same as the mixing distribution, and $P(M = 0) > 0$. Here, the expressions of the reliability metrics $\gamma_1(k)$, MTTF, and $\gamma_2(t, k)$ are

$$\gamma_1(k) = e^{-\mu[F_\theta(\tau+k) - F_\theta(\tau)]},$$

$$\text{MTTF} = \int_0^\infty e^{-\mu[F_\theta(\tau+k)-F_\theta(\tau)]} dk,$$

and

$$\gamma_2(t,k) = e^{-\mu[F_\theta(\tau+t+k)-F_\theta(\tau+t)]}.$$

The joint density of $R, T_{(1)} \le \cdots \le T_{(R)}$ is

$$f_\theta(r, t_{(1)}, \cdots, t_{(r)}) = \mu^r [\prod_{i=1}^r f_\theta(t_{(i)})] e^{-\mu F_\theta(\tau)}. \tag{2}$$

Inferences based on NHPP models can be found in [1, 4, 11, 21, 26, 27].

2.4 Comparison of GOS and NHPP models.

Some differences between the implications and interpretations of GOS and NHPP models are noted. First, NHPP models implicitly assume imperfect debugging (conceptually, they do not allow perfect debugging) as the number of failures in $[0,\infty)$ is unbounded for all values of the parameters μ and θ. The number of errors (initial or remaining) is not of direct interest in NHPP models. Second, the reliability measures $\gamma_1(k)$, MTTF, and $\gamma_2(t,k)$ depend on the parameters (ν and θ) and the data (through R) in a GOS model, but they depend only on the unknown parameters μ and θ (and not the data) in an NHPP model. Thus, it seems that under an NHPP model, the reliability of a software at all points in time are determined when the development of the software is completed. Subsequent debugging activities do not change those reliabilities; the information from testing and debugging only help us estimate them. Actually, an NHPP model does not specify any probability law to describe either how the errors in the software are accessed or how new errors are added during debugging. It specifies the total effect of the initial quality of the software and the debugging activities on the failure process. In contrast, the GOS models describe a probabilistic micro mechanism for failure generation. Also, NHPP models assume independence of the failure process in non-overlapping intervals, which is not true under GOS models. Finally, the parameter θ and the density function $f_\theta(x)$ in an NHPP model may not have physical interpretations.

Although there are basic differences between GOS and NHPP models, the following similarity between inferences based on a GOS model and the corresponding NHPP model may be noted. For given θ, the MLE of μ under an NHPP model is $\hat{\mu} = R/F_\theta(\tau)$, and the MLE of ν (considering ν as continuous) under a GOS model is $\hat{\nu} = R/F_\theta(\tau)$. However, the estimates of μ and ν will be different when θ is unknown and estimated from the data.

The distribution of ν may be interpreted and used in two different ways leading to different interpretations of inferences. It may be used as a mixing distribution as in an NHPP model or as the prior for ν in Bayesian analysis based on a GOS model. In the former case, the distribution of ν is a part of the model whereas in the later case it is our subjective information on a parameter of a model. Some differences of the interpretations based on these two approaches are evident form the expressions of $\gamma_1(k)$, MTTF and $\gamma_2(t,k)$. It can be seen that many calculations in Bayesian analysis based on the a GOS model and an NHPP model would be very similar if the density $f_\theta(x)$ and the prior distribution of θ are same in the two cases, the prior for ν in the GOS model is hierarchical with $\nu \sim$Poisson(μ), $\mu \sim \pi(\mu)$, and in the NHPP model, the mixing distribution is Poisson (μ) and the prior for μ is $\pi(\mu)$.

We also note that a very large number of specific models, some of which are neither GOS nor NHPP, have been proposed in the literature. Detailed discussions of many specific models and their relationships are given in [7, 22, 25, 26, 34, 37].

3   Recapture Debugging

The nature of difficulties in estimating the parameters of a GOS model are exemplified by the Jelinski-Moranda model, the simplest GOS model. There, if one of the two parameters (ν or θ) is known, estimation of the other parameter is easy. For example, if θ is known, $\hat{\nu} = R/F_\theta(\tau)$ is the minimum variance unbiased estimator (MVUE) of ν and it is finite. However, estimating both ν and θ is quite difficult. This suggests that the information about ν and θ in the data is confounded. For further inspection let $Y_i = T_{(i)} - T_{(i-1)}, i \geq 0, T_{(0)} = 0$. Then, observing $R, T_{(1)}, \cdots, T_{(R)}$ is equivalent to observing $R, Y_1, \cdots, Y_R$. Now, note that $Y_1, \cdots, Y_\nu$ are independent exponential random variables and the rate of Y_i, given by $(\nu - i + 1)\theta$, as well as the expected value of R, given by $\nu(1 - e^{-\theta\tau})$, are increasing functions of both ν and θ. So, ν and θ have similar effects on the random observables. In particular, we are likely to observe a large number failures during testing for large ν and small θ, as well as for moderate ν and large θ. Thus, it is difficult to tell from the observed data whether ν is large and θ is small, θ is large and ν is small, or both are moderate.

From the above discussion it appears that for more accurate inferences extra information on one of the two parameters is necessary. Extending the GOS model, if the errors are assumed to be accessed according to iid renewal processes with renewal density $f_\theta(x)$, then extra information on θ would be obtained from not removing the errors during testing and recording all repeated occurrence times of the detected errors. Thus, for increased statistical information, the errors should not be removed and be allowed to be re-accessed during testing. Since it is inconvenient to let the errors remain in the software, in practice, when an error is discovered we may remove it and insert a counter to record the times at which that error would have occurred in the remaining testing time. Nayak [29] proposed this idea and called it recapture debugging. Ross [36] also considered recapture debugging by assuming that efforts for detecting and removing the errors are made only at the end of testing.

A related but different approach, called error seeding, has been proposed earlier (cf. [9]). In that method, some artificial errors, called seeded errors, are inserted deliberately in the software without informing the testing team. At the end of debugging the detected errors are classified as seeded or unseeded errors. Then, the number of errors in the original program is estimated by comparing the numbers of detected seeded and unseeded errors using a hypergeometric model. This approach requires the assumption that detection time distributions of the seeded and unseeded errors are similar. However, in practice, it is difficult to create seeded errors satisfying this assumption. Recapture debugging avoids this difficulty.

The data generated by recapture debugging can be described by R, the number of errors detected during testing, N_i, the number of times the ith detected error is accessed during testing, $i = 1, \cdots, R$, and T_{ij}, the time of the jth access of the ith detected error, $j = 1, \cdots, N_i$. Alternatively, the data can be described (cf. Nayak [30]) as a marked point process $\{W_i, D_i\}$, where, W_i is ith event time (an event is accessing any one of the errors) and $D_i = j$ if the ith event occurs from accessing the jth detected error.

Inferences under the assumption that the common renewal process is a Poisson process with a constant rate λ have been discussed by Nayak [29, 30]. For the fixed time experiment, both the MLE and method of moment estimate of ν are infinite if and only if no error is accessed more than once during testing. Although the probability of this event is small when either ν or $(\lambda \tau)$ is large, statistically it is more meaningful to continue testing until a predetermined number, say b, of repeat accesses occur. In that case, R and the stopping time T are jointly sufficient for ν and λ. The conditional distribution of T given $R = r$ is gamma $(\nu\lambda, r+b)$, and the marginal distribution of R depends only on ν and is given by

$$p_\nu(r) = \nu^{-(r+b)}\nu(\nu - 1) \cdots (\nu - r + 1)c_{r,b}, \quad r = 1, \cdots, \nu,$$

where $c_{r,b}$ is a constant related to Sterling numbers. Specifically, $c_{r,b} = rS(r + b - 1, r)$, where $S(n, r)$ is a Sterling number of the second kind defined as

$$S(n, r) = \frac{1}{n!} \sum_{i=0}^{r} (-1)^{r-i} \binom{r}{i} i^n.$$

Letting $\theta = \nu\lambda$, it is seen that the information in the data about the parameters is separated nicely; R provides information only about ν, and T given R provides information only on θ. Nayak [30] showed that $\hat{\nu} = c_{r,b+1}/c_{r,b}$ is the MVUE of ν. He also presented the MVUE of λ and some functions of ν and λ, and some significance tests. Nayak [29] and Goudie [14] gave stopping rules with the goal of detecting all the errors with a prespecified probability.

Relaxing the assumption that the rates of all errors are equal, i.e., considering the rate to be $\lambda_1, \cdots, \lambda_\nu$, Nayak [30] showed that the event process $\{W_i\}$ is homogeneous Poisson with rate $\Lambda = \sum \lambda_i$, and the event times $\{W_i\}$ and the marks $\{D_i\}$ are independently distributed. He showed that $\sum N_i/\tau$ is the MVUE of Λ and V_1/τ, where V_1 is the number of errors accessed exactly once, is the MVUE of $\sum \lambda_i \exp(-\lambda_i \tau)$, the total expected failure rate of the undetected errors at time τ, but it is difficult to estimate ν and all the λ_i. In the case of unequal rates, Chao, Ma, and Yang [6] proposed two adaptive stopping rules taking cost of testing and penalty due to unremoved errors into account.

Estimation of ν under the homogeneous Poisson process assumption is similar to estimation of the number of classes in a multinomial population based on capture-recapture samples, which has been discussed in [5, 12, 13, 17]. Nayak and Christman [31] discussed this connection and proved that the usual point estimators of ν derived under the assumption that all rates are equal are negatively biased when that assumption fails, and the bias increases as the rates become more heterogeneous.

Inferences for the case where the errors are accessed according to an iid renewal processes with renewal density $f_\theta(x)$ have been discussed by Dewanji, Nayak and Sen [8]. There, the likelihood function is given by

$$L(\nu, \theta | \text{ data}) = \frac{\nu!}{(\nu - r)!} [\bar{F}_\theta(\tau)]^{\nu - r} \prod_{i=1}^{r} [\bar{F}_\theta(l_i) \prod_{j=1}^{n_i} f_\theta(t_{ij})],$$

where $l_i = \tau - t_{in_i}$ and $\bar{F}_\theta(t) = 1 - F_\theta(t)$. It can be seen that for given θ, the MVUE of ν is $R/F_\theta(\tau)$ and the MLE is its integer part. They proposed an iterative method for obtaining the MLE of ν and θ and a conditional maximum likelihood approach, where θ is estimated first by maximizing the conditional likelihood given R, which depends only on θ, and then ν is estimated by maximizing the marginal likelihood of R with θ replaced by its estimate obtained in the earlier step. Dewanji et al. (1995) proved that under certain regularity conditions, asymptotically as $\nu \to \infty$ a large class of estimators (including the MLE and the conditional MLE) of ν and θ converge to the same normal distribution. Thus, asymptotically the MLE and the conditional MLE are equivalent.

4 Discussion

In this paper we have described and compared two general classes of models for analyzing standard debugging data and discussed recapture debugging for generating extra information. Statistical science deals with data analysis as well as data collection. However, statistical research on software reliability estimation has been limited mainly to modeling and analyzing the standard debugging data. We believe the data collection aspect should not be ignored and other debugging methods for generating more useful information should be explored.

It should be noted that, in practice, different parts of the software are tested (e.g., unit testing, module testing, integration testing etc., see [38]) before testing the whole software. However, current statistical models either do not distinguish data coming from different types of testing or analyze data coming only from one type of testing. Analysis of data coming from all types of testing and optimum allocation of resources among different types of testing deserve further research. This can be done only through true collaboration among software engineers and statisticians, taking all relevant information about the nature and structure of the software into account. Significant advances are likely to come from combining statistical principles and methods with software engineering techniques.

Connections between statistical problems in software reliability and the traditional capture-recapture sampling should also be noted. For example, the recapture debugging formulated as a marked point process is quite similar to the standard capture-recapture experiment. The main difference is that in the case of recapture debugging, the captures are made continuously over time with some unknown rate and one needs to estimate the capture rate from the data on capture times.

Knowledge from other areas of applications should be used for deriving methods for analyzing software failure data and also for directing future research.

5 Acknowledgement

The author thanks Professor Nozer D. Singpurwalla and two referees for their helpful comments.

References

1. Achcar, J. A., Dey, D. K., and Niverthi, M. (1997). A Bayesian Approach Using Nonhomogeneous Poisson Process for Software Reliability Models. In *this volume.*

2. Blumenthal, S., and Dahiya, R. C. (1981). Estimating the Binomial Parameter n. *J. Amer. Statist. Assoc.* 76, 903-909.

3. Blumenthal, S., and Marcus, R. (1975). Estimating Population Size With Exponential Failure. *J. Amer. Statist. Assoc.* 70, 913-922.

4. Campodonico, S., and Singpurwalla, N. D. (1995). Inference and Predictions From Poisson Point Processes Incorporating Expert Knowledge. *J. Amer. Statist. Assoc.* 90, 220-226.

5. Chao, A., and Lee, S.-M. (1992). Estimating the Number of Classes via Sample Coverage. *J. Amer. Statist. Assoc.* 87, 210-217.

6. Chao, A., Ma, M.-C., and Yang, M. C. K. (1993). Stopping Rules and Estimation for Recapture Debugging With Unequal Failure Rates. *Biometrika* 80, 193-201.

7. Chen, Y. P., and Singpurwalla, N. D. (1994). A Non-Gaussian Kalman Filter Model for Tracking Software Reliability. *Statistica Sinica* 4, 535-548.

8. Dewanji, A., Nayak, T. K., and Sen, P. K. (1995). Estimating the Number of Components of a System of Superimposed Renewal Processes. Sankhyā, Ser. A 57, 486-499.

9. Duran J. W., and Wiorkowski, J. J. (1981). Capture-Recapture Sampling for Estimating Software Error Content. *IEEE Trans. Software Eng.* SE-7, 147-148.

10. Forman, E. H., and Singpurwalla, N. D. (1977). An Empirical Stopping Rule for Debugging and Testing Computer Software. *J. Amer. Statist. Assoc.* 72, 750-757.

11. Goel, A. L., and Okumoto, K. (1979). Time-Dependent Error Detection Rate Model for Software Reliability and Other Performance Measures. *IEEE Trans. Reliability* R-28, 206-211.

12. Goodman, L. (1949). On the Estimation of the Number of Classes in a Population. *Ann. Math. Statist.* 20, 572-579.

13. Goodman, L. (1953). Sequential Sampling Tagging for Population Size Estimation. *Ann. Math. Statist.* 24, 56-69.

14. Goudie, I. B. J. (1990). A Likelihood-Based Stopping Rule for Recapture Debugging. *Biometrika* 77, 203-206.

15. Goudie, I. B. J., and Goldie, C. M. (1981). Initial Size Estimation for the Linear Pure Death Process. *Biometrika* 68, 543-550.

16. Hall. P. (1994). On the Erratic Behavior of Estimators of N in the Binomial N, p Distribution. *J. Amer. Statist. Assoc.* 89, 344-352.

17. Harris, B. (1968). Statistical Inference in the Classical Occupancy Problem: Unbiased Estimation of the Number of Classes. *J. Amer. Statist. Assoc.* 63, 837-847.

18. Jelinski, Z., and Moranda, P.M. (1972). Software Reliability Research. In *Statistical Computer Performance Evaluation*, ed. W. Freiberger, pp. 465-484. New York, Academic Press.

19. Joe, H., and Reid, N. (1985). Estimating the Number of Faults in a System. *J. Amer. Statist. Assoc.* 80, 222-226.

20. Joe, H., and Reid, N. (1985). On the Software Reliability Models of Jelinski-Moranda and Littlewood. *IEEE Trans. Reliability* R-34, 216-218.

21. Kuo, L., and Yang, T. Y. (1996). Bayesian Computation for Nonhomogeneous Poisson Processes in Software Reliability. *J. Amer. Statist. Assoc.* 91, 763-773.

22. Langberg, N., and Singpurwalla, N. D. (1985). A Unification of Some Software Reliability Models. *SIAM J. Sci. & Statist. Computing*, 6, 781-790.

23. Littlewood, B. (1981). Stochastic Reliability-Growth: A Model for Fault Removal in Computer-Programs and Hardware-Designs. *IEEE Trans. Reliability* R-30, 313-320.

24. Littlewood, B., and Verrall, J. L. (1981). Likelihood Function of a Debugging Model for Computer Software Reliability. *IEEE Trans. Reliability* R-30, 145-148.

25. Miller, D. R. (1986). Exponential Order Statistics Models for Software Reliability. *IEEE Trans. Software Eng.* SE-12, 12-24.

26. Musa, J. D., Iannino, A., and Okumoto, K. (1987). *Software Reliability: Measurement, Prediction, Application.* New York: McGrow-Hill.

27. Musa, J. D., and Okumoto, K. (1984). A Logarithmic Poisson Execution Time Model for Software Reliability Measurement. *Proc. 7th Inter. Conf. Software Eng.*, pp 230-238.

28. Nayak, T. K. (1986). Software Reliability: Statistical Modeling & Estimation. *IEEE Trans. Reliability* R-35, 566-570.

29. Nayak, T. K. (1988). Estimating Population Size by Recapture Debugging. *Biometr- ika* 75, 113-120.

30. Nayak, T. K. (1991). Estimating the Number of Component Processes of a Superimposed Process. *Biometrika* 78, 75-81.

31. Nayak, T. K., and Christman, M. C. (1992). Effect of Unequal Catchability on Estimates of the Number of Classes in a Population. *Scand. J. Statist.* 19, 281-287.

32. Raftery, A. E. (1987). Inference and Prediction for a General Order Statistic Model With Unknown Population Size. *J. Amer. Statist. Assoc.* 82, 1163-1168.

33. Raftery, A. E. (1988). Inference for the Binomial N Parameter: A Hierarchical Bayes Approach. *Biometrika* 75, 223-228.

34. Ries, L. D., and Basu, A. P. (1997). Parameter Estimation in Software Reliability Growth Models. In *this volume.*

35. Rook, P. (ed.) (1990). *Software Reliability Handbook.* New York: Elsevier.

36. Ross, S. M. (1985). Statistical Estimation of Software Reliability. *IEEE Trans. Software Eng.* SE-11, 479-483.

37. Singpurwalla, N. D., and Wilson, S. P. (1994). Software Reliability Modeling. *Int. Statist. Rev.* 62, 289-317.

38. Sommerville, I. (1996). *Software Engineering.* 5th edn. New York: Addison-Wesley.

BIRNBAUM-SAUNDERS-TYPE MODELS FOR SYSTEM STRENGTH ASSUMING MULTIPLICATIVE DAMAGE

W. J. OWEN, W. J. PADGETT

Department of Statistics, University of South Carolina
Columbia, SC 29208
E-mail: owen@stat.sc.edu

Cumulative damage models for tensile strength of composite materials (complex systems) are developed based on a multiplicative damage process and "initial strength" distributions derived from inherent material flaw concepts. The resulting material strength distributions are three-parameter generalizations of the Birnbaum-Saunders distribution and include the effect of "size" of the material specimen on its tensile strength. Estimation and asymptotic distribution theory are investigated, and the model fits to tensile strength data for carbon composite materials are presented. Comparisons of the new models with corresponding additive damage models and with the power-law Weibull model are also given based on the data.

1 Introduction

Most high-performance materials made today are actually complex systems where the strength of the material is a function of many components. For example, a fibrous carbon composite material is a complex system consisting of carbon fibers embedded in a polymer (called the matrix material), and the tensile strength of the system is primarily a function of the strengths of the fibers. In addition, the strength of the material is a stochastic quantity due to random flaws in the components (perhaps from imperfections induced in the manufacturing process), so good statistical models are needed to describe the system strength. This can be difficult since often many standard statistical distributions fail to adequately fit experimental material strength data, particularly in the lower percentiles. This includes the lognormal, the gamma, and the well-known Weibull distributions. The two-parameter Weibull distribution is based on the "weakest link of a chain of L units" argument, and the Weibull cumulative distribution function (CDF) for the tensile strength of a material of length L is given by $F(s) = 1 - \exp[-L(s/\delta)^\rho]$. Here, ρ is the shape parameter and δ is a scale parameter. Several authors have considered different statistical models for tensile strength of fibrous composite materials. These include mixtures of Weibull distributions[1] and mixtures of lognormal distributions[2]. A "brittle fracture" distribution was obtained by Black, Durham and Padgett[3], and competing risks models were considered by Goda and Fukunaga[4] and Wagner[5]. Models that take "clamp effects" of the testing devices into account were developed by Phoenix and Sexsmith[6], Stoner, Edie and Durham[7], and Padgett, Durham and Mason[8].

When a model is being developed for tensile strength of a complex system, it is often desirable to incorporate a "system size" or *gauge length* variable, L, in the model since this has a great effect on the tensile strength. Also, this enables the user to fit experimental observed data at a variety of system sizes at once. For example, the "power law" Weibull model proposed by Padgett, Durham and Mason[6] is a three-parameter extension of the standard Weibull CDF incorporating gauge length using an acceleration model. The power-law Weibull CDF is $F(s;L) = 1 - \exp[-L^\eta(s/\delta)^\rho]$, $s>0$, and the model can be used for fitting over multiple gauge lengths. Durham and Padgett[9] obtained two models based on *additive* cumulative damage processes which begin at the most severe flaw in the material. Their models are three-parameter extensions of the Birnbaum-Saunders distribution that incorporate system size, and could be approximated by a three-parameter extension of the somewhat more tractable inverse Gaussian distribution. The standard two-parameter Birnbaum-Saunders distribution[10] is written as

$$F(s;\alpha,\beta) = \Phi\{ \frac{1}{\alpha}[(\frac{s}{\beta})^{\frac{1}{2}} - (\frac{\beta}{s})^{\frac{1}{2}}]\}, \, s > 0, \, \alpha > 0, \, \beta > 0,$$

where α is a shape parameter and β is a scale parameter.

In this paper, a multiplicative cumulative damage approach is taken to develop two new models for the tensile strength of a material which incorporates the size or gauge length, L, of the material. The models derived here assume that a multiplicative damage process occurs as tensile load on the material is increased until failure. It is widely accepted that the failure of fibrous composites is primarily due to flaws present in the carbon fibers, and these flaws cause a reduction in material strength. The initial damage to the material is assumed to be in the form of the most severe flaw present in the material before any tensile load is applied. The initial damage is quantified by a random variable, X_0, and the fixed (unknown) theoretical strength of the material, Y, is reduced by the random initial damage so that the "initial strength" of the material is assumed to be a random variable W, the appropriate reduction of Y by X_0.

In Section 2, the notion of multiplicative cumulative damage will be defined and two models will be derived. The first model, called the Gaussian Initial Damage Model (GIDM), will assume that the most severe flaw is described by a "flaw process" defined over the system of size L. This accounts for the reduction in system strength, and the flaw process is described by a stationary Brownian motion[11]. This is similar to the approach given by Durham and Padgett[9] where they considered an additive damage model. The second model, called the Weibull Initial Damage Model (WIDM), assumes that the distribution of initial strength, W, is the two-parameter Weibull distribution as given in Weathers[12]. Both of these multiplicative models also result in three-parameter extensions of the Birnbaum-Saunders distribution. In both cases, the third parameter arises due to the incorporation of the known size variable, or gauge length, L. In Section 3, parameter estimation techniques for the GIDM model will be discussed, including "least squares" type estimates and the maximum likelihood estimates (MLEs). Both estimation methods will require numerical root finding techniques. Formulas for estimated percentiles of the strength distribution will be obtained, and asymptotic variances of the MLEs will be given via the estimated Fisher information matrix. In Section 4, the multiplicative cumulative damage models will be fitted to the carbon micro-composite tensile strength data (1000-fiber impregnated tows) of Bader and Priest[13], given in Smith[14]. For the Bader-Priest data, the GIDM will be shown to be superior to the WIDM, superior to the additive damage models given in Durham and Padgett[9], as well as superior to the "power-law" Weibull model proposed by Padgett, Durham and Mason[8]. Finally, using the theory developed in Section 3, confidence intervals for the MLEs for the GIDM will be given, as well as lower bounds on percentiles of the tensile strength.

2   The Multiplicative Cumulative Damage Model

Suppose a composite specimen (or "system") is placed under tensile load that is increased steadily until failure. Assume that:

(*i*) The increasing load is envisioned as being incremented by small, discrete amounts until specimen failure, and the amount of stress that causes failure is observed to be the specimen strength.

(*ii*) Each increment of stress causes a random amount of non-negative damage D, having CDF, F_D.

(*iii*) The "initial damage" to the specimen is in the form of the most severe flaw, and is quantified by a random reduction in the theoretical strength by a random variable $X_0 \geq 1$. The random variable $W = Y / X_0$ represents the "initial strength" of the specimen.

As the tensile stress is incremented, the cumulative damage after $n + 1$ increments is denoted by $X_{n+1} = X_n + D_n X_n$, where $D_n \geq 0$ is the damage to the specimen at increment $n + 1$, $n = 0, 1, 2, ...$ This gives a multiplicative damage model, since the amount of damage at the $(n+1)^{st}$ increment is multiplied by the accumulated amount of damage to that point (see Durham and Padgett[9], and Desmond[15]). Here, the D_n's are independent and identically distributed (i.i.d.) according to the distribution function F_D. Let N denote the number of increments of tensile stress applied to a specimen of strength Y, so that

$$N = \sup_n\{n: X_1 \leq Y, ..., X_{n-1} \leq Y\} = \sup_n\{n: X_1 / X_0 \leq Y / X_0, ..., X_{n-1} / X_0 \leq Y / X_0\}$$

with $N = 1$ if the set is empty.

Conditionally, $P(N > n \mid Y / X_0 = w) = P(X_n / X_0 \leq w)$, since $\{X_n\}$ is a nondecreasing sequence. Thus, the (unconditional) survival probability after n stress increments is just

$$P(N > n) = \int_0^\infty F_n(w)dG_W(w), \tag{1}$$

where $F_n(w) = P(X_n / X_0 \leq w)$ and G_W is the distribution function for the initial strength W. For large n, $F_n(w)$ is approximately a lognormal CDF as follows. Write $D_n = (X_{n+1} - X_n)/X_n$, so that

$$\sum_{i=0}^{n-1} D_i = \sum_{i=0}^{n-1} (X_{i+1} - X_i)/X_i \approx \int_{X_0}^{X_n} u^{-1}du = ln(X_n) - ln(X_0)$$

for large n. Then, by the Central Limit Theorem, X_n / X_0 has an approximate lognormal distribution with parameters $n\mu$ and $n\sigma^2$, where $\mu=E(D)$ and $\sigma^2=Var(D)$, the mean and variance of D, respectively. Thus, for large n,

$$F_n(w) = P(X_n / X_0 \leq w) \approx \Phi[(ln(w) - n\mu)/\sqrt{n}\sigma], \tag{2}$$

where Φ denotes the standard normal CDF. Substituting (2) and an appropriate distribution for the initial strength G_W into (1) gives the probability that the specimen will survive n increments. The next two subsections will consider the two different initial damage models leading to an appropriate G_W as described in the introduction.

2.1   The Gaussian Initial Damage Model

Assume that the unknown theoretical (fixed) strength of the carbon fibrous composite specimen (micro-composite) is Y. Denote the gauge length of the specimen by L. As in Durham and

Padgett[9], let Δ_u denote the (random) amount of initial damage at location u along the length of the micro-composite specimen, $0 \le u \le L$, where $\Delta_0 = C$, a positive constant. Let $M(L) = \max\{\Delta_u : 0 \le u \le L\}$ denote the initial damage in terms of the most severe of the inherent flaws over the material of size L; that is, $M(L)$ gives the random reduction of system strength due to the most severe inherent flaw present in the material before stress is applied. Thus, the "initial strength" of the material is the random variable $W = Y - [M(L) - C] \ge 0$, which is the theoretical strength reduced by an amount due to the initial damage.

Next, assuming that the process $\{\Delta_u : 0 \le u \le L\}$ is a stationary Brownian motion, for a fixed $a > 0$, by Karlin and Taylor[11],

$$P[M(L) - C \ge a] = 2P[\Delta_L - C > a] = \frac{2}{\sqrt{2\pi L}} \int_a^\infty \exp[-x^2/(2L)] \, dx.$$

Then, given the theoretical material strength Y, the CDF of W is given by

$$G_W(w) = P(W \le w) = P[M(L) - C \ge Y - w] = \frac{2}{\sqrt{2\pi L}} \int_{Y-w}^\infty \exp[-x^2/(2L)] \, dx.$$

Thus, the probability density function (PDF) of W is

$$g_W(w) = \frac{2}{\sqrt{2\pi L}} \exp[-(w - Y)^2/(2L)], \, 0 < w < Y. \tag{3}$$

The result (3) will now be used to obtain the strength distribution for fibrous composite specimens of gauge length L. Substituting (2) and (3) into (1) gives the probability of the specimen surviving n increments of stress as

$$P(N > n) \cong \int_0^Y \Phi[(ln(w) - n\mu)/\sqrt{n}\sigma] \frac{2}{\sqrt{2\pi L}} \exp[-(w - Y)^2/(2L)] dw, \tag{4}$$

which is a mixture of lognormal CDFs. In (4), we let $Y - w = \sqrt{n\sigma^2 v}$ to give

$$P(N > n) \cong \int_0^{Y^2/(n\sigma^2)} \Phi[(ln(Y - \sqrt{n\sigma^2 v}) - n\mu)/\sqrt{n}\sigma] \frac{1}{\sqrt{2\pi L/(n\sigma^2)}} v^{-\frac{1}{2}} \exp[-n\sigma^2 v/(2L)] dv, \tag{5}$$

where v represents the value of a gamma random variable, V, with shape parameter $\frac{1}{2}$ and scale parameter $2L/(n\sigma^2)$.

Similar to arguments given in Durham and Padgett[9], the upper limit in the integral (5) can be written as a large multiple of the standard deviation of V, given by $Y^2/(n\sigma^2) = Y^2/(L\sqrt{2})[(L\sqrt{2})/(n\sigma^2)]$. Thus, (5) can be considered as approximately the expectation $E\{\Phi[(ln(Y - \sqrt{n\sigma^2 V}) - n\mu)/\sqrt{n}\sigma]\}$. Furthermore, for very large n, $Y^2/(n\sigma^2)$ is small, so that over the interval of integration, Φ can be considered as approximately linear, allowing the expectation to be taken inside the standard normal CDF Φ. For this formulation, the expected value $E[ln(Y - \sqrt{n\sigma^2 V})]$ is needed. Thus,

$$E[ln(Y - \sqrt{n\sigma^2 V})] = \int_0^\infty ln(Y - \sqrt{n\sigma^2 v}) \frac{1}{\sqrt{2\pi L/(n\sigma^2)}} v^{-\frac{1}{2}} \exp[-n\sigma^2 v/(2L)]dv$$

$$= \int_0^\infty ln(Y - \sqrt{L}x) \frac{1}{\sqrt{2\pi}} \exp[-(x^2/2)]dx \qquad (6)$$

by a change of variable from v to $x = \sqrt{n\sigma^2 v/L}$. To evaluate (6), we will use a two-term Taylor series approximation for $ln(Y - \sqrt{L}x)$ which is $lnY - \sqrt{L}x/Y$, and after substituting this expression, the desired expectation is approximately

$$E[ln(Y - \sqrt{n\sigma^2 V})] \cong lnY - \frac{\sqrt{L}}{Y}\sqrt{2/\pi}.$$

Using these results in (5) gives the survival probability after n increments of stress to be

$$P(N > n) \cong \Phi[(lnY - \frac{\sqrt{L}}{Y}\sqrt{2/\pi})/\sqrt{n}\sigma - \mu\sqrt{n}/\sigma].$$

By letting S be the continuous analog of N, *i.e.* the total applied stress after N increments, and using the reparameterization $\alpha = \sigma$, $\beta = \mu^{-1}$, we can write the tensile strength distribution function for the GIDM as

$$G_S(s;L) = P(S \le s) = \Phi\{ \frac{1}{\alpha}[\frac{\sqrt{s}}{\beta} - (lnY - \frac{\sqrt{L}}{Y}\sqrt{2/\pi})/\sqrt{s}]\}, s > 0. \qquad (7)$$

It is easily seen that in this form, (7) is a three-parameter Birnbaum-Saunders-type distribution, with unknown parameters Y, α and β. The distribution also depends on L, the (known) system size.

2.2 The Weibull Initial Damage Model

Here we will assume that G_W, the distribution for initial strength of the specimen of size L, has the two-parameter Weibull distribution with shape parameter ρ and scale parameter δ. Using this with (2) in (1), our expression for the survival probability of the system becomes, for large n,

$$P(N > n) \cong \int_0^\infty \Phi\{[ln(w) - n\mu]/\sqrt{n}\sigma\} \frac{L\rho}{\delta^\rho} w^{\rho-1} \exp[-L(w/\delta)^\rho]\, dw. \qquad (8)$$

Changing variables in the above integral first to $u = w/(\sqrt{n}\sigma)$ and then to $t = L(\sigma u)^\rho/\delta^\rho$ yields

$$P(N > n) \cong \int_0^\infty \Phi\{[ln(t^{1/\rho}) + ln(\sqrt{n}) + ln(\delta/L^{1/\rho})]/(\sqrt{n}\sigma) - \sqrt{n}\mu/\sigma\}n^{\rho/2}\exp(-n^{\rho/2} t)dt, \qquad (9)$$

which is a mixture with respect to an exponential distribution with mean $n^{-\rho/2}$. Note that (9) is a mathematical expectation with respect to this exponential distribution, and can be approximated (using arguments similar to Durham and Padgett[9], and Weathers[12]) by

$$P(N > n) \cong \Phi\{[Eln(T^{1/\rho}) + ln(\sqrt{n}) + ln(\delta/L^{1/\rho})]/(\sqrt{n}\sigma) - \sqrt{n}\mu/\sigma\}. \qquad (10)$$

Now, it can be shown that $ln(T^{1/\rho})$ is a random variable with an extreme value distribution, so $Eln(T^{1/\rho}) = -(ln\sqrt{n} + .5772/\rho)$, and .5772 is from Euler's constant γ. Thus (10) reduces to

$$P(N \le n) \cong \Phi[\sqrt{n}\mu/\sigma - (\rho ln\delta - lnL - .5772)/(\rho\sqrt{n}\sigma)].$$

Again, letting S be the continuous analog of N, the cumulative distribution function of the system strength is

$$F(s) = P(S \le s) \cong \Phi[\sqrt{s}\mu/\sigma - (\rho ln\delta - lnL - .5772)/(\rho\sigma\sqrt{s})]$$

$$= \Phi\{\frac{1}{\tau}[(\frac{s}{\eta})^{\frac{1}{2}} - \frac{\lambda - lnL}{\sqrt{s}}]\}, \tag{11}$$

which is another three-parameter generalization of the Birnbaum-Saunders distribution indexed by system size or length L. Here, $\tau = \rho\sigma$, $\eta = 1/(\mu\rho)^2$, and $\lambda = \rho ln\delta - .5772$.

The models developed in Sections 2.1 and 2.2 allow for parameter estimation from data observed over more than one value of L, the gauge length. It should be noted that if a single value of L is under consideration, then the third parameter, along with L itself, can be absorbed into the other two parameters. This would result in the usual two-parameter form (1) of the Birnbaum-Saunders distribution. In the next section, we will investigate inference techniques for the GIDM, since this model will be shown in Section 4 to provide a better fit to the Bader-Priest[13] data than the WIDM.

3   Inference for the GIDM

Here, we will suppose that composite specimens at k different gauge lengths are to be tested for their tensile strength, and we will denote the gauge lengths by $L_1, ..., L_k$. At each gauge length L_i, tests are performed for n_i specimens resulting in observed breaking stresses $s_{ij}, j = 1, ..., n_i$, for $i = 1, ..., k$.

3.1   Least Squares-type Estimates

It is possible to "linearize" the GIDM given in (7) by writing it in the form

$$\sqrt{s}\Phi^{-1}[G_S(s;L)] = -\frac{lnY}{\alpha} + \frac{1}{\alpha\beta}s + \frac{\sqrt{2/\pi}}{Y\alpha}\sqrt{L}. \tag{12}$$

Using the observed s_{ij} at each L_i, least squares-type estimates can be obtained for the *functions* of the parameters given in (12) by estimating $G_S(s;L_i)$ for each i by the empirical CDF of the s_{ij}'s. That is, for each i, set

$$\sqrt{s_{ij}}\Phi^{-1}[(j - .5)/n_i] = -\frac{lnY}{\alpha} + \frac{1}{\alpha\beta}s_{ij} + \frac{\sqrt{2/\pi}}{Y\alpha}\sqrt{L_i} , \text{ for } j = 1, ..., n_i. \tag{13}$$

Let b_0, b_1 and b_2 respectively denote the estimates of the intercept and the coefficients of s_{ij} and L_i obtained when the regression of (13) is performed. Estimates (Y^*, α^*, β^*) of (Y, α, β) can be extracted from the least squares estimates (b_0, b_1, b_2) as follows. The estimate Y^* can be obtained by noting that $Y^* ln Y^* = -\sqrt{2/\pi}\, b_0/b_2$, so that Y^* may be found quite easily with a Newton-Raphson root-finding method. Once Y^* has been calculated, simply let $\alpha^* = -ln Y^* / b_0$ and $\beta^* = (b_1 \alpha^*)^{-2}$. Although the estimates (Y^*, α^*, β^*) are not truly the least squares estimates of (Y, α, β), they may be used as initial estimates for solving the likelihood equations by numerical root finding techniques, which will be investigated in the next subsection.

3.2 Maximum Likelihood Estimation

Let $g_s(s;L)$ denote the probability density function of the GIDM given in (7). Thus,

$$g_s(s;L) = \frac{s + \beta(ln Y - \frac{\sqrt{L}}{Y}\sqrt{2/\pi})}{2\sqrt{2\pi}\,\alpha\beta s^{3/2}} \exp\{ -\frac{[s - \beta(ln Y - \frac{\sqrt{L}}{Y}\sqrt{2/\pi})]^2}{2\alpha^2\beta^2 s} \}, \; s > 0. \tag{14}$$

For estimation from data over all of the k gauge lengths, the MLEs of the unknown model parameters in (14) are found by maximizing the likelihood function

$$L(\alpha,\beta,Y) = \prod_{i=1}^{k} \prod_{j=1}^{n_i} g_s(s_{ij}\,; L_i), \tag{15}$$

or equivalently, maximizing the logarithm of (15).

The log-likelihood function for the GIDM is

$$\ell = ln L(\alpha,\beta,Y) = -m\cdot ln(2\sqrt{2\pi}) - m\cdot ln\alpha - m\cdot ln\beta + \sum_{i=1}^{k}\sum_{j=1}^{n_i} ln\left[\frac{s_{ij} + \beta(ln Y - \frac{\sqrt{L_i}}{Y}\sqrt{2/\pi})}{s_{ij}^{3/2}}\right]$$

$$-\frac{1}{2\alpha^2\beta}\sum_{i=1}^{k}\sum_{j=1}^{n_i}\frac{[s_{ij} - \beta(ln Y - \frac{\sqrt{L_i}}{Y}\sqrt{2/\pi})]^2}{s_{ij}}, \tag{16}$$

where $m = \sum_{i=1}^{k} n_i$. The partial derivatives of (16) with respect to each of (α,β,Y) yield three nonlinear likelihood equations. The equation for α can be solved in terms of Y and β, but the equations for Y and β must each be solved numerically, given the values of the other two parameters, as follows:

$$\alpha = \{[\frac{1}{m\beta^2}\sum_{i=1}^{k}\sum_{j=1}^{n_i}\frac{[s_{ij} - \beta(ln Y - \frac{\sqrt{L_i}}{Y}\sqrt{2/\pi})]^2}{s_{ij}}]/m\}^{\frac{1}{2}}, \tag{17.a}$$

$$\beta = [(\sum_{i=1}^{k}\sum_{j=1}^{n_i}[s_{ij} - \beta(ln Y - \frac{\sqrt{L_i}}{Y}\sqrt{2/\pi})])/(\alpha^2\sum_{i=1}^{k}\sum_{j=1}^{n_i}\frac{s_{ij}}{s_{ij} + \beta(ln Y - \frac{\sqrt{L_i}}{Y}\sqrt{2/\pi})})]^{\frac{1}{2}}, \tag{17.b}$$

$$0 = \beta \sum_{i=1}^{k} \left(\frac{1}{Y} + \frac{\sqrt{L_i}}{Y^2} \sqrt{2/\pi} \right) \sum_{j=1}^{n_i} [s_{ij} + \beta(lnY - \frac{\sqrt{L_i}}{Y} \sqrt{2/\pi})]^{-1} + \frac{1}{\alpha^2 \beta} \sum_{i=1}^{k} n_i \left(\frac{1}{Y} + \frac{\sqrt{L_i}}{Y^2} \sqrt{2/\pi} \right)$$

$$- \frac{1}{\alpha^2} \sum_{i=1}^{k} (lnY - \frac{\sqrt{L_i}}{Y} \sqrt{2/\pi})(\frac{1}{Y} + \frac{\sqrt{L_i}}{Y^2} \sqrt{2/\pi}) \sum_{j=1}^{n_i} \frac{1}{s_{ij}}. \tag{17.c}$$

Iteration over these three equations can be performed with a root-finding method until convergence to an approximate solution for the MLEs, $\hat{\alpha}$, $\hat{\beta}$ and $\hat{Y}$. The "least-squares" estimates described in Section 3.1 can be used for starting values for the parameters in the iterative procedure. The estimates from carbon composite material strength data will be given in the Section 4.

The $100 \cdot p^{th}$ percentile, $s_p(L)$, for the strength distribution (7) for a given value of L can be calculated by equating the GIDM with p and by solving for s. This is similar to the procedure given by (13), but the quadratic equation in $\sqrt{s_p(L)}$ (positive root) must be solved. For a specified p, the $100 \cdot p^{th}$ percentile at gauge length L is given by

$$s_p(L) = \frac{\beta^2}{4} \left[\alpha z_p + \sqrt{\alpha^2 z_p^2 + 4(lnY - \frac{\sqrt{L}}{Y} \sqrt{2/\pi})/\beta} \right]^2, \tag{18}$$

where $z_p = \Phi^{-1}(p)$, the $100 \cdot p^{th}$ percentile of the standard normal distribution.

The MLE of $s_p(L)$ can be obtained by substituting the MLEs ($\hat{\alpha}, \hat{\beta}, \hat{Y}$) into (18). Observing that $s_p(L)$ is increasing in Y (α and β fixed), increasing in α (Y and β fixed), and increasing in β (α and Y fixed), a conservative lower confidence bound on $s_p(L)$ can be obtained by substituting the lower confidence bounds (LCBs) for each of the parameters. The (conservative) confidence level for the lower bound of the percentile can be calculated by using a Bonferroni inequality argument. Approximate confidence bounds for the GIDM parameters will be obtained in the next subsection.

3.3 Asymptotic Distribution Theory for the GIDM

For large values of the n_i's (and consequently for large values of m), approximate confidence intervals for α, β and Y can be based on the usual asymptotic normality theory for maximum likelihood estimators. Here, as m goes to infinity, we have that $[I_m(\theta)]^{1/2}(\hat{\theta} - \theta) \to Z$, where Z has the trivariate normal distribution with zero mean vector and identity covariance matrix, $\theta = (\alpha, \beta, Y)'$ and $I_m(\theta)$ is the Fisher information matrix based on m observations. The negative expected values for the six second partial derivatives of the log-likelihood function (16) are complicated but can be obtained. The terms $E(-\partial^2 l/\partial\alpha^2)$, $E(-\partial^2 l/\partial\alpha\partial\beta)$ and $E(-\partial^2 l/\partial\alpha\partial Y)$ can be found explicitly, but the terms $E(-\partial^2 l/\partial\beta^2)$, $E(-\partial^2 l/\partial\beta\partial Y)$ and $E(-\partial^2 l/\partial Y^2)$ require an approximation since they contain intractable integrals that cannot be found in closed form. For illustration, we will present the procedure to obtain the $E(-\partial^2 l/\partial\alpha^2)$ and $E(-\partial^2 l/\partial\beta\partial Y)$ terms.

To simplify the following expressions, we will let $\lambda(Y;L) = lnY - \frac{\sqrt{L}}{Y}\sqrt{2/\pi}$. For calculating $E(-\partial^2 \ell/\partial\alpha^2)$, we obtain

$$\partial^2 \ell/\partial\alpha^2 = m/\alpha^2 - \frac{3}{\alpha^4\beta^2}\sum_{i=1}^{k}\sum_{j=1}^{n_i} s_{ij} + \frac{6}{\alpha^4\beta}\sum_{i=1}^{k}[n_i\lambda(Y;L_i)] - \frac{3}{\alpha^4}\sum_{i=1}^{k}[\lambda(Y;L_i)]^2\sum_{j=1}^{n_i}\frac{1}{s_{ij}}. \quad (19)$$

Using arguments similar to those given in Birnbaum and Saunders[10], it can be shown that for the GIDM,

$$E(S;L) = \beta\lambda(Y;L) + \frac{1}{2}\alpha^2\beta^2 \qquad (20.a)$$

and

$$E(S^{-1};L) = [\beta\lambda(Y;L)]^{-1} + \frac{1}{2}\alpha^2[\lambda(Y;L)]^{-2}. \qquad (20.b)$$

Using (20.a-b) and after simplifying, it is easily seen that the desired negative expectation of (19) is

$$E(-\partial^2\ell/\partial\alpha^2) = -m/\alpha^2 + \frac{3}{\alpha^4\beta^2}\sum_{i=1}^{k} n_i E(S;L_i) - \frac{6}{\alpha^4\beta}\sum_{i=1}^{k}[n_i\lambda(Y;L_i)]$$

$$+ \frac{3}{\alpha^4}\sum_{i=1}^{k}[\lambda(Y;L_i)]^2 n_i E(S^{-1};L_i)$$

$$= 2m/\alpha^2.$$

For calculating $E(-\partial^2\ell/\partial\beta\partial Y)$, we write $\lambda'(Y;L) = \frac{1}{Y} + \frac{\sqrt{L}}{Y^2}\sqrt{2/\pi}$ and observe that

$$\partial^2\ell/\partial\beta\partial Y = \sum_{i=1}^{k}\lambda'(Y;L_i)\sum_{i=1}^{n_i}[s_{ij} + \beta\lambda(Y;L_i)]^{-1} - \frac{1}{\alpha^2\beta^2}\sum_{i=1}^{k} n_i\lambda'(Y;L_i)$$

$$- \beta\sum_{i=1}^{k}\lambda(Y;L_i)\lambda'(Y;L_i)\sum_{i=1}^{n_i}[s_{ij} + \beta\lambda(Y;L_i)]^{-2}. \qquad (21)$$

To solve for the required expectations $E[S + \beta\lambda(Y;L)]^{-1}$ and $E[S + \beta\lambda(Y;L)]^{-2}$, we can employ Laplace's method for approximation of such integrals (see Tierney, Kass and Kadane[16] and Achcar[17]). The Laplace approximations are

$$E[S + \beta\lambda(Y;L)]^{-1} \cong [2\beta\lambda(Y;L)]^{-1} \qquad (22.a)$$

and

$$E[S + \beta\lambda(Y;L)]^{-2} \cong [4\beta^2\lambda^2(Y;L)]^{-1}. \qquad (22.b)$$

Using these approximations and after simplifying, it can be seen that the negative expectation of (21) is

$$E(-\partial^2\ell/\partial\beta\partial Y) \cong -\frac{1}{4\beta}\sum_{i=1}^{k} n_i \frac{\lambda'(Y;L_i)}{\lambda(Y;L_i)} + \frac{1}{\alpha^2\beta^2}\sum_{i=1}^{k} n_i\lambda'(Y;L_i).$$

The expectations given in (20.a-b) and the approximations given in (22.a-b) are also necessary for the other entries in the Fisher information matrix, and they are calculated in the same manner. Without presenting all the derivation details, the six entries in the Fisher information matrix for the GIDM are

$$E(-\partial^2 \ell/\partial \alpha^2) = 2m/\alpha^2 \tag{23.a}$$

$$E(-\partial^2 \ell/\partial \alpha \partial \beta) = m/(\alpha \beta) \tag{23.b}$$

$$E(-\partial^2 \ell/\partial \alpha \partial Y) = -\frac{1}{\alpha} \sum_{i=1}^{k} n_i \frac{\lambda'(Y;L_i)}{\lambda(Y;L_i)} \tag{23.c}$$

$$E(-\partial^2 \ell/\partial \beta^2) \cong 3m/(4\beta^2) + \frac{1}{\alpha^2 \beta^3} \sum_{i=1}^{k} n_i \lambda(Y;L_i) \tag{23.d}$$

$$E(-\partial^2 \ell/\partial \beta \partial Y) \cong -\frac{1}{4\beta} \sum_{i=1}^{k} n_i \frac{\lambda'(Y;L_i)}{\lambda(Y;L_i)} + \frac{1}{\alpha^2 \beta^3} \sum_{i=1}^{k} n_i \lambda'(Y;L_i) \tag{23.e}$$

$$E(-\partial^2 \ell/\partial Y^2) \cong \frac{3}{4} \sum_{i=1}^{k} n_i \frac{[\lambda'(Y;L_i)]^2}{[\lambda(Y;L_i)]^2} + \frac{1}{\alpha^2 \beta} \sum_{i=1}^{k} n_i \frac{[\lambda'(Y;L_i)]^2}{\lambda(Y;L_i)}. \tag{23.f}$$

Using these expressions in $I_m(\alpha,\beta,Y)$, the asymptotic variances of the three MLEs of α, β and Y can be found from the diagonal elements of $I_m^{-1}(\alpha,\beta,Y)$ and estimated by substitution of the MLEs for the three unknown parameters involved. These can be used to calculate approximate confidence bounds on the parameters (using the asymptotic normality result) and hence approximate lower confidence bounds for $s_p(L)$. Real data illustrations of this will furnished in the next section.

4   Estimation From Micro-Composite Strength Data

Bader and Priest[13] obtained strength data (measured in GPa, giga-Pascals) for 1000-fiber impregnated (epoxy resin matrix) tows that were tested under tension in a controlled setting. In their experiment, the tows (micro-composite specimens) were tested at gauge lengths of 20, 50, 150 and 300 mm with 28, 30, 32 and 29 observed specimens at the respective gauge lengths. The amount of tensile stress that caused failure to the specimen of gauge length L was recorded as the strength, and all of these data are given explicitly in Smith[14]. In this section, the GIDM will be used to model the strength distribution of the 1000-fiber tows from the Bader-Priest data. It will be fitted over all four gauge lengths using the methods developed in Section 3. The fit of this model will be compared to the fits of the WIDM (given in Weathers[12]), the additive damage model (given in Durham and Padgett[9]), and the "power-law" Weibull model (given in Padgett, Durham and Mason[8]).

For the estimation of the parameters in (7), starting values were found via the "least-squares" approach described in Section 3.1. For the Bader-Priest data, the starting values were calculated to be $Y^* = 27.485$, $\alpha^* = .1416$ and $\beta^* = .8928$. These starting values were used to begin the iteration over the equations (17.a-c) to yield the MLEs, $\hat{Y} = 26.8055$, $\hat{\alpha} = .13340$ and $\hat{\beta} = .90413$. Note that the MLEs are quite close to the "least-squares" values. The overall mean squared error (MSE) for this fitted model is measured by the average squared distance of the model fit (using the MLEs) to the empirical CDF for *each* of the gauge lengths over all of the $m = 119$ observations. For the GIDM, we calculate this MSE to be 0.005366. For comparison, the MSE for the WIDM is given as 0.009576, the MSE for an additive cumulative damage model (Birnbaum-Saunders form) in Durham and Padgett[9] is .005452, and the MSE for the power-law Weibull was 0.008439. Here, it can be clearly seen that the GIDM is superior to the WIDM, as well as to the additive damage model and to the power-law Weibull model for fitting these carbon micro-composite strength data.

The information matrix entries (23.a-f) were estimated by using the MLEs for Y, α and β. These estimates of (23.a-f) are 13374.767, 986.669, -14.719, 27078.052, 398.664, and 6.086, respectively. The diagonal elements of the inverse of this estimated matrix give the estimated asymptotic variances of the MLEs. We calculate the asymptotic standard deviations (ASD) of the MLEs to be $ASD(\hat{Y}) = 2.558$, $ASD(\hat{\alpha}) = 0.0103$ and $ASD(\hat{\beta}) = 0.03836$. Using the asymptotic normality result stated earlier in this section, approximate 95% confidence intervals for the three parameters are: (21.792, 31.819) for Y, (.1132, .1536) for α and (.8289, .9793) for β. In similar fashion, by calculating asymptotic 96.7% LCBs for each of Y, α and β, and substituting these values into the expression for $s_p(L)$ given by (18), we can obtain conservative 90% LCBs for the $100 \cdot p^{th}$ percentile of the strength distribution (7). These percentile estimates are functions of the gauge length L, and in Table 1 we present point estimates and 90% LCBs of the 10th percentile for the Bader-Priest tow data for each of the gauge lengths. Since increasing the length of a tow has the effect of decreasing its tensile strength, the estimates and LCBs for $s_p(L)$ reported in Table 1 are decreasing with the gauge length as would be expected.

Table 1: Point Estimates and Approximate 90% Lower Confidence Bounds (LCBs) on the 10th Percentile of Tensile Strength for the Bader-Priest Data.

Gauge Length, L (mm)	$\hat{s}_{0.1}(L)$	90% LCB for $s_{0.1}(L)$
20	2.604 GPa	2.263 GPa
50	2.537 GPa	2.187 GPa
150	2.404 GPa	2.038 GPa
300	2.274 GPa	1.892 GPa

In summary, the models given herein based on the multiplicative cumulative damage concept fit the micro-composite tensile strength data quite well, especially the GIDM (model (7)) given in Section 2.1. In particular, the GIDM fits better than its existing competitors, based on MSEs. The models have a sound structural development, and the parameters involved in both the WIDM and the GIDM have specific physical interpretations (it is simple to reparameterize back to μ and σ). Also, it is of great benefit to be able to take into account the "size effect" on the strength of a system, and this method combined with the multiplicative damage model concept seems to work well based on the real data set examined in this section.

Acknowledgements

This work was supported by the National Science Foundation under grant number DMS-9503104.

References

1. C. B. Beetz, *Fiber Science and Technology* **16**, 45-59 (1982).
2. S. H. Own, R.V. Subramanian, and S.C. Saunders, *Journal of Materials Science* **21**, 3912-3920 (1986).
3. C.M. Black, S.D. Durham, and W.J. Padgett, *Communications in Statistics - Simulation and Computation* **19**, 809-825 (1990).
4. K. Goda and H. Fukunaga, *Journal of Materials Science* **21**, 4475-4480 (1986).
5. D.H. Wagner, *Journal of Polymer Science* **B 27**, 115-148 (1989).
6. S.L. Phoenix and R.G. Sexsmith, *Journal of Composite Materials* **6**, 322-337 (1972).
7. E.G. Stoner, D.D. Edie and S.D. Durham, *Journal of Materials Science* **29**, 6561-6574 (1994).
8. W.J. Padgett, S.D. Durham and A.M. Mason, *Journal of Composite Materials* **29**, 1873-1884 (1995).
9. S.D. Durham and W.J. Padgett, *Technometrics* **39**, 34-44 (1997).
10. Z.W. Birnbaum and S.C. Saunders, *Journal of Applied Probability* **6**, 319-327 (1969).
11. S. Karlin and H.M. Taylor, *A First Course in Stochastic Processes* (2nd ed.), New York, Academic Press (1975).
12. P.D. Weathers, *A Multiplicative Damage Model for Strength of Brittle Materials*, Masters Thesis, Department of Statistics, University of South Carolina (1996).
13. S. Bader and A. Priest, *Progress in Science and Engineering of Composites*, eds T. Hayashi, K. Katawa and S. Umekawa, ICCM-IV, Tokyo, 1129-1136 (1982).
14. R.L. Smith, *Reliability Engineering and System Safety*, 55-76 (1991).
15. A.F. Desmond, *The Canadian Journal of Statistics* **13**, 171-183 (1985).
16. L. Tierney, R.E. Kass and J.B. Kadane, *Journal of the American Statistical Association* **84**, 710-716 (1989).
17. J.A. Achcar, *Computational Statistics and Data Analysis* **15**, 367-380 (1993).

SYSTEM-BASED COMPONENT TEST PLANS FOR RELIABILITY INFERENCES

JAYANT RAJGOPAL, MAINAK MAZUMDAR
Department of Industrial Engineering, 1048 Benedum Hall
University of Pittsburgh, Pittsburgh, PA 15261
E-mail: rajgopal@engrng.pitt.edu

This paper addresses the design of system-based component test plans for demonstrating reliability of a series system. There are two primary contributions of this work. First, unlike most of the prior work in this area which has relied on the component failure times being exponentially distributed, this paper examines another common failure time distribution, namely the Weibull distribution and develops a test plan that exploits the relationship between the Weibull and the exponential distributions. Second, it introduces the notion of imperfect interfaces between components within the system which is another issue that all prior work has ignored, and results in plans that also call for system level tests.

1. Introduction

Dedicated programs of testing are an integral part of the process by which new systems are designed and developed. The purpose of such programs is to ensure that the system will satisfy performance criteria, prior to their actual deployment in the field. While there are many different formats that such test plans could follow, the two basic options are (a) system testing, and (b) component testing. With the first option, the entire system is assembled and then tested in order to draw inferences on its reliability, while in the latter case the components that make up the system are tested, and then based on the results of these component tests one draws inferences about the system reliability. Several variations of these two basic options are possible. For instance, one may conduct tests at the level of subassemblies (or subsystems) that are composed of individual components, and that are themselves part of the entire system. Alternatively, one could adopt some combination of system, subsystem and component testing.

The choice of an appropriate test plan depends on the characteristics of the specific system under consideration, as well as cost and feasibility. System level tests are typically more expensive as well as more inconvenient than component tests. There are several reasons for this. First, the instrumentation, test fixtures and facilities, materials and the actual test units all usually cost more for systems than for components. Second, and perhaps most importantly, with component tests the entire system need not be assembled prior to testing. This results in significant savings both with respect to costs as well as the times associated with system development. Third, testing may proceed independently at different times and locations. This is an important issue since components (or subsystems) of many large systems are typically contracted to different organizations who develop these components independently. Fourth, component tests tend to be more informative in that the behavior of individual components can often be understood better than in the case where they are assembled into a larger system.

Finally, component testing goes well with the notions of total quality since the final system is assembled only after a performance guarantee is obtained.

Conversely, system testing is preferable when the component failures are highly dependent on each other since component testing assumes independence of component failure times, and the availability of a mathematical model that expresses system reliability in terms of component reliabilities. Another situation where system testing may be preferable is when the interfaces between components are inherently unreliable; however, in such situations it may still be more economical to use a combination of system and component testing.

In this paper we concentrate on situations where component testing is appropriate. In designing a component test plan the major issue to keep in mind is that the plans must be *system-based*. What this means is that if the plan is not designed with explicit consideration given to the mathematical model that relates the system reliability to component reliabilities, the resulting inferences about system reliability can be very misleading. This point is discussed in [1]. Unfortunately, it is not always considered in practice, and often for testing purposes, the total system reliability is allocated arbitrarily among components. Much of the research in component testing has concentrated on the design of system-based component test plans. Some of the papers that have addressed various issues relating to such plans include references [1] through [11]. The problem of designing optimal test plans leads to a formulation which is a two stage mathematical program. The "inner" stage searches for the maximum probability of an erroneous decision for a given set of test parameters, while the "outer" stage searches for the most desirable set of parameter values among those that yield maximum error probabilities that are within acceptable limits.

2. System-Based Component Test Plans

As with any test plan, the objective is to minimize the total cost of the test program while ensuring that they offer sufficient protection from errors that result from making wrong decisions. Specific examples of reliability test plans can be found in the Department of the Navy document MIL-HDBK-781 [12] which offers protection from the probabilities of Type I error (α) and Type II error (β). In this document values are specified for α and β (with $\alpha+\beta<1$), and for R_0 and R_1 (with $0<R_0<R_1<1$) where a system with reliability less than R_0 over its mission time would be considered definitely unacceptable and a system with reliability higher than R_1 over its mission time would be considered definitely acceptable. Suppose R_S denotes the reliability of the system over its mission time. Then the two constraints to be met by any test plan are:

$$\text{Pr (Plan rejects the system} \mid R_S \geq R_1) \leq \alpha \tag{1}$$
$$\text{Pr (Plan accepts the system} \mid R_S \leq R_0) \leq \beta \tag{2}$$

In general, there may be many different test plans that meet the above constraints. Each such plan could have its own format (test procedure, acceptance rule, etc.) and associated total test cost, and ideally, one would like to select the plan that is the least expensive. Unfortunately, it is not possible to exhaustively list all procedures that will lead to feasible test plans. Therefore, we take the approach of deciding in advance upon a plan format that is plausible for the specific system under consideration, and then searching for plan parameters which minimize total test costs.

3. A Component Test Plan for the Weibull Distribution

Most of the prior work in this area (including the references listed at the end of Section 1) have assumed that the failure times of the components are exponentially distributed. This assumption is justified on the basis of the fact that the exponential distribution is indeed a valid one for many situations and also because it leads to analytical models that are tractable and lead to minimum cost plans. Easterling et al. [1] and Rajgopal and Mazumdar [8] have also examined binomial and gamma failure time distributions, respectively; however, the models in these cases do not possess the same degree of analytical tractability. In this paper we examine under certain assumptions, another failure time distribution, namely the Weibull distribution. We consider this distribution because of its widespread use in reliability applications. It is also analytically more challenging than the exponential distribution and we focus on a series system.

Consider a system of n independent components in series. It is assumed that
- The failure time of component i is a random variable that follows the Weibull distribution with scale parameter b_i and shape parameter a_i.
- The value of b_i is unknown, but that of a_i is known.
- The interfaces between the components are perfect.
- Component failures are mutually independent.
- Without loss of generality, the mission time is equal to one unit of time so that the reliability of the system is given by

$$R_S = \prod_{i=1}^{n} \exp\left(-(1/b_i)^{a_i}\right) = \exp\left(\sum_{i=1}^{n} -(1/b_i)^{a_i}\right) \tag{3}$$

- Two numbers R_0 and R_1 ($0<R_0<R_1<1$) are specified such that the system is considered definitely unacceptable if $R_S \le R_0$ and definitely acceptable if $R_S \ge R_1$.
- Two small fractions α and β corresponding to the maximum acceptable levels of Type I and Type II error respectively are specified, with $\alpha+\beta<1$.

The assumption that a_i is known while b_i is unknown is made in order to make the problem tractable. It may also be justified on the basis of the fact that the parameter a_i determines whether the failure rate is increasing or decreasing (or constant if equal to 1) over time, and usually this characteristic is known for a component with a specific task. On the other hand the scale parameter b_i relates more to the magnitude of the failure rate and is more likely to be unknown. In order to analyze the above system we make use of a property of the Weibull distribution. Specifically, suppose that the time to failure of component i in the system is denoted by W_i and the quantity $(1/b_i)^{a_i}$ is denoted by λ_i (note that the value of λ_i is unknown). Then it follows that the random variable $T_i = W_i^{a_i}$. follows an exponential distribution with mean λ_i^{-1} and the system reliability is given by $R_S = \exp(-\Sigma_i \lambda_i)$.

We use the above property to develop a test plan that is an adaptation of the plan based on the so-called sum rule that has been used in our earlier work with exponentially distributed failure times [8]. With this rule the procedure was to (a) test each component i (with replacement of failed components) for t_i units of time, (b) observe the number of failures X_i observed for component i, (c) compute $X = X_1 + X_2 + ... + X_n$, the total number of failures observed, and (d) accept the system if X does not exceed some integer m, otherwise reject it. The plausibility of the sum rule for series systems has been discussed by Easterling et al. [1], and it has the significant feature of leading to tractable mathematical derivations.

In order to adapt the above test procedure to the case of the components having Weibull distributions we define a test procedure with time parameters t_i for each component i and an acceptance number m. The procedure is as follows: each component i in the system is tested with replacement of failed components and the times to failure are noted as W_{i1}, W_{i2}, W_{i3} etc., where W_{ik} represents the time to failure of the

k^{th} test unit of component i. The statistic $S_{ik} = \sum_{j=1}^{k} W_{ij}^{a_i}$ is computed for each failed unit k and testing of component i is halted as soon as S_{ik} is observed to exceed t_i for some value of k. The quantity X_i is then computed via

$$X_i = Max\left[k \mid \sum_{j=1}^{k} W_{ij}^{a_i} \le t_i \right] \qquad (4)$$

Since each $W_{ij}^{a_i}$ is exponential with mean λ_i^{-1} it follows that X_i is a Poisson random variable with mean $\lambda_i t_i$. (Note that X_i represents the number of "renewals" in t_i units of time when the renewal process is defined in terms of the random variables $W_i^{u_i}$; if the first test unit is still working at time t_i^{1/a_i} then $X_i = 0$.) The quantity $X = X_1 + X_2 + ... + X_n$ which is equal to the total number of renewals observed over all n components is now computed, and if X exceeds m the system is rejected, otherwise it is accepted.

We now address the constraints required to be met by the test plan. First we introduce some notation. Define $F_m(\lambda)$ as the distribution function of a Poisson random variable Y with mean λ, i.e., $F_m(\lambda) = Pr(Y \le m)$. Also, given $0 \le \gamma \le 1$, define $\phi_m(\gamma)$ as the mean of a Poisson random variable Y for which $Pr(Y \le m) = \gamma$, i.e.,

$$F_m\big(\phi_m(\gamma)\big) = \gamma = \exp\big(-\phi_m(\gamma)\big)\left[1 + \phi_m(\gamma) + \frac{\big(\phi_m(\gamma)\big)^2}{2!} + ... + \frac{\big(\phi_m(\gamma)\big)^m}{m!} \right]. \qquad (5)$$

Now consider the Type I error constraint given by (1). Based upon the test plan detailed above and the definition of λ_i this may be stated as

$$Pr\left\{ X \le m \mid \exp(-\sum_i \lambda_i) \ge R_1 \right\} \ge 1 - \alpha \qquad (6)$$

Since each X_i is Poisson with mean $\lambda_i t_i$ it follows that $X = \Sigma_i X_i$ is Poisson with mean $\Sigma_i \lambda_i t_i$. Using (5) we may rewrite (6) as

$$F_m(\Sigma_i \lambda_i t_i) \geq F_m(\phi_m(1-\alpha)) \quad \text{for } \{\lambda \in R^n \mid \exp(-\Sigma_i \lambda_i) \geq R_1, \lambda \geq 0\} \tag{7}$$

Since the Poisson distribution function $F_m(\gamma)$ is strictly decreasing in γ, (7) may be rewritten as

$$\Sigma_i \lambda_i t_i \leq \phi_m(1-\alpha) \text{ for } \{\lambda \in R^n \mid \Sigma_i \lambda_i \leq - \ln R_1, \lambda \geq 0\} \tag{8}$$

In summary (8) states that given the nonnegative parameters $t_1, t_2, ..., t_n$ and the integer m, in order to satisfy the constraint on Type I error the LHS of the inequality in (8) cannot exceed $\phi_m(1-\alpha)$ for <u>any</u> nonnegative vector $\lambda \in R^n$ which also satisfies $\Sigma_i \lambda_i \leq - \ln R_1$. In particular, the <u>maximum</u> value of this LHS across all such λ should be no higher than $\phi_m(1-\alpha)$. Thus constraint (2) reduces to solving the following linear programming subproblem in λ:

$$\text{Maximize } \Sigma_i \lambda_i t_i, \quad \text{st } \{\lambda \in R^n \mid \Sigma_i \lambda_i \leq -\ln R_1, \lambda \geq 0\} \tag{9}$$

and requiring the maximum to be less than or equal to $\phi_m(1-\alpha)$. Along identical lines the Type II error constraint given by (2) reduces to solving the following linear programming subproblem in λ:

$$\text{Minimize } \Sigma_i \lambda_i t_i, \quad \text{st } \{\lambda \in R^n \mid \Sigma_i \lambda_i \geq - \ln R_0, \lambda \geq 0\} \tag{10}$$

and requiring the minimum to be greater than or equal to $\phi_m(\beta)$.

The overall problem is to find nonnegative values of $t_1, t_2, ..., t_n$ and the integer m for which the above feasibility requirements are met and some economic criterion (such as the total test cost) is optimized. Before proceeding further we examine the solutions to subproblems (9) and (10) for given values of m and $t_1, t_2, ..., t_n$. Define

$$i_{max} = \{i \mid t_i = \text{Max}(t_1, t_2, ..., t_n)\} \text{ and } i_{min} = \{i \mid t_i = \text{Min}(t_1, t_2, ..., t_n)\} \tag{11}$$

Consider (9) and (10). It is clear that the optimum solution to (9) occurs at $\lambda_i = -\ln R_1$ for $i = i_{max}$ and $\lambda_i = 0$ for all $i \neq i_{max}$, while that of (10) occurs at $\lambda_i = -\ln R_0$ for $i = i_{min}$ and $\lambda_i = 0$ for all $i \neq i_{min}$. Corresponding objective values are $(-\ln R_1) t_{i_{max}}$ and $(-\ln R_0) t_{i_{min}}$ respectively.

Thus constraints (1) and (2) reduce to $(-\ln R_1) t_{i_{max}} \leq \phi_m(1-\alpha)$ and $(-\ln R_0) t_{i_{min}} \geq \phi_m(\beta)$

respectively, and hence via (11) to the following constraints in the plan parameters m and $t_1, t_2, ..., t_n$.

$$t_i \leq A(m) \quad \text{for } i=1,2,...,n \tag{12}$$
$$t_i \geq B(m) \quad \text{for } i=1,2,...,n, \tag{13}$$

where we define

$$A(m) = \phi_m(1-\alpha)/(-\ln R_1) \tag{14}$$
$$B(m) = \phi_m(\beta) / (-\ln R_0) \tag{15}$$

It is clear that for both (12) and (13) to be satisfied m should satisfy $B(m) \leq A(m)$. This in turn requires $\phi_m(\beta)/(-\ln R_0) \leq \phi_m(1-\alpha)/(-\ln R_1)$, or equivalently, $(-\ln R_1)/(-\ln R_0) \leq \phi_m(1-\alpha)/\phi_m(\beta)$. Since $R_1 > R_0$ the LHS of this last inequality is strictly less than 1, and it has been shown [13] that if $\alpha + \beta < 1$ as assumed, the ratio in the RHS is strictly increasing in m and approaches 1 as m approaches ∞. Therefore we can find $t_1, t_2, ..., t_n$ that satisfy (12) and (13) for all $m \geq m^*$ where m^* is defined as

$$m^* = \text{Min } \{ m |\ (-ln\ R_1)/(-ln\ R_0) \leq \phi_m(1-\alpha)/\phi_m(\beta)\ \} \tag{16}$$

Moreover, although t_i is not equal to the actual time on test for component i it is clear that its test time increases with t_i. Since it is reasonable to assume that test costs are proportional to test times, it follows that in general, one would like to choose t_i to be as low as possible. Thus from (13) the optimum value of t_i for each component will be the same and equal to $B(m)$. Moreover, $B(m)$ strictly increases with m so that the optimum value for m would be m^* as defined by (16). These results are consistent with those obtained for the exponential distribution: for a series system which is only as good as its weakest link, each component has the same optimum value for its test parameter (with the exponential distribution this was the actual time on test).

Unlike with the exponential distribution, there is no clear objective function that can be used to translate the above test times into costs for the test procedure that we have proposed. The total test cost is the sum of the costs for each component, where the cost associated with component i is the product of its time on test and the cost per unit time on test. Unfortunately, with the above analysis for the Weibull distribution, the time on test for component i is a random variable and while its value is related to t_i there is no closed form expression for deriving it. While it has been possible to obtain a well-defined set of parameters for the test plan, it is by no means clear that the test procedure satisfies any "optimum" properties according to well-established statistical criteria. How to combine information on component test data for the purpose of making inferences on system reliability appears to be a valid topic for statistical research.

4. Imperfect Interfaces

A criticism often leveled against component test plans is that it ignores failures that occur at the interfaces between components of the system, e.g., failures in the welding that attaches one component to another or failures in the wiring that connects two components of a system. Practitioners are often uncomfortable with plans that call only for component tests, and often prefer to supplement these with system tests as well. This combination of system and component tests in the context of imperfect interfaces has not received attention in the research literature and in this section we present some results relating to this issue. We restrict ourselves to the case where $a=1$ for all components so that they all have exponentially distributed times to failure and denote the mean for component i by λ_i^{-1}. Now consider a system where the interfaces are also failure prone; specifically, we assume that the failure time of the interfaces taken together is also exponentially distributed with unknown mean λ_I^{-1}. The system reliability here is now given by $R_S = \exp(-\lambda_I - \sum_i \lambda_i)$. The test plan has parameters $t_1, t_2, ...$ t_n, t_S and m, where component i is tested for t_i and the system for t_S units of time. The number of system failures (X_S) and the number of component failures (X_i) are observed and we use the following test procedure: "if $X = X_S + \sum_i X_i$ does not exceed m, the system is accepted, otherwise it is rejected". We use this test procedure mostly for its mathematical advantage. The Type I and Type II error constraints for this test plan may be stated as

$$\Pr\left\{X \le m \,\middle|\, \exp(-\lambda_S - \sum_i \lambda_i) \ge R_1\right\} \ge 1 - \alpha \tag{17}$$

$$\Pr\left\{X \le m \,\middle|\, \exp(-\lambda_S - \sum_i \lambda_i) \le R_0\right\} \ge \beta \tag{18}$$

To develop these, we first note that X is a Poisson random variable with a mean given by $t_S(\lambda_I + \sum_i \lambda_i) + \sum_i \lambda_i t_i = t_S \lambda_I + \sum_i (t_S + t_i)\lambda_i$. Using logic identical to that in Section 3, for a given vector of times, (17) and (18) are equivalent to

$$\left\{\text{Maximum } t_S\lambda_I + \sum_j(t_S + t_i)\lambda_i, \quad \text{st } \{\lambda \in R^{n+1} \mid \lambda_I + \sum_i \lambda_i \le -\ln R_1, \ \lambda \ge 0\}\right\} \le \phi_m(1-\alpha) \tag{19}$$

$$\left\{\text{Minimum } t_S\lambda_I + \sum_j(t_S + t_i)\lambda_i, \quad \text{st } \{\lambda \in R^{n+1} \mid \lambda_I + \sum_i \lambda_i \ge -\ln R_0, \ \lambda \ge 0\}\right\} \ge \phi_m(\beta) \tag{20}$$

and these in turn are equivalent to

$$t_i + t_S \le A(m) \quad \text{for } i = 1, 2, \ldots, n \tag{21}$$

$$t_S \ge B(m) \tag{22}$$

where $A(m)$ and $B(m)$ are given by (14) and (15).

Once again, (21) and (22) are feasible for all $m \ge m^*$, where m^* is given by (16) and it is clear that for feasible m, if test costs are positive and proportional to test times then $t_S = B(m)$, and $t_i = 0$ for all i. The optimum value of m is given by m^*. Thus for a system with imperfect interfaces the optimum strategy is to test only the system (and not the components) for $B(m^*)$ units of time where m^* is given by (16) and accept the system as long as the number of system failures observed is no higher than m^*.

The fact that with imperfect interfaces it is optimal to test only the system is not surprising. This is because in a series system the interfaces are just as important as each of the individual components. Since the interfaces cannot be tested individually, the only option is a system test, which of course obviates the need for individual component tests since the components are implicitly tested as part of the larger system. However, this assumes that one has no knowledge about the *relative* reliability of the interfaces as compared to the components. In practice, this need not be the case, and in general, one would expect the interfaces to be much more reliable than the individual components. Thus one may have an upper bound on the failure rate of the interfaces; this may be in the form of an absolute value or as a function of the individual component failure rates. In such situations, the constraint space of the subproblems defining the LHS of (19) and (20) are different and the solutions to these will not necessarily lead to simple constraints such as (21) and (22). Furthermore, the actual optimum test times will depend on the relative magnitudes of the system and component test costs. Such situations are currently being studied by the authors and some preliminary results may be found in reference [14].

References

1. R. G. Easterling, M. Mazumdar, F. W. Spencer and K. V. Diegert, "System Based Component Test Plans and Operating Characteristics: Binomial Data," *Technometrics*, **33**, 287-298, (1990).
2. S. Gal, "Optimal Test Design for Reliability Demonstration," *Operations Research*, **22**, (1974).
3. M. Mazumdar, "An Optimum Procedure for Component Testing in the Demonstration of Series System Reliability," *IEEE Transactions on Reliability*, **R-26**, 342-345, (1977).
4. M. Mazumdar, "An Optimum Procedure for a Series System with Redundant Subsystems," *Technometrics*, **22**, 23-27, (1980).
5. J. H. Yan and M. Mazumdar, "A Comparison of Several Component Testing Plans for a Series System," *IEEE Transactions on Reliability*, **R-35**, 437-443, (1986).
6. J. H. Yan and M. Mazumdar, "A Comparison of Several Component Testing Plans for a Parallel System," *IEEE Transactions on Reliability*, **R-36**, 419-424, (1987).
7. J. Rajgopal and M. Mazumdar, "A Type-II Censored, Log Test-Time Based Component Testing Procedure for a Parallel System," *IEEE Transactions on Reliability*, **37**, 406-412, (1988).
8. J. Rajgopal and M. Mazumdar, "Designing Component Test Plans for Series System Reliability via Mathematical Programming," *Technometrics* **37** (2), 195-212, (1995).
9. J. Rajgopal and M. Mazumdar, "A System Based Component Test Plan for a Series System, with Type-II Censoring," *IEEE Transactions on Reliability*, **45** (3), 375-378, (1996).
10. K. I. Altinel, "The Design of Optimum Component Test Plans in the Demonstration of a Series System Reliability," *Computational Statistics and Data Analysis* **14** (3), 281-292, (1992).
11. K. I. Altinel, "The Design of Optimum Component Test Plans in the Demonstration of System Reliability," *European Journal of Operational Research*, **78**, 318-333, (1994).
12. MIL-HDBK-781D, Reliability Test Methods, Plans, and Environments for Engineering Development, Qualification, and Production; Department of the Navy, Space and Naval Warfare Systems Command, Washington DC, (1974).
13. J. Rajgopal, M. Mazumdar and T. H. Savits, "Some Properties of the Poisson Distribution with an Application to Reliability Testing," *Probability in the Engineering and Informational Sciences*, **8**, 345-354, (1994).
14. J. Rajgopal, M. Mazumdar and S. V. Majety, "Optimum System-Based Component Test Plan in the Presence of Imperfect Interfaces," Technical Report No. TR 97-3, Department of Industrial Engineering, University of Pittsburgh, Pittsburgh, PA, (1997).

Parameter Estimation in Software Reliability Growth Models

LAWRENCE D. RIES

Department of Statistics, University of Missouri, Columbia, MO 65211
E-mail: ries@stat.missouri.edu

ASIT P. BASU

Department of Statistics, University of Missouri, Columbia, MO 65211
E-mail: basu@stat.missouri.edu

Jelinsky and Moranda (JM, 1972) proposed the first model for relating software inter-failure times. They considered the interfailure times to be exponential variables with intensity proportional to the number of remaining faults in the system. In reaction to the limitations of the JM model, Littlewood and Verrall (LV, 1973) proposed a Bayesian reliability growth model and Goel and Okumoto (GO, 1979) proposed a nonhomogeneous Poisson process model. Langberg and Singpurwalla (1985) showed that these three models could be united under a Bayesian framework and also suggested an extension which we term the LS model. In this paper, we describe these four models and discuss their limitations. We consider parameter estimation in all four models, and in the JM, we suggest some possible alternatives to maximum likelihood estimation.

1   Introduction

The goal of software reliability research is to develop methods for producing software which is more reliable and of higher quality. Toward this goal, many mathematical models for explaining the way software errors are discovered and corrected have been proposed. The market for computer software is quite diverse, and we should not be surprised to find that different models might be better suited for different situations.

Software modeling is undertaken in order to gain information about the state of the software program. Before developing a model, one must consider what information can be derived from the model fitting exercise. General questions include:

- How many (and what kind of) faults currently exist in the software?

- When have we tested the software sufficiently, so that it is "good enough" to release?

- How much faith can we put in our answers to the above?

Any model one chooses to use should be able to answer at least some of the questions of interest to the software developer. Ideally, a model will also reflect several other characteristics, such as:

- It should reasonably reflect reality.

- It should be as simple as possible.

- It should not be excessively difficult or expensive to implement.

Many mathematical models have been proposed for modeling the reliability of computer software, the first of which was by Jelinsky and Moranda (1972). We will review four models, each of which may be viewed, in some sense, as a modification of the Jelinsky-Moranda (JM) model.

2 Specific Models Used in Software Reliability

2.1 Jelinsky and Moranda (JM) Model

The first software reliability model was proposed by Jelinsky and Moranda (1972). It is quite simple conceptually, and has led to several extensions and modifications.

Assumptions

1. The software initially contains an unknown number of faults, N.

2. A detected fault is removed instantaneously, and is corrected without introducing any new faults.

3. Times between failures are independent, exponentially distributed random quantities.

4. All remaining faults contribute the same amount to the software failure intensity.

The failure rate (hazard function) of a software program is taken to be proportional to the number of faults that remain in the code. Specifically, the model implies

$$r(t_i|N,\Lambda) = \Lambda(N - i + 1), i = 1, 2, \ldots, N, \tag{1}$$

where N is the total number of faults in the software at the start of testing, T_i is the ith interfailure time (the time between the $[i-1]$st and ith failures of the software), and Λ is a constant of proportionality. An implication of this is that the ith interfailure time has an exponential density,

$$f(t_i|N,\Lambda) = \Lambda(N - i + 1)\exp(-\Lambda[N - i + 1]t_i).$$

Since the JM model was such an early attempt at formulating a software reliability model, it has naturally met with considerable criticism. Despite this, subsequent modifications and extensions suggest that, at the very least, the model was a good starting point. It continues to receive considerable attention.

Criticism

Exponential distribution. Littlewood (1981) notes that computer scientists often criticize the "memoryless" property of the exponential distribution. Software system failures are thought to be influenced by past events to a greater extent than hardware systems. However, he points out that a distribution more representative of real-life has not been, and may never be, presented.

Equally important faults. Treating all faults as having an equal impact on program reliability seems unrealistic. Faults range from cosmetic (poor choice of screen colors, for example) to "show stoppers" (program locks up). Faults are also less important if they are in little used portions of the code. If a fault causes the program to perform a little-used function in a clumsy, inefficient manner, that may not cause a serious problem. However, if the function is frequently used, the usefulness of the program is more severely hampered.

Perfect repair. Ideally, when a programmer attempts to remove a fault in the program code he will correct the faulty section of code without damaging other sections of code. The assumptions of the JM model require this ideal situation. This may not be an accurate reflection of real life.

Instantaneous repair. Obviously, it is not possible for a programmer to correct a faulty segment of code in zero time. However, if the "clock" is turned off while the repair is taking place, then the data we collect will appear as if there were zero correction time. However, if one considers the economic costs involved with software release then the time needed to correct errors is a factor that should not be ignored.

Alternative formulation

In (1) it is assumed that data collection involves observing n interfailure times. Such a design is not realistic in practice since it requires that one knows a priori that the chosen value of $n \leq N$. A more realistic experiment would be to observe the number of failures that occur in a fixed amount of total time T^*. In this case n becomes a random variable.

In such a situation, we may view the JM model as follows. Consider a population of N software bugs. The times until failure $y_1, y_2, \ldots, y_N$ are iid exponential random variables with mean Λ^{-1}. We observe $\{y_i : y_i \leq T^*, 1 \leq i \leq N\}$. If n number of y's are observed, then the order statistics represent the failure times $y_{(1)} = s_1, y_{(2)} = s_2, \ldots, y_{(n)} = s_n$ and the interfailure times are easily obtained as $t_1 = s_1, t_2 = s_2 - s_1, \ldots, t_n = s_n - s_{n-1}$.

Joe and Reid (1985b) derive the likelihood under this experimental design. If T^* (fixed total testing time, with a random number n of failures occurring) is replaced by $\sum_{i=1}^{n} t_i$ (random total testing time, with fixed n number of failures), the resulting likelihood equations are equivalent.

2.2 Littlewood and Verrall (LV) Model, 1973

Littlewood and Verrall (1973) argue that reliability should not be viewed in terms of fault counting. They model the inter-failure times as exponential random variables with scale Λ_i. Furthermore, they view the problem from a Bayesian perspective, and take a gamma prior for Λ_i with shape parameter α and scale parameter $\psi(i)$, for some function ψ. Specifically,

$$f(t_i | \Lambda_i) = \Lambda_i \exp(-\Lambda_i t_i)$$

and

$$\pi(\Lambda_i | \alpha, \psi(i)) = \frac{[\psi(i)]^{\alpha-1} \Lambda_i^{\alpha-1} \exp(-\psi(i)\Lambda_i)}{\Gamma(\alpha)}.$$

The program hazard rate becomes

$$r(t_i | \alpha, \psi(i)) = \frac{\alpha}{\psi(i) + t_i}.$$

$\psi(i)$ is supposed to describe the quality of the programmer and the difficulty of the programming task. It may be, for example, a linear function of the number of discovered faults, i.e. $\psi(i) = \beta_0 + \beta_1 i$.

Note that the density of T_i is exponential, conditional on Λ_i. We may obtain the unconditional density of T_i (which is actually conditional on the hyperparameters of Λ_i) by noting

$$f(t_i | \alpha, \psi(i)) = \int_0^\infty f(t_i | \Lambda_i) \pi(\Lambda_i) d\Lambda_i = \alpha \left(\frac{\psi(i)}{t_i + \psi(i)} \right)^\alpha \frac{1}{t_i + \psi(i)} \qquad (2)$$

which is a type II Pareto (Lomax) density. (See Johnson, Kotz, and Balakrishnan, 1994, page 575.) The form of the Pareto-II is

$$f(x|a, c) = \frac{ac^a}{(x+c)^{(a+1)}},$$

with $a > 0$, $c > 0$, and $x > 0$. Thus in (2) we have $a = \alpha$ and $c = \psi(i)$.

The LV model shares assumptions 2 (perfect repair) and 3 (exponential interfailure times) with the JM model. Other implicit assumptions depend on the analytical form of $\psi(i)$ used. If $\psi(i) = \beta_0 + \beta_1 i$ is considered, then it must be assumed that the software program contains an infinite number of errors. If, on the other hand, $\psi(i) = \beta(N-i+1)$, then it must be assumed that the number of faults, N, is finite. The effect of remaining faults on the failure intensity also depends on the form of $\psi(i)$.

2.3 Goel and Okumoto (GO) Model, 1979

In contrast to the previous models, Goel and Okumoto (1979) chose to model the number of failures experienced up until time s instead of the time until the occurrence of the next failure. We advise the reader that we are reserving the letters t and T (with appropriate subscripts) to refer to *interfailure* times. When an actual failure time (since time zero) is indicated, we will use the letters s or S (with and without subscripts). Deviations from this notation will be made explicit.

The GO model treats the occurrence of failures as a nonhomogeneous Poisson process (NHPP) mean value function

$$m(s) = a[1 - \exp(-bs)], \ a > 0, \ b > 0,$$

where a represents the total number of expected bugs in the system and b is the fault detection rate.

Note that $m(0) = 0$ (it is expected that no bugs will be discovered before testing begins) and $\lim_{s \to \infty} m(s) = a$ (it is expected that a bugs will be discovered after an infinite amount of testing).

Assumptions

1. The cumulative number of faults detected at time s follows a Poisson distribution.

2. All faults are independent and have the same chance of being detected.

3. All detected faults are removed instantaneously and no new faults are introduced.

4. $m(0) = 0$ and $\lim_{s \to \infty} = a$.

An NHPP may also be defined by its intensity function, which for the GO model is given by

$$\lambda(s) = \frac{dm(s)}{ds} = abe^{-bs}.$$

Goel and Okumoto fit their model to several real data sets. Some of the results are presented in their 1979 paper. They comment that the model seems to give a good fit for other data sets. In the results presented, they compare their model's ability to predict the time of the next software failure with the JM model's ability to do the same. The GO model appears to perform better than the JM model, but the results are not conclusive. It appears that the GO model provides a more conservative prediction of T_{n+1} than the JM model.

2.4 Model Unification

Using a Bayesian framework, Langberg and Singpurwalla (1985) showed that the JM, LV, and GO models could be viewed as the same model. The original JM model proposed an exponential pdf for the ith time between failures,

$$f(t_i | \Lambda, N) = \Lambda(N - i + 1) \exp(-\Lambda(N - i + 1)t_i),$$

where Λ and N were considered fixed, but unknown constants. The three models mentioned above may be obtained by taking a Bayesian approach and assigning prior probability distributions to either Λ or N or both. If a Poisson prior is taken for N ($N \sim \text{poi}(\theta)$) and a degenerate prior for Λ, then the result is the GO model. If a gamma prior is taken for Λ ($\Lambda \sim \text{gam}(\alpha, \psi)$) and N is considered degenerate, then the result is the LV Model. Langberg and Singpurwalla also consider taking prior distributions for both N and Λ (the same priors as above) to obtain a new model (LS). Their work assumes a gamma prior for Λ, but allows for N to follow any discrete prior distribution. A logical example would be to consider $N^* = N - n \sim \text{poi}(\theta)$, where N^* represents the number of errors remaining in the software after n errors have been discovered and removed.

Ries (1995) showed how parameter estimation in the LS model may be accomplished using Gibbs Sampling. He also compared the JM, LV, GO, and LS models using a pre-quential likelihood method proposed by Dawid (1984). He concluded that no model is uniformly superior, and the best model depends on the rate of reliability growth exhibited by the data.

3   Inferences in Software Reliability Models

3.1   Jelinsky-Moranda Model

The JM model proposes that the program hazard rate is proportional to the number of faults remaining in the code. If the program initially contains N faults and testing proceeds until n faults are discovered (i.e., have manifested themselves as program failures) and corrected, then the interfailure times, $T_1, T_2, \ldots, T_n$, will be independent exponential random variables with $T_i \sim e(\Lambda[N - i + 1])$. Note that $E(T_i) = \Lambda^{-1}[N - i + 1]^{-1}$.

Maximum likelihood estimation

Fitting the model may be accomplished by estimating the parameters N and Λ. This may be accomplished through maximum likelihood estimation. The density function of the ith interfailure time is

$$f(t_i|\Lambda, N) = \Lambda(N - i + 1)\exp(-\Lambda[N - i + 1]t_i), t_i \geq 0,$$

so the likelihood function is

$$L(N, \Lambda|t_1, t_2, \ldots, t_n) = \prod_{i=1}^{n} f(t_i|\Lambda, N) = \prod_{i=1}^{n} \Lambda(N - i + 1)\exp(-\Lambda[N - i + 1]t_i). \quad (3)$$

As is frequently the case, it is more convenient to maximize the natural logarithm of the likelihood function,

$$\log[L(N, \Lambda|t_1, t_2, \ldots, t_n)] = n\log(\Lambda) + \sum_{i=1}^{n}\log(N - i + 1) - \Lambda\sum_{i=1}^{n}(N - i + 1)t_i.$$

Taking partial derivatives, we obtain two equations which the MLEs must satisfy,

$$\partial\log(L)/\partial N = \sum_{i=1}^{n}(N - i + 1)^{-1} - \Lambda\sum_{i=1}^{n}t_i = 0$$

and

$$\partial\log(L)/\partial\Lambda = n/\Lambda - \sum_{i=1}^{n}(N - i + 1)t_i = 0.$$

Defining $T = \sum_{i=1}^{n} t_i$ to be the total amount of time spent in testing $W = \sum_{i=1}^{n}(i - 1)t_i$ to be a weighted sum of the interfailure times, and solving the second equation for Λ, we obtain the MLE of Λ in terms of the MLE of N,

$$\hat{\Lambda} = \frac{n}{\hat{N}T - W}. \quad (4)$$

Further substituting the equation for $\hat{\Lambda}$ into the first equation produces

$$\sum_{i=1}^{n}(N - i + 1)^{-1} - \frac{nT}{NT - W} = 0,$$

which may be re-written as

$$\frac{1}{N} + \frac{1}{N-1} + \cdots + \frac{1}{N-n+1} - \frac{n}{N-W/T} = 0. \tag{5}$$

The equation is nonlinear and must be solved numerically. Once $\hat{N}$, the solution of (5), has been calculated, this value may be substituted into (4) to obtain $\hat{\Lambda}$.

One should note that this approach for determining MLEs treats N as a continuous parameter. Naturally, N must be a finite, nonnegative integer. If consideration is restricted to integer estimates for N, then uniqueness may be lost.

While MLEs are very familiar to statisticians, in the JM model they suffer from some severe problems. In particular, the MLE of N may be infinite with substantial probability (Joe and Reid, 1985a). Several authors have investigated problems with MLEs in the JM or related models including Littlewood (1981), Littlewood and Verrall (1981), Goudie and Goldie (1981), and Joe and Reid (1985a,1985b).

Least squares estimation

The method of least squares prescribes obtaining parameter estimates by minimizing the sum of the squared errors between each of the data points and their respective expected values. In the case of the JM model, estimates of N and Λ are chosen so as to minimize

$$Q(N,\Lambda) = \sum_{i=1}^{n} \left(t_i - \frac{1}{\Lambda[N-i+1]} \right)^2.$$

One may obtain the respective minimizers by taking partial derivatives with respect to N and Λ, equating the resulting equations to zero and solving. Differentiating with respect to N yields

$$\frac{\partial Q}{\partial N} = 2 \sum_{i=1}^{n} (t_i \Lambda^{-1}[N-i+1]^{-2} - \Lambda^{-2}[N-i+1]^{-3}) = 0$$

which implies

$$\Lambda = \frac{\sum_{i=1}^{n}(N-i+1)^{-3}}{\sum_{i=1}^{n} t_i(N-i+1)^{-2}}. \tag{6}$$

Similarly, differentiating with respect to Λ produces

$$\frac{\partial Q}{\partial \Lambda} = 2 \sum_{i=1}^{n} (t_i \Lambda^{-2}[N-i+1]^{-1} - \Lambda^{-3}[N-i+1]^{-2}) = 0$$

which implies

$$\Lambda = \frac{\sum_{i=1}^{n}(N-i+1)^{-2}}{\sum_{i=1}^{n} t_i(N-i+1)^{-1}}. \tag{7}$$

By subtracting (7) from (6) we obtain

$$\frac{\sum_{i=1}^{n}(N-i+1)^{-3}}{\sum_{i=1}^{n}t_i(N-i+1)^{-2}} - \frac{\sum_{i=1}^{n}(N-i+1)^{-2}}{\sum_{i=1}^{n}t_i(N-i+1)^{-1}} = 0. \qquad (8)$$

This is a nonlinear equation of a single variable. Solving (8) will produce the non-linear least squares estimate of N and substituting this value into either (6) or (7) will produce the non-linear least squares estimate of Λ.

Joe and Reid (1985) suggest the point estimation of N is not worthwhile, as all reasonable estimators are liable to produce unreasonable estimates. They suggest interval estimation, and suggest using a likelihood interval for this purpose.

Likelihood interval estimation

A *c-fraction likelihood interval for N* is defined to be the set of values for N for which the ratio of the likelihood function at N to the likelihood function at $\hat{N}$ is greater than or equal to a given constant $c \in (0, 1]$. The likelihood interval consists of two endpoints

$$N_1 = \inf\{N \geq n : L(N)/L(\hat{N}) \geq c\}$$

and

$$N_2 = \sup\{N \geq n : L(N)/L(\hat{N}) \geq c\}.$$

Note that both N_1 and N_2 depend on c and the data, $t_1, t_2, \ldots, t_n$, although this dependence is suppressed in the notation.

Joe and Reid (1985a) calculate $P(N_1 \leq N \leq N_2)$ for several cases. They also suggest using the harmonic mean of N_1 and N_2 as a point estimate for N. They note that their estimator,

$$N_{LIE} = \left(\frac{N_1^{-1} + N_2^{-1}}{2}\right)^{-1},$$

does not over-estimate N to the extent that the MLE does. Ries (1995) has done simulation work comparing the MLE, likelihood interval estimator (LIE), and non-linear least squares (NLS) estimators.

3.2 The Littlewood-Verrall Model

If the observed data consists of event times $\underline{t} = (t_1, t_2, \ldots, t_n)$, and we take $\psi(i) = \beta_0 + \beta_1 i$ (which Littlewood [1981] suggests is a good choice), then (2) produces

$$f(\underline{t}|\alpha, \beta_0, \beta_1) = \alpha^n \left(\prod_{i=1}^{n} \frac{\beta_0 + \beta_1 i}{t_i + \beta_0 + \beta_1 i}\right)^{\alpha} \prod_{i=1}^{n} \frac{1}{t_i + \beta_0 + \beta_1 i} = \text{lik}(\alpha, \beta_0, \beta_1|\underline{t}). \qquad (9)$$

Differentiating log(lik) with respect to each parameter and equating to zero produces three equations. The maximum likelihood estimates $\hat{\alpha}, \hat{\beta}_0$, and $\hat{\beta}_1$ satisfy

$$n + \sum_{i=1}^{n} \alpha[\log(\beta_0 + \beta_1 i) - \log(\beta_0 + \beta_1 i + t_i)] = 0,$$

$$\sum_{i=1}^{n} \left[\frac{\alpha}{\beta_0 + \beta_1 i} - \frac{\alpha + 1}{\beta_0 + \beta_1 i + t_i} \right] = 0,$$

and

$$\sum_{i=1}^{n} \left[\frac{\alpha i}{\beta_0 + \beta_1 i} - \frac{(\alpha + 1)i}{\beta_0 + \beta_1 i + t_i} \right] = 0.$$

These equations must be solved through an iterative procedure.

It is not readily evident if either existence or uniqueness of the MLEs of the LV model is guaranteed. Model fitting using *IMSL Math and Stat Library* subroutines to perform nonlinear optimization occasionally produces errors, suggesting possible non-convergence. (In all cases the author has observed, adjusting the starting values of the procedure caused the routine to terminate normally.)

Littlewood and Verrall (1973) consider a parameter estimation procedure based on probability integral transforms. This method is simpler to program and has fewer convergence problems than the MLE computer algorithms. The procedure involves optimizing a goodness of fit statistic between the (parametrically) estimated and empirical distribution functions. While any number of goodness of fit statistics could be optimized, Littlewood and Verrall (1973) suggest the nW^2 statistic (Kendall and Stuart, 1973) is a good choice.

3.3   The Goel-Okumoto Model

In the GO model, if we consider the actual failure times $\underline{s} = (s_1, s_2, \ldots, s_n)$, then the joint density may be written as

$$f(\underline{s}|a, b) = \exp[-m(s_n)] \prod_{i=1}^{n} ab \exp(-bs_i),$$

where $m(s) = a[1 - \exp(-bs)]$ is the mean number of failures expected in the interval $(0, s]$. Rewriting the equation, we have

$$f(\underline{s}|a, b) = \exp\left(a(1 - e^{-bs_n})\right) \prod_{i=1}^{n} ab \exp(-bs_i) = \text{lik}(a, b|\underline{s}). \tag{10}$$

The usual process of taking derivatives of $\text{lik}(a, b|\underline{s})$ and equating to zero implies that the MLEs of a and b must satisfy

$$n/a = 1 - \exp(-bs_n)$$

and

$$n/b = \sum_{i=1}^{n} s_i + as_n \exp(-bs_n).$$

3.4   The Langberg-Singpurwalla Model

We have addressed two experimental designs for software testing. The software may be tested for a fixed amount of time T, and the random number of failures n is observed. The other possibility is to fix n in which case T becomes random. We have concentrated on the latter case (although the analysis is similar for both), and as a result it is necessary to assume *a priori* that $n \leq N$. When employing a Bayesian approach for this case it is more convenient to assign a prior distribution to the number of faults remaining in the software $N^* = N - n$ than to N itself.

Langberg and Singpurwalla (1985) show that by attaching prior distributions for N (or N^*) and Λ, the LV and GO models can be viewed as generalizations of the JM model. Specifically, if Λ is taken to be degenerate and N^* is taken to be Poisson[a], then the GO model is indicated. Similarly, if Λ is taken to be gamma and N is taken to be degenerate, then the LV model can be derived. Langberg and Singpurwalla discuss assigning prior distributions to both parameters. They derive expressions for the joint and marginal posterior distributions of N^* and Λ when a gamma prior is taken for Λ, and N^* may follow any discrete prior distribution which is independent of Λ.

In the latter case, Langberg and Singpurwalla do not consider any specific prior distribution for N^*. A logical combination of the LV and GO models suggests that the combined case utilizes a Poisson prior for N^*. We consider that here and will reference this model as the "LS model."

For clarity, the LS model assumes

$$f(\underline{t}|\Lambda, N^*) = \Lambda^n \frac{(N^* + n)!}{N^*!} \exp(-\Lambda[(N^* + n)T - W]).$$

The prior distributions are

$$\pi(\Lambda|a, b) = \text{gam}(a, b) = \frac{b^a}{\Gamma(a)} \Lambda^{a-1} \exp(-b\Lambda), \ \Lambda \geq 0$$

and

$$\pi(N^*|\theta) = \text{poi}(\theta) = \frac{\theta^{N^*} \exp(-\theta)}{N!}, \ N^* \in \{0, 1, \ldots\}.$$

It follows that the joint density of $(\underline{t}, \Lambda, N^*)$ is given by

$$
\begin{aligned}
p(\underline{t}, \Lambda, N^*|a, b, \theta) &= f(\underline{t}|\Lambda, N^*) \times \pi(\Lambda|a, b) \times \pi(N^*|\theta) \\
&= \Lambda^n \frac{(N^* + n)!}{N^*!} \exp(-\Lambda[(N^* + n)T - W]) \\
&\quad \times \frac{b^a}{\Gamma(a)} \Lambda^{a-1} \exp(-b\Lambda) \times \frac{\theta^{N^*} \exp(-\theta)}{N^*!}.
\end{aligned}
$$

---

[a]Langberg and Singpurwalla develop their results based on an arbitrary discrete prior for N, rather than N^*. Because their results arise from the posterior distributions which are conditional on n, the effect is the same.

In principle, the marginal distribution of the data (which is equivalent to the likelihood function of the hyperparameters) may be obtained from

$$f(\underline{t}|a, b, \theta) = \sum_{N^*=0}^{\infty} \int_{\Lambda=0}^{\infty} p(\underline{t}, \Lambda, N^*|a, b, \theta)d\Lambda = \text{lik}(a, b, \theta|\underline{t}). \qquad (11)$$

Model fitting involves estimation of the hyperparameters a, b, and θ. This may be accomplished by maximizing the above marginal distribution with respect to a, b, and θ. Unfortunately, a closed form expression for $\text{lik}(a, b, \theta|\underline{t})$ is not available. Instead of direct maximization, the Gibbs sampler (Geman and Geman [1984]) may be used to obtain the estimates of a, b, and θ. This is discussed in Ries (1995).

The marginal density for the LS model given by (11) may be re-written as

$$p(t_i|a, b, \theta) = \sum_{N^*=0}^{\infty} (N^* + 1)\frac{b^a}{\Gamma(a)}\frac{e^{-\theta}\theta^{N^*}}{N^*!} \int_{\Lambda} \Lambda^a \exp\left(-\Lambda\left[(N^* + 1)t_i + b\right]\right) d\Lambda.$$

The integrand with respect to Λ is proportional to a $\text{gam}(a + 1, [N^* + 1]t_i + b)$ density, which gives

$$\int_{\Lambda} \Lambda^a \exp\left(-\Lambda\left[(N^* + 1)t_i + b\right]\right) d\Lambda = \frac{\Gamma(a + 1)}{([N^* + 1]t_i + b)^{a+1}}.$$

The marginal density may be re-written again as

$$p(t_i|a, b, \theta) = ab^a e^{-\theta} \sum_{N^*=0}^{\infty} \frac{(N^* + 1)\theta^{N^*}}{N^*!} \left(\frac{1}{[N^* + 1]t_i + b}\right)^{a+1}. \qquad (12)$$

An estimate of $p(t_i|a, b, \theta)$ may be obtained by first estimating a, b, and θ using the Gibbs sampler and then numerically approximating the summation. Moments of the marginal distribution may be obtained by numerically approximating the integral over t_i. In particular, the expected value of t_i may be estimated by

$$E(t_i|a, b, \theta) = \int_{t_i} t_i p(t_i|a, b, \theta)dt_i,$$

where $p(t_i|a, b, \theta)$ is obtained from a numerical evaluation of (12).

It is important to realize that density and moment estimation for the LS model represents a substantial investment in computer time. However, because a closed form solution does not exist for the marginal density (12), a numerical approximation is the only method possible.

4 An Example

John Musa (1978) of AT&T Bell Labs collected several software debugging datasets from programming projects for which he served as supervisor. One of his datasets (which we term *Musa1*) consists of $n = 136$ software interfailure times of a real-time command and control system for in-house use by AT&T. Using Fortran subroutines which we developed, we are able to obtain parameter estimates for each of the four models we have discussed for this dataset.

For the JM model, maximum likelihood estimation produces $\hat{N} = 141.9029$ and $\hat{\Lambda} = 3.4967 \times 10^{-5}$. For the LV model, using Littlewood and Verrall's alternative estimation procedure, we obtain $\hat{\alpha} = 6.2595$, $\hat{\beta}_0 = -22.8616$, and $\hat{\beta}_1 = 48.6758$. The MLEs for the GO model are $\hat{a} = 142.8809$ and $\hat{b} = 3.4204 \times 10^{-5}$. Finally, the hyperparameter estimates for the LS model are $\hat{\theta} = 5.8830$, $\hat{a} = 16.0020$, and $\hat{b} = 513,233.3283$. (The LS estimates were obtained by using $m = 30$ and $k = 10$ in the Gibbs sampler.)

Once parameter estimates have been obtained, they may be used to estimate a characteristic of interest. For example, to estimate $E(T_i)$ we would express the expectation in terms of the parameters and estimate it by substituting the appropriate parameter estimates. Expressions for $\hat{E}(T_i)$ are derived below.

For the JM model, the ith interfailure time follows an exponential distribution with mean $(\Lambda[N - i + 1])^{-1}$. Accordingly, we would predict T_i by replacing the parameters in the mean formula with their MLE's to obtain

$$\hat{E}_{JM}(T_i) = (\hat{\Lambda}[\hat{N} - i + 1])^{-1}. \tag{13}$$

Similarly, predicted values for the LV model may be obtained by utilizing the fact that, under that model, T_i follows a Pareto-II distribution. The density of the Pareto-II is of the form

$$f(x|a, c) = \frac{ac^a}{(x + c)^{(a+1)}}.$$

Its mean is given by

$$E(X) = c/(a - 1).$$

For the LV model $a = \alpha$ and $c = \psi(i) = \beta_0 + \beta_1 i$. Using estimates of α, β_0, and β_1, the usual substitution procedure produces a predicted value of

$$\hat{E}_{LV}(T_i) = \frac{\hat{\beta}_0 + \hat{\beta}_1 i}{\hat{\alpha} - 1}. \tag{14}$$

The GO model treats the cumulative number of software failures that will occur after a total of s units of testing as a nonhomogeneous Poisson process. The distribution of the ith interfailure time T_i for an NHPP is not of a tractable form. It is possible to estimate $E(T_i)$ using a numerical approximation procedure. A more appealing solution, which we consider, is to predict the ith interfailure time as the difference between predictions of the ith and $(i - 1)$st failure times, i.e.

$$\hat{t}_i = \hat{s}_i - \hat{s}_{i-1}.$$

The mean function considered by the GO model is

$$m(s) = a\left[1 - \exp(-bs)\right].$$

It represents the expected number of failures that will occur after s units of total test time. Inverting the function produces the expected amount of time needed to observe i failures,

$$s = m^{-1}(i) = -\frac{1}{b}\log\left(1 - \frac{i}{a}\right).$$

In this manner a prediction of s_i for any i is readily obtainable. Our prediction of the ith interfailure time becomes

$$\hat{E}_{GO}(T_i) = -\frac{1}{b}\left[\log\left(1 - \frac{i}{a}\right) + \log\left(1 - \frac{i-1}{a}\right)\right]. \tag{15}$$

Predicting interfailure times using the LS model is a significant challenge. First, our parameter estimation procedure requires the use of the Gibbs sampler and is computationally intensive. Second, the marginal density of T_i does not have a closed form solution, and must be evaluated numerically. Finally, calculating $\hat{E}_{LS}(T_i)$ requires a numerical integration. Since the integrand must be numerically evaluated, the complexity of the problem increases geometrically.

Using the previously obtained parameter estimates, we may estimate $E(T_{136})$, the time between the 135th and 136th failure of the software, as: 4,143 (under the JM); 1,254 (under the LV); 3,967 (under the GO); and 4,119 (under the LS). The actual observed value in *Musa1* was $t_{136} = 4,116$.

5   Discussion

The JM model was the first model proposed for modeling software interfailure times, and remains to some extent central in the field in that a vastly superior model has not yet been presented. The LV and GO models, while developed independently, can be viewed as extensions of the JM. The LS model is a third, fully Bayesian, extension of the JM. We have discussed parameter estimation procedures for each of these models. The utility of any model lies in its ability to make inferences. Comparisons of these four models are available in Ries (1995).

6   Acknowledgements

This work is supported by the United States Air Force Office of Scientific Research under grant number F49620-97-1-0213. The authors gratefully acknowledge this support.

References

1. Sanjib Basu. Bayesian estimation of the number of undetected errors when both reviewers and errors are heterogeneous, in *this volume*, 1997.

2. A. P. Dawid. Statistical theory: The prequential approach. *Journal of the Royal Statistical Society, series A*, 147:278–292, 1984.

3. S. Geman and D. Geman. Stochastic relaxation, Gibbs distribution and the Bayesian restoration of images. *IEEE Transactions on Pattern Analysis and Machine Intelligence*, 6:721–741, 1984.

4. Amrit L. Goel and Kazu Okumoto. Time-dependent error-detection rate model for software reliability and other performance measures. *IEEE Transactions on Reliability*, R-28:206–211, 1979.

5. I. B. J. Goudie and Charles M. Goldie. Initial size estimation for the linear pure death process. *Biometrika*, 68:543–550, 1981.

6. Z. Jelinsky and P. Moranda. Software reliability research. In Walter Freiberger, editor, *Statistical Computer Performance Evaluation*, pages 465–484. Academic Press, Inc., New York, 1972.

7. H. Joe and N. Reid. Estimating the number of faults in a system. *Journal of the American Statistical Association*, 80:222–226, 1985a.

8. H. Joe and N. Reid. On the software reliability models of Jelinsky-Moranda and Littlewood. *IEEE Transactions on Reliability*, R-34:216–218, 1985b.

9. Norman L. Johnson, Samuel Kotz, and N. Balakrishnan. *Continuous Univariate Distributions, Volume 1*. John Wiley and Sons, Inc., New York, second edition, 1994.

10. M. G. Kendall and A. Stuart. *The Advanced Theory of Statistics, Volume 2: Inference and Relationship*. Hafner Publishing Company, New York, 1973.

11. Naftali Langberg and Nozer D. Singpurwalla. A unification of some software reliability models. *SIAM Journal on Scientific and Statistical Computation*, 6:781–790, 1985.

12. Bev Littlewood. A critique of the Jelinsky-Moranda model for software reliability. In *Annual Reliability and Maintainability Symposium*, pages 357–364, New York, January 1981. IEEE.

13. Bev Littlewood and J. L. Verall. A Bayesian reliability growth model for computer software. *Applied Statistics*, 22:332–346, 1973.

14. Bev Littlewood and J. L. Verall. Likelihood function of a debugging model for computer software reliability. *IEEE Transactions on Reliability*, R-30:145–148, 1981.

15. John D. Musa. Software Reliability Data, Technical Report, Bell Telephone Laboratories, Whippany, NJ, 1978.

16. Tapan K. Nayak. Statistical approaches to modeling and estimating software reliability, in *this volume*, 1997.

17. Lawrence D. Ries. *Software Reliability: Statistical Modeling, Estimation, and Inference*. PhD thesis, University of Missouri - Columbia, 1995.

ANALYSIS OF REPAIRABLE SYSTEMS
—PAST, PRESENT AND FUTURE

Ananda Sen

Oakland University, Rochester, Michigan

The focus of the present article is repairable systems—for which the relevant data comprise either successive times to failure or a sequence of dichotomous success-failure outcomes from successive system configurations or stages. We present a comprehensive compilation of model descriptions and characterizations, as well as discussions of related statistical methodologies for parameter estimation and confidence interval construction. We also identify specific methodological enhancements that are either lacking from or have not been fully integrated into typical applications involving repairable systems.

1 Introduction

Analysis of failure data arising from repairable systems has received considerable attention in the statistical, engineering, computer software, and medical literature over the past three decades. Often times, in many industrial applications, data pertaining to a repairable system is alternatively viewed as the (manifestation of a) 'recurrent event' of some sort. Situations in which individuals or systems in some population experience recurrent events are common in areas such as manufacturing, risk analysis, and clinical trials. In reliability applications, the major thrust of the statistical analysis of repairable-systems data lies in the study of reliability growth, which can be described as follows. At the initial stage of many production processes involving complex systems, prototypes are put into life test under a developmental testing program, corrections or design changes are made at the occurrences of failures, and the modified system is tested again. As this test-redesign-retest sequence contributes to an improvement in the system performance, failures become increasingly sparse at the later stages of testing making it more difficult to assess the *current reliability*, a quantity of utmost importance to reliability engineers. Resolution of this problem is achieved by describing a reliability growth (RG) model which integrates the data from all stages of the process and operating experience while taking proper account of improvements resulting from design changes.

In a medical application, the 'failures' are translated into *times until the occurrence* of a recurrent event (e.g., appearance of tumor) in individuals. Clinical experiments typically consist of a fairly large number of individuals observed over a relatively short period of time. This is also common with databases of manufactured products generating warranty claims. By contrast, most of the field and bench test data for demonstrating product reliability consist of a very small number of prototypes put under test for a fairly long time. The major objective in either situation is to study the rates of recurrence of the events in question, compare different systems, assess the effect of explanatory variables, or predict a future event.

The study of repairable systems exploded in the past decade or so largely due to the recent advances in the computing power and the data analytic methods. Computer-intensive techniques useful for generating data from densities lacking a closed-form brought a renewed thrust into the research in the area of repairable systems, earlier limited to a few smaller-scoped models and methodologies. The purpose of the current article is to present a comprehensive compilation of models and related statistical methodologies associated with recurrent-event data to date. Our emphasis is on the interrelationships between various models and the assumptions that underlie their respective developments. The article is organized as follows.

In Section 2, we present various continuous-time models specifically applicable to time-to-failure data. The entire section is subdivided into categories which list parametric point-process models evolved from various considerations as well as their Bayesian versions. Section 3 describes the extensions to incorporate covariate information. Section 4 is devoted to the discussion of nonparametric estimation and testing procedures under a general time-truncated observation scheme. Models and methodologies for discrete repairable-systems data pertaining to a single-shot operating system are briefed in Section 5. We conclude in Section 6 with the identification of some specific methodological enhancements that are either lacking from or have not been fully integrated into typical repairable-systems applications to date. We shall refrain from any discussions of software reliability models in this article despite the fact that they cover a large body of repairable-systems research. This is due to (1) paucity of space, and (2) availability of recent expository reviews on software reliability models such as the ones by Singpurwalla and Wilson (1994) or Kuo (1997).

2 Continuous Models for Repairable Systems

2.1 NHPP models

In modeling failure data arising from observing a repairable system, the major thrust came from certain empirical findings of Duane (1964). From an examination of time between failure data of a variety of systems such as complex hydromechanical devices, aircraft generators and jet engines in the course of their development, he observed that the cumulative number of failures typically produced a linear relationship with the cumulative operating time t when plotted on a log-log scale. This phenomenon, subsequently referred to as the "learning curve property", was given a concrete stochastic basis by Crow (1974) who assumed that the failures during the development stage of a new system follow a nonhomogeneous Poisson process (NHPP) with an intensity function

$$\lambda_1(t) = (\beta/\theta)(t/\theta)^{\beta-1}. \tag{1}$$

The corresponding cumulative failure rate $\Lambda_1(t) = (t/\theta)^\beta$ is linear on a log-log scale. Incidentally, an NHPP formulation as in (1) was originally proposed by Ascher (1968) in modeling the reliability change of a bad-as-old system, and was later used by Bassin (1973) to obtain optimal overhaul intervals for various machines. Crow (1974), however, was the first to present a unified treatment of the statistical inference related to the model, integrating both time- and failure-truncated sampling schemes.

With $0 < t_1 < t_2 < \ldots < t_n$ denoting the first n successive failure times of a single repairable system, the likelihood function under the failure-truncated scheme (stop testing at the nth failure) assumes the form

$$L(\theta, \beta \mid t_1, \ldots, t_n) = (\beta/\theta)^n \left[\prod_{j=1}^{n} (t_j \mid \theta)^{\beta - 1} \right] \exp\left[(-t_n/\theta)^\beta \right]. \qquad (2)$$

The likelihood function arising from the time-truncated protocol (stop testing at a fixed time τ) is very similar in structure with the t_n in the exponent replaced by truncation-time τ. The maximum likelihood estimates of θ, β from (2) are expressible in the surprisingly simple form

$$\widehat{\beta} = \frac{n}{\sum_{i=1}^{n-1} \ln(t_n/t_i)}, \quad \widehat{\theta} = \frac{t_n}{n^{1/\widehat{\beta}}}.$$

Exact as well as large-sample confidence procedures for the parameters were studied by Crow (1974), Finkelstein (1976). The inference results for the time-truncated sample are similar but more tedious in comparison with the corresponding failure-truncated situation. Crow (1974) detailed formal goodness of fit procedures for the model and applies the model to both simulated as well as real examples. He also studied procedures for testing equality in growth parameters (β's) for $K \geq 2$ independent systems under study. Testing procedures for equality in the scale parameters (θ's) for two or more independent systems were studied by Lee (1980b) under the time-truncated scheme. Bain and Engelhardt (1982) discussed sequential testing procedure for the growth parameter for a single as well as multiple systems.

For a repairable system undergoing a developmental test program, a quantity of extreme importance to reliability engineers is the value of system intensity at the current time, usually the end of a period of data collection. The current value of reliability at the end of the testing stage, typically calculated as a one-to-one function of the current intensity, depicts the reliability of the manufactured system at the operational stage. Exact point and interval estimation of the intensity function were initially studied by Lee and Lee (1978), Bain and Engelhardt (1980), and Crow (1982). Their procedures were based on pivotal properties of certain quantities and were heavily dependent upon the accuracy of Monte Carlo simulation. By contrast, Rigdon and Basu (1988, 1990) investigated estimators in the class of "scale multiples of the MLE of current intensity", based on mean-squared-error criterion. Sen and Khattree (1997) have studied other risk functions and have also derived optimal estimators based on Pitman closeness criterion. Miller (1984) examined the prediction of the future intensity parameter, while Calabria et al (1988) investigated modified maximum likelihood estimators of the failure intensity as well as the expected number of failures in a given time-interval and compared their mean-squared errors with those of the usual MLE's.

In the context of modeling failure data of repairable systems, a different NHPP model, which received some attention in the literature, involves the intensity function

$$\lambda_2(t) = \exp(\alpha_0 + \alpha_1 t), \quad -\infty < \alpha_0, \alpha_1 < \infty. \tag{3}$$

Cox and Lewis (1966) introduced (3) in the process of analyzing failure data of air-conditioning equipment in Boeing 720 aircraft. Although they did not point it out explicitly, the form of intensity in (3) in the reliability growth case ($\alpha_1 < 0$) results in the occurrence of <u>infinite</u> time to first failure with a positive probability. Nevertheless, (3) has been deemed as a useful model to explain a smooth, upward trend in the rate of failure and has found interesting applications (see, e.g., Ascher and Fiengold (1969), Lawless (1982)) in the literature. Lewis and Shedler (1976) developed parametric (maximum likelihood) and nonparametric (kernel type) estimation techniques for generalized Cox-Lewis model with the intensity function

$$\lambda_2^*(t) = \exp\left(\sum_{i=0}^{k} \alpha_i t^i\right).$$

Lee (1980a) integrated both forms λ_1 and λ_2 by introducing an NHPP with intensity function

$$\lambda_{1,2}(t) = (\beta/\theta)(t/\theta)^{\beta-1} \exp(\alpha_1 t), \quad \beta > 0, \ \theta > 0, \ \alpha_1 > 0. \tag{4}$$

For $\beta < 1$, (4) is useful for describing failure data showing an initially decreasing failure rate followed by an increasing trend. Numerous applications of (1) and (3) to various datasets have been documented in the literature. The motivations ranged from investigating the effect of design change on a system to obtaining replacement or overhaul policies which minimize the expected maintenance cost. For a detailed list of the applications as well as a few other less investigated NHPP models applicable to repairable systems, see Ascher and Fiengold (1984).

2.2   Step-intensity models

The NHPP formulation for a repairable system has been deemed to be a realistic reflection of the underlying process governing the failure pattern from a "minimal repair" standpoint. The NHPP with Weibull intensity $\lambda_1(t)$ as in (1) became especially popular due to the (i) existence of markedly simple, closed form expressions for the MLE's, and (ii) availability of exact distributional results for inference purposes. Mathematical tractability, empirical fit to a variety of data vis-a-vis its conformity to the Duane learning curve property made statisticians and engineers ignore some inherent difficulties associated with this model. Finkelstein (1979) pointed out that $\lambda_1(t)$ possesses the rather unrealistic feature that it tends to ∞ as $t \to 0$ and tends to 0 as $t \to \infty$. While this problem is somewhat technical in nature, there is a more serious criticism concerning the principle of modeling. For a system undergoing a test-redesign-retest course of a developmental testing program, "...some provision needs to be present for altering the process of failures when modifications or corrective actions are applied to the system" (Thompson, 1988). Any continually changing intensity function, fixed in advance for all times, ignores this stepwise nature of the process and describes the change in reliability over the entire program irrespective of any fixes or design changes.

Incorporating a step change in the failure pattern at the occurrences of successive failures, Sen and Bhattacharyya (1993) developed a reliability growth model where the sequence of failures are assumed to be governed by a *pure birth process*. The assumption entails that the successive interfailure times are distributed as independent exponential random variables with hazard λ_i at the ith stage of the process. Sen and Bhattacharyya (1993) investigated the specific reliability-growth parameterization

$$\lambda_i = (\mu/\delta)i^{1-\delta}, \quad \mu > 0, \quad \delta \geq 1 \tag{5}$$

which evolved as a reflection of the learning curve property described in Section 2.1. The authors showed that under the failure-truncated sampling scheme, the MLE's of the parameters are jointly asymptotically *singular* normal with quite nonstandard rates of convergence. More recently, Sen (1997) investigated exact as well as large-sample properties of the estimate of current intensity λ_n (say) under the formulation (5).

In the context of reliability growth analysis for systems undergoing development testing programs, Sen and Bhattacharyya (1995) extended (5) to yet another *piecewise exponential* model which accounts for a two-component system with one component experiencing reliability growth and the other remaining unchanged. In principle, the structure is analogous to the setting where a system experiences two sources of failure, namely, the *assignable-cause* and the *inherent* (Barlow and Scheuer, 1966). Maximum likelihood and non-linear least squares estimates of model parameters were examined under the sampling scheme that records the sequence of failure times for a single prototype of the system under the test-redesign-retest cycle. Contrary to the behavior of such estimators in standard situations, several pathologies arose for the model considered by the authors. The most troublesome of them seemed to be the nonexistence of any consistent estimate of a certain parameter. Interestingly, the anomaly vanished in the absence of the *inherent* source of failure. The inherent-cause component added a complexity to the model that made estimation of parameters difficult from the data of a single system, even though testing would have continued over a large number of stages. Evidently, this difficulty can be remedied by allowing sufficient replications in the experiment.

In the context of a developmental program where the times of design modification need not necessarily synchronize with the failure of the system under study, Ebrahimi (1996) described a step-intensity model where failures in different configurations were governed by different Poisson processes. The author considered maximum likelihood estimation of the underlying parameters under various order restrictions evolving from the consideration of a growth in reliability.

2.3 Bayesian models and methodologies

Earlier work in continuous-time models for repairable systems lacked heavily in Bayesian developments. The reason could probably be attributed to the lack of closed-form, analytically tractable solutions to the problems. With the advancement of the statistical computing power and the subsequent breakthroughs in Bayesian computation over the last decade, the situation is rapidly changing. Not only has the Bayesian analysis of repairable systems become a major focus of attention for researchers, but also there is a renewed interest among practitioners and reliability engineers to employ the modern Bayesian methodologies in industrial applications.

One of the earlier Bayesian works in continuous repairable systems (Higgins and Tsokos, 1981) involved (i) approximating $\nu_n/\hat{\nu}_n$ by a chi-square random variable and (ii) imposing a gamma pdf on ν_n, where ν_n, $\hat{\nu}_n$ denote the system intensity and its MLE, respectively, at the nth failure of a repairable system under the Crow (1974) model (1). The authors compared the performance of their (quasi-Bayes) estimator of ν_n with the classical estimator $\hat{\nu}_n$ on the basis of a Monte Carlo simulation study. Guida, Calabria, and Pulcini (1989) first provided a full Bayesian analysis of (1). They considered both informative and noninformative priors on the parameters β, θ and derived the Bayes estimators of the parameters in each case.. Guida et al also supplemented their theoretical findings with numerical comparisons with the MLE's. Their conclusion favored the Bayes estimators for most combinations of the process parameters, especially for small sample sizes. While Guida et al concentrated only on failure-truncated schemes, Kyparisis and Singpurwalla (1985) analyzed both time and failure-truncated data with informative priors on θ and β and derived predictive distributions of future failure times among other things. Calabria et al (1990) undertook the same project and carried the analysis much along the lines of their earlier work. Bar-lev et al (1992) employed Jeffrey's priors on the parameters θ, β and carried out a complete Bayesian analysis of the NHPP model. Major strengths of their presentation were (i) the unification of the two sampling protocols, namely, time truncated and failure truncated, under a single analysis, and (ii) a thorough investigation of statistical inference regarding various quantities of interest, including predictive distributions, posterior distributions of system reliability and the expected number of failures in a future time interval.

A recent work by Kuo and Yang (1996) unified the Bayesian treatment for a NHPP from the computational standpoint. They dichotomized the class of NHPP's in two categories, namely, the ones with finite limiting mean functions and the others which have mean functions increasing to infinity. The former results in a defective distribution for the time to first failure and is sometimes deemed suitable for describing software failures. All the NHPP's we discuss in this article fall in the second category, and so we confine our discussion to those only. Kuo and Yang (1996) exploited the connection between NHPP and record-value statistics to lay out a Bayesian computation scheme based on Gibbs sampling approach. They also addressed the Bayesian inference for the future reliability at a certain time as well as the model selection issues based on conditional predictive densities.

Several researchers have investigated Bayesian approaches incorporated into a step-intensity formulation under the recurrent event context. Littlewood and Verall (1973) have studied a piecewise exponential "reliability-growth" model where the times between $(i-1)^{th}$ and i^{th} failure are modeled as independent exponential variates with failure rate $\lambda(i), i = 1, \ldots, n$. Instead of demanding that the repairs at the $(i-1)$th failure is completely effective in diminishing the failure rate, it is required that this happens probabilistically, i.e., "$\lambda(i)$ is stochastically smaller than $\lambda(i-1)$." For the random failure rates, Littlewood and Verall (1973) chose a family of gamma distributions with the respective scale parameters $\psi(i)$ increasing with stage number. The authors also presented a hierarchical approach where additionally $\psi(i)$'s were assumed to be random. Singpurwalla (1982) extended this to a scheme which included cases where (i) both prior and posterior distributions of $\lambda(i)$ are stochastically ordered, and (ii) the ordering is only with respect to the priors. Erkanli, Mazzuchi and Soyer (1996) imposed an ordered Dirichlet prior on the sequence $R_i \equiv \exp(-\lambda_i)$ and derived various statistical quantities including the predictive density of a future failure time as well as a probabilistic assessment of the number of items to be tested in a certain stage for a prespecified test time τ. Erkanli et al extended their procedure for piecewise Weibull models and take recourse to Gibbs sampling approach in order to conduct a full Bayesian analysis. The major advantages of a Bayesian approach seemed to be in (i) the flexibility of incorporating expert judgment at any stage of the process, (ii) providing a complete description of the test-analyze-and-fix program in an adaptive way and (iii) presenting a natural platform for performing predictive analysis and model selection.

2.4   Miscellaneous models

A couple of recent models in the context of analyzing repairable systems are worth mentioning. One of them is based on a nonhomogeneous gamma process and is intimately related to the NHPP. Berman (1981) introduced a nonhomogeneous gamma process as follows. Consider a Poisson process with intensity function $u(t)$. Imagine that every $\mathcal{K}^{\text{th}}$ event of the process is observed. If the events are thought of as shocks, then the assumption entails a failure at the occurrence of every $\mathcal{K}^{\text{th}}$ shock. The joint pdf of the first n successive failure times $T_1 < T_2 < \ldots < T_n$ assumes the form

$$f(t_1,\ldots,t_n) = \frac{\left[\prod_{i=1}^{n} u(t_i)\{U(t_i) - U(t_{i-1})\}^{\mathcal{K}-1}\right] e^{-U(t_n)}}{\{\Gamma(\mathcal{K})\}^n} \tag{6}$$

with $U(t) = \int_0^t u(x)\,dx$. Clearly, $\mathcal{K} = 1$ reduces (6) to the likelihood of a usual NHPP with intensity $u(t)$, while $u(t) = c$ (constant) defines a renewal process with interfailure times $t_i - t_{i-1}$ distributed as i.i.d. gamma random variables with scale parameter c and shape parameter $\mathcal{K}$. Although lacking a shock model interpretation, (6) defines a valid density function even for non-integer, positive values of $\mathcal{K}$. Black and Rigdon (1996) chose the Weibull form $u(t) = (\beta/\theta)(t/\theta)^{\beta-1}$ for the intensity function in (6) and called it a Modulated Power Law process. It can be shown that $U(T_n) = (T_n/\theta)^\beta$ has a gamma distribution with unit scale parameter and shape parameter $n\mathcal{K}$ and consequently $E(T_n^\beta) = \theta^\beta n\mathcal{K}$, which can again be thought of as a reflection of Duane's learning-curve property. Black and Rigdon (1996) studied maximum likelihood estimation of the parameters and constructed approximate confidence intervals for θ, β, and $\mathcal{K}$ based on the observed matrix of second partial derivatives of the log likelihood. The authors, however, did not report the asymptotic distribution of the MLE's which would involve non-standard manipulation due to the fact that the aforementioned matrix converges in probability to a singular matrix.

Trinner and Phillips (1986) presented an analysis of equipment failure data for a fleet of ships based on a birth-immigration model, where the (unknown) source of failure may be a "birth" or an "immigration." Sweeting and Phillips (1995) demonstrated that the usual MLE asymptotics for time $t \rightarrow \infty$ fails for this model since the information on the immigration parameter turns out to be almost surely finite. Denoting by α and β the immigration and birth parameters, respectively, Sweeting and Phillips (1995) proved that the standard MLE large sample results can be recovered if the asymptotics is considered in the sense $(n\alpha)^{-1}\beta \longrightarrow 0$. The authors presented the theoretical development for the situation where the exact failure times are known and provided numerical support for the extension to the case of interval-censored data.

3   Incorporating covariate information

The discussion of the models and methodologies in this article has so far been limited to lifedata without any covariate information. In the context of observing recurrent events for multiple systems (e.g., recurrence of tumors in subjects), often times it is necessary to incorporate covariate factors describing the system or the environment in which it operates. Regression models for repairable-systems data have been given a comprehensive treatment in Lawless (1987a). The description builds on Cox's proportional hazards model, whereby, an individual i with a $k \times 1$ covariate vector $\boldsymbol{x}_i$ experiences recurrent events occurring according to a nonhomogeneous Poisson process with intensity function

$$\lambda_{\boldsymbol{x}_i}(t) = \lambda_0(t) \exp(\boldsymbol{x}_i'\boldsymbol{\beta}), \ i = 1, \ldots, m,$$

where $\lambda_0(t)$ is an arbitrarily specified baseline function. With $t_{i_1} < t_{i_2} < \ldots < t_{i_{n_i}}$ denoting the successive occurrence times of the events for the ith subject, and parameterizing $\lambda_0(t)$ as $\lambda_0(t, \boldsymbol{\theta})$, the likelihood function becomes

$$L(\boldsymbol{\theta}, \boldsymbol{\beta}) = \prod_{i=1}^{m} \prod_{j=1}^{n_i} \lambda_{\boldsymbol{x}_i}(t_{i_j}) \exp\{-\Lambda_{\boldsymbol{x}_i}(T_i)\}, \tag{7}$$

where T_i is the truncation time for the ith subject and Λ denotes the cumulative (integrated) hazard rate. The likelihood function in (7) can be decomposed as

$$L(\boldsymbol{\theta}, \boldsymbol{\beta}) = \left[\prod_{i=1}^{m} \prod_{j=1}^{n_i} \frac{\lambda_0(t_{i_j}; \boldsymbol{\theta})}{\Lambda_0(T_i; \boldsymbol{\theta})} \right] \prod_{i=1}^{m} \exp\left[-\Lambda_0(T_i, \boldsymbol{\theta}) e^{\boldsymbol{x}_i'\boldsymbol{\beta}} \right] \left[\Lambda_0(T_i; \boldsymbol{\theta}) e^{\boldsymbol{x}_i'\boldsymbol{\beta}} \right]^{n_i}$$

$$\equiv L_1(\boldsymbol{\theta}) L_2(\boldsymbol{\theta}, \boldsymbol{\beta}). \tag{8}$$

Note that $L_1(\boldsymbol{\theta})$ involves $\boldsymbol{\theta}$ only and can be used for inference on $\boldsymbol{\theta}$ alone. On the other hand, $L_2(\boldsymbol{\theta}, \boldsymbol{\beta})$ is simply the likelihood function (apart from the constant) for the Poisson log-linear model, which can be further decomposed to extract Cox's original partial likelihood (Cox, 1972) when T_i's are all equal. Lawless (1987a) presented a semiparametric analysis for (8) with λ_0 unspecified, as well as a fully parametric treatment with $\lambda_0(t)$ assumed to be in the Weibull form $\nu\delta t^{\delta-1}$. Various model checks and tests were presented in this expository article, with applications to the famous dataset due to Gail et al (1980) which presented times to development of mammary tumors in rats distributed in a treatment and a control group. The underlying assumption for the large-sample inference results was the existence of a moderately large number of subjects observed over fairly short periods of time. For an industrial experiment where a relatively long chain of observations is available for every individual system, but a very few prototypes are observed, the direction of asymptotics changes and leaves room for various non-standard results analogous to those described in Section 2.2.

Many recent discussions about both lifetime and stochastic process transition data have focused on modeling and analyzing the effects of so-called unobserved heterogeneity. Lancaster (1979) fitted the model (8) to the unemployment data with $\lambda_0(t) = \nu\delta t^{\delta-1}$ like Lawless, but with ν describing a random frailty parameter. Specifically, Lancaster chose ν to be a gamma random variable to represent all unobservable exogenous variation. This yields what is known in the literature as a negative binomial regression model (Lawless, 1987b) and is deemed to be quite appropriate to describe overdispersion.

Until very recently (see Aroui and Lavergne, 1996), no formal approach for incorporating covariate effects in the step-intensity models described in Section 2.2 has been explored. In the concluding section, we discuss the possibility of utilizing the Generalized Linear Model approach into these models which in our belief, serves as a viable alternative to NHPP regression modeling for repairable-systems data.

4   Nonparametric Methods

Contrary to its discrete counterpart, exploration of nonparametric methods in the continuous-time repairable-systems area had been quite limited until very recently. One of the earlier works in this context was due to Robinson and Dietrich (1987). The scenario assumed by the authors was strictly applicable to the situation of testing new batches of units in successive configurations, characterized by specified time intervals followed by subsequent design changes. The testing proceeds for a fixed number of configurations and the observed data comprise of the (independent) failure times of the respective units during each configuration. Other than requiring mild conditions such as unimodality, no specific distributional assumptions are made about the lifetimes. Nonparametric maximum likelihood estimation of a generic parameter for the life distribution was considered under the order restriction imposed by a growth in reliability. The authors documented simulation results which indicated the superiority of the nonparametric MLE's over some competing parametric estimators with respect to the estimation of failure rate.

Most of the recent nonparametric methods deal with applications in which recurrent events are observed for a fairly large number of independent systems. Formally, system i is observed over the time period $[0, \tau_i]$, during which events (failures) occur at time points $t_{i_1} < t_{i_2} < \ldots < t_{i_{n_i}}$, $i = 1, \ldots, k$. The associated counting processes $\{N_i(t) : t \geq 0\}$, denoting number of failures up to time t, are assumed to be independent, possessing the same cumulative mean function $M(t) =$

$E[N_i(t)]$. Interest lies in the estimation of the $M(t)$ function. The nonparametric MLE of $M(t)$, for $0 \leq t \leq \tau = \max_i(\tau_i)$, is given as

$$\widehat{M}(t) = \sum_{u=0}^{t} \sum_{i=1}^{k} w_i(u)n_i(u),$$

where $n_i(u)$ is observed number of events for the ith system up to time u, and the weight function $w_i(u)$ simply represents the proportion of systems still under observation at time u. $\widehat{M}(t)$ is also popularly known as the Nelson-Aalen estimator in the counting process literature. Nelson (1988, 1995) demonstrated that $\widehat{M}(t)$ is unbiased for $M(t)$ for quite general point processes and constructed approximate confidence limits for $M(t)$. Lawless and Nadeau (1995) derived a robust variance estimate for $\widehat{M}(t)$ and extended Nelson's (1995) work to regression models where the mean function for the ith counting process was specified as:

$$E(N_i(t)) = M_0(t)g_i(t),$$

where M_0 is a baseline function and g_i is a known function involving covariates and regression parameters. The asymptotic distributions (as the number of systems becomes large) of $\widehat{M}_0(t)$ and estimators of regression parameters were derived. Appealing to the general theory of estimating equations, Lawless and Nadeau (1995) demonstrated that these inference results are valid for a quite general class of counting processes. Cook et al (1996) constructed robust tests for the equality of the mean functions $M_i(t)$, $i = 1, \ldots, k$ for the k systems under the setting discussed here. Cook and Lawless (1996) extended these procedures to the longitudinal clinical trial setting where different groups were being monitored continuously at prespecified calendar times.

5 Discrete models for repairable systems

While the continuous-time models discussed thus far are applicable to time-to-failure data, the relevant data for continuous-time models comprise sequences of dichotomous success-failure outcomes from successive system configurations or stages. Numerous reviews examining various classes of discrete models for repairable systems have been scattered in the literature for the past two decades or so. These surveys were smaller-scoped and more restrictive in content compared to the corresponding continuous counterpart. The most comprehensive compilation to date of various discrete models, their characterizations, and interrelationships, is documented in Fries and Sen (1997). Instead of repeating that endeavor, we simply highlight on the major areas of research that have been undertaken in this area.

5.1 Nonparametric formulations

The genesis of the discrete models for repairable systems has its roots in the early attempt to quantify the increase in system reliability attained after implementing post-test design changes. Numerous proposals (nonparametric) for rescoring "observed failures" as "adjusted partial successes" (see, e.g., Drake (1987), Lloyd (1987), Chandler (1988)) were typically involved. Simulations showed, however, that most of the proposed methods were very sensitive to the choice of the failure discounting parameter(s), which was somewhat arbitrary. Another focal point of interest to the reliability researcher and practitioners working in this area has been the generalization evolved from the partitioning of failures in two distinct categories, namely, assignable-cause and inherent. The former corresponds to failure modes that can be eliminated by redesigns or by equipment or operational modifications, while the latter reflects the state of the art and remains unchanged as the system configuration advances. Wolman (1963) initialized work in this area, although his study was confined to the calculation of probabilistic quantities, such as "chance of eliminating all assignable-cause failure modes by trial n." Bresenham (1964) articulated a refinement of the Wolman postulate and, further, developed accompanying MLE procedures. Barlow and Scheuer (1966) followed up with a formal trinomial formulation of the process. The context is a test-analyze-and-fix (TAAF) program conducted over a succession of stages, with improvements attempted after each stage except the last. The likelihood function corresponding to a_k inherent failures, b_k assignable-cause failures, and s_k successes in the n_k observations for stage k, $k = 1, \ldots, K$ is

$$L = \prod_{k=1}^{K} \frac{n_k!}{a_k! b_k! s_k!} q_0^{a_k} q_k^{b_k} (1 - q_0 - q_k)^{s_k},$$

where q_0 is the constant probability of inherent failure and q_k is the probability of an assignable-cause failure at stage k, assumed to decrease with k (i.e., the fixes are effective). Barlow and Scheuer (1966) studied the MLE's $\hat{q}_0$ and $\hat{q}_k$, $k = 1, \ldots, K$ derived under the constraint of having a non-increasing sequence of q_k's.

Based on an extensive simulation study, Olsen (1977) investigated the pros and cons of the Barlow-Scheuer restricted MLE's. Olsen (1977) concluded that these estimators should only be used when (1) a reasonably large sample size is available after the last set of corrections, (2) testing is stopped with the occurrence of assignable-cause failures, (3) corrective actions do not fully remove individual failure modes or introduce new failure modes. It was noted that, in general, the Barlow-Scheuer methodology yielded positive biased estimator $\hat{R}_K = 1 - \hat{q}_0 - \hat{q}_K$ for reliability at the final stage of testing.

Weinrich and Gross (1978) and Mazzuchi and Soyer (1993) investigated Bayesian formulations of Barlow-Scheuer piecewise trinomial model. Weinrich and Gross (1978) modification of the Barlow-Scheuer model entails redistribution of the probability among other causes of inherent and assignable-cause failure, upon removal of an assignable-cause failure mode. Assuming the initial probabilities (success, assignable-cause, inherent-cause) to possess a Dirichlet *pdf*, the posterior *pdf* of stage-k probabilities is updated as either a Dirichlet or a mixture of Dirichlet depending on the extent of knowledge of the amount of initial assignable-cause failure removed. Instead of specifying a prior for the probabilities only at the initial stage, Mazzuchi and Soyer (1993) imposed a Dirichlet prior on the differences of the assignable-cause probabilities from successive stages. This clever manipulation clearly defines a *pdf* for the assignable-cause probabilities decreasing across stages. The major contributions of Mazzuchi and Soyer (1993) formulations were (1) specification of the prior belief in the entire reliability growth pattern, (2) retention of all relevant marginal and posterior *pdf*'s as tractable Dirichlet (or Beta) forms, and (3) flexible framework for elicitation of prior information from experts.

5.2 Parametric models

A large class of discrete parametric models for repairable systems adhere to the representation of the types:

$$R_k = \begin{cases} R_\infty - f(k) \\ \text{or} \\ R_\infty g(k) \end{cases}$$

where R_k is the reliability at the kth stage of testing, R_∞ symbolizes the eventual reliability, and $f(k)$, $g(k)$ are arbitrary functions typically involving unknown parameters. Clearly, $\lim_{k \to \infty} f(k) = 0$ and $\lim_{k \to \infty} g(k) = 1$. Some of the earlier models of this type were extensively studied by Lloyd and Lipow (1964), Chernoff and Woods (1962), Gross and Kamins (1967, 1968). Virene's (1968) 'Gompertz' model is an interesting inclusion in this class in the sense that it was the only model of this form which attempted to characterize S-shaped patterns for reliability growth. The form of the reliability

$$R_k = R_\infty b^{c^k}, \quad 0 < b < 1, \quad 0 < c < 1, \tag{9}$$

however, suffers from the drawback that the curves resulting from (9) always attain 37% of the final reliability R_∞ at their inflection point. Thus, S-shaped trends cannot be obtained unless the initial system reliability is sufficiently depressed, an impractical constraint for many systems. Recently, Kecesioglu et al (1994) modified Virene model by introducing a fourth parameter 'd' in the fashion

$$R_k = (R_\infty - d) + db^{c^k}$$

and investigated accompanying parameter estimation methodologies.

Recent research in discrete reliability models has rendered considerable attention to another class of parametric models intimately linked to the Duane learning curve property described in Section 2.1. Identifying a single trial to a unit time in the continuous case, Crow (1983) developed the following model used extensively by the United States Army:

$$R_k = 1 - \lambda \left[(N_k)^\beta - (N_{k-1})^\beta \right] / n_k, \quad 0 < \lambda < 1, \quad 0 < \beta < 1, \tag{10}$$

where k denotes the configuration number, N_k denotes the cumulative number of test trials until the kth configuration, $N_0 \equiv 0$, and $n_k = N_k - N_{k-1}$. Crow's derivation rested on the underlying assumption that for each stage of testing, the number of observations is fixed and the distribution of outcomes in successes and failures is random. Such a situation can arise, for example, when the system producer delivers batches for sequential phases of testing, all of the scheduled test events are executed, and the results for each stage lead to modifications of the system design. Some estimation procedures for this model were suggested by Crow (1983) and Finkelstein (1983); asymptotic properties were studied by Bhattacharyya, Fries and Johnson (1989), Bhattacharyya and Ghosh (1991), and Johnson (1991).

The development of Crow's model is incompatible with another common strategy of stopping testing as soon as any failure occurs and introducing system redesigns or corrective actions as soon as possible, ie., inverse sampling. This, in particular, is a viable model for destructive testing of very expensive systems, e.g., testing of the launch and landing capabilities of unmanned aerial vehicles. Fries (1993) took the learning curve-property and directly imposed the inverse sampling strategy to derive the following model expressed as:

$$R_k = 1 - \lambda' \left[k^{\beta'} - (k-1)^{\beta'} \right]^{-1}, \quad 0 < \lambda' < 1, \quad \beta' > 1. \tag{11}$$

Recently, Sen and Fries (1997a, b) developed a similar discrete reliability model by first modifying the NHPP Weibull process model (1) to permit changes to the underlying reliability only when redesigns are implemented and then translating to a discrete analog. The resulting model is:

$$R_k = 1 - (\lambda'/\beta')k^{1-\beta'}, \quad 0 < \lambda' < 1, \quad \beta' > 1. \tag{12}$$

The difference between (11) and (12) is $O(k^{-1})$ and consequently, the two models share identical asymptotic properties when the number of configurations becomes large. Sen and Fries (1997a) detailed the inference procedures for (11) and also documented (1997b) the effects of using results from the direct-sampling model (10) to data which is generated from (11) as the underlying process.

5.3 Other models

A number of different types of discrete model have been proposed in the literature which included the time-series model by Singpurwalla (1978), smoothing models by Lloyd and Lipow(1964), Gross (1971), Drake (1987). Most of these models presented alternative, dynamic approaches of modeling discrete data observed for a repairable system but stopped short of presenting any deep investigation of the inferential aspects.

6   Where do we go next?

The intent of the present article was to undertake a journey through the realm of repairable systems and revisit the variety of proposed models and associated statistical methodologies which have been studied in the past years. The research in repairable systems has received a major thrust lately due to (i) the necessity of analyzing complex recurrent-event databases with applications ranging from manufacturing to social sciences, and (ii) immense advances in general methodological and computational research. In an abstract sense, the models can be viewed in one of two ways—as regression and/or curve-fitting procedures generally through some convenient prespecified parameterization, or as semiparametric accounting schemes for tabulating events (either via continuous monitoring or as one-shot operations).

When they fit the test data reasonably well, parametric models often provide more precise estimators in comparison to their nonparametric counterparts. Parametric models also enjoy the relative advantage of directly supporting 'extrapolation.' On the other hand, parametric formulations are limited to a few specific instances of point process models, which may fail to portray the true underlying process governing the sequence of failures. When the interest is centered on the inference regarding the mean failure rate, the robust nonparametric methods (see Section 4) provide satisfactory solutions, although they rely heavily on the existence of a large number of prototypes.

Most developmental testing programs are conducted sequentially. It is therefore of utmost importance to describe a flexible model which incorporates expert's judgment regarding the process at every stage of the program. Bayesian formulations provide direct means for accomplishing this through the prior knowledge which is updated at every configuration. Bayesian theory also provides a natural framework for the selection of "best" model(s) based on the respective predictive powers. Recent advances in Gibbs Sampling and related computational aspects have made Bayesian analysis accessible to the practicing community.

After having traced the history of research in repairable systems from the traditional past to the intersection of computational and methodological advances, one may be curious to know about the possible avenues to travel from this point. In the following paragraphs, we provide a brief discussion of some methodological features that are common in numerous statistical fields but are generally absent from or incompletely treated in typical applications involving repairable-systems data.

Regression methodologies are the logical means by which the influence of multiple identified factors can be addressed, different samples can be compared and an overall growth tend in reliability can be captured. Most of the regression models which have been explored in the context of repairable systems are instances of hazard regression models. Regression models on mean functions have been unified under the umbrella entitled *Generalized Linear models* (GLIM) (McCullagh and Nelder, 1983). In its simplest form, a GLIM is formulated by modeling the independent data $\{y_i\}_{i=1,\dots,n}$ through the family of pdf's $\{f(y_i, \theta_i)\}_{i=1,\dots,n}$ assuming the exponential class structure

$$f(y_i, \theta_i) \propto \exp(\theta_i y_i - b(\theta_i)),$$

$b(\cdot)$ being an arbitrary differentiable function related to the mean of y_i through the equation $\partial b(\theta_i)/\partial \theta_i = E_{\theta_i}(y_i) \equiv \eta_i$. The linearity in the model is imposed by relating the mean to a twice continuously differentiable but otherwise arbitrary function g of a linear combination of unknown parameters β, namely, $\eta_i = g(\beta' x_i)$, x_i's assuming the role of covariates. The log-linear link function $g(x) = \exp(x)$, the logit link function $g(x) = \log[x/(1-x)]$, or the probit link function $g(x) = \Phi^{-1}(x)$, Φ being the cdf of standard normal, are specific instances of g which have received considerable attention in literature. The exponential regression models discussed in Sections 2.2 and 2.3 fall under a GLIM framework, with the successive interfailure times comprising the data and a suitable function of the configuration number acting as the covariate. More general covariate structures and distributional assumptions can be entertained, and one can appeal to the general inference results derived by Fahrmeir and Kaufmann (1985) for assessing the statistical significance of the β parameters. Breslow and Clayton (1993) discussed the inference results for the *generalized linear mixed model* which can be suitably adapted to the recurrent-data situations where the inclusion of a random effect is warranted in order to account for exogenous variation.

Most models for repairable systems incorporate system-level data only, i.e., results for distinct components or subsystems are not separately tracked. One approach to incorporate component-level analyses is to assume an appropriate reliability structure function which can combine the estimates resulting from applying distinct models to each component/subsystem. Such an evaluation strategy may prove to be more efficient in non-standard testing circumstances, e.g., whenever replications of system-level testing may involve (i) different combinations of active or test-instrumented subsystems or (ii) subsystems whose performance histories can be pooled across testing programs.

Because of the nature in which data is accumulated in a study involving a repairable system, sequential procedures seem to provide an effective means of integrating decision-making criteria into the statistical inference. Although the Bayesian paradigm is widely acclaimed for such purposes, its application to the realm of repairable systems has been quite limited. Many of the Bayesian formulations have accommodated its inherent advantages in an incomplete sense—with the incorporation of prior beliefs only or either (i) the initial (stage 1) reliabilities or related parameters or (ii) the perceived likely reliability growth pattern across all stages. Particular insights about specific redesigns and corrective actions implemented throughout the course of the developmental program—for all configurations and for all stages of testing—typically have not been embraced by either approach. For instance, in the perspective (i) the stage-1 posterior pdf assimilates the stage-1 data to provide an updated understanding of stage-2, completely ignoring the types and perceived likely impacts of system reliability in stage 1. To some unknown extent, this shortcoming can partially be offset by future testing outcomes that reflect the incremental adjustments to system reliability induced by the accomplished sequence of redesign efforts. The formulation (ii) certainly entertains a broader scope of prior information, but it also does not utilize any specific knowledge about encountered failure modes and their corresponding fixes. Sequential analysis also creates a natural platform for direct consideration of cost component of the program.

We conclude our discussion with an indication to a new and challenging arena in repairable systems. While modeling time to first failure for *multivariate lifetime data* has receive considerable attention in literature, very little work has been done to incorporate covariate effects, or to extend them to the recurrent-event scenario. Multivariate lifedata arise frequently in biomedical as well as industrial applications. Ghosh and Gelfand (1997) presented the analysis of AIDS patients for whom the responses of interest were multiple, namely, (i) time to first hospitalization after diagnosis, (ii) time to treatment with AZT after diagnosis, and (iii) time to death after diagnosis. In a warranty claim analysis, one may be interested in the true age of a car as well as the number of months in service until sold. Ghosh and Gelfand

(1997) adopted a general approach to model bivariate event data with covariates, in the presence of censoring. They took recourse to a data augmentation technique coupled with a Gibbs sampler step to provide a detailed Bayesian analysis. The authors' work is a promising step in formalizing multivariate regression models with life data, and will hopefully provide useful insights into modeling multivariate recurrent processes.

Acknowledgments:
    The author is indebted to Dr. Arthur Fries and Dr. Mousumi Banerjee for valuable discussions and suggestions which helped the exposition of the article. The author also wishes to extend his gratitude to Ms. Lynette Folken for her diligent service in preparing the manuscript.

References

Aroui, M. and Lavergne, C. (1996), "Generalized linear models in software reliability: parametric and semi-parametric approaches," *IEEE Transactions on Reliability*, **45**, 463–470.

Ascher, H. E. (1968), "Evaluation of repairable system reliability using the bad-as-old concept", *IEEE Transactions on Reliability*, R-**17**, 103–110.

Ascher, H. and Fiengold, H. (1969), "Bad-As-Old Analysis of system failure data," *Annals of Assurance Sciences*, Gordon and Breach, New York, 49–62.

Ascher, H. and Fiengold, H. (1984), *Repairable Systems Reliability*, Marcel Dekker, New York.

Bain, L. J. and Engelhardt, M. (1980), "Inferences on the parameters and current system reliability for a time truncated Weibull process", *Technometrics*, **22**, 421–426.

Bain, L. J. and Engelhardt, M. (1982), "Sequential probability ratio tests for the shape parameter of a NHPP", *IEEE Transactions on Reliability*, R-**31**, 79–83.

Bar-lev, S. K., Lavi, I. and Reiser, B. (1992), "Bayesian inference for the power law process," *Annals of Institute of Statistical Mathematics*, **44**, 623–639.

Barlow, R. E. and Scheuer, E. M. (1966), "Reliability growth during a development testing program", *Technometrics*, **8**, 53–60.

Bassin, W. M. (1973), "A Bayesian optimal overhaul interval model for the Weibull restoration process case," *Journal of the American Statistical Association*, **68**, 575–578.

Berman, M. (1981), "Inhomogeneous and modulated gamma processes," *Biometrika*, **68**, 143–152.

Bhattacharyya, G. K., Fries, A. and Johnson, R. A. (1989), "Properties of continuous analog estimators for a discrete reliability growth model," *IEEE Transactions on Reliability*, **38**, 373–378.

Bhattacharyya, G. K. and Ghosh, J. K. (1991), "Estimation of parameters in a discrete reliability growth model," *J. Statist. Planning and Inference*, **29**, 43–53.

Black, S. E. and Rigdon, S. E. (1996), "Statistical inference for a modulated power law process," *Journal of Quality Technology*, **28**, 81–90.

Bresenham, J. E. (1964), "Reliability Growth Models," *Technical Report No. 74*, Department of Statistics, Stanford University, Stanford, California.

Breslow, N. E. and Clayton, D. G. (1993), "Approximate inference in generalized linear mixed models, *Journal of the American Statistical Association*, **88**, 9–25.

Calabria, R., Guida, M. and Pulcini, G. (1988), "Some modified maximum likelihood estimators for the Weibull process," *Reliabiity Engineering and System Safety*, **23**, 51–58.

Calabria, R., Guida, M. and Pulcini, G. (1990), "Bayes estimation of prediction intervals for a power law process," *Communications in Statistics – Theory Methods*, **19**, 3023–3035.

Chandler, J. D., Jr. (1988), "Estimating reliability growth with discrete reliability models," unpublished M.S. thesis, Naval Postgraduate School, Monterey, California.

Chernoff, H. and Woods, W. M. (1962), *Reliability Growth Models—Analysis and Applications*, Field Memorandum, CEIR, Inc.

Cook, R. J. and Lawless, J. F. (1996), "Interim monitoring of longitudinal comparative studies with recurrent event responses," *Biometrics*, 1311–1323.

Cook, R. J., Lawless, J. F., and Nadeau, C. (1996), "Robust tests for treatment comparisons based on recurrent event responses," *Biometrics*, **52**, 557–571.

Cox, D. R. (1972), "Regression models and life tables (with discussion)," *The Jounal of the Royal Statistical Society*, Series B, **34**, 187–220.

Cox, D. R. and Lewis, P. A. W. (1966), *The Statistical Analysis of Series of Events*, Methuen, London.

Crow, L. H. (1974), "Reliability analysis for complex repairable systems," *Reliability and Biometry*, ed. by F. Proschan and R. J. Serfling, 379–410.

Crow, L. H. (1982), "Confidence interval procedures for the Weibull process with applications to reliability growth," *Technometrics*, **24**, 67–72.

Crow, L. H. (1983), "AMSAA discrete reliability growth model," AMSAA Methodology Office, Note 1–83, Aberdeen Proving Ground, MD.

Drake, J. E. (1987), "Discrete reliability growth models using failure discounting," unpublished M.S. Thesis, Naval Postgraduate School, Monterey, California.

Duane, J. T. (1964), "Learning curve approach to reliability monitoring," *IEEE Transactions on Aerospace*, **2**, 563–566.

Ebrahimi, N. (1996), "How to model reliability-growth when times of design modifications are known," *IEEE Transactions on Reliability*, **45**, 54–58.

Erkanli, A., Mazzuchi, T. A. and Soyer, R. (1997), "Bayesian computations for a class of reliability growth models," submitted for publication.

Fahrmeir, L. and Kaufmann, H. (1985), "Consistency and asymptotic normality of the maximum likelihood estimator in generalized linear models," *The Annals of Statistics*, **13**, 342–368.

Finkelstein, J. M. (1976), "Confidence bounds on the parameters of the Weibullprocess," *Technometrics*, **18**, 115–117.

Finkelstein, J. M. (1983), "A logarithmic reliability growth model for single-mission systems," IEEE Transactions on Reliability, R-**32**, 154–164.

Finkelstein, J. M. (1979), "Starting and limiting values for reliability growth," *IEEE Transactions on Reliability*, **28**, 111–113.

Fries, A. (1993), "Discrete reliability-growth models on learning-curve property," *IEEE Transactions on Reliability*, **42**, 303–306.

Fries, A. and Sen, A. (1997), "A survey of discrete reliability growth models," *IEEE Transactions on Reliability*, **45**, 582–604.

Gail, M. H., Santner, T. J. and Brown, C. C. (1980), "An analysis of comparative carcinogenesis experiments based on multiple times to tumor", *Biometrics*, **36**, 255–266.

Ghosh, S. K. and Gelfand, A. E. (1997), "Latent waiting time models for bivariate event times with censoring," submitted for publication.

Gross, A. J. (1971), "The application of exponential smoothing to reliability assessment," *Technometrics*, **13** 877–883.

Gross, A. J. and Kamins, M. (1967), "Reliability assessment in the presence of reliability growth," *Memorandum RM-5346-PR*, (For United States Air Force Project RAND by) RAND Corporation, Santa Monica, California.

Gross, A. J. and Kamins, M. (1968), "Reliability assessment in the presence of reliability growth," *Ann. Assurance Sciences: Symp. Reliability*, 406–416.

Guida, M., Calabria, R. and Pulcini, G. (1989), "Bayes inference for a non-homogeneous Poisson process with power intensity law," *IEEE Transactions on Reliabiility*. R-**38**, 603–609.

Higgins, J. J. and Tsokos, C. P. (1981), "A quasi-Bayes estimate of the failure intensity of a reliability growth model," *IEEE Transactions on Reliability*, R-**30**, 471–475.

Johnson, R. A. (1991), "Confidence estimation with discrete reliability growth models," *J. Statist. Planning and Inference*, **29**, 87–97.

Kececioglu, D., Jiang, S. and Vassiliou (1994), "The modified Gompertz reliability-growth model," *Proc. Annual Reliability and Maintainability Symp.*, Institute of Electrical and Electronics Engineers, Inc., 160–165.

Kuo, L. (1997), "Software Reliability" (update), *Encyclopedia of Statistical Science*, ed. by S. Kotz, C. B. Read and D. Banks.

Kuo, L. and Yang, T. Y. (1996), "Bayesian computation for nonhomogeneous Poisson processes in software reliability," *Journal of the American Statistical Association*, **91**, 763–773.

Kyparisis, J. and Singpurwalla, N. D. (1985), "Bayesian inference for the Weibull process with applications to assessing software reliability growth and predicting software failures," *Computer Science and Statistics: Proceedings of the Sixteenth Symposium on the Interface* (ed. L. Billard), 57–64, Elsevier Science Publishers B. V. (North Holland), Amsterdam.

Lancaster, T. (1979), "Econometric methods for duration of unemployment," *Econometrica*, **47**, 939–955.

Lawless, J. F. (1982), *Statistical Models and Methods for Lifetime Data*, John Wiley and Sons, New York.

Lawless, J. F. (1987a), Regression methods in Poisson process data," *Journal of the American Statistical Association*, **82**, 308–815.

Lawless, J. F. (1987b), "Negative binomial and mixed regression," *Canadian Jour-*

nal of Statistics, **15**, 209–225.

Lawless, J. F. and Nadeau, J. C. (1995), "Some simple robust methods for the analysis of recurrent events," *Technometrics*, **37**, 158–168.

Lee, L. (1980a), "Testing adequacy of the Weibull and log linear rate models for a Poisson process," *Technometrics*, **22**, 195–199.

Lee, L. (1980b), "Comparing rates of several independent Weibull processes," *Technometrics*,**22**, 427–430.

Lee, L. and Lee, S. K. (1978), "Some results on inference for the Weibull process," *Technometrics*, **20**, 41–45.

Lewis, P. A. W. and Shedler, G. S. (1976), "Statistical analysis of non-stationary series of events in a Data Base system," *IBM Journal of Research and Development*, **20**, 465–482.

Littlewood, B. and Verrall, J. L. (1973), "A Bayesian reliability growth model for computer software," *The Journal of the Royal Statistical Society*, Series C, **22**, 332–346.

Lloyd, D. K. (1987), "Forecasting reliability growth," *Proc. 33rd Annual Technical Meeting*, Institute of Environmental Sciences, 171–175.

Lloyd, D. K. and Lipow, M. (1964), *Reliability: Management, Methods, and Mathematics*, Prentice Hall, New Jersey.

Mazzuchi, T. A. and Soyer, R. (1993), "A Bayes method for assessing product-reliability during development testing," *IEEE Transactions on Reliability*, **42**, 503–510.

McCullagh, P. and Nelder, J. A. (1983), *Generalized Linear Models*, London, Chapman and Hall.

Miller, G. (1984), "Inference on a future reliability parameter with the Weibull process model," *Naval Research Logistics Quarterly*, **31**, 91–96.

Nelson, W. B. (1988), "Graphical analysis of system repair data," *Journal of Quality Technology*, **20**, 24–35.

Nelson, W. B. (1995), "Confidence limits for recurrence data–applied to cost or number of product repairs," *Technometrics*, **37**, 147–157.

Olsen, D. E. (1977), "Estimating reliability growth," *IEEE Transactions on Reliability*, R-**26**, 50–53.

Rigdon, S. E.. and Basu, A. P. (1988), "Estimating the intensity function of a Weibull process at the current time: failure truncated case," *Journal of Statistical Computation and Simulation*, **30**, 17–38.

Rigdon, S. E. and Basu, A. P. (1990), "Estimating the intensity function of a power law process at the current time: time truncated case," *Communications in Statistics–Simulation*, **19**, 1079–1104.

Robinson, D. G. and Dietrich, D. (1987), "A new nonparametric growth model," *IEEE Transactions on Reliability*, R-**36**, 411–418.

Sen, A. (1997), "Reliability estimation in a Duane-based model for repairable systems." Working paper.

Sen, A. and Bhattacharyya, G. K. (1993), "A piecewise exponential model for reliability growth and associated inferences," *Advances in Reliability*, ed. by Asit P. Basu, 331–355.

Sen, A. and Bhattacharyya, G. K. (1995), "A reliability growth model under inherent and assignable-cause failures," *Recent Advances in Reliability*, ed. by N. Balakrishnan, 295–318.

Sen, A. and Fries, A. (1997a), "Estimation in a discrete reliaibility growth model under an inverse sampling scheme," to appear in *Annals of Inst. Statist. Math.*.

Sen, A. and Fries, A. (1997b), "Model misspecification effects within a family of alternative discrete reliability-growth methodologies," submitted for publication.

Sen, A. and Khattree, R. (1997). "On estimating current intensity of a power law process." Working paper.

Singpurwalla, N. D. (1978), "Estimating reliability growth (or deterioration) using time series analysis," *Nav. Res. Log. Quart.* **25**, 1–14.

Singpurwalla, N. D. (1982), "A Bayesian scheme for estimating reliability growth under exponential failure times," *Studies in the Management Science*, **19**, 281–296.

Singpurwalla, N. D. and Wilson, S. P. (1994), "Software reliability modeling," *International Statistical Review*, **62**, 289–317.

Sweeting, T. J. and Phillips, M. J. (1995), "An application of nonstandard asymptotics to the analysis of repairable-systems data," *Technometrics*, **37**, 428–435.

Thompson, W. A. (Jr.) (1988), *Point Process models with applications to safety and reliability*, Chapman and Hall, New York.

Trinner, D. A. and Phillips, M. J. (1986), "The reliability of equipment fitted to a fleet of ships," *Proceedings of the 9th Advances in Reliability Technology Symposium*, Bradford, Warrington, U. K.

Virene, E. P. (1968), "Reliability growth and its upper limit," *Proc. Annual Reliability and Maintainability Symp.*, Institute of Electrical and Electronics Engineers, Inc., 265–270.

Weinrich, M. C. and Gross, A. J. (1978), "The Barlow-Scheuer reliability growth model from a Bayesian viewpoint," *Technometrics* **20**, 249–254.

Wolman, W. (1963), "Problems in System Reliability Analysis," *Statistical Theory of Reliability* (ed. M. Zelen), University of Wisconsin Press, Madison, Wisconsin, 149–160.

QUEUEING NETWORK APPROACH IN RELIABILITY THEORY

KANWAR SEN

Department of Statistics, University of Delhi, Delhi-110007

MANJU AGARWAL

Department of Operational Research, University of Delhi, Delhi-110007

MAITREYEE CHAUDHURI

Department of Operational Research, University of Delhi, Delhi-110007

Analytical modeling of present day high technological complex systems to exploit state of the art technology in their designing and performance/dependability/reliability evaluation is of paramount importance. Analytical modeling techniques can be broadly classified into two different categories: (i) state-space based and (ii) non-state-space based. Product form queueing networks can be classified as non-state-space based methods since their solutions do not require the construction of the underlying state space. As such, they are specially useful in reliability evaluation of complex systems for which the state-space is very large. However, their application in reliability studies has not gained much attention so far. In this paper, an attempt has been made to throw light in this direction and review the work done in reliability theory through this approach.

1 Introduction

The advent of computers and advancement of modern technology has given rise to an enormous increase in complexity of various systems such as production and assembly operations, airline reservations, air-traffic control systems, etc. As such, analytical modeling of these systems to exploit state of the art technology in their designing and performance/dependability/reliability evaluation is of paramount importance.

Analytical modeling techniques can be broadly classified into two different categories: (i) state-space based and (ii) non-state-space based. However, different much sought-after reliability modeling approaches dealing, with state-space of the system such as Markov processes, semi-Markov processes, Markov-renewal processes, and supplementary variable techniques, result in the formidable task of solving a set of probability differential-difference balance equations governing the system behaviour and give results in terms of Laplace-Stieltjes transforms (LSTs). As such, sometimes, one has to feel satisfied with their numerical solutions only. Moreover, as the number of states of a system increases exponentially with increase in both the number of components and complexity, it also becomes very difficult to generate the state space. This difficulty can be overcome considerably by modeling the system as a queueing network (QN). This is so as product form queueing networks can be classified as non-state-space based methods since their solutions do not require the construction of the underlying state space. This is useful in evaluating performance/dependability/reliability measures. Though QNs have been extensively applied in diverse areas such as computer communication networks, time shared and multiprogrammed computer systems, urban traffic control, health systems, etc., Harrison and Patel (1993), their application has not gained much attention in reliability studies so far. In this paper, an attempt has been made to review the work done in reliability theory through this approach to obtain stationary availability of complex redundant repairable systems.

Basically a queueing network consists of an arbitrary number of service centers (i.e. nodes, perhaps with multiple servers at some of them) connected together and each with storage room for queues to form. Customers enter the system at various nodes, queue for service, and upon getting service at a node proceed to some other node with fixed probabilities to receive additional service. Queueing networks can be classified into two types: open and closed (Jackson 1957, 1963).

In closed networks, the number of customers remains constant, i.e., in the sense that no new customers can enter the system and no customer can leave the system. While it is not so in open networks, i.e., number of customers is not constant. Moreover, those systems in which all customers have identical characteristics, i.e., all require the same type of service at a node and have the same routing probabilities, are called single class queueing networks. In some networks, there may be customers of different classes, distinguished by their routing probabilities and service time requirements (Baskett et al. 1975). These are called multiclass queueing neworks (MQNs). Multiclass networks that are closed with respect to some customer types and open with respect to others are said to be mixed networks.

Generally, a node in the queueing network is defined by the service discipline and the service time distribution for each class of customers. The various types of nodes discussed in the literture, Baskett et al. (1975), are:

Type 1. The service discipline is first come-first-served (FCFS) with exponential service time distribution with parameter μ for all the customers at the node. The service rate can be state dependent, say $\mu(j)$ which denotes the service rate with j customers at the node.

Type 2. The service discipline is processor sharing (PS). There is a single server at a node, and each class of customers may have a distinct service time distribution, with service being shared equally amongst all customers in the queue. The service time distributions have rational LSTs.

Type 3. Service discipline is last-come-first-served pre-emptive resume (LCFS-PR). There is a single server at the node, and new arrivals displace the customer in service (if any) to obtain immediate service. The service time distributions have rational LSTs.

Type 4. Infinite server (IS), which provide immediate service to all customers in the queue. Each class of customers may have a distinct service time distribution. The service time distributions have rational LSTs.

The closed queueing network (CQN) modeling can be used to carry out reliability analysis of an n-unit repairable system as the number of units in the system remains constant, i.e., n. This is possible, if the life time of a unit is identified as 'service' in the operating location, a node. To see the application of CQNs in reliability modeling, we describe below in Section 2 the definition and relevant available results on CQNs. In Section 3 we review queueing networks modeling in reliability theory illustrating the use of CQNs in the analysis of complex redundant repairable systems.

2 DEFINITIONS AND RESULTS ON CQNs

2.1 Single Class Gordon Newell Networks (1967)

These networks consist of fixed population, say N, and each of the nodes, say M (in number), is of type 1 described in Section 1. Let n_i denote the number of customers at node i. Obviously, the state space of the network is finite and given by

$$S = \left\{ (n_1, \cdots, n_M) \middle| n_i \geq 0, \sum_{i=1}^{M} n_i = N \right\} \tag{1}$$

The number of elements in the set S can be determined by simple cominatorial arguments and is the same as the number of ways of putting N balls into M cells, i.e.,

$$\binom{N + M - 1}{N} \quad \text{or} \quad \binom{N + M - 1}{M - 1} \qquad \text{(Feller 1985, P.38)}$$

2.2 Multiclass BCMP QNs (1975)

After the CQN of Gorden and Newell (1967), Baskett et al. (1975) have developed and analyzed one of the most general QN, known as the BCMP model. These networks include nodes of all the 4 types described in Section 1. The customers can belong to R different types of job classes. Moreover, these networks also include the possibility of customers changing class from node to node. Thus, these are known as multiclass networks with switching. Through the BCMP model, we can obtain equilibrium state probabilities, and hence other dependability/reliability measures, in explicit computational product form. However, these product form solutions do not enable us to obtain MTSF. As nodes of types 2,3 and 4 need not have exponential service time distributions to maintain the Markov property in the underlying process, it is needed to preserve the memoryless property in service times. This has been possible by using the Cox-phase distribution, C_k, Cox (1955).

The distribution C_k consists of a sequence of k independent exponential phases with service rate μ_i at pharse i $(1 \leq i \leq k)$. After completing service in phase i, the unit either enters phase $i + 1$ with probability a_i or completes service with probability $b_i, b_i = 1 - a_i, i = 1, 2, \cdots, k - 1, b_k = 1$.

The Cox-distribution described above has rational LST:

$$f^*(s) = \sum_{j=1}^{k} b_j A_j \prod_{r=1}^{j} \left(\mu_r / (s + \mu_r) \right) \tag{2}$$

where $A_j = a_1 a_2 \ldots a_{j-1}$.

The mean of the distribution is given by

$$\sum_{i=1}^{k} A_{i-1} b_i \left(\sum_{r=1}^{i} 1/\mu_r \right) \tag{3}$$

The Cox distribution C_k is highly versatile because it generalizes all well known distributions such as the generalized hyper exponential (GHE_k), the generalized Erlang (GE_k), the mixed generalized Erlang (MGE_k), and phase type (pH) and distribution functions whose LSTs are reciprocals of polynomials of degree k (K_k), and also permits CV^2, the square coefficient of variation, to have any value in $[0, \infty]$, Botta et al. (1987). As such, it is vastly applicable in modeling systems like computerss, cummunication networks, and flexible manufacturing systems.

Let node transition probabilities be defined by the set $\{p_{(i,r)(j,s)}\}$, which describe the probability that a class r customer at node i goes next to node j as a class s customer.

2.3 Traffic Equations

Suppose e_{ir} is the relative frequency of visits to node i by class r customers; the e_{ir} satisfy the following traffic equations:

$$e_{js} = \sum_{r=1}^{R} \sum_{i=1}^{M} e_{ir} p_{(i,r)(j,s)}, \text{ for } 1 \leq j \leq M \text{ and } 1 \leq s \leq R. \tag{4}$$

The solution to this set of linear equations can only be determined up to a multiplicative constant.

As is obvious from (4), the number of traffic equations is independent of N and k (the parameter of the Cox distribution C_k) and is at most equal to the product of the number of nodes and the number of job classes in the network.

2.4 Steady State Probabilities

Let

n_{ir} : number of class r customers at node i

$n_i : \sum_{r=1}^{R} n_{ir}$

$n_i : (n_{i1}, n_{i2}, \cdots, n_{ir}, \cdots, n_{iR})$, a state of node i

$n : (n_1, n_2, \cdots, n_M)$, a state of the network

$N_r : \Sigma_{i=1}^{M} n_{ir}$, total number of class r customers in the network

$N_R : (N_1, N_2, \cdots, N_R)$

$N : \Sigma_{r=1}^{R} N_r$, the total number of customers in the network

s_{ir} : mean service time of class r customers at node i

$Y_{ir} = e_{ir} s_{ir}$

For any solution $e_{ir} (i = 1, \cdots, M, r = 1, \cdots, R)$, the general solution of the global balance equation

$p(n)[$ instantaneous transition rate out of state n $]$

$= \Sigma p(n')[$ instantaneous transition rate from n' to $n]$

s.t. $\Sigma_n p(n) = 1$

has the form

$$p(n) = G(N_R) \prod_{i=1}^{M} f_i(n_i) \tag{5}$$

where $G(N_r)$ is the normalization constant that ensures that $\Sigma_n p(n) = 1$ and can be determined by following Bruell and Balbo (1980, pp 61-65). The factor $f_i(n_i)$ depends on the type of node i; and we have

$$f_i(n_i) = \begin{cases} (n_i! / \prod_{k=1}^{n_i} \mu_i(k)) \prod_{r=1}^{R} e_{ir}^{n_{ir}} / n_{ir}! & \text{if node } i \text{ is load} \\ & \text{dependent and has a FCFS,} \\ & \text{PS, or LCFS-PR} \\ & \text{service discipline} \\ n_i! \prod_{r=1}^{R} (e_{ir} s_{ir})^{n_{ir}} / n_{ir}! & \text{if node } i \text{ is load} \\ & \text{independent and has a} \\ & \text{FCFS, PS or LCFS-PR} \\ & \text{service discipline} \\ \prod_{r=1}^{R} (e_{ir} s_{ir}) n_{ir} / n_{ir}! & \text{if node } i \text{ has an IS} \\ & \text{service discilpine} \end{cases} \tag{6}$$

At a FCFS station all customers must have the same mean service time i.e. $s_{i1} = s_{i2} = \cdots = s_{iR}$.

Thus, once the relative rates e_{ir} are determined, the steady-state probabilities p(n) can be obtained as product form solutions.

2.5 Marginal Probability Distribution

The marginal probability of finding exactly n_r customers of class r $(r = 1, 2, \cdots, R)$ at node i is given by (Bruell and Balbo 1980)

$$p_i(n_R, N_R) = f_i(n_R) g_M^i (N_R - n_R) / G(N_R) \quad \text{for } n_R = 0_r, \cdots, N_R \tag{7}$$

where the auxiliary function

$$g_M^i(N_R) = G(N_R) - \sum_{k_R=0_R}^{N_R} \delta(k_R) f_i(k_R) g_M^i(N_R - k_R) \qquad (8)$$

and

$$\delta(k_R) \equiv \begin{cases} 0 \text{ if } k_1 = k_2 = \cdots = k_r = 0 \\ 1 \text{ otherwise} \end{cases}$$

$$\begin{aligned} n_R &= (n_1, n_2, \cdots, n_R) \text{ the vector representing the number of class} \\ r\ (r &= 1, 2, \cdots, R) \text{ customers at a node.} \end{aligned}$$

Values for this auxiliary function are computed iteratively starting with the initial condition

$$g_M^i(O_R) = G(O_R) = 1 \qquad (9)$$

2.6 Throughput

Let $X_{ir}(N_R)$ denote the throughput of class r customers at node i in a network with N_R customers. Then

$$\begin{aligned} X_{ir}(N_R) &= \sum_{n_R=1_r}^{N_R} p_i(n_R, N_R) n_{ir} \mu_i(n_i)/n_i \\ &= e_{ir} G(N_R - 1_r)/G(N_R) \end{aligned} \qquad (10)$$

This result is true for any of the four types of queueing disciplines. It also applies for nodes with load dependent or load independent service rates.

2.7 Utilization

Let $U_{ir}(N_R)$ denote the utilization of node i by class r customers in a network with N_R jobs. (This measure does not apply for an IS node).

For the load independent case:

$$\begin{aligned} U_{ir}(N_R) &= s_{ir} X_{ir}(N_r) \\ &= Y_{ir} G(N_R - 1_r)/G(N_R) \end{aligned} \qquad (11)$$

For the load dependent case:

$$U_{ir}(N_R) = \sum_{n_R=1_r}^{N_R} p_i(n_R, N_R) n_{ir}/n_i \qquad (12)$$

2.8 Mean Queue Length

Let $\bar{n}_{ir}(N_R)$ denote the average number of class r customers at node i in a network with (N_R) jobs. Then,

$$\bar{n}_i(N_R) = \sum_{n_R=1_r}^{N_R} p_i(n_R, N_R)n_r, \text{ FCFS, PS and LCFS-PR nodes} \tag{13}$$

and the total mean queue length may be computed from

$$\bar{n}_i(N_R) = \sum_{r=1}^{R} U_{ir}(N_R)(1 + \bar{n}_i(N_R - 1_r)) \tag{14}$$

with the initial condition $\bar{n}_i(O_R) = 0$.

For an IS node

$$\bar{n}_{ir}(N_R) = X_{ir}(N_R)s_{ir} \tag{15}$$

2.9 Mean Waiting Time

Let $\bar{W}_{ir}(N_R)$ denote the mean waiting time of a customer of class r at node i in a network with N_R jobs. Then

$$\bar{W}_{ir}(N_R) = \bar{n}_{ir}(N_R)/X_{ir}(N_R), \text{ for the FCFS, PS or LCFS-PR nodes} \tag{16}$$

If station i is load independent then the mean waiting time can also be written as a recursion

$$\bar{W}_{ir}(N_R) = s_{ir}(1 + \bar{n}_i(N_R - 1_r)) \tag{17}$$

2.10 Mean Response Time

The mean return time of a customer to an IS node is also referred to as the node's mean response time. Let $T_{ir}(N_R)$ denote the mean response time of a class r job at node i when there are N_R jobs in the closed network.
Then

$$T_{ir}(N_R) = (N_r - \bar{n}_{ir}(N_R))X_{ir}(N_R) \tag{18}$$

where the numerator denotes the mean number of class r jobs in the remainder of the network.

3 Review of CQN Modeling in Reliability

Ravichandaran (1990) has demonstrated the use of single class CQN for 2-unit stand by systems with exponential failure and repair rates. He has shown how through CQN modeling consisting of only type I nodes, the system, even when transit from work place to repair facility and back is exponential r.v., can be analyzed much more easily than with the usual Markov approach. He has also shown how a simple

n-unit parallel system with single repair facility, under the assumption of exponential repairs and K-Erlang lifetimes, can also be modeled as a single class CQN with nodes of types 1 and 4.

Using the single class CQN model Agarwal and Templeton (1990) analyzed a 1-unit repairable system operating under three types of environments changing from one type to another at different exponential rates. Failure time distributions of the system from all types of environments are assumed to be exponential whereas repair time distributions are different and general. In order to account for the effects of environment changes during a repair, along with keeping the model computationally tractable, the repair time distributions are chosen as GE_3 which allow good approximation of a class of distributions with $CV^2\epsilon$ [1/3,1/2]. It is their pioneering work to account for the effect of continuous environment changes which, but for QN modeling could not have been possible in systems with general repair time distributions. Later, Templeton and Agarwal (1990) generalized the model by approximating the repair time distributions by (2,2) series-parallel distributions. Further, Agarwal and Chaudhuri (1995a) developed a more general model of these models for approximating the repair time distributions by C_2 as C_2 covers a much wider class of distributions, i.e. having $CV^2\epsilon[0.5,\infty]$. Recently, in 1997, they have further extended this model by approximating the failure time distributions by C_2.

In (1995b) Agarwal and Chaudhuri analyzed a Markovian complex (K,N):G repairable system with the units having two failure modes, which are diagnosed correctly only with fixed probabilities, thus generalizing the model of (Ibe and Wein 1992) to n-units. The system is modeled as a multiclass CQN with switching to obtain steady-state probabilities and availability. They (1996a) generalized the model by taking the repair time distributions to be general and analyzed the system as a BCMP model on approximating the repair distributions by Cox distributions, C_k. They (1996b,c) generalized this model further by incorporating, respectively, the real life assumptions of non-instantaneous failure-mode diagnosis and the possibility of unsatisfactory repairs resulting in some post-repair. They have numerically computed stationary availability of the system even for N=10, which, otherwise, would have been impossible through Markov modeling. Chaudhuri and Agarwal (1996b) also computed the stationary availability of an n-unit shared-load parallel system with R parallel servers by modeling it as a CQN with load dependent nodes.

We give below three different models by Agarwal and Chaudhuri (1995a) and Chaudhuri and Agarwal (1996a, 1996b) illustrating different types of appropriate queueing network models to obtain stationary availability by using a product from solution.

3.1 A Repairable System Operating under Fluctuating Weather (Agarwal and Chaudhuri (1995a))

This model illustrates the use of single class CQN in steady-state availability analysis of a 1-unit non-Markovian system operating under different changing environments, accounting for the effects of environment changes during a repair, along with keeping the model computationally tractable.

The Model

(1) A one-unit repairable system operates under three types of environments: normal, abnormal 1 and abnormal 2.

(2) The environment can change from one type to another type at different exponential rates.

(3) The failure time distributions for the system operating under all of the environments are exponential and different.

(4) The repair time distributions for the system failed from different environments are different and general and are approximated by Cox distributions C_2.

(5) Repair time of the system is affected by continuous environment changes during its repair. However, a repair does not have to start over from beginning when the environment changes: it is sufficient to continue the current phase of repair with the transition determined by the new environment state.

The system at any instant can be one of the following states:

S_i: the system is operating in environment i, $i = 1, 2, 3$
S_{ij}: the system is under the jth exponential phase of repair in environment
   i, $i = 1, 2, 3$ and $j = 1, 2$.

To find the steady state probabilities, the system can be modeled as a CQN with one customer i.e. N=1. The nine possible states of the system can be viewed as nine nodes of the CQN, with each node having a single server. Obviously, the service time at each node is exponential. The nodes are connected with each other according to the transition probability matrix and hence the traffic equations can be written to get the values of the e_i's.

The solution for the steady state joint probability, P_S, is given by

$$P_S = p_{n_1, n_2, \cdots, n_9} = G(1) \prod_{i=1}^{9} (e_i s_i)^{n_i}$$

As the normalization constraint is

$$\sum_s P_s = 1$$

we get

$$G(1) = [\sum_{i=1}^{9} (s_i s_i)]^{-1}$$

(as there is only one unit in the system).

Hence the steady state probabilites, π_i's, that the unit is in node i, are given by

$$\pi_i = p_{0,0,\cdots,1,0\cdots 0} = e_i s_i / \sum_{i=1}^{9} e_i s_i, i = 1, 2, \cdots, 9 \tag{19}$$

(ith position)

Therefore, the steady state availability

$$= \sum_{i=1}^{3} \pi_i.$$

3.2 A Non-Markovian K-out-of-N:G System (Chaudhuri and Agarwal (1996a))

This model illustrates how the stationary availability of a nonMarkovian complex (K,G):G repairable system, which includes the possibility of incomplete and delayed repairs, can be obtained by modeling it as a BCMP MCQN with switching.

The Model

(1) The system is a K-out-of-N:G system with identical units.

(2) A unit has two s-independent failure modes: 1 and 2, which may not be diagnosed correctly.

(3) The repair, upon failure of a unit, depends on the diagnosis of the failure mode, i.e., if the failure is diagnosed as mode-i, then the unit is repaired for mode-i failure regardless of the true failure mode (i=1,2).

(4) The time to the occurrence of a mode-i failure (called primary failure) is exponentially distributed with rate $\lambda_i (i = 1, 2)$.

(5) The diagnosis time i.e., the time to diagnose the failure mode, is exponentially distributed with rate μ_2.

(6) When a primary mode-i, failure is repaired as the opposite failure mode on wrong diagnosis, the time to occurrence of mode-i failure (i=1,2)(called associated incidental failure) is exponentially distributed with rate $\lambda_3(\lambda_4)$.

(7) The probabilities of correct diagnosis of mode-1 (mode-2) failure, primary or incidental, are p (q), respectively.

(8) If one mode of failure is not correctly diagnosed before the opposite mode occurs, then an incorrect diagnosis of the later does eliminate the former failure.

(9) After each repair, the unit is tested to see whether it meets certain predefined specifications. If it does, it goes for operation, otherwise, undergoes some repair to be called 'post-repair'. The testing time is exponentially distributed with rate μ_4.

(10) The type-i, repair and post-repair time distributions for a unit, assumed to be general are approximated by Cox distributions.

(11) There are two testing centers, one for failure-mode diagnosis and another for repair testing, and two repair facilities, one each for repairs and post-repairs. The repair and post-repair disciplines are assumed to be PS. As, normally, both failure-mode diagnosis and repair-testing take much smaller time in comparison to repairs/post-repairs, failure-mode diagnosis and repair-testing disciplines are assumed to be FCFS.

To do the CQN modeling, we identify different possible states in which a unit can stay and hence bring the analogy of job class of a QN.

The operating states of the unit can be classified as:

S_1 : it is fully up i.e., working in full efficiency and it may fail as primary mode-i with exponential rate λ_i (i=1,2).

S_2 : it is marginally up i.e., operating in less efficiency after getting repair of type-2 on wrong diagnosis of mode-1 failure.

S_3 : it is marginally up i.e., operating in less efficiency after getting repair of type-1 on wrong diagnosis of mode-2 failure.

The failure-mode diagnosis state of a unit can be classified as:
S_{3+j}: The unit is under failure-mode diagnosis upon failure from $S_j, j = 1, 2, 3$.

Depending upon the failure mode of a unit and its subsequent diagnosis, the repair state of the unit can be classified as:

S_7 : The unit is undergoing type-1 repair on correct diagnosis of mode-1 failure, either primary (from S_1) or incidental (from S_2).

S_8 : The unit is undergoing type-1 repair on wrong diagnosis of primary mode-2 failure as mode-1 (from S_1).

S_9 : The unit is undergoing type-1 repair either on correct diagnosis of primary mode-1 failure (from S_3) or wrong diagnosis of primary mode-2 failure (from S_2).

S_{10} : The unit is undergoing type-1 repair on wrong diagnosis of incidental mode-2 failure (from S_3).

S_{11} : The unit is undergoing type-2 repair on correct diagnosis of mode-2 failure, either primary (from S_1) or incidental (from S_3).

S_{12} : The unit is undergoing type-2 repair on wrong diagnosis of primary mode-1 failure as mode 2 (from S_1).

S_{13} : The unit is undergoing type-2 repair either on correct diagnosis of primary mode-2 failure (from S_2) or wrong diagnosis of primary mode-1 failure (from S_2).

S_{14} : The unit is undergoing type-2 repair on wrong diagnosis of incidental mode-1 failure (from S_2).

The repair-testing state of a unit can be classified as:

S_{8+j}: The unit is under testing on repair completion in $S_j, j = 7, 8, \cdots, 14$.

Similarly, the post-repair state of a unit can be classified as:

S_{16+j}: The unit is under post-repair after getting tested in $S_{8+j}, j = 7, 8, \cdots, 14$.

Thus one can visualize the system as a CQN, with 5 nodes and N customers which belong to 30 different classes. A unit is in node (the 'operating' node), node 2 (the 'failure-mode diagnosis' node), node 3 (the 'repair' node), node 4 (the 'post-repair' node), according as it is in the operating states $\{S_1, S_2, S_3\}$, diagnosis states $\{S_4, S_5, S_6\}$, repairing states $\{S_7, S_8, \cdots, S_{14}\}$, testing states $\{S_{15}, S_{16}, \cdots, S_{22}\}$ or post-repair states $\{S_{23}, S_{24}, \cdots, S_{30}\}$, respectively.

From the state transitions of a unit discussed above, it is clear that, while a unit can change class membership from any 3 job classes at node 2 to any of the 8 job classes at node 3 and from any of these 8 job classes at node 5 to any of the 3 job classes at node 1, the customers go from node 1 to node 2 and from node 3 to node 4 without changing their class membership. Thus, the number of effective different job classes of customers in the QN is only 11.

The lifetime of a unit in a particular upstate at node 1 is identified as the service time of a customer belonging to the corresponding class at node 1. Since, at any instant of time any number of units our of N can be in an operating state, node 1, may be visualized as a service node with infinite servers. Thus, according to the assumptions of failure times, the service times for class $r(= 1, 2, 3)$ customers at node 1 are exponential. Node 2 and 4, both are single server, with service discipline FCFS and exponential service rates. Further, nodes 3 and 5 are single server with service discipline PS and have the Cox-phase service times.

To see the advantage of CQN modeling of the system over Markov modeling (MM), we give below a detailed comparsion in tabular form.

Table: Number of nodes, job classes/system states and traffic
equations/balance equations in CQN Modeling/Markov Modeling(MM)

	CQN	MM
N=1, k=1		
Nodes	5	-
job classes/system states*	11	30
Traffice eqs./balance eqs.	30	30
N=2, k=2		
Nodes	5	-
job classes/system states*	11	$\binom{47}{2} = 1081$
Traffic eqs./balance eqs.	30	1081
N=10, k=2		
Nodes	5	-
job classes/system states*	11	$\binom{55}{10} = 29248649430$
Traffic eqs./balance eqs.	30	29248649430

*The formula $\binom{N+r-1}{N}$ (Feller, 1985) is used where N is the number of units and r is the number
of possible states of a unit (each of the 16 repair states will be bifurcated into 2 states resulting
in $r=46$ states).

3.3 A Shared-load (K,N): G System with Parallel Repairs
(Chaudhuri and Agarwal 1996b)

This model illustrates the use of CQN where the service rate at a node is load
dependent and has multiple servers.

The Model

(1) The system considered consists of N identical units with exponential failures
and repairs.

(2) Failure of a unit increases the failure rate of the remaining operating units.

(3) There are R repair men. The repair discipline is FCFS.

As at any time point, a unit can be either in operating state or in down state,
the system can be visualized as a single class CQN. The network consists of two
nodes, the 'operating' node and the 'repairing' node with N customers. The service
discipline at the 'operating' node is with exponential service times. As the system
is shared-load, the service rate changes with the change in the number of customers
at the 'operating' node. The service discipline at the 'repairing' node is FCFS
and there are R identical exponential servers. The service time distributions are
identical and exponential. The relative arrival e_i are given by $e_1 = e_2 = c$ (say).

Let the service times of customers at node 1 be exponential with rates α or
$\alpha_i (i = 1, 2 \cdots N - 1)$, respectively, according as whether all units are operating or
i units have failed ($\alpha < \alpha_i < \alpha_{i+1}, i = 1, 2 \cdots N - 1$).

i units have failed ($\alpha < \alpha_i < \alpha_{i+1}, i = 1, 2 \cdots N - 1$).

Thus for node 1 (the operating node), from(6),

$$f_1(n) = (e_1 s_1)^n / n! = \begin{cases} c^n (1/\alpha_{N-n})^n / n!, N - k \leq n < N \\ c^n (1/\alpha)^n / n!, n = N \end{cases}$$

and for node 2 (the repairing node), from (6),

$$f_2(n) = \begin{cases} (e_2 s_2)^n / n! \text{ if } 0 \leq n \leq R \\ (e_2 s_2)^n / R! R^{n-R} \text{ if } n \geq R \end{cases}$$

Marginal probabilites $p_i(n_i, N)(i = 1, 2)$ can be obtained using (7) and hence the availability.

It may be mentioned here that, if the failure of a unit does not affect the failure rates of the remaining operating units, i.e., $\alpha_i = \alpha(i = 1, 2, \cdots, N - 1)$, then the system becomes a simple K-out of N:G system with R repairmen. The availability so obtained coincides with that given in John G. Rau (1970, p, 252).

References

Agarwal, M., and Chaudhuri, M. (1997), "Availability Analysis of a Non-Markovian System Operating Under Fluctuating Environment", presented at 84th Indian Science Congress, Delhi, Jan 3-8.

Agarwal, M., and Chaudhuri, M. (1996a), "Queueing Network Approach to Availability Analysis of a K-out-of-N:G system with Partially Observable Failures and Coxian Repair Time Distributions", A.C. Borthakur and H. Choudhaury (eds.), *Probability Models and Statistics*: A.J. Medhi Festschrift, New Age International Publishers, New Delhi, 211-226.

Agarwal, M., and Chaudhuri, M. (1996b), "A Closed Queueing Network Model for Availability Analysis of a K-out-of N:G System with Cox-phase Delayed Repairs", (revised version submitted to EJOR).

Agarwal, M., and Chaudhuri, M. (1996c), "A Closed Queueing Network Model in Availability Analysis of a K-out-of N:G System with Two Types of Failure and Cox-phase Repairs" (communicated).

Agarwal, M., and Chaudhuri, M. (1995a), "A Repairable System Operating under Fluctuating Weather with Coxian-2 Repair-time Distributions", (communicated).

Agarwal, M., and Chaudhuri, M. (1995b), "Queueing Network Modeling for Availability Analysis of a K-out-of-N:G System with Partially Observable Failures", *Journal of Indian Statistical Association*, **33**, 107-121.

Agarwal, M., and Templeton, J.G.C. (1990), "Stochastic Analysis of a Repairable System under Fluctuating Weather", *International Journal of Systems Science*, **21**, 2019-2028.

Baskett, F., Chandy, K.M., Mutz, R.R., and Palacios, F.G.(1975), "Open Closed and Mixed Networks of Queues with Different Classes of Customers", *Journal of the Association for Computing Machinery*, **22**, 248-260.

Botta, R.F., Harris, C.M., Marchal, W.G. (1987), "Characterization of Generalized Hyperexponential Distribution Functions", *Communications of Statistics-Stochastic Models*, **3**, 115-148.

Bruell, S.C., and Balbo, G. (1980), *Computational Algorithms for Closed Queueing Networks*, North Holland.

Chaudhuri, M., and Agarwal, M. (1996a), " A Multiclass Closed Queueing Network Model to Study Availability of a Non-Markovian K-out-of-N:G System with Failure Mode and Repair Analysis", (communicated).

Chaudhuri, M., and Agarwal, M. (1996b), "Queueing Network Approach to Availability Analysis of a Shared-load (K,N):G System with Parallel Repairs", presented at Int. Conf. on Quality Improvement Through Statistical Methods, Cochin, Dec. 29-31.

Cox, D.R. (1955), "A Use of Complex Probabilities in the Theory of Stochastic Processes", *Proceedings of Cambridge Philosophical Society*, **51**, 311-319.

Feller, W. (1985), An *Introduction to Probability Theory and Its Applications*, New Delhi: Wiley Eastern.

Gordon, W.J., and Newell, G.P. (1967), "Closed Queueing Systems with Exponential Servers", *Operations Reserach*, **15**, 254-265.

Harrison, P.G. and Patel, N.M. (1993), *Performance Modelling of Communication Networks and Computer Architectures*, Addison Wesley.

Ibe, O.C., and Wein, A.S. (1992), "Availability of Systems with Partially Observable Failures", *IEEE Transactions on Reliability*, **41**, 92-96.

Jackson, J.R. (1957), "Networks of Waiting Lines", *Operations Research*, **5**, 518-522.

Jackson, J.R. (1963), "Jobshop Like Queueing Systems", *Management Science*, **10**, 131-142.

Rau, J.G. (1970), *Optimization and Probability in Systems Engineering*, New York: Van Nostrand Reinhold.

Ravichandran, N. (1990), *Stochastic Methods in Reliability Theory*, New Delhi: Wiley Eastern.

Templeton, J.G.C. and Agarwal, M. (1990), " A Repairable System with Fluctuating Weather and Series-parallel Repair Time Distributions", *SCIMA*, **19**, 79-91.

ROLE OF STOCHASTIC ORDERINGS
IN SPARE ALLOCATION IN SYSTEMS

HARSHINDER SINGH
Department of Statistics
Panjab University
Chandigarh

Allocation of spare components in order to optimize the lifetime of the system with respect to a suitable criterion is of considerable interest in reliability, engineering, industry and defense. In this review article we emphasize the role of stochastic orderings in active spare allocation problems considered in the literature.

1 Introduction.

The problem of allocating a spare component in a system in such a way as to optimize, in some sense, the life time of the system is an important problem in reliability theory having important applications in engineering, industry and defense. There are mainly two types of spares (redundancy) that are used, namely (i)active (parallel), where the spare is allocated in parallel to the component and starts functioning at the same time as the component and (ii) standby, where the spare starts working only when it actually replaces the failed component. Problems of optimal allocation of spares have posed many interesting theoretical problems in probability. For references see Barlow and Proschan (1975), El-Neweihi et al (1986), Boland et al (1988, 1991, 1992, 1994), Shaked and Shanthikumar (1992), Boland and El-Neweihi(1995), Singh and Misra (1994), Singh and Singh (1996, 1997).

Consider a simple example where T_1 and T_2 are two nonnegative random variables denoting the lifetimes of two components in a series system. Assume that life time of a spare component is T which is available for active allocation to one of the two components in order to increase the reliability of the system. There are two choices available: one is to assign the spare in parallel to the component with life time T_1 in which case the life time of the resulting system is

$$X = \min(\max(T_1, T), T_2)$$

The other choice is to assign the spare component to the component with life time T_2, in which case the resulting system has life time

$$Y = \min(T_1, \max(T_2, T))$$

We shall prefer the first choice to the second if X is larger (in some sense) than Y. X and Y are random variables. What do we mean that X is larger than Y? Here comes the role of stochastic orderings in spare allocation in systems. In Section 2, we define the common stochastic orderings between two random variables namely the expectation ordering, usual stochastic ordering, hazard rate ordering and the likelihood ratio ordering. In Sections 3-6, we review literature on different problems of active spare allocation emphasizing the role of stochastic orderings in these problems.

In Section 3 we discuss the role of usual stochastic ordering and hazard rate ordering in a problem of allocation of spare to a k-out-of-n system. Section 4 emphasizes the role of usual stochastic ordering, hazard rate ordering and likelihood ratio ordering in componentwise versus systemwise spare allocation. In Section 5 we discuss relevance of usual stochastic ordering and hazard rate ordering in allocation of resources to nodes of a series system. Some unsolved problems in these areas of spare allocations have also been pointed out in the sections.

2   Stochastic Orderings.

Let X and Y be two random variables having cumulative distribution functions F(x) and G(x) respectively. The simplest way of defining X larger than Y is the expectation ordering.

2.1   Definition.

X is said to be larger than Y in expectation ordering (written as $X \overset{E}{\geq} Y$) if $E(X) \geq E(Y)$.

Expectation ordering is a complete ordering, that is, for any two random variables X and Y either $X \overset{E}{\geq} Y$ or $Y \overset{E}{\geq} X$. However it is based on only two numbers and therefore it is often not very informative. In addition, the means sometimes do not exist. Comparing distribution functions is more informative.

2.2   Definition.

X is said to be larger than Y in usual stochastic ordering(written as $X \overset{st}{\geq} Y$) if

$$P[X > x] \geq P[Y > x] \text{ for every real x}$$

or equivalently

$$F(x) \leq G(x) \text{ for every real x}$$

An important characterization of the usual stochastic order is the following theorem:

Theorem 2.1. Two random variables X and Y satisfy $X \overset{st}{\geq} Y$ if and only if there exist two random variables X^* and Y^* defined on the same probability space such that X^* has same distribution as X, Y^* has the same distribution as Y and $P[X^* \geq Y^*]=1$.

Another important characterization result is given by the following theorem:

Theorem 2.2. $X \overset{st}{\geq} Y$ if and only if $E[\phi(X)] \geq E[\phi(Y)]$ for all increasing functions ϕ.

In reliability and survival analysis, apart from the survival function, the hazard rate function also plays an important role. Let X be an absolutely continuous nonnegative random variable with probability density function f(x) and survival function $\overline{F}(x) = 1\text{-}$ F(x). The hazard rate function of X is defined as:

$$\lambda(x) = \frac{f(x)}{\overline{F}(x)}$$

for all those x for which $\overline{F}(x) > 0$ (see Barlow and Proschan(1975)).
Clearly the higher the hazard rate, the smaller X should be. This is the motivation behind defining the hazard rate ordering.

2.3 Definition.

Let X and Y be absolutely continuous nonnegative random variables with hazard rate functions $\lambda(x)$ and $\beta(x)$ respectively. X is said to be larger than Y in the hazard rate ordering (written as $X \overset{hr}{\geq} Y$) if

$$\lambda(x) \leq \beta(x)$$

for all those x for which $\lambda(x)$ and $\beta(x)$ are defined.

The following theorem characterizes the hazard rate ordering.

Theorem 2.3(a).$X \overset{hr}{\geq} Y$ if and only if $\frac{\overline{F}(x)}{\overline{G}(x)}$ is an increasing function of x.

(b) $X \overset{hr}{\geq} Y$ if and only if

$$(X - t)/X > t \overset{st}{\geq} (Y - t)/Y > t \ \ \text{for every} \ \ t \geq 0.$$

Thus hr ordering is stronger than the st ordering. More stronger than hr ordering is the likelihood ratio ordering.

2.4 Definition.

Let X and Y be random variables with probability density functions f(x) and g(x) respectively. X is said to be larger than Y in the likelihood ratio ordering(written as $X \overset{lr}{\geq} Y$) if $\frac{f(x)}{g(x)}$ is an increasing function of x.

lr ordering does not have any clear intuitive justification. But in many situations they are easy to verify.

We have the following chain of implications for these orderings

$$X \overset{lr}{\geq} Y \implies X \overset{hr}{\geq} Y \implies X \overset{st}{\geq} Y \implies X \overset{E}{\geq} Y$$

For a detailed study of these orderings see Ross(1983), Singh (1989), Deshpande et al(1990), Shaked and Shanthikumar(1994).

3   Usual Stochastic and Hazard Rate Ordering in Allocation of a Spare to a k-out-of-n System.

We reconsider the problem active spare allocation to a series system of two components I and II with lifetimes T_1 and T_2 respectively. Let T_1 and T_2 be independent and let $T_2 \overset{st}{\geq} T_1$. Let life time T of a spare component be independent of T_1 and T_2. Then with respect to the usual stochastic ordering it is more favorable to make the spare allocation with stochastically weaker component I than with component II , that is,

$$\min(\max(T_1, T), T_2) \overset{st}{\geq} \min(T_1, \max(T_2, T))$$

More generally, let $T_1, T_2, \ldots, T_n$ denote the independent lifetimes of a k-out-of-n system where

$$T_n \overset{st}{\geq} T_{n-1} \overset{st}{\geq} \ldots \overset{st}{\geq} T_1$$

Suppose T is the lifetime of an independent spare, then

$$\tau_k^{(1)} \overset{st}{\geq} \tau_k^{(2)} \overset{st}{\geq} \ldots \overset{st}{\geq} \tau_k^{(n)}$$

for all $k = 1, 2, \ldots, n$, where $\tau_k^{(i)}$ denotes the life time of the system when the spare is allocated to component i with lifetime T_i. Thus with respect to the usual stochastic ordering, it is most favorable to allocate the spare component with the stochastically weakest component when the lifetimes of the components of the system are independent and ordered with respect to the usual stochastic ordering. For proofs and more details of these results see Boland et al (1992).

In view of the importance of hazard rates it is of interest to know whether the results similar to above hold when we compare life times of the resulting systems with respect to the hazard rate ordering. Curiously these do not hold as brought out by the following example:

Example. Let T_1, T_2 and T have independent exponential distributions with means 10/3, 10, and 2 respectively. The hazard rate functions $\lambda_{T_1}(t)$ and $\lambda_{T_2}(t)$ of T_1 and T_2 are given by

$$\lambda_{T_2}(t) = 1/10 < 3/10 = \lambda_{T_1}(t), \quad \text{for all } t > 0.$$

Thus $T_2 \overset{hr}{\geq} T_1$. However, plots of hazard rate functions of $X = \min(\max(T_1, T), T_2)$ and $Y = \min(T_1, \max(T_2, T))$ satisfy

$$\lambda_X(t) \leq \lambda_Y(t), t < 3.65; \lambda_X(t) > \lambda_Y(t), t > 3.65$$

Thus X and Y are not comparable with respect to the hazard rate ordering. So allocation of an idependent spare to the weaker component with respect to hr ordering does not always optimize the life time of the series system of two independent components with respect to the hazard rate ordering. Singh and Misra(1994) gave sufficient conditions for X to be larger than Y in hr ordering when T_1, T_2 and T have independent exponential distributions. Unlike the case of stochastic ordering here the distribution of the spare component plays a role in the comparisons of the resulting two systems with respect to hr ordering. Sufficient conditions for X to be larger than Y in hr ordering when T_1, T_2 and T have arbitrary distributions still needs to be investigated.

4   Usual Stochastic, Hazard Rate and Likelihood Ratio Orderings in Componentwise Versus Systemwise Spare Allocation.

Again consider a simple series system of two independent components with lifetimes T_1 and T_2 respectively. Suppose two spares with life times T_1^* and T_2^* are available for active spare allocation. Let V be the lifetime of the system when the two spares are allocated componentwise and let W be the lifetime of the system when these are allocated systemwise, that is

$$V = \min(\max(T_1, T_1^*), \max(T_2, T_2^*))$$

$$W = \max(\min(T_1, T_2), \min(T_1^*, T_2^*))$$

Assuming that T_1, T_2, T_1^*, T_2^* are independent,

$$V \overset{st}{\geq} W$$

More generally for any coherent system with structure function $\tau(x_1, x_2, \ldots, x_n)$

$$\tau(\max(T_1, T_1^*), \max(T_2, T_2^*), \ldots, \max(T_n, T_n^*))$$

$$\overset{st}{\geq} \max(\tau(T_1, T_2, \ldots, T_n), \tau(T_1^*, T_2^*, \ldots, T_n^*))$$

where T_i's and T_i^*'s are independent (see Barlow and Proschan(1975) for proof). Thus componentwise active spare allocation is superior to systemwise spare allocation with respect to the usual stochastic ordering.

Recently Boland and El-Neweihi (1995) investigated this problem in the hazard rate ordering. They demonstrated through an example that for independent but nonidentical spares and components it does not hold always. For a 2-out-of-n system when component and spares lifetimes are independent and identically distributed, they showed that component wise spare allocation is more favorable than systemwise spare allocation in the hazard rate ordering and conjectured that it shall hold for a general k-out-of-n system. Singh and Singh(1997) established their conjecure. In fact they proved a stronger result that componentwise spare allocation is better than systemwise allocation in the likelihood ratio ordering which is stronger than the hazard rate ordering. Thus if $T_i, T_i^*, i = 1, 2, \ldots, n$ are independent and identically distributed, then

$$\tau_{k/n}(\max(T_1, T_1^*), \ldots, \max(T_n, T_n^*))$$

$$\overset{lr}{\geq} \max(\tau_{k/n}(T_1, \ldots, T_n), \tau_{k/n}(T_1^*, \ldots, T_n^*)),$$

where $\tau_{k/n}(x_1, \ldots, x_n)$ denotes the structure function of a k-out-of-n system.

Boland and El-Neweihi (1995) had also established that active spare allocation at the component level is superior to the system level in the hazard rate ordering for a series system when the components and spares are matching ie T_i and T_i^* have the same distribution for $i = 1, 2, \ldots, n$ ($T_i's$ may not be identically distributed) Whether the generalization of this holds for a k-out-of-n system is still an open question.

5   Usual Stochastic and Hazard Rate Orderings in Allocation of Resources to Nodes of a Series System.

Suppose we have a series system of m independent and identical components. Let K spares identical to the components be available for allocation to the components. Consider the problem of active allocation of these K spares to the m components of the system. Assume that lifetimes of components and spares are independently distributed. Let k_i ($0 \leq k_i \leq K$) spares be allocated to the ith component. Let $\mathbf{k} = (k_1, k_2, \ldots, k_m)$. Denote the life time of the resulting system by $T^S(\mathbf{k})$. Shaked and Shanthikumar (1992) showed that the survival function of $T^S(\mathbf{k})$ is a Schur concave function of $\mathbf{k}$ (for definitions of Schur convex and Schur concave functions see Marshal and Olkin(1979)). Thus if K =mk, then the balanced allocation (ie k_i = k for all $i = 1, 2, \ldots, m$) optimizes $T^S(\mathbf{k})$ with respect to the usual stochastic ordering. If K =mk does not hold for any integer k, the allocation which is balanced as much as the restrictions allow maximizes $T^S(\mathbf{k})$ with respect to the ususal stochastic order. Singh and Singh (1996) investigated whether the allocation which optimizes the life time of the system with respect to the usual stochastic ordering also optimizes it with respect to the more strong hazard rate ordering. They showed that the hazard rate function of $T^S(\mathbf{k})$ is a Schur convex function of $\mathbf{k}$, which establishes that balanced allocation of resources as much as the restrictions allow maximizes $T^S(\mathbf{k})$ with respect to the hazard rate ordering. Extension of this result to a k-out-of-n system is yet to be investigated.

References

1. Barlow, R.E. and Proschan, F. (1975). *Statistical Theory of Reliability and Life Testing.* Holt Rinehart and Winston. Inc.

2. Boland, P.J., El-Neweihi, E. and Proschan, F. (1988). Active redundancy allocation in coherent systems. *Probability in Engineering and Information Sciences*, **2**, 343-353.

3. Boland, P.J., El-Neweihi, E. and Proschan, F. (1991). Redundancy importance and allocation of spares in coherent systems. *Journal of Statistical Planning and Inference*, **29**, 55-66.

4. Boland, P.J., El-Neweihi, E. and Proschan, F. (1992). Stochastic order for redundancy allocation in series and parallel systems. *Advances in Applied Probability*, **24**, 161-171.

5. Boland, P.J., El-Neweihi, E. and Proschan, F. (1994). Applications of the hazard rate ordering in reliability and order statistics. *Journal of Applied Probability*, **31**, 180-192.

6. Boland, P.J. and El-Neweihi, E. (1995). Component redundancy versus system redundancy in the hazard rate ordering. *IEEE Transaction on Reliability*, **8**, 614-619.

7. Deshpande, J.V., Singh, H., Bagai, I. and Jain K. (1990). Some partial orderings describing positive ageing. *Stochastic Models*, **6**, 471-482.

8. El-Neweihi, E., Proschan, F. and Sethuraman, J. (1986). Optimal allocation of components in parallel-series and series-parallel systems. *Journal of Applied Probability*, **23**, 770-777.

9. Marshall, A. and Olkin, I. (1979). *Inqualities: Theory of Majorization and its Applications*. Academic Press, New-York.

10. Ross, S.M. (1983). *Stochastic Processes*. John Wiley and Sons, New York.

11. Shaked, M. and Shanthikumar, J.G. (1992). Optimal allocation of resources to nodes of series and parallel systems. *Advances in Applied Probability* , **24**, 894-914.

12. Shaked, M. and Shanthikumar, J.G. (1994). *Stochastic Orders and Their Applications*. Academic Press.

13. Singh, H. (1989). On partial orderings of life distributions. *Naval Research Logistics*, **24**, 103-110.

14. Singh, H. and Misra, N. (1994). On redundancy allocations in systems. *Journal of Applied Probability*, **31**, 1004-1014.

15. Singh, H. and Singh, R.S. (1997). Optimal allocation of resources to nodes of series systems with respect to failure rate ordering. *Naval Research Logisitics*, **44**, 147-152.

16. Singh, H. and Singh, R.S. (1997). On allocation of spares at component level versus system level. *Journal of Applied Probability* , **34**, 283-287.

COMPLETE CONFOUNDING IN A TWO-COMPONENT SERIES SYSTEM: STATUS OF BIVARIATE EXPONENTIALS

Bikas K Sinha and Gopal Chaudhuri

Stat-Math Unit, Indian Statistical Institute

203, B.T. Road, Calcutta:700035, India.

Our purpose in this article is to examine the concept of complete confounding as applicable to the case of Bivariate Exponential (BVE) distributions in the study of reliability theory. We borrow the concept from an interesting piece of work of Sethuraman (1965) and verify the same for some well-known and some new BVE distributions. The results seem to be interesting, though a complete characterization of such BVE distributions would require careful study. We believe this study will generate interest among probabilists as well as reliability theorists.

1   Introduction

We consider a system with 2 components connected in series, and we assume that the life distributions of the components form a bivariate density $f(x,y), 0 < x, y < \infty$. In such a situation, the competing risk analysis of data starts with the basic premises that once the system has failed, not only we are in a position to assess the life of the system but also we can identify the particular component which is responsible for the disaster.

Given this type of background information, we can do all kinds of usual data analysis (see, for example, Guess et al (1991), Miyakawa (1984), Usher et al (1988), Usher (1996), Chaudhuri(1997)). In this paper we are concerned with some probabilistic features of data arising out of the study of the above system. This closely follows the work in Sethuraman(1965). See also Sinha and Roy(1992).

If we do not assume any masking effect, then the recorded data set will have the following feature :

Serial Number : 1, 2, 3,...,n

Life observed : ....

Component responsible :................

A little reflection will show that we have incidentally generated three different data sets : those for which component 1 is responsible, those labeled with the second component, and the combined or global one.

What if we work out the means of the three data sets separately and look at the limitting values of such means when the number of systems put on test tends to infinity ? It is clear that the limitting values will be :

(1)$E(X|X < Y)$

(2)$E(Y|Y < X)$

(3)$E(U), where U = min(X, Y)$.

It follows from Sethuraman(1965) and Sinha and Roy(1992) that all the three quantities listed above are identical whenever the components are independent and the survival functions $\overline{F}_1$ and $\overline{F}_2$ are log-linearly related.

This raises an important question as to the characterization of life distributions of two independent components for which

(i) *conditional distribution of $X|X < Y$*
(ii) *conditional distribution of $Y|Y < X$*
(iii) *distribution of $U = min(X, Y)$*

are identical. When this holds, the life distributions are said to be **completely confounded**. The log-linear relation between $\overline{F}_1$ and $\overline{F}_2$ is a consequence of such a requirement ((i)-(iii)) and clearly, this implies equality of (1), (2), and (3) listed above.

In the next section, we intend to illustrate several examples of bivariate exponential distributions (BVEs) for which complete confounding (in the sense described above) holds.

As a general rule, it is evident that the distribution of U is a mixture of those of $X|X < Y$ and $Y|Y < X$ with mixing proportions given by $P[X < Y]$ and $P[Y < X]$, respectively. Hence complete confounding holds if and only if the two mixing distributions are equal or, equivalently, either one of them is identical to that of the mixture U. In situations where X and Y are exchangeable, this holds trivially and hence complete confounding is immediate.

2   Bivariate Exponentials (BVEs)

A random variable X is said to be exponentially distributed with parameter λ, abbreviated as $exp(\lambda)$ when the pdf of X is

$$f(x, \lambda) = \frac{1}{\lambda} exp(-\frac{x}{\lambda}), x > 0, \lambda > 0. \tag{2.1}$$

It follows that $E(X)=\lambda, Var(X)=\lambda^2$. Further, for any $s, t > 0$, $\overline{F}(s + t) = \overline{F}(s)\overline{F}(t)$ where $\overline{F}(.) = 1 - F(.)$. This is known as the loss of memory property (LMP) which is a charactistic property of exponential distribution. A bivariate exponential distribution relates to the joint distribution of two random variables X and Y such that each marginal distribution is exponential or at least a mixture (finite) of exponentials. We refer to Johnson and Kotz (1972) and Marshall and Olkin(1967) for details. Bivariate exponentials have been studied by Gumbel (1960), Freund (1960), Marshall and Olkin(1967), and Block and Basu(1974) among others.

Gumbel's BVEs:

(a) $f_\theta(x, y) = e^{(-x-y-\theta xy)}\{(1 + \theta x)(1 + \theta y) - \theta\}, x, y > 0, 0 \leq \theta \leq 1$ (2.2)

(b) $f_\alpha(x, y) = e^{-(x+y)}\{1 + \alpha(2e^{-x} - 1)(2e^{-y} - 1)\}, x, y > 0, -1 \leq \alpha \leq 1$ (2.3)

Freund's BVE:

$$f_\theta(x, y) = \begin{cases} \alpha\beta' e^{-\beta' y - (\alpha+\beta-\beta')x}, 0 \leq x \leq y \\ \alpha'\beta e^{-\alpha' x - (\alpha+\beta-\alpha')y}, 0 \leq y \leq x. \end{cases} \tag{2.4}$$

Marshall-Olkin BVE:

$$\overline{F}_\theta(x,y) = e^{-\lambda_1 x - \lambda_2 y - \lambda_{12} max(x,y)}, x,y > 0, \theta = (\lambda_1, \lambda_2, \lambda_{12}) > 0 \qquad (2.5)$$

Block-Basu BVE:

The joint density is given by:

$$f(x,y) = \begin{cases} \frac{\lambda\lambda_1(\lambda_2+\lambda_{12})}{\lambda_1+\lambda_2}e^{-\lambda_1 x - (\lambda_2+\lambda_{12})y}, 0 < x < y < \infty \\ \\ \frac{\lambda_2\lambda(\lambda_1+\lambda_{12})}{\lambda_1+\lambda_2}e^{-(\lambda_1+\lambda_{12})x - \lambda_2 y}, 0 < y < x < \infty. \end{cases} \qquad (2.6)$$

Sinha-Sinha BVE Model I(1995):

$$f(x,y) = \begin{cases} \frac{\lambda\lambda_2}{(\lambda_2+\lambda_3)^2}e^{-\lambda_2 x - (\lambda_3+\lambda_4)y}[(\lambda_2+\lambda_3)(\lambda_3+\lambda_4) - \lambda\lambda_3 e^{-\lambda_2 y}]. \\ (1-e^{-\lambda_2 x})^\gamma(1-e^{-\lambda_2 y})^{-1-\gamma}, 0 < x < y, \\ \frac{\lambda\lambda_3}{(\lambda_2+\lambda_3)^2}e^{-\lambda_3 y - (\lambda_2+\lambda_4)x}[(\lambda_2+\lambda_3)(\lambda_2+\lambda_4) - \lambda\lambda_2 e^{-\lambda_3 x}]. \\ (1-e^{-\lambda_3 y})^\gamma(1-e^{-\lambda_3 x})^{-1-\gamma}, 0 < y < x. \end{cases} \qquad (2.7)$$

For $\lambda_4=0$, X and Y are independent. Otherwise, they are dependent but marginally, both are exponentially distributed where $\lambda = \lambda_2 + \lambda_3 + \lambda_4$, $\gamma = \lambda_4(\lambda_2 + \lambda_3)^{-1}$, and $\lambda_2, \lambda_3, \lambda_4 > 0$.

Sinha-Sinha BVE Model II(1995):

$$f(x,y) = \begin{cases} \frac{f_1(x)f_2(y)}{(F_2(y)-F_1(y))}e^{-\int_y^x \frac{dF_2(t)}{F_2(t)-F_1(t)}}, 0 < y < x < \infty, \\ \\ 0, otherwise. \end{cases} \qquad (2.8)$$

This yields: X has marginal density f_1 and Y has marginal density f_2. When $f_1 \equiv exp(\sigma_1)$ and $f_2 \equiv exp(\sigma_2)$, f(x,y) assumes the form:

$$f_{\sigma_1,\sigma_2}(x,y) = \frac{e^{-x\sigma_1^{-1} - y\sigma_2^{-1}}}{\sigma_1\sigma_2(e^{-y\sigma_1^{-1}} - e^{-y\sigma_2^{-1}})}[\frac{1 - e^{-y(\sigma_2^{-1} - \sigma_1^{-1})}}{1 - e^{-x(\sigma_2^{-1} - \sigma_1^{-1})}}]^{\sigma_2}$$

Using (2.9), we may derive the following joint density :

$$\phi(x,y) = \pi f_{\sigma_1,\sigma_2}(x,y) + (1-\pi)f_{\lambda_2,\lambda_1}(y,x), 0 < x,y < \infty, 0 < \pi < 1. \qquad (2.10)$$

3   Status of BVE distributions

Below we propose to examine the BVE distributions mentioned in Section 2 for verifying the property of complete confounding. It turns out that the verification is in the affirmative in each case except for Sinha-Sinha Model II. The illustrative examples also have some special features. Some give rise to exchangeable distributions, whereas some others yield complete confounding because of identical nature of the two conditional distributions. The special feature of Sinha-Sinha Model II is that complete confounding may hold even if the two distributions are very much different.

Gumbel's BVE:

For both the forms of Gumbel's BVE distributions, it follows that X and Y are exchangeable and hence, complete confounding holds trivially. In particular, for the distribution in (a), we have

$$P[X > t | X < Y] = P[Y > t | Y < X] = P[U > t] = e^{-t(2+\theta t)} \tag{3.1}$$

and

$$E(X|X < Y) = E(Y|Y < X) = E(U) = \frac{1}{2} + \frac{1}{2}\sum_{j=1}^{\infty}(-1)^j \frac{2j!}{j!}\frac{\theta^j}{2^{2j}}. \tag{3.2}$$

Also for that in (b), we have

$$P[X > t | X < Y] = P[Y > t | Y < X] = e^{-2t} + 2\alpha[(e^{-t} - e^{-2t})^2 + e^{-3t} - \frac{1}{2}e^{-2t} - \frac{1}{2}e^{-4t}]. \tag{3.3}$$

Further,

$$E(X|X < Y) = E[Y|Y < X] = E(U) = \frac{1}{2} + \frac{\alpha}{12}. \tag{3.4}$$

Freund's BVE:

For this distribution,

$$P[X > t | X < Y] = P[Y > t | Y < X] = P[U > t] = e^{-(\alpha+\beta)t}. \tag{3.5}$$

Consequently,

$$E(X|X < Y) = E(Y|Y < X) = E(U) = (\alpha + \beta)^{-1}. \tag{3.6}$$

Hence complete confounding holds for Freund's BVE model.

However, the marginals of X and Y are generally different (unless $\alpha = \beta$ and $\alpha' = \beta'$) and each is a mixture of exponentials.

Block-Basu Model:

We have,

$$P[X > t | X < Y] = P[Y > t | Y < X] = P[U > t] = e^{-\lambda t}. \tag{3.7}$$

Hence complete confounding holds for Block-Basu model. The marginals are different unless $\lambda_1 = \lambda_2$.

Sinha-Sinha Model I:

Here

$$P[X > t | X < Y] = e^{-\lambda t} = P[Y > t | Y < X]. \tag{3.8}$$

Therefore, complete confounding holds. The marginals are exponentials with parameters $\lambda_2 + \lambda_4$ and $\lambda_3 + \lambda_4$, respectively. Since this model is relatively new, we provide below a quick proof of (3.8). Not to obscure any essential step of reasoning, we proceed through the following steps:

Step 1

$$P[X < Y] = \frac{\lambda_2}{\lambda_2 + \lambda_3} \tag{3.9}$$

Step 2

$$P[X > t | X < Y] = \frac{P[t < X < Y]}{P[X < Y]} = \frac{\lambda_2 + \lambda_3}{\lambda_2} P[t < X < Y] \tag{3.10}$$

Step 3

$$\int_t^y e^{-\lambda_2 x}(1 - e^{\lambda_2 x})^\gamma dx = \frac{1}{\lambda_2(1 + \gamma)}[(1 - e^{-\lambda_2 y})^{\gamma+1} - (1 - e^{-\lambda_2 t})^{\gamma+1}] \tag{3.11}$$

Step 4

$$P[t < X < Y] = \int_t^\infty [\int_t^y f(x,y)dx]dy \tag{3.12}$$

Step 5

$$\begin{aligned} J(t) &= \int_t^\infty e^{-(\lambda_3+\lambda_4)y}[(\lambda_2 + \lambda_3)(\lambda_3 + \lambda_4) - \lambda\lambda_3 e^{-\lambda_3 y}](1 - e^{-\lambda_2 y})^{-\gamma-1}dy \\ &= (\lambda_2 + \lambda_3)e^{-(\lambda_3+\lambda_4)t}(1 - e^{-\lambda_2 t})^{-\gamma} \end{aligned} \tag{3.13}$$

(using integration by parts)

Step 6

$$\begin{aligned} I(t) &= \int_t^\infty e^{-(\lambda_3+\lambda_4)y}[(\lambda_2 + \lambda_3)(\lambda_3 + \lambda_4) - \lambda\lambda_3 e^{-\lambda_3 y}] \\ &\quad [(1 - e^{-\lambda_2 y})^{\gamma+1} - (1 - e^{-\lambda_2 t})^{\gamma+1}](1 - e^{-\lambda_2 y})^{-\gamma-1}dy \\ &= \lambda_2 e^{-\lambda t} \end{aligned} \tag{3.14}$$

Step 7

$$P[X > t | X < Y] = \frac{1}{\lambda_2}I(t) = e^{-\lambda t} \tag{3.15}$$

Sinha-Sinha Model II:

We have

$$P[X > t | X < Y] = e^{-\frac{t}{\lambda_1}} \tag{3.16}$$

and

$$P[Y > t | Y < X] = e^{-\frac{t}{\sigma_2}}. \tag{3.17}$$

Therefore, complete confounding fails unless $\lambda_1 = \sigma_2$. It is clear that both the marginals are mixtures of exponentials.

References

1. Block, H.W. and Basu, A.P. (1974). A continuous bivariate exponential extension.*J. Amer. Statist. Assoc.*,69,1031-1037.
2. Chaudhuri, G.(1997).Effect of aging in competing risk model under random masking,*Submitted*.
3. Freund, R.J. (1961). A bivariate extension of the exponential distribution. *J. Amer. Statist. Assoc.*,56,971-977.
4. Guess M.F.,J.S. Usher and T.J. Hodgson (1991). Estimating system and component reliabilities under partial information on cause of failure.*J. Statist. Plann. Inference*,29,75-85.
5. Gumbell, E.J. (1960). Bivariate exponential distribution. *J. Amer. Statist. Assoc.* 55,698-707.
6. Johnson, N.L. and Kotz, S.(1972).Continuous multivariate distributions.Wiley.
7. Marshall, A.W. and Olkin, I.(1967).A multivariate exponential distribution.*J. Amer. Statist. Assoc.*, 62,30-40.
8. Miyakawa, M.(1984). Analysis of incomplete data in competing risk model.*IEEE Trans. on Reliability*, 33,293-296.
9. Sarkar,S.K.(1987). A continuous bivariate exponential distribution.*J. Amer. Statist. Assoc.*,82,667-675.
10. Sethuraman, J.(1965). On a characterization of the three limiting types of the extreme. it Sankhyā,Ser A,,27,357-364.
11. Sinha, B.K.(1991). A statistical procedure to evaluate clean up standards: A Followup Study. Statistical Policy Branch,Envirnmental Protection Agency.U.S.A.
12. Sinha, B.K. and Roy, D. (1992). Characterization of probability laws using paired comparison models,*Order Statistics and Nonparamertics: Theory and Applications*, P.K.Sen and I.A.Salama(Editors),Elsevier Science Publishers.
13. Sinha, B.K. and Sinha, B.K.(1995). An application of bivariate exponential models and related inference,*Jour. Statist. Planning and Inference*
14. Usher, J.S. and T.J. Hodgson (1988). Maximum likelihood analysis of component reliability using masked system life data. *IEEE Trans. on Reliabilitty*,37,550-555.
15. Usher, J.S.(1996). Weibull component reliability prediction in presence of masked data. *IEEE Trans. on Reliability*, 45,229-232.

Distribution-Free Tests for Two Crossing Cumulative Incidence Functions

Ram C. Tiwari and Jyoti N. Zalkikar

University of North Carolina at Charlotte and Florida International University

In the setting of competing risks model with two causes of failure, there exist real situations where the cause specific cumulative incidence functions cross at a point. In this paper, we propose the estimation of the crossing point and obtain a large sample confidence interval for it employing the kernel approach. Distribution free tests for testing the equality of the two cumulative incidence functions against the alternatives that these two cross are given. Finally, an application and power comparison of the tests with an existing test in the literature is considered.

1 Introduction

A competing risks model refers to a situation where a system or an organism is exposed to two or more causes of failure or death, but its eventual failure or death is attributed to precisely one of the causes. Associated with each cause of failure there is a nonnegative random variable which represents the latent or coceptual failure time of a system whose failure is attributed to only one of the causes. In the simultaneous presence of all causes only the smallest of such nonnegative random variables is in fact observable, together with the actual cause of the failure. Data like this arises in many fields including medicine and industry.

In this paper, we consider the competing risks model with two causes of failure. One often has dependent causes of failure in many physical situations, i.e., the theoretical lifetime of an individual failing from one cause may be correlated with the theoretical lifetime of the same individual failing from a different cause. Thus, in order to allow for such dependencies, the joint distribution of the theoretical lifetimes associated with an individual must be multivariate in nature. Let us assume that the lifetimes of the unit under the two dependent risks are denoted by T_1^0 and T_2^0. Also, assume that $P(T_1^0 = T_2^0) = 0$. We only observe (T, δ), where $T = \min(T_1^0, T_2^0)$ is the time of failure and $\delta = 2 - I(T_1^0 \leq T_2^0)$ is the cause of failure. Here $I(A)$ is the indicator function of the event A. Thus the observed data is in the form of (T, δ) for each observed item. It is well known that the joint and marginal distributions of T_1^0 and T_2^0 are not identifiable on the basis of the observable random variables (T, δ) unless T_1^0 and T_2^0 are independent. For more on this aspect of nonidentifiability see Cox (1959), Tsiastis (1975), and Kalbfleisch and Prentice (1980), among others. Because of the nonidentifiability inherent in the competing risks model, the independence of T_1^0 and T_2^0 cannot be tested on the basis of the data on the pair (T, δ) and must be assumed *a priori* on the basis of the physical or biological process leading to the failure of the unit.

In situations where it is not realistic to assume the independence of T_1^0 and T_2^0, a comparison of the marginal distributions of T_1^0 and T_2^0 is not meaningful as these functions may not represent the distributions of T_1^0 and T_2^0 in any practical situation. See Gail (1975), Prentice et al. (1978), and Prentice and Kalbfleisch (1988), among others, for further discussion on this issue. It is however possible to compare the two causes of failures within the environment in which the two are acting simultaneously on the basis of the competing risks data by comparing their *cause specific hazard rates* (CSHRs) or their *cumulative incidence functions* (CIFs). The cumulative incidence function corresponding to cause j $(j = 1, 2)$ is defined by

$$F_j(t) = P(T \leq t, \delta = j).\tag{1}$$

Define

$$F(t) = F_1(t) + F_2(t)\tag{2}$$

to be the cumulative distribution of T. Its survival function is given by $S(t) = 1 - F(t)$. We assume that the F_j are absolutely continuous and have subdensities f_j with respect to the Lebesgue measure. The cause specific hazard rate (CSHR) of cause j is given by

$$h_j(t) = \frac{f_j(t)}{S(t)}.\tag{3}$$

The overall hazard rate of failure is then $h(t) = h_1(t) + h_2(t)$. CSHRs provide detailed information on the extent of each type of risks at each time t. In models where the causes of failures are independent, the h_j reduces to the hazard rate corresponding to the marginal distribution of failure due to cause j. Prentice et al. (1978) emphasize that only those quantities which are expressible in terms of CSHRs are estimable and can be estimated by the competing risks data even if risks are dependent. Note that we can express F_j as

$$F_j(t) = \int_0^t h_j(u)S(u)du.\tag{4}$$

Hence the problem of testing the null hypothesis $H_0 : h_1(t) = h_2(t)$ for all t, against the one-tailed alternative(s) $H_1 : h_1(t) \leq (\geq) h_2(t)$ for all t, with strict inequality for some t, is equivalent to testing $H_0 : F_1(t) = F_2(t)$, $t \geq 0$ versus $H_2 : F_1(t) \leq (\geq)F_2(t), t \geq 0$, with strict inequality for some t. Note that there may be no reason to expect a priori that the CSHRs are all equal (except, say, when they represent identical components in a series system), but this is a natural choice of null hypothesis for the ordered alternatives H_1 and H_2.

Various tests procedures based on the competing risks data have been developed for testing the equality of two or more (cause specific) hazard rates against the ordered alternative when the underlying risks are independent, as well when they are dependent. See, for example, Bagai, Deshpandé, and Kochar (1989, 1990), Froda(1987), Gray (1988), Neuhaus (1991), and Yip and Lam (1992,1993) for the competing risks models with independent risks; and Aly, Kochar, and McKeague (1994), Aras and Deshpandé (1992), Deshpandé (1989), Dykstra, Kochar, and Robertson (1995), Sen (1979), and Sun and Tiwari (1995, 1996) for the case of the competing risks models with dependent risks. For a detailed discussion on these test procedures see the review article of Kochar (1995).

Let $C_A = \{(F_1, F_2) : F_1$ and F_2 are CIFs; there exists a unique $x^* \in (0, \infty)$ such that $F_1(x) > F_2(x)$ for $x < x^*$ and $F_1(x) < F_2(x)$ for $x > x^*\}$. Define,

$C_B = \{(F_1, F_2) : (F_2, F_1) \in C_A\}$, and $C_{A \cup B} = \{(F_1, F_2) : (F_1, F_2) \in C_A \cup C_B\}$. In Section 2, we develop distribution-free tests for testing H_0 versus the alternatives $H_1^A : (F_1, F_2) \in C_A$, $H_1^B : (F_1, F_2) \in C_B$, or $H_1^{AB} : (F_1, F_2) \in C_{A \cup B}$. Clearly under the alternative hypothesis H_1^A (H_1^B), there exists some (life-)time x^* such that F_1 crosses F_2 at x^* from above (below). Data of this type where the cumulative incidence functions corresponding to two causes of failure cross one another are available in medical and engineering studies. For example, the plots in Figure 1 of the empirical CIFs for Hoel's (1972) data given in Table 3 suggest that $(F_1, F_2) \in C_A$. Also the empirical CIFs corresponding to failure code 5 and failure code 17 for Nelson's (1982) data consisting of the lifetimes of small appliances satisfy such an alternative. When F_1 and F_2 are distribution functions the tests for H_0 versus H_1^A, H_1^B or H_1^{AB} based on independent samples from F_1 and F_2 have been studied by Hawkins and Kochar (1991). However, to the authors knowledge, there are no such test procedures developed in the context of a competing risks model with two dependent risks.

Let

$$\psi(t) = F_1(t) - F_2(t), t \geq 0, \tag{5}$$

and define

$$\varphi(t) = \int_0^t \psi(x) dF(x) - \int_t^\infty \psi(x) dF(x), t \geq 0. \tag{6}$$

Clearly $\varphi(t) = 0$ for all t under H_0. It is easy to see by differentiation that under H_1^A, $\varphi(t)$ is increasing in $t < x^*$ and decreasing in $t > x^*$, with $\varphi(x^*) = \sup\{\varphi(t) : t \geq 0\} > 0$. Similarly, under H_1^B, $\varphi(t)$ is decreasing (increasing) in $t < x^*$ ($t > x^*$) and $\varphi(x^*) = \inf\{\varphi(t) : t \geq 0\} < 0$. A point estimate and an asymptotically distribution-free confidence interval estimate of the crossing point x^* is obtained in Section 3.

2 Distribution-free tests

Let $(T_1, \delta_1), ..., (T_n, \delta_n)$ be n iid copies of (T, δ). Define

$$F_{jn}(t) = \frac{1}{n} \sum_{i=1}^n I(T_i \leq t, \delta_i = j), j = 1, 2, \tag{7}$$

$$F_n(t) = F_{1n}(t) + F_{2n}(t), \tag{8}$$

$$\psi_n(t) = F_{1n}(t) - F_{2n}(t), and \tag{9}$$

$$\varphi_n(t) = \int_0^{t-} \psi_n(x) dF_n(x) - \int_t^\infty \psi_n(x) dF_n(x). \tag{10}$$

Consider the process

$$v_n(t) = \sqrt{n}(\psi_n(t) - \psi(t)), 0 \leq t < \infty. \tag{11}$$

Let $D[a, b]$ be the Skorohod space of cadlag functions on $[a, b]$, and $v(.)$ be the mean zero Gaussian processes with the following covariance functions for $s \le t$:

$$cov(v(s), v(t)) = F(s) - \psi(s)\psi(t). \tag{12}$$

Then we have the following.

Theorem 2.1 *The process $v_n(t)$ converges weakly to $v(t)$ on $D[0, \infty]$.*

Proof. Since $\psi_n(t) = F_{1n}(t) - F_{2n}(t)$, applying the central limit theorem, it can be shown that the finite dimensional distributions of $v_n(t)$ converges to those of $v(t)$. The tightness condition for $v_n(t)$ follows from the tightness for $u_{1n}(t) = \sqrt{n}(F_{1n}(t) - F_1(t))$ and $u_{2n}(t) = \sqrt{n}(F_{2n}(t) - F_2(t))$, which in turn follow from

$$E\{| u_{1n}(t) - u_{1n}(t_1) |^2 | u_{1n}(t_2) - u_{1n}(t) |^2\} \le 6(F_1(t) - F_1(t_1))(F_1(t_2) - F_1(t)),$$

for $t_1 \le t \le t_2$ and similarly for $u_{2n}(t)$, by using the inequality (13.17) and Theorem 15.6 of Billingsley (1968). This completes the proof.

Define

$$W_n(t) = \sqrt{n}(\varphi_n(t) - \varphi(t)), W(t) = \int_0^t v(x)dF(x) - \int_t^\infty v(x)dF(x), 0 \le t < \infty. \tag{13}$$

Lemma 2.1 *Under $H_0, W_n(t)$ converges weakly to $W(t)$ on $D[0, \infty]$, where $W(.)$, defined in (13), is the mean zero Gaussian process with covariance, for $s \le t$, given by*

$$cov(W(s), W(t)) = \tfrac{1}{3} + \tfrac{1}{3}(F^3(t) - F^3(s)) - (F^2(t) + F^2(s)) + 2F^2(s)F(t).$$

Proof. Note that, under H_0,

$$\begin{aligned} W_n(t) &= \sqrt{n}\int_0^{t-} (\psi_n(x) - \psi(x))dF_n(x) - \sqrt{n}\int_t^\infty (\psi_n(x) - \psi(x))dF_n(x) \\ &= \int_0^{t-} v_n(x)dF(x) - \int_t^\infty v_n(x)dF(x) + o_p(1) \end{aligned}$$

is a continuous functional on $D[0, \infty]$. Hence it follows from the continuous mapping theorem (Billingsley, 1968) that the process $\{W_n(t) : 0 \le t < \infty\}$ converges weakly to $\{W(t) : 0 \le t < \infty\}$ on $D[0, \infty]$. Using (2.6), the covariance kernel, for $s \le t$, is given by

$$\begin{aligned} cov(W(s), W(t)) = cov(&\int_0^s v(x)dF(x), \int_0^t v(x)dF(x)) \\ -cov(&\int_0^s v(x)dF(x), \int_t^\infty v(x)dF(x)) \\ -cov(&\int_s^\infty v(x)dF(x), \int_0^t v(x)dF(x)) \\ +cov(&\int_s^\infty v(x)dF(x), \int_t^\infty v(x)dF(x)) \end{aligned} \tag{14}$$

$$= \frac{1}{3} + \frac{1}{3}(F^3(t) - F^3(s)) - (F^2(t) + F^2(s)) + 2F^2(s)F(t).$$

Clearly the covariance kernel of $W(.)$ depends on the (unknown) distribution function F. In order to develop distribution free tests for testing H_0 versus the hypothesis H_1^A, H_1^B, or H_1^{AB}, for $0 < p < 1$, define

$$\varphi^{(0)}(p) = \int_0^{F^{-1}(p)} \psi(x)dF(x) - \int_{F^{-1}(p)}^{\infty} \psi(x)dF(x) \text{ and} \qquad (15)$$
$$\varphi_n^{(0)}(p) = \int_0^{F^{-1}(p)} \psi_n(x)dF(x) - \int_{F^{-1}(p)}^{\infty} \psi_n(x)dF(x).$$

Note that

$$F_{jn}(x) = \frac{1}{n}\sum_{i=1}^{n} I(U_i \le F(x), \delta_i = j),$$

where $U_i = F(T_i)$. Set $u = F(x)$, and define

$$W_n^0(u) = \frac{1}{n}\sum_{i=1}^{n} I(U_i \le u, \delta_i = 1) - \frac{1}{n}\sum_{i=1}^{n} I(U_i \le u, \delta_i = 2), 0 \le u \le 1. \qquad (16)$$

Then

$$\varphi_n^{(0)}(p) = \int_0^p W_n^0(u)du - \int_p^1 W_n^0(u)du = (\phi W_n^0)(p), 0 \le p \le 1, \qquad (17)$$

where the map $\phi : D[0, 1] \to D[0, 1]$, defined by

$$(\phi h)(p) = \int_0^p h(u)du - \int_p^1 h(u)du, \qquad (18)$$

is Skorohod continuous. In view of the weak convergence of the process $v_n(.)$ in (2.5), under H_0, the process $\{\sqrt{n}W_n^0(u) : 0 \le u \le 1\}$ converges weakly to $\{W^0(u) : 0 \le u \le 1\}$ on $D[0, 1]$ with covariance kernel given by $\text{cov}(W^0(u), W^0(v)) = u$, for $u \le v$. Hence, using the continuous mapping theorem

$$(\phi W_n^0) \to^d (\phi W^0) \text{ as } n \to \infty \qquad (19)$$

on $D[0, 1]$, where in view of (15) the covariance kernel given by, for $0 \le u \le v \le 1$,

$$\text{cov}(\phi W^0(u), \phi W^0(v)) = \frac{1}{3} + \frac{1}{3}(v^3 - u^3) - (v^2 + u^2) + 2u^2v. \qquad (20)$$

Let $\{Z(u) : 0 \le u \le 1\}$ be a mean zero Gaussian process with the covariance kernel given by (20). We have the following.

Theorem 2.2 *Under H_0, as $n \to \infty$,*

$$Z_n^A = \sup\{\sqrt{n}\varphi_n(t) : t \ge 0\} \to^d Z^A = \sup\{Z(u) : 0 \le u \le 1\};$$

$$Z_n^B = \inf\{\sqrt{n}\varphi_n(t) : t \ge 0\} \to Z^B = \inf\{Z(u) : 0 \le u \le 1\};$$

$$Z_n^{AB} = \sup\{\sqrt{n} \mid \varphi_n(t) \mid : t \ge 0\} \to^d Z^{AB} = \sup\{\mid Z(u) \mid : 0 \le u \le 1\}.$$

For $0 < \alpha < 1$, let Z_α^A, Z_α^B, and Z_α^{AB} denote the 100α quantiles of Z^A, Z^B, and Z^{AB}, respectively. Since $Z^B = -Z^A$, so that $Z_{1-\alpha}^A = -Z_\alpha^B$. The approximate values of Z_α^A, Z_α^B, and Z_α^{AB} are obtained based on Monte Carlo simulation of the process $Z(.)$ consisting of 5000 realizations using the Cholesky method on a grid of 70 points on $[0, 1]$. These are given in Table 1.

Table 1: Approximate critical values for tests.

α	Z_α^A	Z_α^B	Z_α^{AB}
0.90	0.954	-0.124	0.958
0.95	1.134	-0.076	1.137
0.99	1.486	-0.024	1.490

To our knowledge, there are no other tests in the literature designed specifically to detect any of the alternatives H_1^A, H_1^B, or H_1^{AB}, in the framework of competing risks. However, it may be possible to view an omnibus Kolmogorov- Smirnov type test of Aly et al. (1994) as competitor to test $H_0 : F_1(t) = F_2(t)$, $t \geq 0$. Aly et al. (1994) showed that under H_0,

$$\sqrt{n}D_n = \sup\{| \psi_n(t) |: t \geq 0\}$$

is distributed as $\sup\{| W(t) |: t \geq 0\}$, where $\{W(t) : t \geq 0\}$ is a standard Brownian motion. This test (referred to as AKM test) rejects H_0 in favor of arbitrary departures from H_0 for large values of $\sqrt{n}D_n$. Table 2 gives the results of a small Monte Carlo study comparing the power of our test Z_n^A with that of AKM test. All powers are estimates based on 10000 Monte Carlo trials at the nominal 0.05 significance level. The data is generated from a bivariate Weibull (BVW) distribution mentioned briefly in Marshall and Olkin (1967) and discussed in more detail in Moeschberger(1974). The form of the survival function is as follows.

$$
\begin{aligned}
S(t_1, t_2) &= P(T_1^0 > t_1, T_2^0 > t_2) \\
&= \exp\{-\lambda_1 t_1^{c_1} - \lambda_2 t_2^{c_2} - \lambda_{12} \max(t_1^{c_1}, t_2^{c_2})\}; \quad t_1, t_2 > 0.
\end{aligned}
$$

The parameter λ_{12} measures the dependence between the two risks. To generate the data from this bivariate distribution, we first use an algorithm due to Friday and Patil (1977). We use their Corollary 3.4 to generate observations on random variables (X_1, X_2) that have Marshall and Olkin's (1967) bivariate exponential distribution and then make a simple transformation $T_1^0 = X_1^{1/c_1}$ and $T_2^0 = X_2^{1/c_2}$. For several choices of the parameters $\lambda_1, \lambda_2, c_1, c_2$ and $\lambda_{12} = 0.1$, (F_1, F_2) are plotted in Figure 2. It is clear that for all of these choices $(F_1, F_2) \in C_A$. The computations in Table 2 are based on large sample critical values for both Z_n^A and AKM tests and sample size $n = 100$. The table shows that Z_n^A test has higher power than the AKM test, except for those choices of parameter values that result in the two CIFs that are very close to each other for $t < x^*$. This reflects the fact that when the two CIFs are very close to each other for $t < x^*$, it is harder to detect that they belong to class C_A than it is to detect arbitrary deviation from H_0. These observations were also made (but not reported here) for other choices of the sample size and for the dependence parameter $\lambda_{12} = 0.5$.

Table 2: Power comparison for H_0 versus H_1^A

$(T_1^0, T_2^0) \sim BVW(\lambda_1, c_1, \lambda_2, c_2, \lambda_{12} = 0.1)$
(nominal significance level is 0.05 and $n = 100$)

		Estimated Power		
(c_1, c_2)	(λ_1, λ_2)	Z_n^A	AKM	Figure 2
$(0.7, 4)$	$(.5, .5)$	.9975	.9889	$(a1)$
	$(.5, .6)$	.9951	.9853	$(a2)$
	$(.5, .9)$	.9851	.9713	$(a3)$
	$(.5, 1.5)$	.9802	.9707	$(a4)$
$(1, 2)$	$(.5, .5)$	.4178	.3313	$(b1)$
	$(.5, .6)$	.2673	.271	$(b2)$
	$(.5, .9)$	.1242	.4437	$(b3)$
	$(.5, 1.5)$	.4515	.8674	$(b4)$
$(1, 4)$	$(.5, .5)$	.9465	.9008	$(c1)$
	$(.5, .6)$	.9223	.8736	$(c2)$
	$(.5, .9)$	.8468	.8217	$(c3)$
	$(.5, 1.5)$	.7978	.8408	$(c4)$
$(3, 4)$	$(.5, .5)$	.1778	.1116	$(d1)$
	$(.5, .6)$	.0754	.1059	$(d2)$
	$(.5, .9)$	.1996	.5049	$(d3)$
	$(.5, 1.5)$	.8147	.9703	$(d4)$

Hoel (1972) presented a set of data consisting of the lifetimes of RFM strain male mice that had received a radiation dose of 300 rads at 5-6 weeks of age and were kept in a conventional laboratory environment. He uses this information to illustrate his representation of cohort mortality data. There are essentially two types of cancer mortality, namely thymic lymphoma and reticulum cell sarcoma that are of interest. Twenty two mice died due to first cause, and 38 died due to second cause. The data for this example appears in Table 3. To illustrate our procedure with this data, we do not need to assume that the two diseases are independent of one another. Plots of estimates of the cumulative incidence functions in Figure 2 suggest that $(F_1, F_2) \in C_A$. The value of the test statistic $Z_n^A = 1.257$, when compared to the approximate critical value results in rejection of H_0 in favor of H_1^A at 5% level of significance.

Table 3: Cancer Mortality data for RFM strain male mice

Thymic Lymphoma				Reticulum Cell Sarcoma					
159	189	191	198	317	318	399	495	525	536
200	207	220	235	549	552	554	557	558	571
245	250	256	261	586	594	596	605	612	621
265	266	280	343	628	631	636	643	647	648
356	383	403	414	649	661	663	666	670	695
428	432	697	700	705	705	712	713	738	748
									753

3 Estimation of the change point

We estimate the change point x^* by $\hat{x}^*$ which maximizes $\varphi_n(t)$:

$$\hat{x}^* = \min\{t : \varphi_n(t) = \max[\varphi_n(s) : s \geq 0]\}, \tag{21}$$

where $\varphi_n(t)$ can be expressed as

$$\varphi_n(t) = n^{-1} \sum_{k:T_{(k)}<t} \psi_n(T_{(k)}) - n^{-1} \sum_{k:T_{(k)}\geq t} \psi_n(T_{(k)}) \tag{22}$$

with $T_{(1)} < .... < T_{(n)}$, the order statistics of $(T_1, ..., T_n)$. Let

$$p^* = F_1(x^*) = F_2(x^*). \tag{23}$$

Then proceeding along the lines of Hawkins and Kochar (1991), it follows that if $(F_1, F_2) \in C_A$, then $\hat{x}^* \to x^*$ a. s. as $n \to \infty$. That is, $\hat{x}^*$ is strongly consistent for x^*. Note that $2p^* = F_1(x^*) + F_2(x^*) = F(x^*), 0 \leq 2p \leq 1$. From a direct computation or substituting $\text{var}(nF_n(x^*)) = n(2p^*(1-2p^*))$, $\text{var}(nF_{jn}(x^*)) = n(p^*(1-p^*))$ into the following

$$\text{var}(nF_n(x^*)) = \text{var}(nF_{1n}(x^*)) + \text{var}(nF_{2n}(x^*)) + 2\text{cov}(nF_{1n}(x^*), nF_{2n}(x^*)),$$

gives $\text{cov}(F_{1n}(x^*), F_{2n}(x^*)) = -p^{*^2}/n$. Hence, $\text{var}(\sqrt{n}\psi_n(x^*)) = n[\frac{2p^*(1-p^*)}{n} + \frac{2p^{*^2}}{n}] = 2p^*$. We thus have,

$$\sqrt{n}(\psi_n(x^*) - \psi(x^*)) \to^d N(0, 2p^*) as n \to \infty. \tag{24}$$

Using the consistency of $\hat{x}^*$ and arguments similar to the ones given in Hawkins and Kochar (1991), it can be shown that $\sqrt{n}(\psi(\hat{x}^*) - \psi_n(x^*)) = o_p(1)$. Hence using the delta-method from (24),we have the following.

Theorem 3.1 *If $(F_1, F_2) \in C_A$, then $\sqrt{n}(\hat{x}^* - x^*) \to^d N(0, \frac{2p^*}{[f_1(x^*)-f_2(x^*)]^2})$ as $n \to \infty$.*

A point estimate of p^* is given by

$$\hat{p}^* = \frac{1}{2}F_n(\hat{x}^*). \tag{25}$$

Note that for $(F_1, F_2) \in C_A$,

$$0 \leq F_1(x) - F_2(x) = F_1(x^*) - F_2(x^*) + (f_1(x^*) - f_2(x^*))(x - x^*) + o(x - x^*)^2.$$

Hence, for $x < x^*$, $f_1(x^*) < f_2(x^*)$. From Theorem 3.1, a $100(1 - \alpha)\%$ confidence interval for x^* is given by

$$I_1(x^*; \alpha) = (\hat{x}^* - z_{\alpha/2}\frac{\sqrt{2p^*}}{\sqrt{n} \mid f_1(x^*) - f_2(x^*) \mid}, \hat{x}^* + z_{\alpha/2}\frac{\sqrt{2p^*}}{\sqrt{n} \mid f_1(x^*) - f_2(x^*) \mid}), \tag{26}$$

if $p^*, f_1(x^*)$ and $f_2(x^*)$ are known. Here z_α is $100(1 - \alpha)$ quantile of $N(0, 1)$. We construct an asymptotically distribution-free confidence interval using smooth estimates of f_j given as follows. Let $K(.)$ be the bounded kernel function with badwidth b_n that has integral 1. We estimate f_j by

$$f_{jn}(x) = \frac{1}{b_n} \int K(\frac{x-u}{b_n}) dF_{jn}(u), j = 1, 2. \tag{27}$$

Let $I_2^1(x^*; \alpha)$ and $I_2^2(x^*; \alpha)$ be the $100(1-\alpha)\%$ confidence intervals for x^* obtained from (26) by replacing f_j by its estimate f_{jn} using biweight kernel $K(y) = \frac{15}{16}(1-y^2), | y | < 1$ and normal density kernel on the interval $(-1, 1)$, respectively. Scott's (1979) normal reference rule is used to obtain window width for both kernel as $b_n = 3.49\hat{\sigma}n^{-1/3}$, where $\hat{\sigma} = \min(S, IQR/1.349)$. In this popular choice of $\hat{\sigma}$, S is the sample standard deviation and IQR is the interquantile range. More sophisticated choices for $\hat{\sigma}$ and for b_n with some optimality properties are possible, see for example Wand (1997). However, for our application these choices perform fairly well. Table 4 gives Monte Carlo estimates (based on 10000 trials) of the coverage probabilities and average lengths of $I_2^1(x^*; 0.05)$ and $I_2^2(x^*; 0.05)$ for $(F_1, F_2) \in C_A$. These are compared with the length $2(1.96)(2p^*/n)^{1/2} | f_1(x^*) - f_2(x^*) |$ and Monte Carlo estimates of the coverage probability of the interval $I_1(x^*; 0.05)$. The data is generated from $BVW(\lambda_1 = 0.5, c_1 = 1, \lambda_2 = 0.9, c_2 = 2, \lambda_{12} = 0.1)$ for which the true values of x^* and p^* are 0.8 and 0.326, respectively. We observe that both I_2^1 and I_2^2 compare well with I_1 in terms of coverage probability with length of I_2^1 being considerably shorter than I_2^2. However, the coverage probability of I_1 is lower than 0.95 even for very large sample size such as 300. To investigate this a little further, we plotted the sampling distribution of $\hat{x}^*$ in Figure 3, using 1000 replications for those sample sizes considered in Table 4. This figure shows that although convergence to normal distribution is quite satisfactory even for sample size 60, the standard deviation (sd) of the sampling distribution is always larger than asymptotic standard deviation (asd) even for sample size 300 with the ratio sd/asd varying from 1.2 to 1.08. When we construct confidence intervals $\hat{x}^* \pm 1.96(sd)$, the coverage probabilities are as expected for all the sample sizes considered. This observed negative bias in the asd, resulting in lower coverage probabilities in Table 4, stems from the fact that second and higher order terms in the Taylor series expansion are ignored when using the delta method in the proof of Theorem 3.1. Our calculations showed that, for as large a sample size as 1000, the ratio sd/asd is 1.0195 with coverage probability of I_1 improving to 0.93; with length of I_1 being 0.343. In this case, confidence intervals based on kernel estimates perform very well with coverage probabilities and lengths of I_2^1 and I_2^2 being 0.915, 0.432, and 0.926, 0.441, respectively. Since the convergence to the asd could be extremely slow as shown in our example of $BVW(\lambda_1 = 0.5, c_1 = 1, \lambda_2 = 0.9, c_2 = 2, \lambda_{12} = 0.1)$ distribution, the authors would like to suggest caution while interpreting the confidence level of the kernel based confidence intervals for real data. As a note, we would like to mention that inclusion of the second order term in the Taylor series expansion in the derivation of asd in Theorem 3.1 would lead to correction in the negative bias to a considerable degree, but we have not made any attempt to study this problem at this stage.

Table 4: Performance of confidence intervals $I_1(x^*; 0.05)$ and $I_2(x^*; 0.05)$

n	Coverage Probability			Length		
	$I_1(x^*;.05)$	$I_2^1(x^*;.05)$	$I_2^2(x^*;.05)$	$I_1(x;0.05)$	I_2^1	I_2^2
60	.903	.898	.920	1.40	4.94	7.33
100	.910	.897	.922	1.08	3.66	5.04
200	.910	.898	.920	0.77	1.84	2.74
300	.910	.893	.918	0.63	1.41	3.51

For the data in Table 3, the estimated change point $\hat{x}^*$ is 643 with 95% (caution !) confidence intervals $I_2^1(x^*; 0.05) = (585.87, 700.13)$ and $I_2^2(x^*; 0.05) = (570.69, 715.31)$.

References

Aras, G. and Deshpandé, J. V. (1992), Statistical analysis of dependent competing risks, *Statistics and Decisions*, 10, 323-336.

Aly, E. A. A., Kochar, S. C. and McKague, I. W. (1994), Some tests for comparing cumulative incidence functions and cause-specific hazard rates, *J. Amer. Statist. Assoc.*, 89, 994-999.

Bagai, I., Deshpandé, J. V. and Kochar, S. C. (1989), A distribution-free test for the equality of failure rates due to two competing risks, *Commun. Statist. Part A - Theory and Methods*, 18, 107-120.

——————————————————————(1990), Distribution-free tests for stochastic ordering among two independent risks, *Biometrika*, 76, 775-778.

Billingsley, P. (1968), *Convergence of Probability Measures*, Wiley, New York.

Cox, D. R. (1959), The analysis of exponentially distributed lifetimes with two types of failures, *J.R.S.S. Ser. B*, 21, 411-421.

Deshpandé, J. V. (1989), A test for bivariate symmetry of dependent competing risks. Technical Report.

Dykstra, R., Kochar, S. and robertson, T. (1995), Likelihood based inference for cause specific hazard rates under order restrictions, *J. Multi. Anal.*, 54, 263-274.

Friday, D. S. and Patil, G. P. (1977), A bivariate exponential model with applications to reliability and computer generation of random variables, In *Theory and Applications of reliability (eds. C. P. Tsokos and I. N. Shimi)*, vol. 1, 527 - 549, Academic Press, New York.

Froda, S. (1987), A signed rank test for censored paired lifetimes, *Commun. Statist. Part A-Theory and Methods*, 16, 3497-3517.

Gray, R. J. (1988), A class of of k-sample tests for comparing the cumulative incidence of a competing risk, *Ann. Statist.*, 16, 1141-1154.

Hawkins, D. L. and Kochar, S. C. (1991), Inference for the crossing point of two continuous cdf's, *Ann. Statist.*, 19, 1626-1638.

Hoel, D. G. (1972), A representation of mortality data by competing risks, *Biometrics*, 28, 475-488.

Kalbfleisch, J. D. and Prentice, R. L. (1980), *The Statistical Analysis of Failure Time Data*, Wiley, New York.

Kochar, S. C. (1995), A review of some distribution-free tests for the equality of cause specific hazard rates, In *Analysis of Censored Data, IMS Lecture Notes-Monograph Series (eds. H.L. Koul and J. V. Deshpandé)*, 27, 147-162.

Marshall, A. W. and Olkin, I. (1967), A multivariate exponential distribution, *J. Amer. Statist. Assoc.*, 63, 30-44.

Moeschberger, M. L. (1974), Life tests under dependent competing causes of failure, *Technometrics*, 16, 39-47.

Nelson, W. (1982), *Applied Life data Analysis*, Wiley, New York.

Neuhaus, G. (1991), Some linear and nonlinear rank tests for competing risks models, *Commun. Statist. Part A-Theory and Methods*, 20, 667-701.

Prentice, R. L. and Kalbfleisch, J. D. (1988), Reply to Slud, Byar and Schatzkin, *Biometrics*, 44, 1205.

Prentice, R. L., Kalbfleisch, J. D., Peterson, A. V., Flourney, N., Farewell, V. T. and Breslow, N. E. (1978), The Analysis of failure times in the presence of competing risks, *Biometrics*, 34, 541-554.

Scott, D. W. (1979), On optimal and data-based histograms, *Biometrika*, 66, 605-610.

Sen, P. K. (1979), Nonparametric tests for interchangeability under competing risks, In *Contributions to Statistics, Jaroslav Hajek Memorial Volume* (J. Jureckova, ed.), 211-228, Reidel, Dordrecht.

Sun, Y. and Tiwari, R. C. (1995), Comparing cause-specific hazard rates of a competing risks model with censored data, In *Analysis of Censored Data, IMS Lecture Notes-Monograph Series*, 27, 255-270.

————————————————(1996), Comparing cumulative incidence functions of a competing risks model. Technical Report.

Tsiatis, A. (1975), A nonidentifiability aspect of the problem of competing risks, *Proc. Natl. Acad. Sci.*, 72, 20-22.

Wand, M. P. (1997), Data-based choice of histrogram binwidth, *The American Statistician*, 51, 59-64.

Yip, P. and Lam, K. F. (1993), A class of non-parametric test for the equality of failure rates in a competing risks model, *Commun. Statist. Part A- Theory and Methods*, 21, 2541-2556.

————————————————(1993), A multivariate nonparametric test for the equality of failure rates in a competing risks model, *Commun. Statist. Part A- Theory and Methods*, 22, 3199-3222.

Ordinary and Bayesian Approach to Life Testing Using the Extreme Value Distribution

Chris P. Tsokos
Department of Mathematics
University of South Florida
Tampa, FL 33620-5700

Abstract

We shall review the extreme value probability density function to characterize the behavior of life testing data as a failure model. We shall derive the minimum variance unbiased estimate of the reliability function, $R(t)$, along with Bayesian estimates. Bayesian estimates will be obtained of $R(t)$ under different priors and mean square error loss. Analytical difficulties will be discussed, along with sensitivity of the choice of a loss function.

1 INTRODUCTION

The model under consideration in this paper will be the modified extreme value distribution, given by

$$f(x; \theta) = \frac{1}{\theta} \exp\left\{ -\frac{1}{\theta}\left[e^x - 1 \right] + x \right\}, \quad \begin{matrix} x > 0 \\ \theta > 0 \end{matrix}. \tag{1.1}$$

The study of extreme values, aside from its mathematical interest, has become increasingly important in recent years and has many physical applications, see Gumbel (1958).

Here we shall be concerned with using the extreme value theory in life testing. The usual life test envisages n components subjected to some specified operating environment. If times to failure are recorded, an estimate of MTBF (mean time between failure) can be made after all n components have failed. Testing time and cost may be cut down by terminating the test after some specified amount of time has elapsed or by stopping the test after a pre-assigned fraction of the components have failed. Suppose, however, that the life test is designed in the following manner. We divide all available test components into N groups of n components each. A group is removed from the test after the first failure in that group. The test is terminated after each group has contributed a first failure. The data of the test then consist of times to first failure. If N and n are large enough, we can then employ extreme-value statistics to ascertain with some preassigned confidence a time interval during which no component will fail. In many cases such information is of greater value than an estimate of MTBF, particularly in one-shot devices such as missles. In addition, test time will usually be shortened and the test components that have not failed will still be available for use, Godbold, J.H. and C.P. Tsokos (1970).

In the study of life-testing and reliability analysis one important approach has been to consider an underlying life distribution and to find suitable estimators of the parameters of that distribution. Also it is of importance to obtain estimators of $R(t)$, the reliability function. In the present study we shall discuss two kinds of estimators for the parameter θ and the reliability function $R(t)$ using the modified extreme value distribution as our model. We shall consider the uniformly minimum variance unbiased estimator, MVUE, and a Bayes' estimator. For practical reasons a relevant problem would be to get an unbiased estimator of reliability as otherwise, especially in complex systems, the cumulative effect of bias might be quite considerable and a system might prove unsatisfactory during operation time. However, if we have available some information about the behavior of the parameter, a Bayesian approach would be more appropriate. A Bayesian analysis implies the use of suitable prior information in association with Bayes' theorem and rests on the exploitation of such information as well as the belief that a parameter is not merely an unknown fixed quantity but rather a random variable with some prior distribution and the choice of a loss function. The prior distributions we shall consider will be the general uniform, the exponential, and the inverted gamma. Also, we shall use the popular mean square loss function.

Several authors, including Barton (1961), Laurent (1963), Pugh (1963), Tate (1959) and Basu (1964), among others, have considered the problem of obtaining unbiased estimators of reliability of the modified extreme value failure model or of other related functions using different density functions as probabilistic models and have made use of the Rao-Blackwell and Lehmann-Scheffe theory to derive minimum variance unbiased estimators. Barton (1961), has considered the binomial, the Poisson, and the normal density functions. Pugh (1963), has found the minimum variance unbiased estimator of reliability for the one-parameter exponential distribution. Laurent (1963), and Tate (1959), have considered the problem for the two parameter exponential distribution. Basu (1964), has considered the Weibull and the gamma distributions. Using the above studies as a basis, we shall determine the minimum variance unbiased estimate of reliability for the modified extreme value distribution.

2   PRELIMINARY REMARKS

Following the approach of Basu (1964), let $\underline{x} = (x_1, x_2, \ldots, x_n)$ be a random sample of size n from a population with density $f(x; \theta)$, where θ is a d-dimensional parameter vector $(d \geq 1)$, and let $\underline{x}_0 = \left(x_{(1)} \leq x_{(2)} \leq \ldots \leq x_{(n)}\right)$ be the corresponding order statistics, where x is the failure time of a system. The reliability at time t of a system whose life follows the probability law $f(x; \theta)$ is given by

$$R(t) = \Pr(X \geq t) = \int\limits_{t}^{\infty} f(x; \theta) \, dx. \tag{2.1}$$

Let $\widehat{\theta}$ be a complete sufficient statistic, assuming one exists, for estimating θ and let its probability density function be given by $h\left(\widehat{\theta}\right)$. The random sample $\underline{x}$ may possibly be thought of as two independent components, ξ and $v = (v_1, v_2, \ldots, v_{n-1})$ of sizes 1 and $(n-1)$, respectively, where ξ may be any one of the x_i's and $\underline{v}$ comprises the remaining $(n-1)$ x_i's for $i = 1, 2, \ldots, n$. If $\widehat{\theta}^*$ is the minimum variance unbiased estimator of θ from $\underline{v}$, we may find the joint density of ξ and $\widehat{\theta}^*$ and hence that of ξ and $\widehat{\theta}$, from which the conditional distribution $g\left(\xi \mid \widehat{\theta}\right)$ is obtained. Let an unbiased estimator of $R(t)$ be $I_t(\xi)$ where $I_t(\cdot)$ is the indicator function of the set $[t, \infty]$; hence by the Rao-Blackwell and Lehman-Scheffé theorems, the unique minimum variance unbiased estimator of $R(t)$ is given by

$$R^*(t) = E\left[I_t(\xi) \mid \widehat{\theta}\right] = \int_t^\infty g\left(\xi \mid \widehat{\theta}\right) d\xi. \tag{2.2}$$

This method of finding an unbiased estimator of reliability can be used even if no complete sufficient statistic for θ is available, provided $\widehat{\theta}$ is a sufficient statistic. In that case the estimate of reliability will no longer be unique.

3   RESULTS

The modified extreme value probability density function is given by (1.1), and hence the likelihood function for a random sample of size n is given by

$$L\left(\theta; x_1, \ldots, x_n\right) = \frac{1}{\theta^n} \exp -\frac{\sum_{i=1}^n e^{x_i} - n}{\theta} e^{\sum_{i=1}^n x_i}. \tag{3.1}$$

The maximum likelihood estimator for θ is given by

$$\widehat{\theta}_{\text{M.L.}} = \frac{\sum_{i=1}^n e^{x_i} - n}{n}. \tag{3.2}$$

That, $\widehat{\theta}_{\text{M.L.}}$ is also a sufficient statistic can easly be seen by applying the Neyman factorization criterion. The distribution of the maximum likelihood estimator of θ has been derived, Godhold, J.H. and C.P. Tsokos, (1970). That is, it can be shown that

$$h\left(\widehat{\theta}; \theta, n\right) = \frac{1}{[\Gamma(n)]\left[\frac{\theta}{n}\right]^n} \widehat{\theta}^{n-1} e^{-\widehat{\theta}/[\theta/n]}. \tag{3.3}$$

Note that the density of $\widehat{\theta}$ given by (3.3) is a gamma density with parameters n and θ/n. Since the gamma density is complete, $\widehat{\theta}$ is therefore a complete sufficient statistic. It can also be shown that it is also unbiased. The joint probability density functions of $\widehat{\xi}$ and $\widehat{\theta}$ can be shown to be

$$g\left(\xi, \widehat{\theta}; \theta, n\right) = \frac{n^2 \left[\sum_{i=1}^{n-1} \exp(v_i) - n + 1\right]^{n-2} \cdot \left[\sum_{i=1}^{n-1} \exp(v_i)\right]}{\theta^n [\Gamma(n)] \cdot \left[\exp(\xi) + \sum_{i=1}^{n-1} \exp(v_i)\right]^{n-1}}$$

$$\cdot \exp\left[-\frac{1}{\theta}\left\{\sum_{i=1}^{n-1} \exp(v_i) - n + \exp(\xi)\right\} + \xi\right] \tag{3.4}$$

where

$$h(\xi; \theta) = \frac{1}{\theta} \exp\left[-\frac{1}{\theta}\{\exp(\xi) - 1\} + \xi\right].$$

The conditional density of ξ given $\widehat{\theta}$ is

$$g\left(\xi \mid \widehat{\theta}\right) = n\left[\sum_{i=1}^{n-1} \exp(v_i) - n + 1\right]^{n-2} \cdot \left[\sum_{i=1}^{n-1} \exp(v_i)\right]$$

$$\cdot \frac{\exp(\xi)}{\left[\exp(\xi) + \sum_{i=1}^{n-1} \exp(v_i)\right] \cdot \left[\exp(\xi) + \sum_{i=1}^{n-1} \exp(v_i) - n\right]^{n-1}}. \quad (3.5)$$

Thus, the minimum variance unbiased estimator of the reliability function, $R(t)$, is given by

$$R^*(t) = n\left[\sum_{i=1}^{n-1} \exp(v_i) - n + 1\right]^{n-2} \cdot \left[\sum_{i=1}^{n-1} \exp(v_i)\right]$$

$$\cdot \int_t^\infty \frac{\exp(\xi)\, d\xi}{\left[\exp(\xi) + \sum_{i=1}^{n-1} \exp(v_i)\right] \cdot \left[\exp(\xi) + \sum_{i=1}^{n-1} \exp(v_i) - n\right]^{n-1}}. \quad (3.6)$$

The integral in (3.6) will be evaluated using numerical techniques to obtain MVUE of $R(t)$.

4 BAYES' ESTIMATORS OF RELIABILITY

Here we shall consider a Bayesian analysis of the extreme value distribution using the general uniform, the exponential, and the inverted gamma as prior probability densities of the parameter and under the influence of mean square loss.

When θ is assumed to be a random variable, we shall examine the problem for each of the following three prior densities of θ:

(i) A general uniform density given by

$$g(\theta; \alpha, \beta) = \frac{(a-1)(\alpha\beta)^{a-1}}{\beta^{a-1} - \alpha^{a-1}} \cdot \frac{1}{\theta^a}, (a < \alpha \le \theta \le \beta) \quad (4.1)$$

which for $a = 0$ reduces to the uniform density on $[\alpha, \beta]$.

(ii) The exponential density

$$g(\theta; \lambda) = \frac{1}{\lambda} \exp\left[-\frac{\theta}{\lambda}\right], (0 < \theta < \infty, \lambda > 0). \quad (4.2)$$

(iii) The inverted gamma density

$$g(\theta; \mu, \nu) = \frac{\exp\left(-\frac{\mu}{\theta}\right) \cdot (\mu/\theta)^{\nu+1}}{\mu\Gamma(\nu)}, (0 < \theta < \infty; \mu, \nu > 0). \quad (4.3)$$

The uniform prior density of θ is surely a realistic choice if one considers the possibility of some prior information concerning the range of the parameter. The inverted gamma prior will give rise to a posterior density which is also an inverted gamma; thus the property of closure under sampling is realized.

For the modified extreme value probability failure distribution, the reliability function is given by

$$R(t) = \exp\left[-\frac{1}{\theta}\{\exp(t) - 1\}\right].$$ (4.4)

Let $(x_1, x_2, \ldots, x_r) = \underline{x}$ denote the observed ordered lifetimes of the test items. The distribution of the first r out of a total of n order statistics is given by

$$g(y_1, y_2, \ldots, y_r) = \frac{n!}{(n-r)!} \prod_{i=1}^{r} [f(y_i)] [1 - F(y_r)]^{n-r}.$$ (4.5)

Hence, the likelihood of our sample of ordered values is given by

$$l(\underline{x} \mid \theta) = \frac{n!}{(n-r)!} \prod_{i=1}^{r} [f(x_i \mid \theta)] [1 - F(x_r \mid \theta)]^{n-r}$$

$$= \frac{n!}{(n-r)!} \left(\frac{1}{\theta}\right)^r \exp\left[-\frac{1}{\theta}(n-r)(\exp(x_r) - 1)\right] \prod_{i=1}^{r} \exp\left[-\frac{1}{\theta}(\exp(x_i) - 1) + x_i\right].$$

Let

$$S_r = \prod_{i=1}^{r} [\exp(x_i) - 1] + (n-r)[\exp(x_r) - 1].$$

Then

$$1(\underline{x} \mid \theta) = \frac{n!}{(n-r)!} \left(\frac{1}{\theta}\right)^r \left[\prod_{i=1}^{r} \exp(x_i)\right] \cdot \exp\left(-\frac{1}{\theta} S_r\right).$$ (4.6)

4.1 General Uniform Prior

For a general uniform density of θ we have

$$h(\theta \mid \underline{x}) = \frac{1(\underline{x} \mid \theta) \cdot g(\theta; \alpha, \beta)}{\int_{\alpha}^{\beta} 1(\underline{x} \mid \theta) \cdot g(\theta; \alpha, \beta) \, d\theta}$$

$$= \frac{\frac{n!}{(n-r)!} \left[\left(\frac{1}{\theta}\right)^r \left\{\prod_{i=1}^{r} \exp(x_i)\right\} \exp\{-S_r/\theta\}\right] \cdot \frac{(a-1)(\alpha\beta)^{a-1}}{\theta^a[\beta^{a-1} - \alpha^{a-1}]}}{\frac{n!}{(n-r)!} \cdot \frac{(a-1)(\alpha\beta)^{a-1}}{[\beta^{a-1} - \alpha^{a-1}]} \cdot \left[\prod_{i=1}^{r} \exp(x_i)\right] \int_{\alpha}^{\beta} \frac{1}{\theta^{r+a}} \exp(-S_r/\theta) \, d\theta}$$

$$= \frac{\left[\frac{1}{\theta^{r+a}}\right] \exp(-S_r/\theta)}{\int_{\alpha}^{\beta} \frac{1}{\theta^{r+a}} \exp(-S_r/\theta) \, d\theta}.$$ (4.7)

Using the incomplete gamma function

$$\gamma(n, z) = \int_0^\zeta \exp(-t) \cdot t^{n-1} \, dt$$

we can write

$$
\begin{aligned}
h(\theta \mid \underline{x}) &= \frac{\frac{1}{\theta^{r+a}} \exp(-S_r/\theta)}{\frac{1}{S_r^{r+a-1}} \cdot \gamma^*(r+a-1, S_r)} \\
&= \frac{\left[\frac{S_r^{r+a-1}}{\theta^{r+a}}\right] \exp(-S_r/\theta)}{\gamma^*(r+a-1, S_r)},
\end{aligned}
\tag{4.8}
$$

where for brevity $\gamma^*(k, y) = \gamma(k, y/\alpha) - \gamma(k, y/\beta)$. The Bayes' estimator of θ with respect to squared error loss is given by the posterior mean. Thus,

$$
\begin{aligned}
\widehat{\theta}_B &= E(\theta \mid x) = \int_\alpha^\beta \theta \cdot h(\theta \mid x) \, d\theta \\
&= \int_\alpha^\beta \theta \cdot \frac{\left[\frac{S_r^{r+a-1}}{\theta^{r+a}}\right] \exp(-S_r/\theta)}{\gamma^*(r+a-1, S_r)} \, d\theta \\
&= \frac{S_r^{r+a-1}}{\gamma^*(r+a-1, S_r)} \int_\alpha^\beta \frac{\exp(-S_r/\theta)}{\theta^{r+a-1}} \, d\theta \\
\widehat{\theta}_B &= \frac{\gamma^*(r+a-2, S_r)}{\gamma^*(r+a-1, S_r)} \cdot S_r.
\end{aligned}
\tag{4.9}
$$

The Var $(\theta \mid \underline{x})$ is given by

$$
\begin{aligned}
\text{Var}(\theta \mid \underline{x}) &= \frac{S_r^2 \{\gamma^*(r+a-3, S_r)\,\gamma^*(r+a-1, S_r)\}}{[\gamma^*(r+a-1, S_r)]^2} \\
&- \frac{[\gamma^*(r+a-2, S_r)]^2}{[\gamma^2(r+a-1, S_r)]^2}.
\end{aligned}
\tag{4.10}
$$

The Bayes' estimator of the reliability function for squared error loss is given by the posterior mean:

$$E[R(t) \mid \underline{x}] = \int_\alpha^\beta R(t)\, h(\theta \mid \underline{x}) \, d\theta$$

or

$$\widehat{R(t)}_B = \int_\alpha^\beta \exp\left[-\frac{1}{\theta}\{\exp(t) - 1\}\right] \cdot \frac{\frac{S_r^{r+a-1}}{\theta^{r+a}} \exp(-S_r/\theta)}{\gamma^*(r+a-1, S_r)} \, d\theta$$

$$= \frac{S_r^{r+a-1}}{\gamma^*(r+a-1,S_r)} \int_{\alpha}^{\beta} \frac{1}{\theta^{r+a}} \exp\left[-\frac{1}{\theta}\{S_r + \exp(t) - 1\}\right] d\theta$$

$$\widehat{R(t)}_B = \frac{\gamma^*[r+a-1,S_r+\exp(t)-1]}{\gamma^*[r+a-1,S_r]} \cdot \left[1 + \frac{\exp(t)-1}{S_r}\right]^{1-r-a}. \qquad (4.11)$$

The Var $(R(t))$ is given by the following expression:

$$\text{Var}\,[R(t)\mid \underline{x}] = \frac{1}{[\gamma^*(r+a-1,S_r)]^2}$$

$$\left\{ \frac{\gamma^*(r+a-1,S_r)\,\gamma^*(r+a-1,S_r+2\,[\exp(t)-1])}{\left[1 + \frac{2[\exp(t)-1]}{S_r}\right]^{r+a-1}} \right.$$

$$\left. - \frac{[\gamma^*(r+a-1,S_r+[\exp(t)-1])]^2}{\left[1 + \frac{\exp(t)-1}{S_r}\right]^{2(r+a-1)}} \right\}.$$

It is quite realistic to expect many situations in which the experimenter has no knowledge on the prior density of θ. This reality may be handled by letting $a = 0$ and $(\alpha, \beta) \to (0, \infty)$ in the general uniform density. The result is a diffuse prior over the positive real line. Under these assertions, we have

$$\widehat{\theta}_B = \frac{S_r}{r-2}, \quad r > 2$$

$$\text{Var}\,(\theta \mid \underline{x}) = \frac{S_r^2}{(r-2)^2(r-3)}, \quad r > 3$$

$$\widehat{R(t)}_B = \left\{1 + \frac{\exp(t)-1}{S_r}\right\}^{1-r}$$

and

$$\text{Var}\,[R(t)\mid \underline{x}] = \left\{1 + \frac{2\,[\exp(t)-1]}{S_r}\right\}^{1-r} - \left\{1 + \frac{\exp(t)-1}{S_r}\right\}^{2(1-r)}.$$

4.2   Exponential Prior Distribution

Assuming the prior density of θ is given by the exponential, then we have

$$h(\theta\mid \underline{x}) = \frac{\frac{n!}{(n-r)!}\left(\frac{1}{\theta}\right)^r \left[\prod_{i=1}^{r} \exp(x_i)\right] \exp(-S_r/\theta) \cdot \frac{\exp(-\theta/\lambda)}{\lambda}}{\int_0^{\infty} \frac{n!}{(n-r)!}\left(\frac{1}{\theta}\right)^r \left[\prod_{i=1}^{r} \exp(x_i)\right] \exp(-S_r/\theta) \cdot \frac{\exp(-\theta/\lambda)}{\lambda}\,d\theta}$$

$$= \frac{\frac{1}{\theta^r}\exp\left[(-S_r/\theta) - (\theta/\lambda)\right]}{\int_0^{\infty} \frac{1}{\theta^r}\exp\left[(-S_r/\theta) - (\theta/\lambda)\right] d\theta}. \qquad (4.12)$$

The denominator of (4.12) can be evaluated using the relation

$$K_\nu(az) = \left(\frac{1}{2}\right) a^\nu \int_0^\infty \frac{\exp\left[\frac{1}{2}zt - \left(a^2 z/2t\right)\right]}{t^{\nu+1}} \, dt,$$

where $K_\nu(az)$ is the modified Bessel function of the third kind of order ν as given by Erdélyi, et al. (1953). We make the substitutions

$$a^2 z/2 = S_r, \quad z/2 = 1/\lambda, \text{ and } \nu + 1 = r,$$

giving

$$K_\nu(az) = \frac{1}{2} a^\nu \int_0^\infty \frac{\exp\left(-zt/2 - a^2 z/2t\right)}{t^{\nu+1}} \, dt$$

$$K_{r-1}(az) = \frac{1}{2} a^{r-1} \int_0^\infty \frac{\exp\left(-\theta/\lambda - S_r/\theta\right)}{\theta^r} \, d\theta$$

$$\frac{2K_{r-1}(az)}{a^{r-1}} = \int_0^\infty \frac{\exp\left(-\theta/\lambda - S_r/\theta\right)}{\theta^r} \, d\theta.$$

Now to solve for a and z we have

$$a = \sqrt{\lambda S_r} \text{ and } z = 2/\lambda, \text{ and hence } az = 2\sqrt{S_r/\lambda}.$$

Thus

$$\frac{2K_{r-1}(az)}{a^{r-1}} = \frac{2K_{r-1}\left(2\sqrt{S_r/\lambda}\right)}{(\lambda S_r)^{(r-1)}/2}$$

or

$$\int_0^\infty \frac{\exp\left(-S_r/\theta - \theta/\lambda\right)}{\theta^r} \, d\theta = \frac{2K_{r-1}\left(2\sqrt{S_r/\lambda}\right)}{(\lambda S_r)^{(r-1)}/2}.$$

This enables us to write the posterior density of θ as follows:

$$h\left(\theta \mid \underline{x}\right) = \frac{(\lambda S_r)^{(r-1)}/2 \cdot \frac{\exp(-S_r/\theta - \theta/\lambda)}{\theta^r}}{2K_{r-1}\left(2\sqrt{S_r/\lambda}\right)}. \tag{4.13}$$

The posterior mean is given by

$$E\left(\theta \mid \underline{x}\right) = \int_0^\infty \theta h\left(\theta \mid \underline{x}\right) d\theta$$

$$= \int_0^\infty \theta \left[\frac{(\lambda S_r)^{\frac{r-1}{2}} \cdot \left(\frac{1}{\theta^r}\right) \exp\left[-S_r/\theta - \theta/\lambda\right]}{2K_{r-1}\left(2\sqrt{S_r/\lambda}\right)}\right] d\theta$$

$$= \frac{(\lambda S_r)^{\frac{r-1}{2}}}{2K_{r-1}\left(2\sqrt{S_r/\lambda}\right)} \int_0^\infty \frac{\exp\left[-S_r/\theta - \theta/\lambda\right]}{\theta^{r-1}}\, d\theta$$

$$= \frac{(\lambda S_r)^{\frac{r-1}{2}}}{2K_{r-1}\left(2\sqrt{S_r/\lambda}\right)} \cdot \frac{2K_{r-2}\left(2\sqrt{S_r/\lambda}\right)}{(\lambda S_r)^{\frac{r-2}{2}}},$$

or

$$E\left(\theta \mid \underline{x}\right) = (\lambda S_r)^{\frac{1}{2}} \cdot \frac{K_{r-2}\left(2\sqrt{S_r/\lambda}\right)}{K_{r-1}\left(2\sqrt{S_r/\lambda}\right)}.$$

The $\mathrm{Var}\left(\theta \mid \underline{x}\right)$ is given by

$$\mathrm{Var}\left(\theta \mid \underline{x}\right) = \frac{\lambda S_r}{K_{r-1}\left(2\sqrt{S_r/\lambda}\right)} \left[K_{r-3}\left(2\sqrt{S_r/\lambda}\right) - \frac{K_{r-2}^2\left(2\sqrt{S_r/\lambda}\right)}{K_{r-1}\left(2\sqrt{S_r/\lambda}\right)}\right].$$

Also, one can calculate the $\mathrm{Var}\left(R(t) \mid \underline{x}\right)$ to be

$$\mathrm{Var}\left[R(t) \mid \underline{x}\right] = \frac{K_{r-1}^2\left(2\sqrt{S_r + \exp(t) - 1/\lambda}\right)}{K_{r-1}\left(2\sqrt{S_r/\lambda}\right)} \cdot \frac{1}{\left[1 + \frac{\exp(t)-1}{S_r}\right]^{r-1}}.$$

4.3 Inverted Gamma Prior Distribution

If the prior density of θ is the inverted gamma, we have for the posterior density of the following:

$$h\left(\theta \mid \underline{x}\right) = \frac{\frac{n!}{(n-r)!}\left(\frac{1}{\theta}\right)^r \left[\prod_{i=1}^r \exp\left(x_i\right)\right] \exp\left(-S_r/\theta\right)\left[\frac{(\mu/\theta)^{\nu+1}\exp(-\mu/\theta)}{\mu\Gamma(\nu)}\right]}{\int_0^\infty \frac{n!}{(n-r)!}\left(\frac{1}{\theta}\right)^r \left[\prod_{i=1}^r \exp\left(x_i\right)\right] \exp\left(-S_r/\theta\right)\left[\frac{\exp(-\mu/\theta)(\mu/\theta)^{\nu+1}}{\mu\Gamma(\nu)}\right] d\theta}$$

$$= \frac{\frac{1}{\theta^r}\exp\left(-S_r/\theta\right)\exp\left(-\mu/\theta\right)(\mu/\theta)^{\nu+1}}{\int_0^\infty \frac{1}{\theta^r}\exp\left(-S_r/\theta\right)\exp(-\mu/\theta)(\mu/\theta)^{\nu+1} d\theta}$$

$$= \frac{\frac{1}{\theta^{r+\nu+1}}\exp\left[-\frac{1}{\theta}\left\{S_r + \mu\right\}\right]}{\int_0^\infty \frac{1}{\theta^{r+\nu+1}}\exp\left[-\frac{1}{\theta}\left\{S_r + \mu\right\}\right] d\theta}. \tag{4.14}$$

The above posterior density can be reduced to

$$h\left(\theta \mid \underline{x}\right) = \frac{\frac{\exp\left[-\frac{1}{\theta}(S_r+\mu)\right]}{\theta^{r+\nu+1}}}{\frac{\Gamma(r+\nu)}{(S_r+\mu)^{r+\nu}}}$$

$$= \frac{\left(\frac{S_r+\mu}{\theta}\right)^{r+\nu+1}\exp\left[-\frac{1}{\theta}\left(S_r + \mu\right)\right]}{(S_r + \mu)\,\Gamma(r + \nu)}. \tag{4.15}$$

This is also an inverted gamma with parameters $(S_r + \mu)$ and $(r + \nu)$. Therefore, the inverted gamma prior density is the natural conjugate family of prior densities for the scale parameter θ of the extreme value distribution. The posterior mean is given by

$$E(\theta \mid \underline{x}) = \int_0^\infty \theta h(\theta \mid \underline{x}) \, d\theta$$

$$= \int_0^\infty \theta \left[\frac{\left(\frac{(S_r + \mu)}{\theta} \right)^{r+\nu+1} \exp\left[-\frac{1}{\theta}(S_r + \mu) \right]}{(S_r + \mu) \Gamma(r + \nu)} \right] d\theta$$

$$= \frac{(S_r + \mu)^{r+\nu}}{\Gamma(r + \nu)} \int_0^\infty \frac{\exp\left[-\frac{1}{\theta}(S_r + \mu) \right]}{\theta^{r+\nu}} \, d\theta.$$

Using the fact that

$$\int_0^\infty \frac{(S_r + \mu)^{r+\nu}}{\Gamma(r + \nu)} \cdot \frac{\exp\left[-\frac{1}{\theta}(S_r + \mu) \right]}{\theta^{r+\nu+1}} \, d\theta = 1$$

$$\int_0^\infty \frac{\exp\left[-\frac{1}{\theta}(S_r + \mu) \right]}{\theta^{r+\nu+1}} \, d\theta = \frac{\Gamma(r + \nu)}{(S_r + \mu)^{r+\nu}}$$

$$\int_0^\infty \frac{\exp\left[-\frac{1}{\theta}(S_r + \mu) \right]}{\theta^{r+\nu}} \, d\theta = \frac{\Gamma(r + \nu - 1)}{(S_r + \mu)^{r+\nu-1}},$$

we have as our posterior mean

$$E(\theta \mid \underline{x}) = \frac{(S_r + \mu)^{r+\nu}}{\Gamma(r + \nu)} \cdot \frac{\Gamma(r + \nu - 1)}{(S_r + \mu)^{r+\nu-1}}$$

and

$$\widehat{\theta}_B = \frac{S_r + \mu}{r + \nu - 1}; r + \nu > 1. \tag{4.16}$$

The variance of the Bayes estimate of θ is given below:

$$\text{Var}\,[\theta \mid \underline{x}] = \frac{(S_r + \mu)^2}{(r + \nu - 1)^2 (r + \nu - 2)}.$$

Similarly, the variance of the Bayesian estimate of $R(t)$ is given by

$$\text{Var}\,[R(t) \mid \underline{x}] = \frac{1}{\left[1 + \frac{2(\exp(t) - 1)}{S_r + \mu} \right]^{r+\nu}} - \frac{1}{\left[1 + \frac{\exp(t) - 1}{S_r + \mu} \right]^{2(r+\nu)}}.$$

Note the above results can be generalized to any life-testing model of the form

$$f(x;\theta) = \left[\frac{\varphi(k)}{\theta}\right] \exp\left\{-\frac{1}{\theta}[g(x)]^{\varphi(k)}\right\} \cdot g'(x); \quad \begin{array}{l} \theta > 0 \\ x \geq \ell \geq 0 \\ -\infty < k < \infty \end{array}$$

where $g(x)$ is any strictly monotone increasing function, $\varphi(k)$ is any non-zero function, and ℓ is the greatest lower bound of x. The cumulative distribution function is given by

$$F(x;\theta) = -\int_{\ell}^{x} \exp\left\{-\frac{1}{\theta}[g(z)]^{\varphi(k)}\right\} \left[-\frac{\varphi(k)}{\theta}g'(z)\,dz\right],$$

which can be reduced to the following form

$$F(x) = 1 - \exp\left\{-\frac{1}{\theta}[g(x)]^{\varphi(k)}\right\}.$$

Thus, the reliability function can be written as

$$R(t) = \exp\left\{-\frac{1}{\theta}[g(t)]^{\varphi(k)}\right\}.$$

The likelihood function of ordered failures is given by

$$l\left(\underline{x}\mid\theta\right) = \frac{n!}{(n-r)!}\left[\frac{\varphi(k)}{\theta}\right]^{r}\left[\prod_{i=1}^{r}g'(x_i)\right]\exp\left(-S_r/\theta\right)$$

with

$$S_r = \sum_{i=1}^{r}[g(x_i)]^{\varphi(k)} + (n-r)[g(x_r)]^{\varphi(k)}.$$

For uniform prior, the Bayes estimate of θ is given by

$$\widehat{\theta}_B = \frac{\gamma^*(r+a-2,S_r)}{\gamma^*(r+a-1,S_r)},$$

and its variance by

$$\mathrm{Var}\left(\theta\mid\underline{x}\right) = \frac{S_r^2\left\{\gamma^*(r+a-3,S_r)\,\gamma^*(r+a-1,S_r) - [\gamma^*(r+a-2,S_r)]^2\right\}}{[\gamma^*(r+a-1,S_r)]^2}.$$

The Bayesian estimate of $R(t)$ and its variance is given by

$$R(t)_B = \frac{\gamma^*\left[r+a-1,S_r + (g(t))^{\varphi(k)}\right]}{\gamma^*(r+a-1,S_r)}\left[1 + \frac{[g(t)]^{\varphi(k)}}{S_r}\right]^{1-r-a}.$$

and

$$\text{Var}\,[R(t)\mid \underline{x}] \;=\; \frac{1}{[\gamma^*\,(r+a-1,S_r)]^2}$$

$$\cdot \;\frac{\{\gamma^*\,(r+a-1,S_r)\,\gamma^*\,(r+a-1,S_r+2)\,[g(t)]^{\varphi(k)}\}}{\left[1+\frac{2[g(t)]^{\varphi(t)}}{S_r}\right]^{r+a-1}}$$

$$\frac{\left[\gamma^*\,(r+a-1,S_r+[g(t)]^{\varphi(k)})\right]^2}{\left[1+\frac{2[g(t)]^{\varphi(t)}}{S_r}\right]^{2(r+a-1)}}.$$

Similar results can be obtained from the other priors that we studied under square error loss function.

5   Sensitivity of the Loss Function

Recently, Tang, H. and C.P. Tsokos (1997) have studied the effect of the loss function on the Bayesian estimate of the parameter and reliability function of the extreme value failure model. Keeping the same prior probability densities that we used in the present study, we obtained Bayesian estimates of θ and $R(t)$ for the following loss functions.

Linear Loss Function:

$$L_{LP}\left(\hat{\theta},\theta\right) = \begin{cases} p\left|\hat{\theta}-\theta\right|, & \hat{\theta}\le\theta \\[2mm] (1-p)\left|\hat{\theta}-\theta\right|, & \hat{\theta}>\theta \end{cases} \quad 0<p<1\;.$$

Harris' Loss Function:

$$L_H\left(\hat{\theta},\theta\right) = \left|\frac{1}{1-\hat{\theta}}-\frac{1}{1-\theta}\right|^k, \quad k=2.$$

Higgins-Tsokos Loss Function (1980):

$$L_{H-T}\left(\hat{\theta},\theta\right) = \frac{f_1 e^{f_2\left(\hat{\theta}-\theta\right)} + f_2 e^{f_1\left(\hat{\theta}-\theta\right)}}{f_1+f_2} - 1, \quad f_1,f_2>0.$$

The results were compared with those obtained in using the popular mean squared error.

In some cases we were able to obtain analytical tractible estimates in other cases very complicated theoretical expressions from which we could not compare our findings with those obtained in the present study. However, numerical results indicate that both Bayesian estimates, parameter and reliability, are sensitive with respect to the choice of the loss function.

6   Numerical Simulation

By means of a Monte Carlo simulation we compared the various estimates we have obtained in the present study. For each prior probability distribution, we consider five hundred repetitions of the experiment. In each repetition, there are ten randomly-generated lifetimes from the failure model, the extreme value distributions and a randomly-generated value of the parameter θ from its prior distribution. Using the generated data, we calculated the M.V.U.E. and the Bayes' estimate of θ. Also, the corresponding estimates of the reliability function, $R(t)$, with time increments of 0.1 was obtained. In each case, the exact reliability is calculated and from this result the square error loss is obtained. For each repetition, we calculated the average squared error loss for each of the two reliability estimators. For each successive repetition after the first, the averages of that repetition are added to the previous averages.

The ratio of the accumulated averages of squared error loss of the Bayes' estimator and M.V.U.E. was calculated for each repetition. That is, this ratio gives the accumulated averages of square error losses.

Representative graphical examples of the simulated results are given by Figures 1–3. The results of the 400th and 500th experiments for uniform prior distribution with θ equal to 219 and 293, M.V.U.E. of θ is 192 and 253, Bayes' estimator of θ is 206 and 334, accumulated average of squared error loss for minimum variance of estimates of reliability is 2.85 and 3.5, accumulated average of squared error loss for Bayes' estimates of reliability is 1.3 and 1.8 and the ratio of accumulated averages of square loss is 0.46 and 0.45, respectively. The exact, Bayes' estimator and M.V.U.E. of $R(t)$ are given by Figure 1.

For the 400th and 500th experiments for exponential prior with θ equal to 10.7 and 2.1, M.V.U.E. of θ is 10.7 and 3.7, accumulated average of squared error loss for minimum variance estimate of reliability is 1.8 and 2.8, accumulated average of square error loss of Bayes' estimator of reliability is 3.9 and 4.9 and the ratio of accumulated averages of square loss is 0.46 and 0.47, respectively. The exact Bayes' and M.V.U.E. of $R(t)$ is given by Figure 2.

For the 400th and 500th experiments for inverted gamma prior with θ equal to 3.2 and 24, M.V.U.E. of θ is 3.9 and 21.3, Bayes' estimator for θ is 5.4 and 21.2, accumulated average of squared error loss for minimum variance estimate of reliability is 1.8 and 2.2, accumulated average of squared errror loss for Bayes' estimator of reliability is 3.6 and 4.7 and the ratio of accumulated averages of squared loss is 0.5 and 0.48, respectively. Figure 3 gives the exact Bayes' and M.V.U.E. of $R(t)$.

As a result of this extensive simulation we have observed that the ratio of the accumulated averages of squared error losses tends to be between 0.40 and 0.5 as the number of experiments increases. This behavior suggests that if data is available about the behavior of the parameter θ in the failure model, then the square error loss will be reduced by using the Bayes' estimator of $R(t)$ rather than the unbiased M.V.U.E. Thus, as expected, the Bayes' estimator of reliability is better than the unbiased minimum variance estimator.

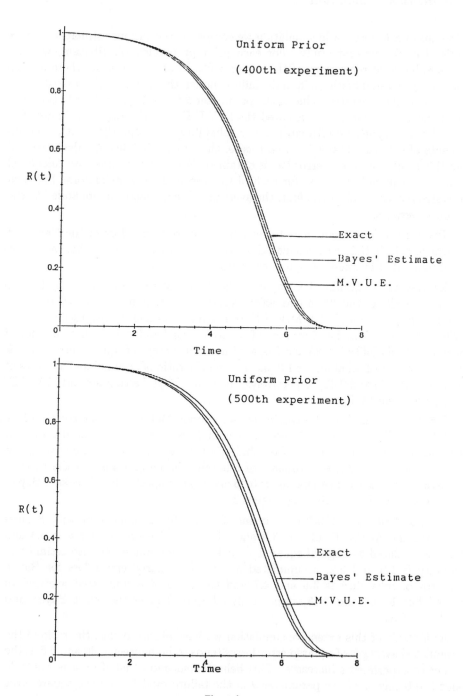

Figure 1

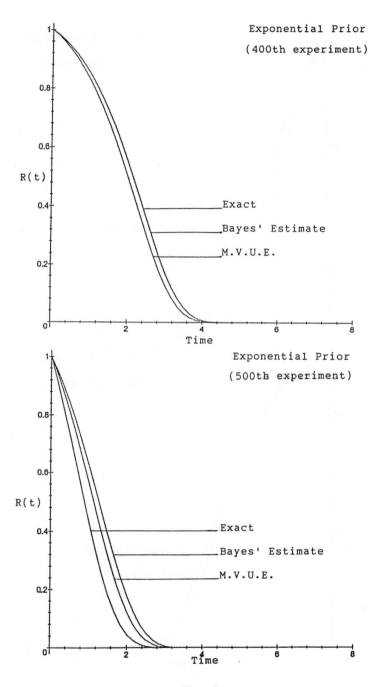

Figure 2

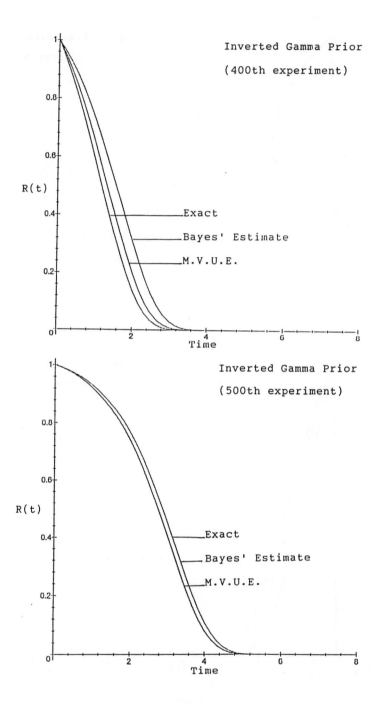

Figure 3

References

1. Barton, D.E., "Unbiased Estimation of a Set of Probabilities," *Biometrika* 48, 227–229, (1961).
2. Basu, A.P., "Estimates of Reliability for Some Distributions Useful in Life Testing," *Technometrics* 6, 215–219, (1964).
3. Erdélyi, A., et. al, Higher Transcendental Functions, Volume II, McGraw-Hill Book Co., Inc., New York, (1953).
4. Godbold, J. and C.P. Tsokos, The Extreme Value Distribution as a Life Testing Model, VPI & SU Technical Report, Sept. (1970).
5. Gumbel, E.J., Statistics of Extremes, Columbia University Press, New York, (1958).
6. Higgins, J. and C.P. Tsokos, "A Study of the Effect of Loss Function on Bayes Estimates of Failure Intensity, MTBF and Reliability," *J. Applied Math. and Computations* 6, 145–166, (1980).
7. Laurent, A.G., "Conditional Distribution of Order Statistics of the Exponential Model," *Ann. Math. Statistist.* 34, 652–657, (1963).
8. Tang, H. and C.P. Tsokos, "Bayes Estimates of Reliability Under Different Loss Functions," to appear, (1997).
9. Tate, R.F., "Unbiased Estimation: Function of Location and Scale Parameters," *Ann. Math. Statist.* 30, 341–366, (1959).

RELIABILITY MODELLING AND INFERENCE THROUGH FATIGUE CRACK GROWTH

CATHAL D. WALSH, SIMON P. WILSON

Department of Statistics, Trinity College, Dublin 2, Ireland

E-mail: Simon.Wilson@tcd.ie

In many structures and components, failure occurs because of the unstable propagation of cracks through them. These cracks initiate and grow as a result of imposed stresses and are also a function of the environment and material properties. In this paper, we discuss the various approaches to modelling crack propagation and illustrate how reliability models may be derived from them. Emphasis is placed on stochastic models, and an example is given of how reliability predictions from crack length data may be made.

Key Words: crack growth, fatigue, Monte Carlo Markov Chain, Paris-Erdogan equation, predictive distribution, stochastic process.

1 Introduction

When a metal structure or component is subject to a cyclic load, our experience is that it must eventually fail. The failure is the result of *fatigue*, which is the deterioration in a material's ability to carry its intended load. It is well documented that fatigue, and so failure, of a metallic structure or component occurs because of the propagation of cracks through it, caused by the repeated cyclic stresses and strains imposed on the material. This propagation is also a function of several other internal and external factors: size of load, microstructual properties of the material, temperature and humidity, for example. For the reliability modeller, the propagation of such *fatigue cracks* can therefore be a very important factor in assessing lifetime.

In terms of predicting reliability, there are two strategies. The simplest is to propose a model for the lifetime directly, usually in terms of the number of stress cycles to failure. The second is to model the underlying physical causes of failure, that is the growth of cracks, and by modelling those derive a reliability model. It is this second strategy that our paper describes.

Whatever model is chosen, our aim will be to specify the length of a crack after N cycles of the load, which we denote $a(N)$. Given this, the lifetime of a component is easily defined. Since, near failure, the crack grows very quickly, one can specify lifetime as the time at which the length of a crack first exceeds a suitable threshold length A_{th}. Thus the time to failure N_f is

$$N_f = \inf\{N \mid a(N) \geq A_{th}\} \tag{1}$$

and, under the assumption that $a(N)$ is non-decreasing, the reliability function is

$$P(N_f > N) = P(a(N) \leq A_{th}). \tag{2}$$

The academic interest in modelling fatigue crack growth has been and is considerable, with researchers in mechanical engineering, chemistry, material science, applied mathematics and physics, as well as reliability theory, all contributing. In this paper, we review some of the vast literature that has been generated, with an emphasis on its use in reliability prediction. We concentrate particularly on stochastic modelling, and the role of statistical methods in the analysis of fatigue crack data.

The paper is organised as follows. In the next section, we give a brief description of the process of crack growth. In Section 3, we look at some of the modelling stategies that have been proposed. In Section 4, we conclude with an example of reliability prediction from short crack data, which illustrate the possibilities that new simulation methods might have for conducting inference with the more complex models.

2   The Process of Crack Growth

Broadly speaking, one can consider the process of crack propagation to have five phases. In the first *dormant* phase, the crack has not formed. Then there is *nucleation*, where the crack is initially formed by the sliding of atomic layers in the material, usually as a result of dislocations in the atomic structure and local stress concentrations. After this is the *micro* or *short crack growth* phase, where the crack grows rather haphazardly up to about 1mm in length. This is followed by the *macro-crack growth* phase, where the crack continues to propagate before finally its growth rate increases dramatically and *failure* of the component occurs. The final failure phase occurs very quickly, relative to the other phases, so can be ignored as a factor in determining reliability.

Out of the many cracks that nucleate and become micro-cracks in a specimen, usually there is only one that becomes dominant and causes failure. The fact that only one macro-crack is important makes the modelling of this phase rather easier than the others; it is only necesssary to model this dominant crack. Also, there are many situations where the macro-crack phase is the longest phase in the lifetime of the specimen. Thus, most of the considerable body of work on crack propagation is devoted to macro-crack modelling. In the micro-crack case, one often attempts to model the largest crack only. The micro-crack phase can form a very sizeable proportion of the failure time, particularly in situations of relatively low stress levels where lifetimes are long.

Many models for fatigue crack growth have been proposed. A good overview, particularly of stochastic models, can be found in Sobczyk and Spencer [1].

3   Models for Crack Growth

3.1   Deterministic Modelling

A deterministic model of fatigue crack growth may be considered to be a mathematical system which will allow one to make accurate predictions about the lifetime of a material or structure, given information about the material properties, details of the geometry, and the actions to which it is subject. A deterministic model would suggest that if one could specify the various parameters exactly, then one would get an exact prediction for the lifetime.

In reality, we know that there is a large amount of scatter in observed lifetime data for similar materials under similar conditions. Such scatter will be addressed in the probabilistic adaptations of some of the deterministic models.

One may divide the different types of model into two categories; namely those which are empirical, based on observed data and constructed to fit the data, and those that are more theoretical, based on some physical reasoning, or mechanism which is known to affect the lifetime.

Empirical Models

Laboratory tests on materials and observation of structures indicate that the length of the largest crack is dependent on factors such as the stresses, material properties, temperature, chemical properties to which the sample or structure is subjected. Two equations which describe crack growth, and fit well with the observed data, are the Paris-Erdogan and Forman equations.

Paris-Erdogan Equation   The Paris-Erdogan equation is derived from empirical considerations, and has no real theoretical basis. The equation models the relation between crack velocity and an abstract quantity called the *stress intensity range*, which describes the magnitude of the stress at the crack tip. This range is denoted ΔK and is usually defined as $\Delta K = Q\Delta\sigma\sqrt{a}$, where the constant Q reflects the crack geometry and $\Delta\sigma$ is the range in stress per cycle.

The form of the Paris-Erdogan equation is

$$\frac{da}{dN} = C(\Delta K)^n,$$
$$a(0) = A_0. \tag{3}$$

We note that C and n are regarded as material constants, but that they also depend upon factors such as frequency, temperature and stress ratio.

The Paris-Erdogan equation gives good results for long cracks when the material constants are known, but we note that a large effort is required to determine them, since they are functions of many variables.

Forman Equation   The stress ratio $R = \sigma_{min}/\sigma_{max}$ (ratio of minimum to maximum stress in a cycle) has been shown to have an important effect on crack growth, according to Bannantine *et al.*[2]. The Forman equation accounts for the stress ratio, and has the following form:

$$\frac{da}{dN} = \frac{C(\Delta K)^n}{(1-R)K_c - \Delta K}, \tag{4}$$

where K_c is a critical level for the stress intensity, corresponding to unstable fracture.

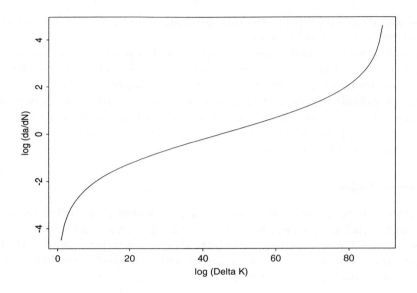

Figure 1: The Forman equation

In Figure 1, we graph this function on a log / log scale, from which we observe two limits:

$$\Delta K \to 0 \;\Rightarrow\; \frac{da}{dN} \to 0,$$

and

$$\Delta K \to (1-R)K_c \;\Rightarrow\; \frac{da}{dN} \to \infty.$$

We see from this that there is a lower limit to ΔK, below which crack growth is negligible, and an upper limit, around which growth is explosive.

Finally, with regard to the Forman equation, we simply note that it has been shown to give good results for crack growth in aluminium alloys and steels, but not in many other materials.

Theoretical Models – Fracture Mechanics

While the Forman equation might yield some nice physical insights, one can tackle the whole problem of predicting lifetime directly, using the stress intensity factor and the idea of *fracture toughness*. This requires the application of fracture mechanics, the details of which are beyond the scope of this paper, and which, in any case, are well covered in numerous texts such as Rolfe and Barsom[3]. It is sufficient to note that the application of fracture mechanics, with the appropriate examination of the crack geometry, nature of stress applied, and length of crack to date, along with experimental data for materials can give us an idea of when failure may be likely to occur.

Linear fracture mechanics is the simplest form, where a linear relationship between stress on the material and the resulting strain (or deformation) is assumed. While it is valid in many of the situations of interest, it is certainly not valid in them all. There are many structures where a material, under normal operating conditions, will experience a load which will cause the material to yield and deform locally. Regions where this occurs are called plastic zones. Where this arises, it is called elastic-plastic loading; it is more common with lighter modern materials.

One way of modelling elastic-plastic loading is the path independent J integral proposed by Rice [4] which describes the stress-strain field at the tip of the crack under such loading. Since the integral can be taken far from the crack tip and then information about the crack tip deduced, this allows us to extend fracture mechanics from linear elastic to elastic-plastic behaviour. Again we will not discuss the details here, but allow the reader to refer to texts such as Sobczyk and Spencer [1] and Rolfe and Barsom [3].

3.2   Stochastic Modelling

Using a stochastic model has several advantages. First, crack growth shows considerable variation in behaviour, even within the same material and under the same conditions. Thus, there would appear to be some underlying randomness in crack propagation. Secondly, components in the field are often subjected to a random load and environment, factors that are very important to its reliability and which cannot be readily modelled deterministically.

The simplest form of stochastic model for crack growth is to assume that crack length data is described by a common probability model such as the Weibull or lognormal, parameterised in some ad hoc way by N. As with many other reliability applications, these are also used to model time to failure directly, and can fit lifetime data quite well, particularly in situations of high stress. The first stochastic models were of this type (see Birnbaum and Saunders [5] and Weibull [6]). The failure rate is also used, with the rate derived by consideration of the fatigue process (see Yang and Heer [7], for example). However, most stochastic models are formed by "randomising" a deterministic model, thus taking account of theoretical or empirical characteristics of the growth process.

The ways in which the randomisation occurs are many and varied. Suppose we have a deterministic model for $a(N)$, which we may write as $a(N, \theta)$, where θ are model parameters. Some possibilities are:

- Assume random model parameters θ. Since a is a function of θ, a distribution on crack length is implied. Since N_f is a function of a and therefore of θ, a lifetime distribution is also implied.

- Take some non-decreasing stochastic process (such as a birth process) indexed by N and specify process parameters such that the expected value of the process is $a(N)$.

- If $a(N)$ has been defined in terms of a differential equation, one can form an equivalent stochastic differential equation whose solution is a stochastic process model for $a(N)$.

A Random Parameter Model

A simple random model for $a(N)$ is to take the Paris Erdogan model and assign probability distributions to the parameters C, n and A_0. Here, we give another example, this time from the theory of ceramic materials (Paluszny and Nicholls[8]). For ceramics, it is more relevant to assume the material is under a constant stress σ, instead of a cyclic one, in which case one indexes growth by time t and the equivalent to the Paris-Erdogan equation is

$$\frac{da}{dt} = v_0 K^n, \tag{5}$$

where $K = Q\sigma\sqrt{a}$ is the stress intensity, and v_0 and n are material specific parameters.

The approach to finding the lifetime distribution is different, as it is based on modelling the strength of the specimen rather than crack length. The strength $S(t)$ (or maximum load that the specimen can bear) at time t is related to crack length by

$$S(t) = \sqrt{\frac{B}{\pi a(t)}}, \tag{6}$$

for a material constant B; this equation is derived in Griffith's seminal work[9]. So, as $a(t)$ increases, the strength decreases and failure occurs as soon as $S(t) < \sigma$:

$$T_f = \inf\{t \,|\, S(t) < \sigma\}. \tag{7}$$

Using Equations 5, 6 and 7, and given that $n > 10$ for ceramic materials, one can show that the lifetime is approximately

$$T_f \approx \frac{2}{(n-2)\, v_0\, B^n\, \sigma^n\, (\sqrt{A_0})^{n-2}}. \tag{8}$$

The model is randomised by considering the initial strength $S(0)$ — which is related to the initial crack length by Equation 6 — to be Weibull distributed. The implied distribution on T_f is also Weibull, thus justifying one of the most common reliability models.

The assumption of a Weibull distribution for the initial strength is justified by an "extreme value" argument; the initial strength of a specimen is governed by the size of the largest initial crack, and it is assumed that this largest crack is distributed according to one of the three extreme value distributions for the maximum of a set of normalised independent and identically distributed random variables (see Galambos[10]). If this maximum is a particular one of the three possible, then the initial strength is Weibull; see Singpurwalla et. al.[11] for more details.

Wiederhorn and Fuller[12] have shown this model to be consistent with data on the failure of an alumina ceramic. A model, formed by randomising parameters, for the growth of short cracks in metals is considered in Section 4.

A Stochastic Process Model

A representative stochastic process model is that of Sobczyk and Trębicki [13]; they also index crack length by time t instead of cycles and state that

$$a(t, \theta) = A_0 + \sum_{i=1}^{N(t)} X_i(\theta). \tag{9}$$

The idea is that the crack has some initial length A_0, then grows in increments of size $X_i(\theta)$, with the number of such increments by time t being $N(t)$. The idea that a crack does not grow continuously but in short bursts has been observed in many materials. The authors assume a Poisson process for $N(t)$ and quote experimental evidence (Kogaiev and Liebiedinskij [14]) for the use of the exponential distribution for the X_i. Assuming independent increments, the density for crack length is then

$$f_{a(t)}(a \mid A_0, \Lambda(t), \mu) = e^{-\Lambda(t) - \mu(a - A_0)} \sum_{k=0}^{\infty} \frac{(\Lambda(t)\mu)^{k+1}(a - A_0)^k}{k!\,(k+1)!}, \quad a > A_0, \tag{10}$$

with a point probability mass $e^{-\lambda t}$ at $a = A_0$, and that for the lifetime T_f is

$$f_{T_f}(t \mid A_0, \Lambda(t), \mu, A_{th}) = \lambda(t) e^{-\Lambda(t) - \mu(A_{th} - A_0)} \sum_{k=0}^{\infty} \frac{(\Lambda(t)\mu(A_{th} - A_0))^k}{(k!)^2}, \tag{11}$$

where $\Lambda(t)$ is the mean value function of the Poisson process, $\lambda(t) = \Lambda'(t)$ and $1/\mu$ is the mean of X_i.

In Figure 2 the density in Equation 11 is plotted, for $\lambda(t) = 1$ (a homogeneous Poisson process), $\mu = 1$, $A_0 = 0.02$, and three threshold lengths: $A_{th} = 5$, 10 and 20.

This stochastic process has not been related to any physical model for crack growth. This can be done in several ways. For example, one could set $\mu = 1$ and define $\Lambda(t)$ to be the solution to the Paris-Erdogan equation. In this way, the expected value of $a(t)$ is $\Lambda(t)$.

A Stochastic Differential Equation Model

If we take the Paris-Erdogan equation and a stochastic process $X(N, \theta)$, one can form a stochastic model as follows:

$$\frac{da}{dN} = C(\Delta K)^n X(N, \theta), \tag{12}$$
$$a(0) = A_0;$$

the reasoning is that deviation from the equation over time can be described by a random process $X(N, \theta)$. The solution to the above stochastic differential equation will be a stochastic process model for $a(t)$.

This approach can be difficult to make operational, as solving Equation 12 is not possible for many processes. Typically, $X(N, \theta)$ is taken to be Gaussian white noise,

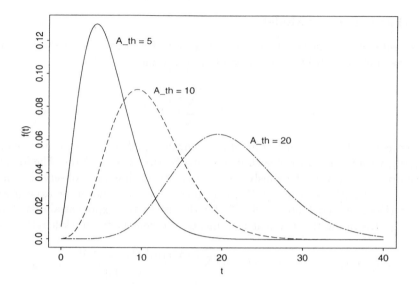

Figure 2: Lifetime density for the model of Sobczyk and Trębicki

which produces a tractable solution, although this process has the disadvantage of being physically unrealisable, and one must ensure the variance in the process is small enough so that the chance of $X(N)$ being negative is negligible.

Sobczyk[15] uses the theory of stochastic differential equations to solve Equation 12 with $X(N, \theta)$ as Gaussian white noise with mean value μ_X and variance σ_X^2. If $n = 2$ then the process $a(N)$ satisfies

$$a(N) = A_0 \exp[C(Q \Delta\sigma)^n (N\mu_X + \sigma_X(W(N) - W(0)))], \quad (13)$$

where $W(N)$ is the Wiener process. Since $W(N)$ is normally distributed, this implies that $a(N)$ is lognormally distributed. If $n > 2$ then the solution is a little more complex, but one can show that the density of $a(N)$ is

$$f_{a(N)}(a) = \frac{m - 1}{a^m \sqrt{2\pi N(m - 1)C(Q\Delta\sigma)^n \sigma_X}}$$
$$\times \exp\left[-\frac{(a^{1-m} + N(m - 1)C(Q\Delta\sigma)^n \mu_X - A_0^{1-m})^2}{2N(m - 1)^2(C(Q\Delta\sigma)^n)^2 \sigma_X^2}\right], \quad (14)$$

where $m = n/2$.

The reliability function can only be obtained in the usual manner if $n \leq 2$. If $n > 2$ then the solution for $a(N)$ is only defined on an interval of random length $(0, N_e)$. When $N = N_e$ then the solution explodes to infinity, and we interpret N_e as the time at which unstable crack growth occurs and the material fails; i.e. $N_e = N_f$.

For the case $n > 2$, Sobczyk shows that N_f has an inverse Gaussian distribution:

$$f_{N_f}(N) = \frac{C_1}{\sqrt{2\pi N}} \exp\left[-\frac{(C_1 - C_2 N)^2}{2N}\right],\tag{15}$$

where

$$C_1 = [A_0^{m-1}(m-1)C(Q\Delta\sigma)^n \sigma_X]^{-1}, \text{ and}$$
$$C_2 = \frac{\mu_X}{\sigma_X}.\tag{16}$$

4 Statistical Analysis of Fatigue Crack Data

In many papers that describe a stochastic model, the authors will include a section on statistical inference with the model, and make use of all the usual statistical estimation techniques. In the case of the Paris-Erdogan equation, parameter estimation is often done by regressing log growth against log crack length, which is a linear relationship under the equation, and then fitting C and n by least squares. Maximum likelihood is also commonly used. Bayesian techniques are all but absent, certainly in the engineering literature, although where they are used, they are not called as such (see the final chapter of Sobczyk and Spencer[1], for example). There are some analyses in the statistical literature, such as Palettas and Goel[16].

In this section, we conclude the paper with an example of the analysis of a set of micro-cracks observed in a cast iron. We start the example with a description of a model for the growth of this collection of micro-cracks. One aim of the example is to show that one can make use of recently developed Monte Carlo techniques to conduct inference with such models.

4.1 A Hierarchical Model for Micro-crack Propagation

Micro-crack propagation has been shown to obey Paris-Erdogan, apart from one key point; at certain short lengths the rate of growth can slow dramatically. This widely observed phenomenon is caused by the crack encountering a boundary between grains in the material microstructure; at such a boundary, growth rate is slowed by a factor that depends on local conditions. In this way, some micro-cracks propagate to become macro-cracks with hardly any delay, while others may be held back for some time or even stopped altogether. The difference in the length at which cracks slow down is due to the different distance that the cracks progress before hitting a grain boundary.

The effect of a grain boundary is modelled by taking the Paris-Erdogan equation and multiplying the right hand side by a factor that accounts for the local conditions at the first grain boundary. An example of such an equation is

$$\frac{da}{dN} = C(\Delta K)^n \left\{1 - \phi \exp(-m(a - D)^2)\right\},\tag{17}$$
$$a(0) = A_0,$$

where $\phi \in [0, 1]$ and $m > 0$ are local property parameters and D is the distance from crack nucleation point to first grain boundary. The size of the slowdown in

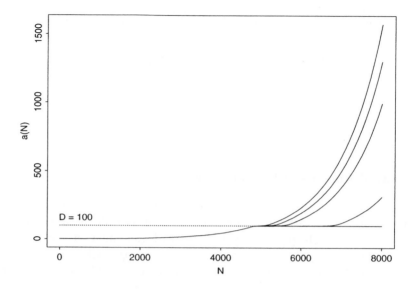

Figure 3: Solution to Equation 17 for different values of ϕ

growth at the grain boundary is governed by ϕ, assumed to lie between 0 and 1. If $\phi = 0$ then there is no local effect and the crack moves according to Paris-Erdogan. If $\phi = 1$ then when $a = D$, that is when the crack hits the first grain boundary, $da/dN = 0$ and the crack is stopped. The parameter m is a scaling factor.

The solution for $a(N)$ from Equation 17 is shown in Figure 3 for five values of ϕ from 0.6 (the fastest growing), 0.7, 0.8, 0.9 and 1 (the slowest). Other parameter values are: $C = 0.2$, $\Delta\sigma = 10$, $n = 1$, $m = 0.02$, $Q = 1$ and $D = 80$. Observe that the crack growth does indeed slow around the value of D, with the extent of the slowdown depending on the size of ϕ.

This is a deterministic model for a single crack. One probabilistic scheme for the collection of micro-cracks in a specimen is the hierarchical model, where the parameters are randomised and assumed exchangeable. We assume a collection of K micro-cracks , and define $A_i(N)$ to be the length of crack i after N cycles. The local conditions at each crack are described by the local parameters m_i, D_i and ϕ_i, for $i = 1, \ldots, K$, which are assumed exchangeable. Conditional on the local parameters, the deterministic model of Equation 17 now gives a solution $a_i(N)$ for the length of crack i, and we assume that $A_i(N)$ is a Gaussian distribution with mean $a_i(N)$ and variance $\sigma^2 a_i(N)^2$. The use of multiplicative error for $A_i(N)$ is necessary because crack lengths vary over several orders of magnitude with time.

Somewhat arbitrarily, we assume the $\log(m_i)$'s are independent draws from a normal distribution conditional on its mean M and variance σ_m^2, whereas the logit(ϕ_i)'s are also normally distributed with mean Φ and variance σ_ϕ^2. Taylor and Wilson [17] derive a distribution for the D_i's from geometrical considerations, which we do not describe here.

Given a collection of cracks, the lifetime distribution is a function of the largest crack, since

$$P(N_f > N) = P(\max(A_1(N), \dots, A_K(N)) \le A_{th}).\tag{18}$$

4.2 Bayesian Inference and Prediction

The statistical inference that we are aiming at will be Bayesian, and we assume vague priors on the hyperparameters: normals with large variance for M and Φ, and inverse gamma with scale and shape parameters 0.5 for σ_m^2, σ_ϕ^2 and σ^2.

Typically, the data will consist of the lengths of K cracks measured at a set of times $N_1, \dots, N_L$. We denote a_{ij} as the observed length of crack i at N_j cycles, and let $\mathcal{A}$ denote the entire data set:

$$\mathcal{A} = \{a_{ij} \mid i = 1, \dots, K; j = 1, \dots, L\}.\tag{19}$$

If we denote the set of all model parameters by Θ, then the aim of the inference is to calculate the joint posterior distribution, $f(\Theta \mid \mathcal{A})$, as well as any marginal posterior and posterior predictive distributions, such as the posterior reliability $P(N_f > N \mid \mathcal{A})$.

The posterior distributions of the model parameters are not in any closed form, and must be obtained indirectly by a Monte Carlo Markov Chain method, in this case the Gibb's sampler. One problem with implementing this scheme is that $a_i(t)$ is a highly non-linear function of the first level parameters m_i, ϕ_i and D_i, which can produce convergence problems for the sampler. Wakefield [18] uses the Metropolis algorithm in similar circumstances, for an application in pharmacology, with very long runs to achieve convergence.

For those familiar with MCMC methods, the posterior distributions are approximated by a kernel density estimate technique. In the case of the posterior reliability prediction, this is derived using the conditional independence assumptions of the model. First, we note that

$$\begin{aligned}
P(N_f > N | \mathcal{A}) &= P(\max\{A_1(N), \dots, A_K(N)\} \le A_{th} \mid \mathcal{A})\\
&= P(A_1(N) \le A_{th}, \dots, A_K(N) \le A_{th} \mid \mathcal{A})\\
&= \int \left[\prod_{i=1}^{K} P(A_i(N) < A_{th} \mid m_i, \phi_i, Di, \sigma^2) \right.\\
&\qquad \left. \times\, f(m_i, \phi_i, D_i \mid \mathcal{A})\, dm_i\, d\phi_i\, dD_i \right] f(\sigma^2 \mid \mathcal{A})\, d\sigma^2\\
&= E_{(m_i, \phi_i, D_i, \sigma^2)} \left[\prod_{i=1}^{K} P(A_i(N) < A_{th} \mid m_i, \phi_i, Di, \sigma^2) \right].
\end{aligned}\tag{20}$$

Since the conditional distribution $P(A_i(N) < A_{th} \mid m_i, \phi_i, Di, \sigma^2)$ is Gaussian, it is readily evaluated. This expectation is then approximated by a sample mean, using the simulated values from the Gibb's sampler, so the posterior reliability function

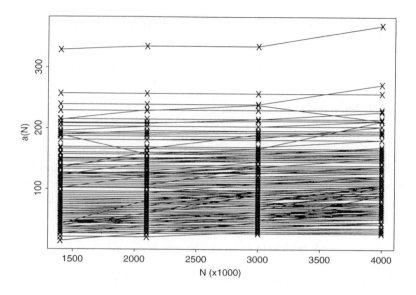

Figure 4: Micro-crack data from a specimen of steel

is

$$P(N_f > N \mid \mathcal{A}) \approx \frac{1}{J} \sum_{j=1}^{J} \prod_{i=1}^{K} P(A_i(N) \le A_f \mid m_i(j), \phi_i(j), D_i(j), \sigma^2(j)), \quad (21)$$

where we assume that we have sampled J values of each parameter from its posterior distribution, the jth sampled value denoted $m_i(j)$ etc.

Figure 4 shows the observed lengths of 190 micro-cracks in a specimen of a grey cast iron subjected to a constant frequency cyclic stress of range 140 kPa. The lengths were observed at four points only, which means that our model is overparameterised as regards estimation of each individual crack's properties (since there are 3 parameters per crack). So we concentrate on the 6 global parameters and the reliability prediction. Figure 5 shows the two predictions for the reliability, given threshold lengths $A_{th} = 1000$ and $A_{th} = 1500$. Observe from Figure 4 that there is one dominant crack that is larger than the others, so the reliability prediction is almost entirely dependent on the predicted growth of this crack alone.

References

1. K. Sobczyk and B. F. Spencer. *Random fatigue: from data to theory.* Academic Press, San Diego, 1992.
2. J. A. Bannantine, J. J. Comer, and J. L. Handrock. *Fundamentals of metal fatigue analysis.* Prentice-Hall, Englewood Cliffs, New Jersey, 1990.
3. S. T. Rolfe and J. M. Barsom. *Fracture and fatigue control in structures.* Prentice-Hall, Englewood Cliffs, New Jersey, 1977.

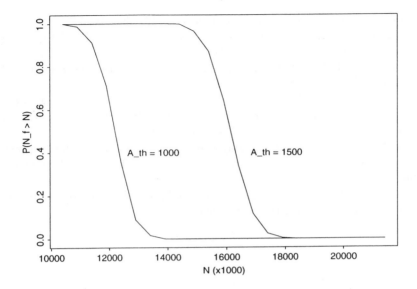

Figure 5: The estimated reliability of the steel specimen, for both $A_{th} = 1000$ and $A_{th} = 1500$

4. J. R. Rice. A path independent integral and the approximate analysis of strain concentration by notches and cracks. *Journal of Applied mechanics, Transactions ASME*, 35, 1968.

5. Z. W. Birnbaum and S. C. Saunders. Estimation for a family of life distributions with application to fatigue. *J. Appl. Prob.*, 6:328–337, 1969.

6. W. Weibull. A statistical distribution function of wide applicability. *ASME J. Appl. Mech.*, 18:293–297, 1951.

7. J. N. Yang and E. Heer. Reliability of randomly excited structures. *AIAA J.*, 9:1262–1268, 1971.

8. A. Paluszny and P. F. Nicholls. Predicting time-dependent reliability of ceramic motors. In J. J. Burke, E. N. Lenoe, and R. N. Katz, editors, *Ceramics for high performance applications - II*, pages 95–112. Brook Hill, Chestnut Hill, Mass., 1978.

9. A. A. Griffith. The phenomenon of rupture and flow in solids. *Phil. Trans. Roy. Soc. London, Ser. A*, 221:163–198, 1921.

10. J. Galambos. *The asymptotic theory of extreme order statistics*. John Wiley & Sons, New York, 1978.

11. N. D. Singpurwalla, S. P. Wilson, and E. R. Fuller. Statistical aspects of failure processes in ceramics. In J. M. Bernardo, J. O. Berger, A. P. Dawid, and A. F. M. Smith, editors, *Bayesian Statistics 5*. Oxford University Press, 1996.

12. S. M. Wiederhorn and E. R. Fuller. Structural reliability of ceramic materials. *Mat. Sci. Eng.*, 71:169–186, 1985.

13. K. Sobczyk and J. Trębicki. Modelling of random fatigue by cumulative jump

processes. *Engineering Fracture Mechanics*, 34:477–493, 1989.

14. V. H. Kogaiev and S. G. Liebiedinskij. Probabilistic model of fatigue crack growth. *Mashinoviedinije*, 4:78–83, 1983.

15. K. Sobczyk. Modelling of random fatigue data. *Engineering Fracture Mechanics*, 24:609–623, 1986.

16. P. N. Palettas and P. K. Goel. Bayesian modelling for fatigue crack curves. In J. P. Klein and P. K. Goel, editors, *Survival Analysis: State of the Art*, pages 153–170. Kluwer Academic Publishers, 1992.

17. D. Taylor and S. P. Wilson. Bayesian inference for micro-crack growth. Technical Report 96/03, Department of Statistics, Trinity College Dublin, 1996.

18. J. Wakefield. The Bayesian analysis of population pharmacokinetic models. *J. Amer. Statist. Assoc.*, 91:62–75, 1996.

FAULT TREE REDUCTION FOR RELIABILITY ANALYSIS AND IMPROVEMENT

Dr. M. XIE, Dr. K. C. TAN and K. H. GOH
Department of Industrial and Systems Engineering,
National University of Singapore,
Kent Ridge, Singapore 119260

Fault tree analysis is a technique widely used in the study of the reliability of complex systems. Most of the studies carried out are related to how to construct a fault tree, and how to determine the top event probability. Fault trees are usually very complex after the construction phase and almost all fault trees can be reduced so that further analysis can be made easier. In this paper some principles of fault tree reduction are presented. By reducing the fault trees to a much more graphically simplified configuration, analysis of the fault trees can be carried out with less effort. Examples are given to demonstrate the usefulness of the proposed reduction technique in both the qualitative and quantitative fault tree analysis. The MOCUS algorithm is for example used as a basis for comparison between non-reduced fault trees and reduced fault trees on the reduction in steps required to obtain the minimal cut sets. Furthermore, using the reduced fault tree, a simple ranking method for basic events in reliability improvement analysis is presented. The ranking method has the advantage of being accurate compared with various importance measures but computationally much easier, and in addition it can easily be applied to fault trees with repeated events and fault trees with undeveloped events.

1   Introduction

Fault tree analysis (FTA) is a reliability engineering technique with application in almost all engineering fields. It is especially useful for complex and safety critical systems. However, the analysis of large fault trees is usually tedious. Obtaining cut sets on a personal computer can be very slow, and sometimes completely impractical. Furthermore it is difficult to understand the physical importance of a large number of minimal cut sets and path sets.

In practice, almost all fault trees can be reduced after its initial construction. Hence, introducing fault tree reduction will greatly simplify the fault trees and this will in turn cut down on the number of gates used. Another useful application of fault tree reduction is that repeated events can often be removed. As such, minimal cut sets can be easily obtained, and FTA can be conducted with less effort making it more useful for practical application.

Table 1: Fault tree reduction and optimisation.

Techniques	
Modularization	[1 - 4]
Binary decision diagrams	[5 - 8]
Bit manipulation	[9, 10]
Truncation methodology and approximations	[11 - 13]
Partitioning cut sets	[14]
Object-oriented programming	[15, 16]

Many techniques were developed to help in speeding up the reduction and optimisation process. Table 1 provides a list of the fault tree reduction and optimisation techniques available.

The above mentioned methods to fault tree reduction are difficult to apply and are mostly approximations. With regard to this difficulty, some simple principles for fault tree optimisation or reduction are presented. Fault trees can be highly reduced based on these principles. This greatly facilitates both qualitative and quantitative FTA.

This paper begins by studying some simple fault tree reduction principles. Some examples are presented using these principles. Fault tree reduction should be used before applying the MOCUS algorithm so that the number of steps can be greatly reduced. Also, a simple ranking approach based on counting the number of AND gates that leads to the top event using the reduced fault tree is introduced.

2 Principles for Fault Tree Reduction

To reduce the fault tree to a simpler configuration, three principles are proposed here. The first principle simplifies fault trees by removing consecutive AND and OR gates. The second and third principles simplify fault trees by combining repeated events under different gates. As shown in later examples, the successful application of these three principles will reduce fault trees to a much more manageable proportion so that analysis can be carried out easily.

2.1 Principle 1 - Removing consecutive AND and OR gates

A common sight in most fault trees is repeated AND or OR gates. When there is a repetition of AND or OR gates, the repeated gates can be reduced to only one AND or OR gate. Figures 1a and 1b show a fault tree with repeated OR gates and the reduced configuration. An important point to note here is that the gates used in the derivation of these two principles are interchangeable. That is, OR gates can be replaced with AND gates and the resulting reduction will still be the same.

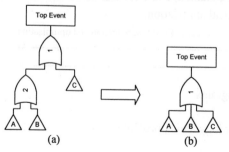

(a)   (b)

Figure 1: Fault tree with repeated gates and the equivalent reduced configuration.

Therefore, fault trees having repeated AND or OR gates can be reduced to having one AND or OR gate only when Principle 1 is applied. The proof for the equivalence between (a) and (b) in Figure 1 is straight-forward and hence it is omitted here.

2.2 Principle 2 - Combining repeated events under different gates at the same level

A common sight among most fault trees is repeated events. In some instances, repeated events may occur under different gates. Thus, repeated events can be combined as shown in Figure 2a and 2b.

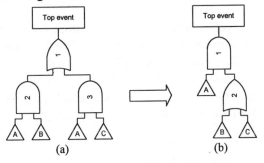

Figure 2: Fault tree having repeated events and the equivalent reduced configuration.

2.3 Principle 3 - Combining repeated events under different gates at the same level (with extra events)

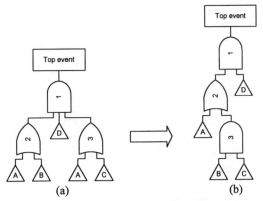

Figure 3: Fault tree having repeated events and the equivalent reduced configuration.

When there is an additional event or gate D as shown in Figure 3a, the fault tree can be reduced to the configuration as shown in Figure 3b. Although in this case the number of gates remain the same, the repeated event is eliminated. Fault trees without repeated event is much easier to deal with than those with repeated events. Hence, this reduction is a useful one. It can be easily shown that (a) and (b) in Figure 3 are equivalent.

It can be seen from the above that fault trees having repeated events under different gates can be reduced to a fault tree without repeated events. The number of gates can also be reduced. This implies that analysis of the fault tree can be greatly simplified.

3 Some Examples of the Application

In this section, two examples will be shown to illustrate the usefulness and ease of application of the fault tree reduction technique. The first example, as shown in Figure 4, is extracted from a recent paper by Dai [17] on the study on the residual life prediction for pressurised tubes of an industrial furnace operated at elevated temperature. Note that there are a number of repeated basic events in this fault tree (basic events 1, 2, 3 and 4). The fault tree consists of 12 gates.

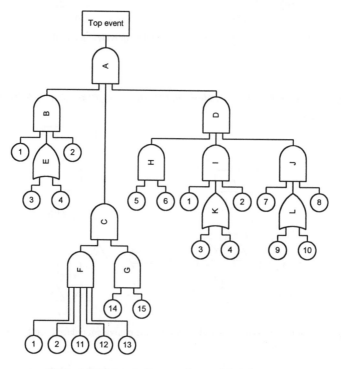

Figure 4: Fault tree for furnace tube possible failure [17].

To illustrate the different sequences of how the reduction procedure proceeds, Figure 5 shows a blow-up of the different phases of fault tree reduction. Note that along the branch of the fault tree with gates *A*, *C*, and *F*, there exist basic events 1 and 2 also. During this stage of the fault tree reduction, there exist two sets of repeated basic events (basic events 1 and 2), and therefore only one set shall remain.

At the end of the reduction, the resulting fault tree should have no consecutive AND or OR gates. By applying the first principle of fault tree reduction, the fault tree will eventually be reduced to the configuration as shown in Figure 6. Note that the fault tree contains two levels now.

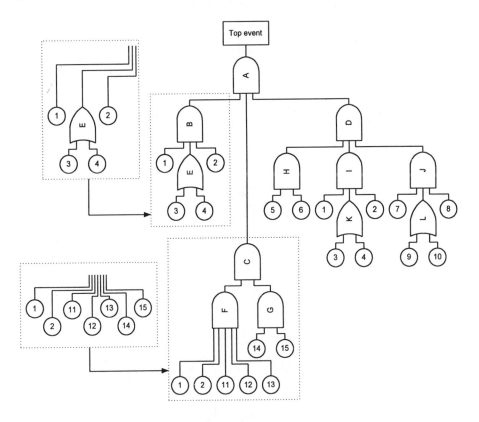

Figure 5: Summary of reduction procedure.

To illustrate the use of the other reduction principles, another example is extracted from a recent paper by Teng and Ho [18] on reliability analysis for the design of an inflator. Figure 7 shows the fault tree on inadequate system output. Note that instead of consecutive AND gates, the fault tree has now a number of consecutive OR gates, which is common in practice.

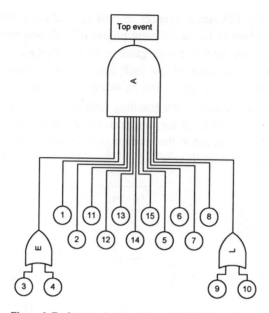

Figure 6: Fault tree reduction after applying first principle.

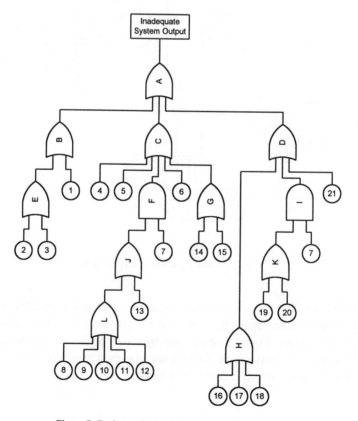

Figure 7: Fault tree for inadequate system output [18].

In the above fault tree, there is again a repeated basic event (basic event 7). However in this case, basic event 7 is not connected to the consecutive OR gates. Therefore, this implies that the repeated basic event will not lead to the same gate.

After applying the first principle of fault tree reduction technique, the fault tree takes on the following configuration as shown in Figure 8.

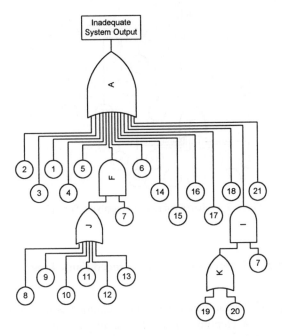

Figure 8: Fault tree reduced after applying Principle 1.

Notice from Figure 8 that gates F and I are at the same level and having at least one common incoming basic event (basic event 7). By applying Principle 3 of the fault tree reduction technique, the fault tree can be simplified further by combining gates F and I into a single AND gate with basic events 19 and 20 leading to gate J. The above manipulation is shown in Figure 9.

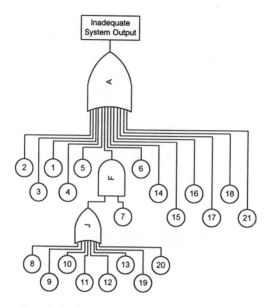

Figure 9: Fault tree reduced after applying Principle 3.

4 Simplifying MOCUS Algorithm Using Reduced Fault Tree

A clear advantage of this fault tree reduction technique is the ease with which cut sets can be obtained. A common mode of obtaining minimal cut sets is to use the MOCUS algorithm [19, 20]. This can be stated as follows:
1.   Alphabetise each gate and number each basic event;
2.   Locate the uppermost gate in the first row and column of a matrix;
3.   Iterate either of the fundamental permutations a) or b) below in a top-down fashion.
     a)   Replace OR gates by a vertical arrangement of the input to the gates, and increase the cut sets;
     b)   Replace AND gates by a horizontal arrangement of the input to the gates, and enlarge the size of the cut sets; and
4.   When all gates are replaced by basic events, obtain the minimal cut sets by removing the supersets.

Consider the second example used in Section 3 again, and apply the MOCUS algorithm on a non-reduced fault tree. The following steps as shown in Table 2 were used to obtain the cut sets.

Table 2: Steps for obtaining the cut sets.

Steps	1	2	3	4	5	6	7
	A	B	1	1	1	1	1
		C	E	2	2	2	2
		D	C	3	3	3	3
			D	C	4	4	4
				D	5	5	5
					6	6	6
					F	J, 7	L, 7
					G	G	13, 7
					D	D	G
							D

Steps	8	9	10	11	12	13
	1	1	1	1	1	1
	2	2	2	2	2	2
	3	3	3	3	3	3
	4	4	4	4	4	4
	5	5	5	5	5	5
	6	6	6	6	6	6
	8, 7	8, 7	8, 7	8, 7	8, 7	8, 7
	9,7	9,7	9,7	9,7	9,7	9,7
	10, 7	10, 7	10, 7	10, 7	10, 7	10, 7
	11, 7	11, 7	11, 7	11, 7	11, 7	11, 7
	12, 7	12, 7	12, 7	12, 7	12, 7	12, 7
	13, 7	13, 7	13, 7	13, 7	13, 7	13, 7
	G	14	14	14	14	14
	D	15	15	15	15	15
		D	H	16	16	16
			I	17	17	17
			21	18	18	18
				I	K, 7	19, 7
				21	21	20, 7
						21

Therefore, from the above example, it can be observed that there are a total of twenty cut sets. The twenty cut sets are {1}, {2}, {3}, {4}, {5}, {6}, {8, 7}, {9, 7}, {10, 7}, {11, 7}, {12, 7}, {13, 7}, {14}, {15}, {16}, {17}, {18}, {19, 7}, {20, 7}, {21}.

The fault tree reduction technique is now applied to the same fault tree, and the MOCUS algorithm is again employed to derive the cut sets. Using the MOCUS algorithm, the following steps as shown in Table 3 were used to obtain the cut sets.

Table 3: Steps for obtaining the cut sets using the reduced fault tree.

Steps	1	2	3	4
	A	1	1	1
		2	2	2
		3	3	3
		4	4	4
		5	5	5
		6	6	6
		F	J, 7	8, 7
		14	14	9, 7
		15	15	10, 7
		16	16	11, 7
		17	17	12, 7
		18	18	13, 7
		21	21	19,7
				20,7
				14
				15
				16
				17
				18
				21

The exact cut sets were obtained using the reduced fault tree. However, as can be seen from the above example, the steps taken to derive all the cut sets were greatly reduced.

From the examples presented, it can be seen that the reduction principles proposed are easy to apply and greatly simplify fault trees for further analysis. This new reduction methodology consists of three principles. The first principle simplifies fault trees by removing consecutive AND and OR gates. The second and third principles simplify fault trees by combining repeated events under different gates. From the example used, it was also demonstrated that by applying the new fault tree reduction techniques, minimal cut sets can be obtained much faster using the MOCUS algorithm.

5 An Application to Ranking Basic Events

In FTA, AND gates imply that there are redundancies in the system. In this way, basic events under an AND gate can be viewed as less important since it is dependent on the other basic events' occurrence for the intermediate event to occur. As the number of AND gates needed to relate the basic event to the top event increases, the basic event decreases in importance.

5.1 The ranking based on the number of AND gates

The fundamental concept behind this new approach is to add up the number of AND gates leading up to the top event from each basic event. An important point to note before application of this new ranking approach is to reduce the fault tree first. The procedure leading to the successive execution of the new method is as follows:

1.   Group basic events into categories according to the number of AND gates leading to the top event;
2.   Basic events with fewer AND gates leading to the top event will take precedence over basic events with higher number of AND gates leading to the top event;
3.   Within each group, basic events that are linked by OR gates should be treated with equal significance although they may appear along different levels of the fault tree; and
4.   Within each group, basic events with the same reliability and occur nearer to the top event will have a higher influence on the top event compared to basic events that are connected by AND gates that are further down the fault tree.

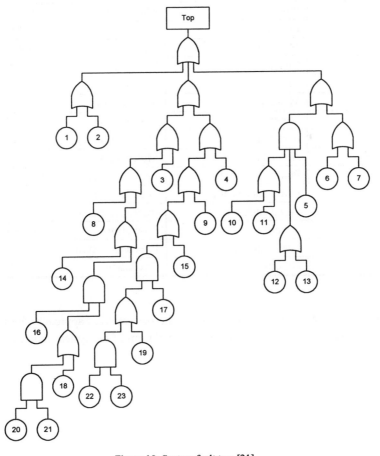

Figure 10: System fault tree [21].

Table 4: The failure data of the gasoline pump.

Comp. No.	Designation	Failure mode	Reliab.
1	Computer	Wrong output	0.9995
2	Display	Display faulty	0.9995
3	Gear transmission 1	Incorrect trans.	0.998
4	Gear transmission 2	Incorrect trans.	0.998
5	Drop cone	Leakage	0.999
6	Mast 1	Leakage	0.98
7	Mast 2	Leakage	0.98
8	Pulse sensor 1	Malfunction	0.995
9	Pulse sensor 2	Malfunction	0.995
10	Pilot cone	Leakage	0.999
11	Main cone	Leakage	0.95
12	Check valve 1	Leakage	0.999
13	Check valve 2	Leakage	0.999
14	Meter 1	Malfunction	0.995
15	Meter 2	Malfunction	0.995
16	Air separator 1	Malfunction	0.99
17	Air separator 2	Malfunction	0.99
18	Pump suck seal 1	Leakage (air)	0.99
19	Pump suck seal 2	Leakage (air)	0.99
20	Pump suck pipe 1	Improper install.	0.999
21	Gasoline	Cavitation	0.999
22	Pump suck pipe 2	Improper install.	0.999
23	Gasoline	Cavitation	0.999

In order to obtain the ranking through the new ranking approach, the fault tree has to be reduced. Figure 11 shows the reduced form of the system fault tree.

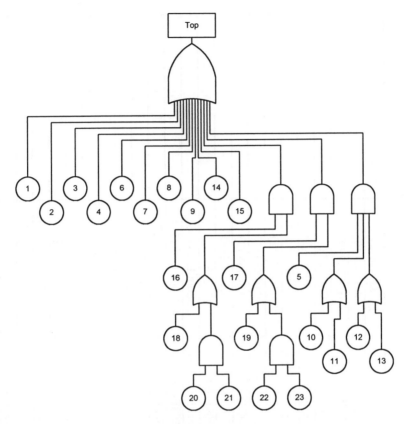

Figure 11: Reduced fault tree.

5.2 A comparison with the existing basic event importance

Given the reliability data, the importance and ranking of the basic events are calculated and shown in Table 5. In this particular example, the importance measures used include *Birnbaum's* basic event importance, *Criticality* importance measure, *Fussell-Vesely* cut set importance, and *Improvement potential* (see Appendix for the definition of the different importance measures).

Table 5: Importance calculations for all basic events.

Comp. No.	I^B (rank)	I^{CR} (rank)	I^{VF} (rank)	I^{IP} (rank)	Proposed classification(rank)
1	0.93700 (9)	0.00050 (9)	0.00787 (9)	0.00047 (9)	I (9)
2	0.93700 (9)	0.00050 (9)	0.00787 (9)	0.00047 (9)	I (9)
3	0.93800 (7)	0.00200 (7)	0.03147 (7)	0.00187 (7)	I (7)
4	0.93800 (7)	0.00200 (7)	0.03147 (7)	0.00187 (7)	I (7)
5	9.54000E-05 (15)	1.01874E-07 (15)	1.60502E-06 (15)	9.53000E-08 (15)	II(16)
6	0.95600 (1)	0.02042 (1)	0.31471 (1)	0.01870 (1)	I (1)
7	0.95600 (1)	0.02042 (1)	0.31471 (1)	0.01870 (1)	I (1)
8	0.94100 (3)	0.00502 (3)	0.07868 (3)	0.00468 (3)	I (3)
9	0.94100 (3)	0.00502 (3)	0.07868 (3)	0.00468 (3)	I (3)
10	1.78000E-06 (23)	1.90080E-09 (23)	3.14709E-08 (23)	1.78000E-09 (23)	II (16)
11	1.87000E-06 (16)	9.98452E-08 (16)	1.57355E-06 (16)	8.88000E-08 (16)	II (11)
12	4.77000E-05 (17)	5.09371E-08 (17)	8.02508E-07 (17)	4.76000E-08 (17)	II (16)
13	4.77000E-05 (17)	5.09371E-08 (17)	8.02508E-07 (17)	4.76000E-08 (17)	II (16)
14	0.94100 (3)	0.00502 (3)	0.07868 (3)	0.00468 (3)	I (3)
15	0.94100 (3)	0.00502 (3)	0.07868 (3)	0.00468 (3)	I (3)
16	9.36600E-03 (11)	1.00016E-04 (11)	1.57370E-03 (11)	9.27300E-05 (11)	II (12)
17	9.36600E-03 (11)	1.00016E-04 (11)	1.57370E-03 (11)	9.27300E-05 (11)	II (12)
18	9.36500E-03 (13)	1.00005E-04 (13)	1.57355E-03 (13)	9.27200E-05 (13)	II (12)
19	9.36500E-03 (13)	1.00005E-04 (13)	1.57355E-03 (13)	9.27200E-05 (13)	II (12)
20	9.27000E-06 (19)	9.89900E-09 (19)	1.57355E-07 (19)	9.26200E-09 (19)	III (20)
21	9.27000E-06 (19)	9.89900E-09 (19)	1.57355E-07 (19)	9.26200E-09 (19)	III (20)
22	9.27000E-06 (19)	9.89900E-09 (19)	1.57355E-07 (19)	9.26200E-09 (19)	III (20)
23	9.27000E-06 (19)	9.89900E-09 (19)	1.57355E-07 (19)	9.26200E-09 (19)	III (20)

Upon reducing the fault tree, the basic events are arranged in order of their significance. For the case of *Birnbaum's* basic event importance, basic events 6, 7 are the most significant. This is followed by basic events 8, 9, 14, 15, and so forth. It can be seen from Table 5 that according to the new ranking approach, the ranking of basic events with only one OR gate follows similarly to those of the other importance measures. The remaining basic events, although they have been ranked differently, are quite insignificant to require special attention.

6 Discussions and Conclusions

A point to note here is that the proposed classification refers to the grouping of basic events according to the number of AND gates that leads to the top event. For example classification I refers to those basic events with no AND gates leading to the top event. In the previous example, it was observed that by applying the four importance measures, the same ranking for all basic events was achieved. However on application of the new ranking approach, slight differences in the rankings have been obtained. This however is no cause for alarm. In a practical sense, only the most significant basic events require special attention. Therefore, it can be safely said that the new ranking approach has been able to achieve similar results when compared to the other importance measures on the basis of the most significant basic events.

FTA is a very useful tool in the analysis of system reliability. However almost all fault trees encountered are rather huge and complex. This makes analysis rather tedious and time consuming. By applying the reduction principles, almost all fault trees can be greatly simplified and thus cut down on calculation effort.

From the examples presented, it is demonstrated that the structures of the fault trees greatly simplifies application of the reduction principles. Successful reduction of the fault trees implies that analysis time of the fault trees using algorithms such as MOCUS will be greatly reduced.

By applying the new ranking approach after employing the fault tree reduction principles, it was found from the examples and case study that similar rankings were obtained when compared against the other commonly used importance measures. Although not all basic events have achieved exact ranking as the other basic events, it can be seen that the most significant basic events have achieved identical ranking as the other importance ranking. In practical sense, the identification of the few most significant basic events outweighs the correct ranking of all the basic events as there will only be limited resources for system improvement. Therefore, it is hopeful that through this identification and improvement of the worst performers, the system reliability will grow in time.

Appendix (Commonly used importance measures)

In this section, a brief introduction of the importance measures used is provided.

1 Birnbaum's basic event importance

Birnbaum's measure of importance of event *i* is defined as

$$I_i^B = \frac{\partial h(\mathbf{p})}{\partial p_i} \, . \tag{1}$$

It is the rate of system improvement for an improvement to a basic event [22, 23].

2 Criticality basic event importance

The *Criticality* measure is used to identify the event that most probably caused the system to fail. This approach is suitable after a system failure has occurred [24].

$$I_i^{CR} = \frac{I^B \cdot (1 - p_i)}{1 - h(\mathbf{p})} \tag{2}$$

3 Fussell-Vesely cut set importance

The *Fussell-Vesely* measure is given by

$$I_i^{VF} = \frac{P(D_i \cap C)}{P(C)}, \tag{3}$$

where

$$D_i = E_1^i \cup E_2^i \cup \cdots \cup E_{m_i}^i, \tag{4}$$

where m_i denote the number of minimal cut sets that contain event *i*, and E_j^i denote that the minimal cut set *j* among those containing event *i* is failed for $i = 1, 2, \ldots, n$ and $j = 1, 2, \ldots, m_i$ [25].

4 Improvement potential

The *Improvement potential* can be used in cases when knowledge of how the system reliability increases with replacement of a component by a perfect component is known. That is

$$I_i^{IP} = h(1_i, \mathbf{p}) - h(\mathbf{p}), \tag{5}$$

where

$$h(1_i, \mathbf{p}) = h(p_1, p_2, \ldots, 1_i, \ldots, p_{n-1}, p_n). \tag{6}$$

The *Improvement potential*, like the *Birnbaum's* basic event importance measure, is applied to instances where system failure has yet to occur, and it is used to identify events that should be improved to increase system reliability [26, 27].

References

1. Chatterjee, P., Modularization of fault trees: a method to reduce the cost of analysis. *Reliability and Fault Tree Analysis*, eds. Barlow, R. E., Fussell, J. B. and Singpurwalla, N. D., SIAM, Philadelphia, 1975.

2. Kohda, T., Henley, E. J. and Inoue, K., Finding modules in fault trees. *IEEE Transactions on Reliability*. **38** 165-76 (1989).

3. Cheng, Y. L. and Yuan, J., Structured fault tree synthesis based on system decomposition. *Reliability Engineering and System Safety*. **50** 109-20 (1995).

4. Contini, S., A new hybrid method for fault tree analysis. *Reliability Engineering and System Safety*. **49** 13-21 (1995).

5. Bryant, R., Graph based algorithms for Boolean function manipulation. *IEEE Transactions on Computer*. **35** 677-91 (1987).

6. Brace, K., Rudell, R. and Bryant, R., Efficient implementation of a BDD package. Paper presented at the 27th *ACM/IEEE Design and Automation Conference*. IEEE, (1990).

7. Rauzy, A., New algorithm for fault trees analysis. *Reliability Engineering and System Safety*. **40** 203-11 (1993).

8. Coudert, O. and Madre, J. C., MetaPrime: An interactive fault-tree analyser. *IEEE Transactions on Reliability*. **43** 121-27 (1994).

9. Chakib, K., An improved minimal cut set algorithm. *Department of Industrial Technology, University of Bradford, UK*.

10. Wheeler, D. B., Hsuan, J. S., Duersch, R. R. and Roe, G. M., Fault tree analysis using bit manipulation. *IEEE Transactions on Reliability*. **26** 95-98 (1977).

11. Modarres, M. and Dezfall, H., A truncation methodology for evaluating large fault trees. *IEEE Transactions on Reliability*. **33** 325-28 (1984).

12. Takaragi, K., Sasaki, R. and Shingai, S., An algorithm for obtaining simplified prime implicant sets in fault-tree and event-tree analysis. *IEEE Transactions on Reliability*. **32** 386-89 (1983).

13. Huang, P. and Li, Z., The measure of effect in an expert system for automatic generation of fault trees. *IEEE Transactions on Reliability*. **41** 57-62 (1992).

14. Rosenberg, L., Algorithm for finding minimal cut sets in a fault tree. *Reliability Engineering and System Safety*. **53** 67-71 (1996).

15. Patterson-Hine, F. A. and Koen, B. V., direct evaluation of fault trees using object-oriented programming techniques. *IEEE Transactions on Reliability*. **38** 186-92 (1989).

16. Stefix, M. and Bobrow, D. G., Object-oriented programming: themes and variations. *The AI Magazine*. **VI** 40-62 (1985).

17. Dai, S. H., Study on the residual life prediction for pressurised tubes of an industrial furnace operated at elevated temperature by using the method of extrema of fuzzy functions. *International Journal of Pressure Vessels and Piping.* **63** 111-17 (1995).

18. Teng, S. H. and Ho, S. Y., Reliability analysis for the design of an inflator. *Quality and Reliability Engineering International,* **11** 203-14 (1995).

19. Fussell, J. B., Henry, E. B. and Marshall, N. H., MOCUS - A computer program to obtain minimal cut sets from fault tree. ANCR - 1156. 1974.

20. Fussell, J. B. and Vesely, W. E., A new methodology for obtaining cut sets for fault trees. *Trans. Amer. Nuc. Soc.* **15** 794 (1972).

21. Shen, K., Some approaches to system reliability improvement in engineering design - by component importance measures and stress-strength modelling in the conceptual and embodiment phases. ISRN LUTMDN/TMKT -- 90/1014 - SE.

22. Birnbaum, Z. W., On the importance of different components in a multicomponent system. *Multivariate analysis II,* 581-92 (1969).

23. Barlow, R. E. and Proschan, F., Importance of system components and fault tree analysis. Operations Research Center, University of California, Report ORC 74-3, 1974.

24. Henley, E. J. and Kumamoto, H., *Reliability Engineering and Risk Assessment.* Prentice-Hall, New Jersey, 1992.

25. Fussell, B. J., How to hand-calculate system reliability characteristics. *IEEE Transactions on Reliability,* **24** (1973).

26. Xie, M. and Shen, K., The increase of reliability of k-out-of-n system through improving a component. *Reliability Engineering and System Safety,* **26** 189-95 (1989).

27. Xie, M. and Shen, K., On ranking of system components with respect to different improvement actions. *Microelectronics Reliability,* **29** 159-64 (1989).

CONTRIBUTING AUTHOR INDEX

KEYWORD INDEX